Advances in Thermodynamics

Advances in Thermodynamics

Volume 1: **C_{7+} Fraction Characterization**

Volume 2: **Fluctuation Theory of Mixtures**

ADVANCES IN THERMODYNAMICS
VOLUME 2

Fluctuation Theory of Mixtures

EDITED BY
Enrico Matteoli and G. Ali Mansoori

Taylor & Francis
New York • Bristol, PA • Washington D.C. • London

USA	Publishing Office:	Taylor & Francis New York Inc. 79 Madison Ave., New York, NY 10016-7892
	Sales Office:	Taylor & Francis Inc. 1900 Frost Road, Bristol, PA 19007-1598
UK		Taylor & Francis Ltd. 4 John St., London WC1N 2ET

Fluctuation Theory of Mixtures

First published 1990
Printed in the United States of America

Library of Congress Cataloging in Publication Data

Fluctuation theory of mixtures / edited by Enrico Matteoli and G. Ali Mansoori.
p. cm. - - (Advances in thermodynamics ; v. 2)
Includes index.
ISBN 0-8448-1630-2
1. Mixtures, G. A. (G. Ali) 2. Solution (Chemistry) I. Matteoli, Enrico. II. Mansoori, G. A. (G. Ali) III. Series.
QD541.F525 1990
541.3'4--dc20 90-36337
CIP

Contents

(§) Reproduced from the original magazine.

List of Contributors

K. Arakawa
Research Institute of Applied Electricity, Hokkaido University, Japan

A. Ben-Naim
Department of Physical Chemistry, The Hebrew University of Jerusalem, Israel

A. A. Chialvo
Department of Chemical Engineering, Princeton University, U. S. A.

M. Diaz Peña and R. G. Rubio
Departamento de Quìmica Fìsica, Universidad Complutense, Spain

H. D. Cochran
Chemical Technology Division, Oak Ridge National Laboratory, U. S. A.

E. Z. Hamad and G. A. Mansoori
Department of Chemical Engineering, University of Illinois at Chicago, U. S. A.

D. A. Jonah
Department of Mathematics, University of Sierra Leone, Sierra Leone

Y. Kataoka
Department of Chemistry, Kyoto University, Japan

T. Kato
Department of Chemistry, Tokio Metropolitan University, Japan

J. G. Kirkwood and F. P. Buff§
California Institute of Technology, U.S.A.

L. L. Lee and D. M. Pfund
Department of Chemical Engineering, University of Oklahoma, U. S. A.

L. Lepori and E. Matteoli
N. R. C., Institute of Quantum Chemistry and Molecular Energetics, Italy

R. M. Mazo§
University of Chicago, Ill., U.S.A.

K. E. Newman
Department of Chemistry, The King's College, Canada

J. P. O'Connell
Department of Chemical Engineering, University of Virginia, U. S. A.

D. Patterson
Department of Chemistry, McGill University, Canada

E. Peacock-López
Department of Chemistry, Willams College, U. S. A.

F. J. Pearson and G. S. Rushbrooke§
Physics Department, University of Durham, U.K.

L. Woodhouse
Department of Chemistry, University California, U. S. A.

(§) Their contribution is the reproduction of a paper published before.

Introduction

In the last decade, after about a quarter of a century in which it has been almost ignored, the Fluctuation theory of mixtures, as developed by Kirkwood and Buff in 1951, has attracted the attention of more and more investigators, as demonstrated by the still increasing large number of papers recently published in international scientific magazines.

The topics treated in this literature range from the generalization and theoretical developments of the theory, to its applications in experimental physical chemistry and chemical engineering.

The applications which have been implemented are due to the recent theoretical developments which, on one hand, have allowed a more complete modelization of mixtures and consequently a better prediction and a more accurate calculation of thermodynamic quantities of mixtures, such as activity coefficients, partial molar volumes, solubility and phase behaviour in general; and, on the other, have promoted a deeper understanding of the microscopic structure of mixtures in terms of correlation functions, local composition, interfacial composition and size of clusters.

It is now certain that the theory is still susceptible of further important developments which will involve both its predictive potentiality as well as its capability in characterizing the local environment of molecules in mixtures. For this reason, we believe, the publication of a volume containing the state of the art of the fluctuation theory approach to mixtures will be welcome by all scientists who work in the general areas of basic and applied physical chemistry and of chemical process engineering.

To this purpose, as well as to draw attention to a topical research field which lends itself to interesting developments in different directions, we have here collected a series of articles written by investigators who have been and still are very active in this area.

After the first chapter, where the main conceptions and equations of statistical mechanics relevant to the subsequent articles are reviewed, three papers published in the fifties by Kirkwood and Buff, by Pearson and Rushbrooke, and by Mazo, respectively, are fully reproduced from the original magazines. These articles contain the fundamental equations and relationships among thermodynamic properties and correlation functions integrals of the fluctuation theory, and are therefore essential references for all chapters of the volume.

In O'Connell's paper, a new and more convenient formalism of the theory based on the integrals of the direct correlation functions is developed and used to obtain thermodynamic properties of pure and mixed liquids as well as of supercritical and ionizing substances in liquids.

In the contribution by Cochran, Lee and Pfund, the supercritical solution behaviour is studied and explained in terms of the fluctuation theory; expressions are obtained for the size of solvent cluster about a solute molecule, degree of solute aggregation and other thermodynamic properties for dilute solutions.

Hamad and Mansoori propose a new fluctuation theory approach which produces analytical expression for total and partial molar properties of mixtures at both infinite dilution and finite concentration. Vapour-liquid and liquid-liquid phase equilibria are predicted accurately by a new theory based on newly developed exact relations among the mixture direct correlation function integrals.

The behaviour of the activity coefficients in binary mixtures in relation to their values at infinite dilution and their limiting slopes is exhamined in Chialvo's paper in terms of the linear combination of the fluctuation integrals, $\Delta = G_{11}+G_{22}-2G_{12}$. Also, new correlations for the activity coefficients are derived.

New general relationships among the mixture direct correlation function integrals are proposed in the paper by Lepori, Matteoli, Hamad and Mansoori. Their validity is demonstrated with the aid of the values of these integrals for symmetric and asymmetric mixtures calculated from experimental data. A simple closure expression is also derived.

In Ben-Naim's paper, the procedure based on the inversion of the Kirkwood-Buff theory for the calculation of the fluctuation integrals from thermodynamic quantities is outlined and its use in characterizing mixtures on a molecular level is pointed out. A new theory of preferential solvation is also developed.

Other applications which exploit the possibility of calculating the fluctuation integrals from scattering data or from experimental thermodynamic quantities are dealt with by Kato, by Rubio, Diaz Peña and Patterson, and by Matteoli and Lepori, respectively. The behaviour with concentration of either the fluctuation integrals or of the concentration fluctuation, a quantity connected to the linear combination $G_{11}+G_{22}-2G_{12}$, is used to get insight into the local environment of molecules in mixtures, to identify association phenomena, to test mixture models based on local compositions, and to explain a peculiar behaviour of the excess heat capacity of a number of mixtures.

An example of the fluctuation integral calculation via integration of the radial distribution function is shown in the paper by Kataoka. The radial distribution function is obtained by Montecarlo and molecular dynamics studies for a 2-dimensional model of water. It is shown that the calculated thermodynamic and dynamical properties are in qualitative agreement with alcohol+water mixtures.

The concentration fluctuations, related to the fluctuation integrals, are calculated in Arakawa's paper for liquid alkali alloys through a theoretical model and successfully compared with experimental data.

Peacock-López and Woodhouse use the fluctuation-dissipation theory to analyze a multicomponent system arbitrarily away from equilibrium, and the mass and energy fluctuations around stationary states are discussed. An expression for the mutual diffusion coefficient for binary non-electrolyte mixtures is obtained and the theory is shown to give a good qualitative and quantitative description of experimental behaviour.

The paper by Jonah presents a perturbation theory approach to the infinitely diluted solute chemical potential, with the solvent component as the reference. It deals mainly with the second order perturbation term, which is related to fluctuations. Applications to low-pressure vapour-liquid equilibria and solution thermodynamics, such as the calculation of Henry's constant and solid solubilities in supercritical fluids, are discussed.

In the last chapter, an extensive treatment of electrolyte solutions by means of the Kirkwood-Buff theory is presented by Newman. Starting from the theory equations for non-electrolyte solutions, new relations are derived for a variety of thermodynamic properties for solutions of salts which are conform to the ion-addivity rule and infinite dilution behaviour. The theory is used to interpret partial molal volumes, free energies of transfer and activity coefficients of electrolytes in solution.

Enrico Matteoli
Institute of Quantum Chemistry
and Molecular Energetics, NRC
Via Risorgimento, 35
56126 Pisa, Italy

G. Ali Mansoori
Department of Chemical Engineering
University of Illinois at Chicago
Box 4348
Chicago, Ill., USA

Fluctuation Theory of Mixtures
(A Statistical Mechanical Background)

G. A. Mansoori
Department of Chemical Engineering
University of Illinois
Chicago, IL 60680

Enrico Matteoli
Istituto di Chimica Quantistica ed Energetica
Molecolare del CNR, Via Risorgimento 35
56126 Pisa, Italy

ABSTRACT
The statistical mechanical basis of the fluctuation theory of mixtures is reviewed. An overview of the statistical mechanical relations between the microscopic properties of a system and its macroscopic properties is presented. The distribution functions in equilibrium systems are defined and the relations between distribution functions and thermodynamic properties for pure fluids and fluid mixtures are reported. The available approximations to distribution functions are presented. Definitions of the direct correlation functions, direct correlation function integrals, and their relations with fluctuation integrals are reported.

INTRODUCTION
Scientists in their efforts to develop new ideas and engineers in their endeavors to build and improve industrial processes and efficient utilization of the available resources, deal with variety of physical, chemical, and biological phenomena. This requires detailed knowledge of the thermodynamic and transport properties of the substances being considered. In thermodynamics, the effects of pressure, temperature, and mixing with other substances on thermodynamic properties are of major concern.

Experimental measurements of thermodynamic properties can not cover all the needs of science and industry. They are costly and time consuming. The large number of substances and the wide range of conditions we have to deal with make it rather difficult to rely completely on experimental measurements. The problem becomes more complicated for mixtures, because the number of measurements needed grows rapidly with the number of components in a given mixture. Because of this problem, it becomes necessary to minimize the amount of experimental measurements by developing theoretical models that could be capable of correlating the existing data and extrapolating to regions where none or little experimental data are available.

The early work in this field relied mainly on empirical approaches. For example, Kamerlingh Onnes (1901) developed the empirical version of the virial equation of state. Later, it was shown that this equation had a solid theoretical basis. The van der Waals equation (1873) was derived from intuitive ideas. This equation was then proved to be theoretically sound (Prins, 1974). It has been also modified, both theoretically and empirically, by many workers.

The theoretical tools needed to describe the behavior of real fluids at equilibrium were developed in the late nineteenth century by Boltzmann and Gibbs. They laid the foundation for statistical mechanics. However, the astronomical amount of calculations needed to get the thermodynamic properties from molecular data delayed the use of statistical mechanics in practical problems for some time. The work of Kirkwood and others in the early part of the twentieth century opened the door for applying statistical mechanics in practical thermodynamic problems (Haile and Mansoori, 1983).

In their effort to model the behavior of real mixtures, scientists and engineers follow two methods: activity coefficient models, and equations of state approach. The early models of the activity coefficients were based on composition power series expansions of the excess Gibbs free energy. Margules and van Laar equations (Walas, 1985) are examples of such expansions. Local composition models have been used to obtain activity coefficient expressions by a number of workers (Wilson, 1964; Renon and Prausnitz, 1968). The connection between local composition and statistical mechanics was later shown in a number of ways (Kemeny and Rasmussen, 1981; Hu et al., 1983; Mansoori and Ely, 1985b). Abrams and Prausnitz (1975) applied statistical mechanics directly to obtain expressions for the activity coefficients.

Statistical mechanical equations of state can be obtained in a number of ways. One route is to make an assumption that allows evaluating analytically the integrals in the partition function. The lattice model does this by assuming that the molecules can move in only defined volumes around their equilibrium positions. This assumption makes the integrals in the partition function independent and as a result each integral can be evaluated separately. This approach, however, is successful in describing the behavior of solids but not fluids.

Another approach to obtaining equations of state from statistical mechanics is to assume the partition function to be divided into repulsive and attractive parts. This implies that the same division holds for the equation of state. Usually the repulsive part is taken as that of a hard body, where the potential is infinite inside the core of the particles and zero outside. Accurate statistical mechanical expressions for the properties of such particle systems are available for both pure fluids and mixtures. The attractive part is usually treated as a perturbation expansion around the repulsive part.

The distribution function approach to the equilibrium properties of substances relates the relative locations of particles in a given substance to the thermodynamic properties of the substance. The distribution functions are normalized probability densities for the locations of n particles at given intermolecular distances from a central particle. Information about the distribution functions of particles in real fluids are not quite available. As a result, approximations to the distribution functions are used to get useful practical results. In the perturbation and the variational theories the distribution functions of a reference fluid (usually a hard-sphere fluid) are used rather than the real functions. These two theories usually require the knowledge of the potential energy functions between the molecules. This limits their use to simple fluids for which intermolecular potential energy functions and data are available.

Modeling of the behavior of complex mixtures can be accomplished more effectively provided one assumes the properties of the pure components are known. A large amount of experimental data has been collected for the widely used pure fluids, and has been put into the format of empirical equations of state. The contribution of statistical mechanics can be more effective in predicting mixture properties from the existing pure fluid equations of state.

CLASSICAL AND STATISTICAL THERMODYNAMICS

Collection of particles (atoms or molecules) possess both microscopic and macroscopic properties. The thermodynamic properties, such as pressure, internal energy and entropy are macroscopic properties. Through classical thermodynamics we can derive relations which allow calculating one thermodynamic property from another. Such relations have the advantage of being applicable to all kinds of systems at equilibrium. Techniques of statistical mechanics are the needed tools for predicting the behavior of a given substance from microscopic information or microscopic energy levels. The energy levels, in turn, can be obtained from the knowledge of the potential energy between the particles forming the system and from the system volume. The energy levels are obtained from quantum mechanics, usually through the Schrödinger equation. The energy levels are the eigenvalues of this equation. However, it is usually difficult to obtain a solution to the Schrödinger equation. In fact the Schrödinger equation has not been solved exactly for systems consisting of more than two particles, except when the potential energy function is assumed to be zero everywhere. In real systems the number of particles is of the order of 10^{23}. Such a large number of interacting particles makes it extremely difficult to obtain the quantum mechanical energy levels.

Assuming the energy levels are available, the macroscopic properties can then be obtained from the partition function. In the canonical ensemble where the temperature, the volume and the number of particles of each component in the system, N_i, are fixed, the partition function, Q, for a c-component system is:

$$Q(T,V,N) = \sum e^{-\varepsilon_j/kT}, \tag{1}$$

where k is Boltzmann constant, T is the absolute temperature, V is the system volume, ε_j is the energy of the particles at level j and $\mathbf{N}=N_1, N_2, \ldots, N_c$. The summation in Equation (1) is carried over all the molecules in all energy levels. The canonical partition function, Q, is related to the macroscopic Helmholtz free energy of the system, A, by

$$A = -kT \ln Q(T,V,N). \tag{2}$$

The relation between the microscopic and the macroscopic properties can be obtained for other kinds of ensembles as well. For example, in the grand canonical ensemble, where the temperature, the volume and the chemical potential of each component in the system are fixed, the partition function, Ξ is:

$$\Xi(T,V,\mu) = \sum_{N_1} \cdots \sum_{N_c} Q(T,V,N)\, e^{\mu.N/kT}, \tag{3}$$

where $\mathbf{N}=\mu_1, \mu_2, \ldots, \mu_c$ and the scalar product $\mu.\mathbf{N}$ is:

$$\mu.\mathbf{N} = \sum_{k=1}^{c} \mu_k N_k. \tag{4}$$

The relation between the grand canonical ensemble partition function and the system pressure, P, is

$$PV = kT \ln \Xi(T,V,\mu). \tag{5}$$

Equations (1) and (3) are completely general for equilibrium systems. However, as mentioned above it is very difficult to get the quantum mechanical energy levels for real systems. To avoid this problem, one makes use of the observation that the spacings between the translational energy levels for most molecules away from absolute zero are very small. This means that one can treat the translational energy of the molecules as continuous rather than discrete without introducing a detectable error. Since this is not generally the case for the rotational and vibrational energy levels for molecules with internal motion, it is advantageous to separate the contributions of these two motions from the contribution of the translational motion.

To achieve the separation between internal and translational motions it is usually assumed that the internal motion of a molecule does not interfere appreciably with the motion of the center of its mass. The partition function separates, then, into two parts. The first part, which is called the configurational partition function, represents the motion of the centers of mass in the potential field created by the interaction between molecules. The second part, which is called the internal partition function, represents the contribution from the internal motions of the molecules and it is equal to the internal partition function of an ideal gas under the same conditions. The small energy gap between consecutive energy levels for all molecules, except the lightest, such as H_2 and He, allows replacing the summation in Equation (1) by integration over the phase space. Thus obtaining the semi-classical partition function as:

$$Q(T,V,N) = Q_{int} \prod^{c} (N_i!\, \lambda_i^{3N_i})^{-1} \int_V \cdots \int_V e^{-\Phi/kT}\, d\mathbf{r}_1 \ldots d\mathbf{r}_N, \tag{6}$$

where Q_{int} is the internal partition function, $\prod$ denotes the product over i, λ_i is the De Broglie thermal wavelength of component i,

$$\lambda_i = h/(2\pi m_i kT), \tag{7}$$

h is Planck's constant, m_i is the mass of molecule i, Φ is the total potential energy of the system and $\mathbf{r}_i$ is the position vector of molecule i. All the integrals in Equation (6) are carried over the system volume.

DISTRIBUTION FUNCTIONS

The partition function can be evaluated, at least in principle, by carrying out the integrations in Equation (6) for a substance with known potential function. However, this task is rather difficult because of the very large number of molecules involved in real systems. A more convenient statistical mechanical formulation is based on the concept of distribution functions. The probability, $P^{(N)}$, of finding molecule 1 in volume element $d\mathbf{r}_1$ at $\mathbf{r}_1$, molecule 2 in volume element $d\mathbf{r}_2$ at $\mathbf{r}_2, \ldots$, and molecule N in volume element $d\mathbf{r}_N$ at $\mathbf{r}_N$ is given by (Hill, 1956)

$$P^{(N)}\, d\mathbf{r}_1 \ldots d\mathbf{r}_N = e^{-\Phi/kT} d\mathbf{r}_1 \ldots d\mathbf{r}_N / Z(T,V,N), \tag{8}$$

where Z(T,V,N) is the configurational integral,

$$Z(T,V,N) = \int \ldots \int e^{-\Phi/kT}\, d\mathbf{r}_1 \ldots d\mathbf{r}_N. \tag{9}$$

Usually we are interested in the relative position of two molecules, irrespective of the location of the other molecules in the system. This can be obtained by integrating Equation (8) over the positions of all molecules except those which we are interested in. This leads to the definition of the distribution function, $\rho^{(2)}{}_{ij}(\mathbf{r}_1,\mathbf{r}_2)$, which gives the probability of finding a molecule of type i in $d\mathbf{r}_1$ at $\mathbf{r}_1$ and molecule of type j in $d\mathbf{r}_2$ at $\mathbf{r}_2$,

$$\rho^{(2)}{}_{ij}(\mathbf{r}_1,\mathbf{r}_2) = N_i\,(N_j - \delta_{ij}) \int \ldots \int e^{-\Phi/kT}\, d\mathbf{r}_3 \ldots d\mathbf{r}_N / Z(T,V,N), \tag{10}$$

where δ_{ij} is the Kronecker delta. Note that $\rho^{(2)}{}_{ij}(\mathbf{r}_1,\mathbf{r}_2)$ depends on temperature, density, and composition in addition to $\mathbf{r}_1$ and $\mathbf{r}_2$. For molecules which interact with radially symmetric potential functions $\rho^{(2)}{}_{ij}$ depends only on the distance between the centers of masses,

$$\rho^{(2)}{}_{ij}(\mathbf{r}_1,\mathbf{r}_2) = \rho^{(2)}{}_{ij}(r_{12}); \quad r_{12} = |\mathbf{r}_1 - \mathbf{r}_2|.$$

In the limit of ideal gas ($\Phi/kT \to 0$) the distribution function $\rho^{(2)}{}_{ij}(\mathbf{r}_1,\mathbf{r}_2)$ approaches the value $N_i\,(N_j - \delta_{ij})/V^2$. This suggests defining the pair (or radial) distribution function, $g_{ij}(r)$, by

$$g_{ij}(r) = \rho^{(2)}{}_{ij}(r) V^2/(N_i N_j), \tag{11}$$

which approaches $1-\delta_{ij}/N_j$ in the ideal gas limit. Combining Equations (10) and (11) gives

$$g_{ij}(r) = V^2(1-\delta_{ij}/N_j) \int \ldots \int e^{-\Phi/kT}\, d\mathbf{r}_3 \ldots d\mathbf{r}_N / Z(T,V,N). \tag{12}$$

A similar definition for the distribution function exists in the grand canonical ensemble as the following

$$g_{ij}(r) = V^2/(N_i N_j) \sum_{N_i \geq i} \sum_{N_j \geq j} \cdots \sum_{N_c}^{c} \pi (N!_k) e^{\mu . N/kT} \int \cdots \int e^{-\Phi/kT} dr_3 \ldots dr_N \Big/ \Xi(T,V,\mu) \qquad (13)$$

The pair (or radial) distribution function in the grand canonical ensemble approaches unity in the limit of ideal gas.

RELATIONS BETWEEN THERMODYNAMIC PROPERTIES AND g(r)

The relation between g(r) and the internal energy, U, can be derived from the definition of the internal energy in terms of the canonical ensemble partition function:

$$U - U^\circ = kT^2 \left[\partial \ln Z(T,V,N)/\partial T\right]_{N,V} , \qquad (14)$$

where U° is the ideal gas internal energy. Substituting for Z(T,V,N) from Equation (9) we get

$$U - U^\circ = \int \cdots \int e^{-\Phi}/kT \, dr_1 \ldots dr_N \Big/ Z(T,V,N). \qquad (15)$$

For pure fluids under the assumption of pairwise additivity of intermolecular interactions one can write Φ as the summation of N(N-1)/2 potential energy terms

$$\Phi = \sum_{i=1}^{N} \sum_{j<i}^{N} \phi(r_{ij}). \qquad (16)$$

Upon substitution of Equation (16) into Equation (15) the relation between the internal energy and the radial distribution function for pure fluids is obtained:

$$U - U^\circ = 1/2 \, N \rho \int_0^\infty \phi(r) \, g(r) \; 4\pi \, r^2 \, dr, \qquad (17)$$

where ρ is the molecular number density, N/V. Equation (17) can also be derived from the grand canonical ensemble distribution function. For mixtures of c-components Equation (17) takes the form

$$U - U^\circ = 1/2 \, N \rho \sum_{i=1}^{c} \sum_{j=1}^{c} x_i \, x_j \int_0^\infty \rho_{ij}(r) \, g_{ij}(r) \; 4\pi \, r^2 \, dr, \qquad (18)$$

where x_i is the mole fraction of component i and ϕ_{ij} is the pair intermolecular potential energy function for molecules of type i and j. For pure fluids the pressure is given in terms of g(r) by (Hill, 1956)

$$P = kT\rho - \rho^2/6 \int_0^\infty r\,\phi'(r)\,g(r)\;4\pi\,r^2\,dr. \tag{19}$$

For mixtures, Equation (18) takes the following form

$$P = kT\rho - \rho^2/6 \sum_{i=1}^{c}\sum_{j=1}^{c} x_i\,x_j \int_0^\infty r\,\phi'_{ij}(r)\,g_{ij}(r)\;4\pi\,r^2\,dr, \tag{20}$$

where ϕ' is the derivative of the potential function with respect to the intermolecular distance r. The assumption of pairwise additivity of the potential energy function is not generally needed in deriving the internal energy and pressure equations. However, definition of the functional form of the intermolecular potential energy function is necessary for these two equations. A relation between the isothermal compressibility, $\kappa_T=-(1/v)(\partial v/\partial P)_T$, and g(r) can be derived without the need to know the functional form of the intermolecular potential. This relation is obtained from the grand canonical ensemble radial distribution function (Hill, 1956)

$$\rho\,kT\kappa_T = 1 + \rho \int_0^\infty [\,g(r) - 1\,]\;4\pi\,r^2\,dr. \tag{21}$$

The corresponding relation for mixtures was derived by Kirkwood and Buff (1951)

$$\rho\;kT\kappa_T = \frac{|B|}{\rho \sum\sum x_i\,x_j\,|B|_{ij}}, \tag{22}$$

where $|B|_{ij}$ symbolizes the cofactor of the element B_{ij} in the c x c matrix B and $|B|$ is the determinant of B. The elements, B_{ij}, of the matrix are:

$$B_{ij} = \rho\,x_i\,[\delta_{ij} + \rho\,x_j G_{ij}], \tag{23}$$

where G_{ij} is called the radial distribution function integral, the fluctuation integral, and the Kirkwood-Buff integral. It is defined by the following expression

$$G_{ij} = \int_0^\infty [g_{ij}(r) - 1] 4\pi r^2\,dr. \tag{24}$$

In addition to the mixture isothermal compressibility, the Kirkwood-Buff solution theory gives the partial molar volumes and the derivatives of the chemical potential with respect to mole fractions in terms of the fluctuation integrals. Details of this theory are discussed in the last section of this chapter.

DIRECT CORRELATION FUNCTIONS

In the compressibility equation, Equation (21), the term [g(r)-1] in the integrand is known as the total correlation function. This is because it gives the total influence of molecule 1 on molecule 2 at a distance r. Ornstein and Zernike (1914) suggested dividing the total correlation function, h(r)=g(r)-1, into two parts, a direct part and an indirect part. They defined the direct correlation function, c (r), by

$$c(r_{12}) = h(r_{12}) - \rho \int c(r_{13})h(r_{23})dr_3. \tag{25}$$

The compressibility equation, (21), can be expressed in terms of c (r) by using Fourier transforms (McQuarrie, 1975). The resulting expression is:

$$(kT)^{-1} [\partial P/\partial \rho]_T = 1 - \rho \int c(r)dr. \tag{26}$$

For mixtures the direct correlation function of the pair ij is defined by

$$c_{ij}(r_{12}) = h_{ij}(r_{12}) - \sum_{k=1}^{c} \rho_k \int c_{ik}(r_{13})h_{jk}(r_{23})dr_3, \tag{27}$$

where

$$h_{ij}(r) = g_{ij}(r) - 1 \tag{28}$$

and $\rho_k = \rho x_k$.

DISTRIBUTION FUNCTION APPROXIMATIONS

Obtaining thermodynamic properties from the relations given in the above sections requires knowledge of the distribution functions. Evaluating the distribution functions from microscopic information, such as the intermolecular potential energy, is not easier than evaluating the partition function. However, the distribution functions are easier to deal with because of their physical interpretation. Next we will look into some of the approximations which were suggested to allow calculating the distribution functions without having to perform the large number of integrations in Equations (12) and (13).

In the previous section three basic relations between thermodynamic properties of mixtures and distribution functions were derived. Those relations are the energy equation, (18), the pressure equation, (20), and the compressibility equation, (22). The pressure and energy equations, in the form they are reported here, are based on two assumptions: 1) Classical mechanics and 2) Pairwise additivity of the total intermolecular potential energy. In addition, they are not valid for fluids with angle-dependent potentials in their present form. For such fluids the dependence of the pair distribution function on the various angles needs to be known. The compressibility equation is free from these assumptions. The dependence of the distribution function on the various angles is not necessary for obtaining thermodynamic properties. The fact that the compressibility equation is free from the assumption of any form for the potential energy function is obvious from its derivation. However, its validity for quantum fluids is not so obvious. This was shown by Buff and Brout (1955).

Historically, the first approximation that allows calculating the distribution functions was suggested by Kirkwood. This assumption is called the superposition approximation (Hill, 1956):

$$\rho^{(3)}(\mathbf{r}_1,\mathbf{r}_2,\mathbf{r}_3) = \rho^{(2)}(\mathbf{r}_1,\mathbf{r}_2)\,\rho^{(2)}(\mathbf{r}_1,\mathbf{r}_3)\,\rho^{(2)}(\mathbf{r}_2,\mathbf{r}_3). \tag{29}$$

The motivation for making this approximation is the existence of exact equations (within the pairwise additivity assumption) for n-molecule distribution functions in terms of (n+1)-molecule functions which were first derived by Yvon (1935). For n=2 the equation is (Boublik, 1980):

$$\begin{aligned} kT\partial \ln\rho^{(2)}(\mathbf{r}_1,\mathbf{r}_2)/\partial\mathbf{r}_1 &= -\,\partial\rho(\mathbf{r}_1,\mathbf{r}_2)/\partial\mathbf{r}_1 \\ &\quad -\int \partial\rho(\mathbf{r}_1,\mathbf{r}_3)/\partial\mathbf{r}_1\,\rho^{(3)}(\mathbf{r}_1,\mathbf{r}_2,\mathbf{r}_3)/\rho^{(2)}(\mathbf{r}_1,\mathbf{r}_2)\,d\mathbf{r}_3. \end{aligned} \tag{30}$$

Combining equations (29) and (30) yields the Yvon-Born-Green integro-differential equation (Yvon, 1935; Born and Green, 1946). A similar equation was derived based on the notion of the coupling parameter (Kirkwood and Monroe, 1941; Kirkwood and Boggs, 1942). Unfortunately, both equations are nonlinear and difficult to solve, even numerically. Salpeter (1958) derived another exact relation between the two-body and three-body distribution functions. However, this relation is of limited practical use because of its infinite number of terms which are related to irreducible clusters (Salpeter, 1958).

The method of topological reduction was also utilized to derive integral equations for the distribution functions. In this approach a relation between the total and the direct correlation functions is assumed, in addition to the relation which defines the direct correlation function. The Percus-Yevick (Percus and Yevick, 1958) and the hypernetted chain (van Leeuwen et al., 1959) integral equations are derived using this method. They are based on the assumptions:

$$c\,(r) = g(r)[1 - e^{\phi/kT}] \qquad \text{(Percus-Yevick)} \qquad (31)$$

$$c\,(r) = g(r) - \ln[g(r)] - 1 - \phi/kT \qquad \text{(hypernetted chain)} \qquad (32)$$

The distribution functions c (r) and g (r) can be solved for by combining either Equations (31) or (32) with the definition of c (r), Equation (27). The extension of these equations to mixtures is straightforward.

Except for the simplest models of pair interactions, such as the hard-sphere model, mixture calculations using the integral equations approach are rather complicated. This problem and the fact that for real fluids the potential energy functions are not generally well understood makes the above approach unsuitable for getting real fluid mixture properties. For the time being, it seems that the best approach is to start from known pure fluid properties (in the form of equations of state, for example) and develop mixture models that utilize these properties in predicting mixture behavior. Some of the models which have been successful in this field are the conformal solution theory and some versions of the perturbation and variational theories (Lucas, 1986; Rowlinson, 1982).

The conformal solution theory (Brown, 1957; Massih and Mansoori, 1983) presents another alternative for calculating mixture properties. The basic assumption in this theory is that all species interact by potential functions which have the same functional forms. They differ only in the values of the potential parameters. Radial distribution functions of mixtures are approximated by that of a pure reference fluid with appropriate scalings of distance, temperature and density. Van der Waals mixture theory is based on the conformal solution theory. A number of other conformal solution theories have been derived. The mean density approximation (Mansoori and Leland, 1972), and the density expansion theory (Mansoori and Ely, 1985a) are among those suggested.

In the perturbation theory (McQuarrie and Katz, 1966; Barker and Henderson, 1967; Weeks, et al., 1971) one divides the intermolecular potential energy function into a reference part and a perturbation part. The reference part represents a potential model for which the thermodynamic properties are known, such as the hard-sphere model. The variational theory (Mansoori and Canfield, 1969; Mansoori and Leland, 1970; Hamad and Mansoori, 1987) provides inequalities which give least upper bound and highest lower bound to the Helmholtz free energy. In both theories a reference system, for which thermodynamic properties and radial distribution functions are known, is needed. Mixture calculations based on these theories, although simpler than the integral equations approach, are still lengthy and usually no closed form expressions can be obtained from them.

FLUCTUATION THEORY OF MIXTURES

The Kirkwood-Buff solution theory is useful in analysis and prediction of properties of asymmetric and/or highly polar mixtures for which there is limited or no knowledge about their intermolecular potential energy functions (Landau and Lifshitz, 1980). The basic relation of the Kirkwood-Buff solution theory is derived using the grand canonical ensemble theory and it is as follows (Kirkwood and Buff, 1951):

$$kT[\partial\langle N_i\rangle/\partial\mu_j]_{V,T,\mu i\neq j} = \langle N_i\rangle\langle N_j\rangle G_{ij}/V + \delta_{ij}\langle N_i\rangle \quad (33)$$

In this expression k is the Boltzmann constant, T is the absolute temperature, V is the total volume, $\langle N_i\rangle$ is the average number of particles of type i in the grand canonical ensemble, μ_j is the chemical potential of component j, δ_{ij} is the Kronecker delta and G_{ij} is defined by Equation (24). For binary mixtures, Equation (33) reduces to the following equation:

$$[\partial\mu_1/\partial x_1]_{P,T} = kT/[x_1\{1+x_1x_2\rho(G_{11}+G_{22}-2G_{12})\}]. \quad (34)$$

Provided information is available about G_{ij} integrals, defined by Equation (24), Equation (34) can be used to calculate chemical potentials of components of a mixture. Also, the information about G_{ij} can provide us with the means of calculating other properties of a mixture such as the isothermal compressibility κ_T, as given by Equation (22), and the partial molar volumes v_i as given by the following expression:

$$\rho v_i = \sum_{j=1}^{c} x_j |B|_{ij} \Big/ \sum_{j=1}^{c}\sum_{k=1}^{c} x_j x_k |B|_{jk}, \quad (35)$$

where c is the number of components in the mixture, B is a cXc matrix with elements, B_{ij}, defined by Equation (23). For binary mixtures the above relation reduce to the following equation:

$$v_1 = \frac{1 + x_2\rho(G_{22}-G_{12})}{\rho\,[1+x_1x_2\rho(G_{11}+G_{22}-2G_{12})]} \quad (36)$$

The above equations have been studied extensively for the case of infinitely dilute solutions (Ben-Naim, 1974 and 1980). The major difficulty in utilizing the fluctuation theory in mixture property calculations is the lack of knowledge about the fluctuation integrals, G_{ij}'s. For solutions with finite compositions expansions of the G_{ij}'s in powers of concentration are available (Buff and Brout, 1955; Buff and Schindler, 1958). However, the coefficients in these expansions are given in terms of third, fourth and higher order correlation functions. Since little is known about correlation functions of order higher than two, the expansions are in the time being of limited practical use.

The Kirkwood-Buff fluctuation theory has been used by Mazo (1958) as a starting point for the development of a theory based on the perturbation expansion of the excess free energy of mixtures. It has been applied to classical mixtures as well as to quantum mechanical isotope mixtures.

Theoretical calculation of G_{ij} integrals requires the knowledge about the radial distribution functions, g_{ij}. The radial distribution functions can generally be calculated using the theory of intermolecular potential energy functions in the context of a partition function. However, for complex mixtures, such as mixtures consisting of asymmetric, highly polar, and associating molecules the intermolecular potential energy functions are not very well known. Also, the existing techniques of calculating radial distribution functions from the knowledge of intermolecular potential energy functions require extensive numerical calculations (McQuarrie, 1975).

Pearson and Rushbrooke (1957) reformulated the fluctuation theory expressions, Equations 34-36 in terms of direct correlation function integrals, C_{ij}, as defined by

$$C_{ij} = \int_V c_{ij}(\mathbf{r})d\mathbf{r}. \tag{37}$$

In this equation $c_{ij}(\mathbf{r})$ is the direct correlation function, which is related to the radial distribution function as defined by Equations (27) and (28). The chemical potential is related to the direct correlation function integrals as follows (O'Connell, 1971):

$$(kT)^{-1}\,[\partial\mu_i/\partial\rho_j]_{T,\rho k\neq j} = \delta_{ij}/\rho_i - C_{ij} \tag{38}$$

For example, for binary mixtures the relation between the chemical potential and the direct correlation function integrals will reduce to the following equation:

$$x_1[\partial\mu_1/\partial x_1]_{P,T} = kT\frac{1-x_1\rho C_{11}-x_2\rho C_{22}+x_1x_2\rho^2(C_{11}C_{22}-C_{12}^2)}{1-\Sigma\Sigma x_i x_j \rho C_{ij}} \tag{39}$$

In terms of the direct correlation function integrals the isothermal compressibility and the the partial molar volume become (Pearson and Rushbrooke, 1957; O'Connell, 1971).

$$(\rho kT\kappa)^{-1} = 1-\rho\sum_{i=1}^{c}\sum_{j=1}^{c} x_i x_j C_{ij} \tag{40}$$

and

$$\rho\overline{v}_i = \left(1-\rho\sum_{j=1}^{c} x_j C_{ij}\right) \Big/ \left(1-\rho\sum_{j=1}^{c}\sum_{k=1}^{c} x_j x_k C_{jk}\right) \tag{41}$$

According to Equations (38)-(41) the Kirkwood-Buff expression is reformulated with respect to another molecular function, c_{ij} rather than g_{ij}. It should be pointed out that the following general relation in matrix form holds between the fluctuation integrals, G_{ij}'s, and the direct correlation function integrals, C_{ij}'s (O'Connell, 1971),

$$\underline{\underline{G}} = \underline{\underline{C}} + \underline{\underline{G}}\,\underline{\underline{X}}\,\underline{\underline{C}} \tag{42}$$

where $\underline{\underline{G}}$ is a cxc matrix with elements G_{ij} as given by Eq. (24), $\underline{\underline{C}}$ is a cxc matrix with elements C_{ij} given by Eq. (37), and $\underline{\underline{X}}$ is a diagonal matrix with elements $X_{ij}=\langle Ni/N\rangle$. Eq. (42) reduces to the following expressions in a binary mixture

$$\rho G_{ii} = \frac{\rho C_{ii}-(1-x_i)\rho^2(C_{11}C_{22}-C_{12}^2)}{1-x_1\rho C_{11}-x_2\rho C_{22}+x_1x_2\rho^2(C_{11}C_{22}-C_{12}^2)}, \quad i=1,2 \tag{43}$$

$$\rho G_{12} = \frac{\rho C_{12}}{1-x_1\rho C_{11}-x_2\rho C_{22}+x_1x_2\rho^2(C_{11}C_{22}-C_{12}^2)}, \tag{44}$$

or

$$\rho C_{ii} = \frac{\rho G_{ii}+(1-x_i)\rho^2(G_{11}G_{22}-G_{12}^2)}{1+x_1\rho G_{11}+x_2\rho G_{22}+x_1x_2\rho^2(G_{11}G_{22}-G_{12}^2)}, \quad i=1,2 \tag{45}$$

$$\rho C_{12} = \frac{\rho G_{12}}{1+x_1\rho G_{11}+x_2\rho G_{22}+x_1x_2\rho^2(G_{11}G_{22}-G_{12}^2)} \, . \quad (46)$$

The equations and conceptions illustrated in the previous sections constitute a useful basis for a better comprehension of the various subjects presented in the following chapters of this volume. The books and articles quoted in the present paper are the most appropriate for those readers who look for a deeper mastering of the matter.

ACKNOWLEDGMENT. This work has been supported by the NATO Advanced Study Program.

REFERENCES

Abrams, D.S. and J.M. Prausnitz. 1975. Statistical thermodynamics of liquid mixtures: A new expression for the excess Gibbs energy of partly or completely miscible systems. AIChE J. 21: 116-128.

Barker, J.A. and D. Henderson. 1967. Perturbation theory and equation of state for fluids: The square-well potential. J. Chem. Phys. 47: 2856-2861.

Ben-Naim, A. 1974. Water and Aqueous Solutions. New York: Plenum Press.

Ben-Naim, A. 1980. Hydrophobic Interactions. New York: Plenum Press.

Born, M. and H.S. Green. 1946. A general kinetic theory of liquids. I: The molecular distribution functions. Proc. R. Soc. London Ser. A 188: 10-18.

Boublik, T., I. Nezbeda and K. Hlavaty. 1980. Statistical thermodynamics of simple liquid mixtures. Amsterdam: Elsevier Scientific Publ. Co.

Brown, W.B. 1957. The second-order theory of conformal solutions. Proc. R. Soc. London, Ser A 240: 561-580.

Buff, F.P. and R. Brout. 1955. Molecular formulation of thermodynamic functions encountered in solution theory. J. Chem. Phys. 23: 458-465.

Buff, F.P. and F.M. Schindler. 1958. Small perturbations in solution theory. J. Chem. Phys. 29: 1075-1081.

Haile, J.M. and G.A. Mansoori (eds.). 1983. Molecular based study of fluids. Adv. Chem. Series 204. Washington, DC.: ACS.

Hamad, E.Z. and G.A. Mansoori. 1987. Variational theory of mixtures. Fluid Phase Equilib. 37: 255-285.

Hill, T.L. 1956. Statistical mechanics. New York: McGraw-Hill Book Co., Inc.

Hu, Y., E.G. Azevedo, and J.M. Prausnitz. 1983. The molecular basis for local compositions in liquid mixture models. Fluid Phase Equilib. 13: 351-360.

Kamerlingh Onnes, H. 1901. Expression of the equation of state of gases and liquids by means of series. Comm. Phys. Lab. Leiden n.71: 136-158.

Kemeny, S. and P. Rasmussen. 1981. A derivation of local composition expressions from partition functions. Fluid Phase Equilib. 7: 197-203.

Kirkwood, J.G. and E. Monroe. 1941. Statistical mechanics of fusion. J. Chem. Phys. 9: 514-526.

Kirkwood, J.G. and E.M. Boggs. 1942. The radial distribution function in liquids. J. Chem. Phys. 10: 394-402.

Kirkwood, J.G. and F.P. Buff. 1951. The statistical mechanical theory of solutions: I. J. Chem. Phys. 19: 774-777.

Landau, L. and E. M. Lifshitz. 1980. Course of theoretical physics. Vol. 5, Part 1. Oxford: Pergamon Press.

Lucas, K. 1986. Applied statistical thermodynamics. Berlin: Springer Verlag.

Mansoori, G.A. and F.B. Canfield. 1970. Perturbation and variational approaches to equilibrium thermodynamics of gases, liquids and phase transitions. Ind. Eng. Chem. 62: 12-29.

Mansoori, G.A. and T.W. Leland. 1970. Variational approach to the equilibrium properties of simple fluid mixtures, III. J. Chem. Phys. 53: 1931-1936.

Mansoori, G.A. and T.W. Leland. 1972. Statistical thermodynamics of mixtures: A new version for the theory of conformal solution. J. Chem. Soc., Faraday Trans. II 68: 320-344.

Mansoori, G.A. and J.F. Ely. 1985a. Density expansion (DEX) mixing rules: Thermodynamic modeling of supercritical extraction. J. Chem. Phys. 82: 406-413.

Mansoori, G.A. and J.F. Ely. 1985b. Statistical Mechanical theory of local compositions. Fluid Phase Equilib. 22: 253-275.

Massih, A.R. and G.A. Mansoori. 1983. Conformal-solution theory of polar fluids: The statistical mechanical basis of shape factors. Fluid Phase Equilib. 10: 57-72.

Mazo, R.M. 1958. On the statistical mechanical theory of solution. J. Chem. Phys. 29: 1122-1128.

McQuarrie, D.A. and J.L. Katz. 1966. High-temperature equation of state. J. Chem. Phys. 44: 2393-2397.

McQuarrie, D.A. 1975. Statistical mechanics. New York: Harper and Row.

O'Connell, J.P. 1971. Thermodynamic properties of solutions based on correlation functions. Molec. Phys. 20: 27-33.

Ornstein, L.S. and F. Zernike. 1914. Fortuitous density changes and opalescence at the critical point of an element. Verslag Akad. Wetenschappen 23: 582.

Pearson, F.J. and G.S. Rushbrooke. 1957. On the theory of binary fluid mixtures. Proc. Roy. Soc. Edinb. A 64: 305-317.

Percus, J.K. and G.J. Yevick. 1958. Analysis of classical statistical mechanics by means of collective coordinates. Phys. Rev. 110: 1-13.

Prins C. (Editor). 1973. Proc. of the van der Waals centennial Conference on statistical Mechanics. Amsterdam, 27-31 August 1973. Amsterdam: North-Holland Publishing Co.

Renon, H. and J.M. Prausnitz. 1968. Local compositions in thermodynamics excess functions for liquid mixtures. AIChE J. 14: 135-144.

Rowlinson, J.S. and F.L. Swinton. 1982. Liquids and Liquid Mixtures. 3rd Ed. Wolborn, Mass.: Butterworths.

Salpeter, E.E. 1958. On Mayer's theory of cluster expansions. Ann. Phys. 5: 183-223.

Van der Waals, J. D. 1873. Over de continuiteit van den gasen vloeistoftoestand. Ph.D. thesis. Leiden.

Van Leeuwen, J.M., J. Groeneveld, and J. de Boer. 1959. New method for the calculation of the pair correlation function. I. Physica 25: 792-808.

Walas, S.M. 1985. Phase Equilibria in Chemical Engineering. Boston: Butterworths.

Weeks, J.D., D. Chandler and H.C. Andersen. Role of repulsive forces in determining the equilibrium structure of simple liquids. J. Chem. Phys. 54: 5236-5247.

Wilson, G.M. 1964. Vapor-liquid equilibrium. XI: A new expression for the excess free energy of mixing. J. Am. Chem. Soc. 86: 127-130.

Yvon, J. 1935. Actualities Scientifiques et Industriel. Paris: Herman et Cie.

The Statistical Mechanical Theory of Solutions. I

John G. Kirkwood and Frank P. Buff*
Gates and Crellin Laboratories of Chemistry
California Institute of Technology
Pasadena, CA†

A general statistical mechanical theory of solutions is developed with the aid of the theory of composition fluctuations in the grand canonical ensemble. It is shown that the derivatives of the chemical potentials and osmotic pressure with respect to concentrations, the partial molar volumes, and compressibility may be expressed in terms of integrals of the radial distribution functions of the several types of molecular pairs present in the solution. Explicit coefficients of a q-fraction expansion of the thermodynamic variables are presented in a detailed treatment of the two-component system.

I

THIS paper will be concerned with the development of a general statistical mechanical theory of solutions which is applicable to all types of intermolecular interaction and is valid both classically and quantum mechanically. The theory of the grand canonical ensemble is employed on the one hand to relate composition fluctuations to derivatives of the chemical potentials of the components and on the other hand to relate them to integrals of the radial distribution functions of the several types of molecular pairs present in the solution. When the composition fluctuations are themselves eliminated from these relations, expressions for the derivatives of the chemical potentials and osmotic pressure with respect to concentrations, the partial molar volumes, and compressibility are obtained in terms of integrals of the radial distribution functions. For short-range intermolecular forces, these integrals may be developed in power series in the concentrations of the solutes, the coefficients of which involve distribution functions of sets n of molecules at infinite dilution in the solvent. The resulting expansion for the osmotic pressure is identical with that given earlier by McMillan[1] and Mayer in their valuable paper on solution theory.

In Part II of this article, following a discussion of notation and the definition of various molecular distribution functions, we present a review of the theory of composition fluctuations in the grand ensemble. The results are subsequently employed to evaluate important thermodynamic variables of a multicomponent system in terms of integrals of the radial distribution functions. The theory provides an alternative to the familiar charging process encountered in electrolyte theory, while in the absence of long-range intermolecular forces, expansions of the excess chemical potentials, partial molar volumes and compressibility in terms of the q-fractions of the solute components may be obtained.

In Part III of the article the two-component system will be treated in detail. Here, anticipating the results of a future publication, we present the dominant q-fraction expansion coefficients of the thermodynamic variables.

II

We first wish to formulate the relations between average densities in systems of molecules and the fluctuations in density. We shall treat explicitly those relations involving the number density in the configuration space of molecular pairs. The results may be generalized without difficulty to densities in the configuration space of sets of n molecules and fluctuations of higher order. We consider an open region of volume v, which is a part of a system of infinite extent, and we select an example of the grand ensemble representing its statistical behavior, for which v contains exactly $N_1, \cdots N_\nu$ molecules of the ν-species of the multicomponent system. For specified configurations, $\mathbf{R}_{i\alpha}$, of all of the molecules, the singlet density $\nu_\alpha^{(1)}(\mathbf{R}_1)$ of species α at a point $\mathbf{R}_1$ in v and the density $\nu_{\alpha\beta}^{(2)}(\mathbf{R}_1, \mathbf{R}_2)$ of ordered pairs of molecules of species $\alpha\beta$ at a point $\mathbf{R}_1$

* AEC Postdoctoral Fellow, 1949–1950. Present address: Department of Chemistry, University of Rochester, Rochester, New York.

† Contribution No. 1548.

[1] W. G. McMillan, Jr., and J. E. Mayer, J. Chem. Phys. 13, 276 (1945).

Reproduced by permission of the American Institute of Physics from J. Chem. Phys. 19, 774-777, 1951.

and species β at $\mathbf{R}_2$ are given by

$$\nu_\alpha^{(1)}(\mathbf{R}_1)=\sum_{i_\alpha=1}^{N_\alpha}\delta(\mathbf{R}_{i_\alpha}-\mathbf{R}_1),$$

$$\nu_{\alpha\beta}^{(2)}(\mathbf{R}_1,\mathbf{R}_2)=\sum_{\substack{i_\alpha=1\\ i_\alpha\neq k_\alpha}}^{N_\alpha}\sum_{k_\beta=1}^{N_\beta}\delta(\mathbf{R}_{i_\alpha}-\mathbf{R}_1)\delta(\mathbf{R}_{k_\beta}-\mathbf{R}_2), \tag{1}$$

where $\delta(\mathbf{R}_{i_\alpha}-\mathbf{R}_1)$ is the three-dimensional Dirac delta-function. These densities possess by virtue of their definitions the following integrals,

$$\int^v \nu_\alpha^{(1)}(\mathbf{R}_1)dv_1=N_\alpha,$$

$$\int^v\int^v \nu_{\alpha\beta}^{(2)}(\mathbf{R}_1,\mathbf{R}_2)dv_1dv_2=N_\alpha N_\beta-N_\alpha\delta_{\alpha\beta}, \tag{2}$$

where $\delta_{\alpha\beta}$ is the Kronecker delta, unity for $\alpha=\beta$ and zero for $\alpha\neq\beta$. We denote by $\rho_\alpha^{(1)}(\mathbf{R}_1)$ and $\rho_{\alpha\beta}^{(2)}(\mathbf{R}_1,\mathbf{R}_2)$ the average number densities in the singlet and pair spaces,

$$\rho_\alpha^{(1)}(\mathbf{R}_1)=\langle\nu_\alpha^{(1)}\rangle_{\text{Av}},$$

$$\rho_{\alpha\beta}^{(2)}(\mathbf{R}_1,\mathbf{R}_2)=\langle\nu_{\alpha\beta}^{(2)}\rangle_{\text{Av}}, \tag{3}$$

where the averages are understood to be taken over all numbers of molecules $N_1\cdots N_\nu$ and over the accessible phase space of each set of molecules with the probability distribution function, P, of the grand canonical ensemble,

$$P=\exp([\Omega+\sum_{\alpha=1}^{\nu}N_\alpha\mu_\alpha-H_{N_1\ldots N_\nu}]/kT), \tag{4}$$

where μ_α is the chemical potential of species α, per molecule, $H_{N_1\ldots N_\nu}$ is the hamiltonian of the set of molecules, $N_1\cdots N_\nu$, and Ω is equal to $-pv$ to within terms of thermodynamically negligible order of magnitude. Calculation of the mean values of both sides of Eqs. (3) yields the following expressions for the integrals of the mean densities,

$$\int^v \rho_\alpha^{(1)}(\mathbf{R}_1)dv_1=\langle N_\alpha\rangle_{\text{Av}},$$

$$\int^v\int^v \rho_{\alpha\beta}^{(2)}(\mathbf{R}_1,\mathbf{R}_2)dv_1dv_2=\langle N_\alpha N_\beta\rangle_{\text{Av}}-\delta_{\alpha\beta}\langle N_\alpha\rangle_{\text{Av}}, \tag{5}$$

$$\int^v\int^v \{\rho_{\alpha\beta}^{(2)}(\mathbf{R}_1,\mathbf{R}_2)-\rho_\alpha^{(1)}(\mathbf{R}_1)\rho_\beta^{(1)}(\mathbf{R}_2)\}dv_1dv_2$$
$$=[\langle N_\alpha N_\beta\rangle_{\text{Av}}-\langle N_\alpha\rangle_{\text{Av}}\langle N_\beta\rangle_{\text{Av}}]-\delta_{\alpha\beta}\langle N_\alpha\rangle_{\text{Av}}.$$

For a fluid system, liquid solution, or gas mixture, the mean densities have the form

$$\rho_\alpha^{(1)}(\mathbf{R}_1)=\langle N_\alpha\rangle_{\text{Av}}/v=c_\alpha,$$

$$\rho_{\alpha\beta}^{(2)}(\mathbf{R}_1,\mathbf{R}_2)=c_\alpha c_\beta g_{\alpha\beta}^{(2)}(R), \tag{6}$$

$$R=(\mathbf{R}_2-\mathbf{R}_1),$$

where c_α is bulk molecular concentration of species α, and $g_{\alpha\beta}^{(2)}(R)$ is the radial distribution function of species α and β, dependent only on the scalar distance R between the pair of type $\alpha\beta$. With the use of Eq. (6), the last of Eq. (5) assumes the form

$$\int[g_{\alpha\beta}^{(2)}(R)-1]dv=v\frac{\langle N_\alpha N_\beta\rangle_{\text{Av}}-\langle N_\alpha\rangle_{\text{Av}}\langle N_\beta\rangle_{\text{Av}}}{\langle N_\alpha\rangle_{\text{Av}}\langle N_\beta\rangle_{\text{Av}}}-\frac{\delta_{\alpha\beta}}{c_\alpha}, \tag{7}$$

where the integral extends over all relative coordinates of a pair of type $\alpha\beta$. These are the desired relations between the radial distribution functions and the density fluctuations, which form the basis of the present theory. It is to be remarked that although the integral extends only over the relative centers of mass of the molecules of the pair, it has not been assumed that they possess only translational degrees of freedom. The effect of hindered relative rotation for molecules possessing internal degrees of freedom is implicit in the radial distribution functions $g_{\alpha\beta}^{(2)}(R)$, which is to be interpreted as the integral of a more general pair correlation function over all relative orientations of the pair. It is the task of a detailed molecular theory of the molecular distribution functions to determine the implicit influence of hindered rotation, which in many cases is of decisive importance on $g_{\alpha\beta}^{(2)}$.

Next we recall that the composition fluctuations of a thermodynamically open system of volume v are directly related to thermodynamic variables by means of the Gibbs grand canonical ensemble.[2–4] The well-known relations are

$$\langle N_\alpha N_\beta\rangle_{\text{Av}}-\langle N_\alpha\rangle_{\text{Av}}\langle N_\beta\rangle_{\text{Av}}=|A|_{\alpha\beta}/|A|,$$

$$|A|=|A_{\alpha\beta}|;\quad A_{\alpha\beta}=(1/kT)(\partial\mu_\alpha/\partial N_\beta)_{T,V,N_\gamma}, \tag{8}$$

where $|A|_{\alpha\beta}$ symbolizes the cofactor of $A_{\alpha\beta}$ in the determinant $|A|$, μ_α is the chemical potential, per molecule, of species α, T is the thermodynamic temperature, and k is Boltzmann's constant.

Elimination of the composition fluctuations from Eq. (8) immediately leads to expressions for useful thermodynamic variables. Upon taking the inverse of Eq. (8) we obtain for an infinite system, with the use of Eq. (7),

$$(1/kT)(\partial\mu_\alpha/\partial N_\beta)_{T,V,N_\gamma}=|B|_{\alpha\beta}/v|B|,$$

$$B_{\alpha\beta}=c_\alpha\delta_{\alpha\beta}+c_\alpha c_\beta G_{\alpha\beta}, \tag{9}$$

$$G_{\alpha\beta}=\int[g_{\alpha\beta}^{(2)}(R)-1]dv.$$

In order to obtain expressions for variables measured at the constant pressure p we employ the Gibbs-Duhem

[2] H. C. Brinkman and J. J. Hermans, J. Chem. Phys. 17, 574 (1949).
[3] J. G. Kirkwood and R. J. Goldberg, J. Chem. Phys. 18, 54 (1950).
[4] W. H. Stockmayer, J. Chem. Phys. 18, 58 (1950).

equation,

$$\sum_{\alpha=1}^{\nu} N_\alpha \left(\frac{\partial \mu_\alpha}{\partial N_\beta}\right)_{T,p,N_\gamma} = 0 \qquad (10)$$

and the mathematical relation

$$(\partial \mu_\alpha/\partial N_\beta)_{T,V,N_\gamma} = (\partial \mu_\alpha/\partial N_\beta)_{T,p,N_\gamma} + \bar{v}_\alpha \bar{v}_\beta/\kappa v, \qquad (11)$$

where $\bar{v}_\alpha$ is the partial molar volume, per molecule, of species α and κ is the compressibility of the solution. Use of Eqs. (9), (10), and (11) leads after a straightforward transformation to the desired expressions for the partial molar volumes, compressibility, and derivatives of the chemical potentials,

$$(v/kT)(\partial \mu_\alpha/\partial N_\beta)_{T,p,N_\gamma} = |\Delta|_{\alpha\beta}/c_\alpha c_\beta|\Delta|; \qquad \alpha,\ \beta = 2, \cdots \nu, \qquad (12)$$

$$\Delta_{\alpha\beta} = (\delta_{\alpha\beta}/c_\alpha) + (1/c_1) + G_{\alpha\beta} + G_{11} - G_{1\alpha} - G_{1\beta},$$

$$\bar{v}_\alpha = \sum_{\beta=1}^{\nu} c_\beta |B|_{\alpha\beta} \Big/ \sum_{\beta,\gamma=1}^{\nu} c_\beta c_\gamma |B|_{\beta\gamma}, \qquad (13)$$

$$\kappa kT = |B| \Big/ \sum_{\alpha,\beta=1}^{\nu} c_\alpha c_\beta |B|_{\alpha\beta}. \qquad (14)$$

The derivative of the osmotic pressure π of a solution, solvent 1 and solutes 2, $\cdots$ may be calculated in a similar manner from Eq. (9) and the thermodynamic equation,

$$\left(\frac{\partial \pi}{\partial c_\alpha}\right)_{T,\mu_1,c_\gamma} = \sum_{\beta=2}^{\nu} c_\beta \left(\frac{\partial \mu_\beta}{\partial c_\alpha}\right)_{T,\mu_1,c_\gamma} \qquad (15)$$

with the result

$$\left(\frac{\partial \pi}{\partial c_\alpha}\right)_{T,\mu_1,c_\gamma} = kT \sum_{\beta=2}^{\nu} \frac{c_\beta |B'|_{\alpha\beta}}{|B'|}, \qquad (16)$$

$$B_{\alpha\beta}' = c_\alpha \delta_{\alpha\beta} + c_\alpha c_\beta G_{\alpha\beta}.$$

It will be noted that this expression does not contain explicitly solvent-solvent and solvent-solute interaction terms.

III

We shall now present a more detailed treatment of the two component system, solvent 1 and solute 2. In this case Eqs. (12), (13), (14), and (16) may be exhibited in the form

$$\bar{v}_1 = \frac{1 + (G_{22} - G_{12})c_2}{c_1 + c_2 + c_1 c_2 (G_{11} + G_{22} - 2G_{12})}, \qquad (17)$$

$$c_1 \bar{v}_1 + c_2 \bar{v}_2 = 1,$$

$$\kappa kT = \frac{1 + G_{11}c_1 + G_{22}c_2 + (G_{11}G_{22} - G_{12})^2 c_1 c_2}{c_1 + c_2 + (G_{11} + G_{22} - 2G_{12})c_1 c_2}, \qquad (18)$$

$$(\partial \pi/\partial c_2)_{T,\mu_1} = kT/(1 + G_{22}c_2), \qquad (19)$$

$$\frac{1}{kT}\left(\frac{\partial \mu_2}{\partial c_2}\right)_{T,p} = \frac{1}{c_2} + \frac{G_{12} - G_{22}}{1 + c_2(G_{22} - G_{12})}, \qquad (20)$$

$$\frac{1}{kT}\left(\frac{\partial \mu_1}{\partial c_1}\right)_{T,p} = \frac{1}{c_1} + \frac{G_{12} - G_{11}}{1 + c_1(G_{11} - G_{12})}, \qquad (21)$$

$$\frac{1}{kT}\left(\frac{\partial \mu_2}{\partial x_2}\right)_{T,p} = \frac{1}{x_2} + \frac{c_1(2G_{12} - G_{11} - G_{22})}{1 + c_1 x_2 (G_{11} + G_{22} - 2G_{12})}, \qquad (22)$$

where $x_\alpha = N_\alpha/(N_1 + N_2)$, is the mole fraction of component α.

The preceding relations are completely general, and it is of interest to remark that in electrolyte theory Eq. (20) provides an alternative to the usual charging process. For example, the well-known limiting law immediately follows when G_{22} is calculated by means of the Debye-Huckel radial distribution function. In the absence of long-range intermolecular forces, the integrals $G_{\alpha\beta}$ may be developed in power series in c_2 the coefficients of which involve distribution functions of sets n of molecules at infinite dilution in the solvent. This development, which leads to the solution theory of McMillan and Mayer[1] as a special case, constitutes a generalization of the methods of Part II and will be described in a forthcoming paper. In the remainder of this article these results will be anticipated, and we shall give the explicit expansion coefficients involving the cluster integrals of sets of 3 molecules. The following additional integrals are encountered:

$$\begin{aligned} vG_{\alpha\beta\gamma} = (1/c_\alpha c_\beta c_\gamma) \int\int\int \{ &\rho_{\alpha\beta\gamma}^{(3)}(\mathbf{R}_1, \mathbf{R}_2, \mathbf{R}_3) \\ &- \rho_{\alpha\beta}^{(2)}(\mathbf{R}_1, \mathbf{R}_2)\rho_\gamma^{(1)}(\mathbf{R}_3) \\ &+ 2\rho_\alpha^{(1)}(\mathbf{R}_1)\rho_\beta^{(1)}(\mathbf{R}_2)\rho_\gamma^{(1)}(\mathbf{R}_3) \\ &- \rho_{\alpha\gamma}^{(2)}(\mathbf{R}_1, \mathbf{R}_2)\rho_\beta^{(1)}(\mathbf{R}_3) \\ &- \rho_{\beta\gamma}^{(2)}(\mathbf{R}_1, \mathbf{R}_2)\rho_\alpha^{(1)}(\mathbf{R}_3)\} dv_1 dv_2 dv_3, \end{aligned} \qquad (23)$$

where $\rho_{\alpha\beta\gamma}^{(3)}$ is the mean density of ordered triplets which when expressed in terms of the generalized correlation functions of the centers of mass of the molecules simplify for a fluid system,

$$\begin{aligned} G_{\alpha\beta\gamma} = \int\int\int \{ &2 + g_{\alpha\beta\gamma}^{(3)}(\mathbf{R}_{12}, \mathbf{R}_{13}) - g_{\alpha\beta}^{(2)}(R_{12}) \\ &- g_{\alpha\gamma}^{(2)}(R_{13}) - g_{\beta\gamma}^{(2)}(R_{23})\} dv_{12} dv_{13}, \end{aligned} \qquad (24)$$

$$\rho_{\alpha\beta\gamma}^{(3)} = c_\alpha c_\beta c_\gamma g_{\alpha\beta\gamma}^{(3)}.$$

It is found that the osmotic pressure π is given by

$$\pi = c_2 kT\{1 - \tfrac{1}{2}G_{22}^0 c_2 - \tfrac{1}{3}(G_{222}^0 - 3G_{22}^{02})c_2^2 + \cdots\}, \qquad (25)$$

where the superscript 0 denotes that the cluster integrals are evaluated in the infinitely dilute solvent. Similar expansions in powers of c_2 may be obtained for the other variables, but in most applications it is more convenient to employ expansions in powers of the q_2-fraction which is defined by the relation,

$$q_2 = x_2/(x_1 + m x_2),$$
$$m = \bar{v}_2^0/\bar{v}_1^0; \quad \bar{v}_2^0 = \lim_{c_2=0} \bar{v}_2 = \bar{v}_1^0 + G_{11}^0 - G_{22}^0. \tag{26}$$

The theory leads to the following expansions for the chemical potentials:

$$\mu_2 = kT \log q_2 + \mu_2^E + \mu_2^0(T, p),$$
$$\mu_2^0 = \lim_{q_2=0} [\mu_2 - kT \log q_2], \tag{27}$$

$$\mu_1 = kT \log(1 - m q_2) + \mu_1^E + \mu_1^0(T, p), \tag{28}$$

where

$$\mu_2^E/kT = a_1 q_2 + a_2 q_2^2 + \cdots, \tag{29}$$

$$\mu_1^E/kT = b_1 q_2 + b_2 q_2^2 + b_3 q_2^3 + \cdots, \tag{30}$$

and

$$a_1 = (G_{12}^0 - G_{22}^0)/\bar{v}_1^0,$$
$$a_2 = (\tfrac{1}{2}\bar{v}_1^{02})[3G_{122}^0 + G_{111}^0 - 3G_{112}^0 - G_{222}^0 + 3(G_{22}^0 - G_{12}^0)^2 - (G_{11}^0 - G_{12}^0)^2 + (G_{11}^0 - G_{12}^0)(G_{22}^0 - G_{12}^0) + \bar{v}_1^0(G_{11}^0 + G_{22}^0 - 2G_{12}^0)],$$
$$b_1 = (G_{11}^0 - G_{12}^0)/\bar{v}_1^0,$$
$$b_2 = (\tfrac{1}{2}\bar{v}_1^{02})[(G_{11}^0 - G_{12}^0)^2 + \bar{v}_1^0(G_{11}^0 + G_{22}^0 - 2G_{12}^0)],$$
$$b_3 = (\tfrac{1}{3}\bar{v}_1^{03})\{(G_{11}^0 - G_{12}^0)^3 + \bar{v}_1^0[3(G_{112}^0 - G_{221}^0) + 3(G_{11}^0 - G_{12}^0)^2 - 3(G_{22}^0 - G_{12}^0)^2 + G_{222}^0 - G_{111}^0]\}.$$

Similarly, the partial molar volumes are given by

$$\bar{v}_2 = \bar{v}_2^0 + d_1 q_2 + \cdots, \tag{31}$$

$$\bar{v}_1 = \bar{v}_1^0 - \tfrac{1}{2} d_1 q_2^2 + \cdots, \tag{32}$$

where

$$\bar{v}_2^0 = \bar{v}_1^0 + G_{11}^0 - G_{12}^0,$$
$$d_1 = 2G_{12}^0 - G_{11}^0 - G_{22}^0 + (1/\bar{v}_1^0)[2G_{12}^0 G_{22}^0 + 2G_{112}^0 + G_{11}^{02} - 2G_{12}^{02} - G_{122}^0 - G_{111}^0 - G_{11}^0 G_{22}^0].$$

Finally, the compressibility is found to be

$$\kappa kT = \kappa^0 kT + e_1 q_2 + \cdots, \tag{33}$$

where

$$\kappa^0 kT = \bar{v}_1^0 + G_{11}^0,$$
$$e_1 = G_{12}^0 - G_{11}^0 + (1/\bar{v}_1^0)[G_{112}^0 + G_{11}^{02} - G_{111}^0 - G_{12}^{02}].$$

While mole fraction expansions are not as useful as the preceding expressions, for the sake of completeness we present the expansion for the chemical potential

$$\mu_1/kT = \log x_1 + f_2 x_2^2 + f_3 x_2^3 + \cdots + \mu_1^0/kT, \tag{34}$$

where

$$f_2 = [G_{11}^0 + G_{22}^0 - 2G_{12}^0]/2\bar{v}_1^0,$$
$$f_3 = (\tfrac{1}{3}\bar{v}_1^{02})[3G_{112}^0 + 3(G_{11}^0 - G_{12}^0)^2 - G_{111}^0 - 3G_{221}^0 - 3(G_{22}^0 - G_{12}^0)^2 + G_{222}^0 + 3(G_{12}^0 - G_{11}^0)(G_{11}^0 + G_{22}^0 - 2G_{12}^0)].$$

APPENDIX

Equation (8) is readily derived from the grand partition function,

$$e^{\Omega'} = \sum_{N_1 \cdots N_\nu = 0}^{\infty} \prod_{\alpha=1}^{\nu} \lambda_\alpha^{N_\alpha} \exp\left[-\frac{A(N_1 \cdots N_\nu, T_1 v)}{kT}\right],$$
$$\Omega' = pv/kT; \quad \lambda_\alpha = \exp(\mu_\alpha/kT), \tag{A1}$$

where $A(N_1 \cdots N_\nu, T_1 v)$ is the Helmholtz free energy of a member of the grand ensemble. Repeated differentiation of Ω with respect to the absolute activity λ_α leads to mean values of the composition fluctuations:

$$\lambda_\alpha(\partial\Omega'/\partial\lambda_\alpha)_{T,v,\lambda_\gamma} = \langle N_\alpha \rangle_{\mathrm{Av}},$$
$$\lambda_\alpha\lambda_\beta(\partial^2\Omega'/\partial\lambda_\alpha\partial\lambda_\beta)_{T,v,\lambda_\gamma} + \lambda_\alpha(\partial\Omega'/\partial\lambda_\alpha)_{T,v,\lambda_\gamma}\delta_{\alpha\beta} = \langle N_\alpha N_\beta \rangle_{\mathrm{Av}} - \langle N_\alpha \rangle_{\mathrm{Av}} \langle N_\beta \rangle_{\mathrm{Av}}. \tag{A2}$$

Next by equating the mathematical identity, we obtain

$$\left(\frac{\partial^2\Omega'}{\partial N_\alpha \partial N_\beta}\right)_{T,v,N_\gamma} = \frac{1}{(kT)^2} \sum_{\sigma,\rho=1}^{\nu} \left[\lambda_\sigma\lambda_\rho\left(\frac{\partial^2\Omega'}{\partial\lambda_\sigma\partial\lambda_\rho}\right)_{T,v,N_\gamma} + \lambda_\sigma\left(\frac{\partial\Omega'}{\partial\lambda_\sigma}\right)_{T,v,N_\gamma}\delta_{\sigma\rho}\right]\left(\frac{\partial\mu_\sigma}{\partial N_\alpha}\right)_{T,v,N_\gamma}\left(\frac{\partial\mu_\rho}{\partial N_\beta}\right)_{T,v,N_\gamma} + \frac{1}{kT}\sum_{\sigma=1}^{\nu}\lambda_\sigma\left(\frac{\partial\Omega'}{\partial\lambda_\sigma}\right)_{T,v,N_\gamma}\left(\frac{\partial^2\mu_\sigma}{\partial N_\alpha\partial N_\beta}\right)_{T,v,N_\gamma} \tag{A3}$$

to the result obtained by differentiating the Gibbs-Duhem equation

$$kT\left(\frac{\partial^2\Omega'}{\partial N_\alpha\partial N_\beta}\right)_{T,v,N_\gamma} = \left(\frac{\partial\mu_\alpha}{\partial N_\beta}\right)_{T,v,N_\gamma} + \sum_{\sigma=1}^{\nu} N_\sigma\left(\frac{\partial^2\mu_\sigma}{\partial N_\alpha\partial N_\beta}\right)_{T,v,N_\gamma}. \tag{A4}$$

We finally obtain with use of Eq. (A2)

$$\frac{1}{kT}\left(\frac{\partial\mu_\alpha}{\partial N_\beta}\right)_{T,v,N_\gamma} = \sum_{\sigma,\rho=1}^{\nu}\frac{1}{k^2T^2}[\langle N_\sigma N_\rho\rangle_{\mathrm{Av}} - \langle N_\sigma\rangle_{\mathrm{Av}}\langle N_\rho\rangle_{\mathrm{Av}}] \times \left(\frac{\partial\mu_\sigma}{\partial N_\beta}\right)_{T,v,N_\gamma}\left(\frac{\partial\mu_\rho}{\partial N_\alpha}\right)_{T,v,N_\gamma}, \quad \alpha, \beta = 1, \cdots \nu. \tag{A5}$$

This set of equations is equivalent to the matrix equation,

$$A = ABA,$$
$$A_{\alpha\beta} = A_{\beta\alpha} = (1/kT)(\partial\mu_\alpha/\partial N_\beta)_{T,v,N_\gamma},$$
$$B_{\sigma\rho} = B_{\rho\sigma} = \langle N_\sigma N_\rho\rangle_{\mathrm{Av}} - \langle N_\sigma\rangle_{\mathrm{Av}}\langle N_\rho\rangle_{\mathrm{Av}}. \tag{A6}$$

with the solution,

$$B = A^{-1}; \quad B_{\sigma\rho} = |A|_{\sigma\rho}/|A|.$$

On the Theory of Binary Fluid Mixtures

F. J. Pearson*† and G. S. Rushbrooke
Physics Department, King's College
Newcastle upon Tyne, UK

Synopsis

A number of analogues to the simple fluid compressibility equation are deduced by considering fluctuations in a binary mixture; and their simplest expressions are found to be in terms of the binary mixture direct correlation functions. The accuracy of these results is tested with the aid of the appropriate extension of the approximate Born-Green theory, which facilitates the demonstration of consistency with the first three terms of the virial expansion. The problem of the evaluation of the corresponding radial distribution functions by means of X-ray or optical scattering is taken as far as the determination, in principle, of a concentration correlation function from observations of critical opalescence.

1. Introduction

Two of the many ways of approaching the theory of fluids are based upon the partition, and radial distribution, functions, respectively, of statistical mechanics. Their formal equivalence has been demonstrated by Rushbrooke and Scoins (1953), who considered, in particular, the equations for the pressure and the isothermal compressibility of a simple fluid, both of which involved a single radial distribution function capable of experimental determination by means of X-ray scattering, and related to the direct and indirect correlation functions of Ornstein and Zernike (1914), Zernike (1916), Klein and Tisza (1949) and Rosenfeld (1951). In the case of a binary fluid mixture of A and B systems, the pressure equation has already been given by, for example, Green (1952), in terms of three radial distribution functions $g_{AA}(r)$, $g_{BB}(r)$ and $g_{AB}(r)=g_{BA}(r)$. The corresponding compressibility equation will be discussed in this paper, appropriate direct and indirect correlation functions will be introduced, and a partial answer will be given to the problem of demonstrating that the resulting equations are consistent with the partition function approach epitomized by the virial expansion for a binary mixture given by Mayer (1939). This information will facilitate discussion of X-ray scattering and critical opalescence.

* This forms part of the author's Thesis for the degree of Doctor of Philosophy of the University of Durham, June 1955.

† Present address: Department of Physics, City of Birmingham College of Advanced Technology.

The simple fluid compressibility equation arises from the equivalence of expressions for the mean square fluctuation in the total number of systems in terms of thermodynamic, and radial distribution, functions respectively. The thermodynamics of binary fluctuations have already been considered by Landau and Lifshitz (1938) in the case of dilute solutions, by Kirkwood and Goldberg (1950) and Stockmayer (1950) in connection with the problem of critical opalescence later investigated by Furth and Williams (1954) and Quantie (1954), and by Munster (1953) for multi-component mixtures, but none of this previous work is sufficiently detailed for the present purpose, nor has the relation between fluctuations and binary mixture distribution functions been investigated before. Both aspects of the problem therefore require discussion.

2. The Thermodynamics of Binary Fluctuations

We consider fluctuations in a sub-assembly of N_A A systems and N_B B systems occupying volume V within a larger assembly, which is, by hypothesis, in equilibrium. The probability $P(\Delta S)$ of certain thermodynamic co-ordinates having specified values can be expressed in terms of the temperature T, pressure p and chemical potentials μ_A and μ_B of the A and B systems. It then follows that

$$P(\Delta S) \propto \exp\left[-R(E, T)\right], \tag{2.1}$$

where

$$R(E, T) = (2kT)^{-1}[\Delta S \Delta T - \Delta p \Delta V + \Delta\mu_A \Delta N_A + \Delta\mu_B \Delta N_B], \tag{2.2}$$

a straightforward extension of equation (59.9) of Landau and Lifshitz (1938). Here $R(E, T)$ is essentially positive and capable of expansion in terms of the Helmholtz free energy F, the Gibbs' function or the enthalpy. For example, using F, we obtain

$$\begin{aligned} 2kTR = & -\left(\frac{\partial^2 F}{\partial T^2}\right)_{VN_AN_B}(\Delta T)^2 + \left(\frac{\partial^2 F}{\partial V^2}\right)_{TN_AN_B}(\Delta V)^2 \\ & + \left(\frac{\partial^2 F}{\partial V \partial N_A} + \frac{\partial^2 F}{\partial N_A \partial V}\right)_{TN_B} \Delta V \Delta N_A + \left(\frac{\partial^2 F}{\partial V \partial N_B} + \frac{\partial^2 F}{\partial N_B \partial V}\right)_{TN_A} \Delta V \Delta N_B \\ & + \left(\frac{\partial^2 F}{\partial N_A^2}\right)_{TVN_B}(\Delta N_A)^2 + \left(\frac{\partial^2 F}{\partial N_A \partial N_B} + \frac{\partial^2 F}{\partial N_B \partial N_A}\right)_{TV} \Delta N_A \Delta N_B \\ & + \left(\frac{\partial^2 F}{\partial N_B^2}\right)_{TVN_A}(\Delta N_B)^2. \end{aligned} \tag{2.3}$$

Equations (2.1), (2.2) and (2.3) lead immediately to multi-variate Gaussian distributions for various fluctuations. They can be simplified by replacing extensive by intensive variables. For example, if the densities $\rho_A = N_A/V$ and $\rho_B = N_B/V$ are introduced into (2.3), it can be shown that

$$R = \frac{1}{2kT}\left[-\left(\frac{\partial^2 F}{\partial T^2}\right)_{VN_AN_B} (\Delta T)^2 + V^2\left\{\left(\frac{\partial \mu_A}{\partial N_A}\right)_{TVN_B} (\Delta \rho_A)^2 + \left[\left(\frac{\partial \mu_A}{\partial N_B}\right)_{TVN_A} + \left(\frac{\partial \mu_B}{\partial N_A}\right)_{TVN_B}\right] \Delta\rho_A \Delta\rho_B + \left(\frac{\partial \mu_B}{\partial N_B}\right)_{TVN_A} (\Delta\rho_B)^2 \right\}\right], \tag{2.4}$$

whence fluctuations in number density form a bivariant Gaussian distribution from which it follows that

$$\langle (\Delta N)^2 \rangle_{TV} = V^2 \langle (\Delta\rho)^2 \rangle = \frac{kT\left\{\dfrac{\partial \mu_A}{\partial N_A} - \dfrac{\partial \mu_A}{\partial N_B} - \dfrac{\partial \mu_B}{\partial N_A} + \dfrac{\partial \mu_B}{\partial N_B}\right\}_{TV}}{\dfrac{\partial(\mu_A, \mu_B)}{\partial(N_A, N_B)_{TV}}}, \tag{2.5}$$

where $\langle (\Delta N)^2 \rangle = \langle N^2 \rangle - \langle N \rangle^2$, the bracket $\langle\ \rangle$ denoting a mean value and the suffices indicating that $\langle (\Delta N)^2 \rangle_{TV}$ is independent of fluctuations in temperature, and is taken, by hypothesis, at constant volume. Furthermore, the introduction into (2.3) of $\rho = N/V$ and the concentration $x = x_A = N_A/N$ leads to

$$R = \frac{1}{2kT}\left[-\left(\frac{\partial^2 F}{\partial T^2}\right)_{VN_AN_B} (\Delta T)^2 + \frac{V}{\rho}\left(\frac{\partial p}{\partial \rho}\right)_{xT} (\Delta\rho)^2 + \frac{2V}{\rho}\left(\frac{\partial p}{\partial x}\right)_{\rho T} \Delta\rho\Delta x + V\rho\left(\frac{\partial \mu_A}{\partial x} - \frac{\partial \mu_B}{\partial x}\right)_{\rho T} (\Delta x)^2 \right], \tag{2.6}$$

from which

$$\langle (\Delta N)^2 \rangle_{TV} = V^2 \langle (\Delta\rho)^2 \rangle = NkT\left(\frac{\partial \rho}{\partial p}\right)_{xT} \left[1 + \frac{\dfrac{1}{\rho^2}\left(\dfrac{\partial p}{\partial x}\right)_{\rho T}\left(\dfrac{\partial \rho}{\partial x}\right)_{pT}}{\left(\dfrac{\partial \mu_A}{\partial x} - \dfrac{\partial \mu_B}{\partial x}\right)_{\rho T}} \right]^{-1}, \tag{2.7}$$

$$\langle (\Delta x)^2 \rangle = \frac{kT}{N\left(\dfrac{\partial \mu_A}{\partial x} - \dfrac{\partial \mu_B}{\partial x}\right)_{pT}} \left[1 + \frac{\dfrac{1}{\rho^2}\left(\dfrac{\partial p}{\partial x}\right)_{\rho T}\left(\dfrac{\partial \rho}{\partial x}\right)_{pT}}{\left(\dfrac{\partial \mu_A}{\partial x} - \dfrac{\partial \mu_B}{\partial x}\right)_{\rho T}} \right]^{-1}, \tag{2.8}$$

$$\langle \Delta N \Delta x \rangle_{TV} = V \langle \Delta \rho \Delta x \rangle = V\left(\frac{\partial \rho}{\partial x}\right)_{pT} \langle (\Delta x)^2 \rangle. \tag{2.9}$$

It can be shown that equations (2.5) and (2.7) for $\langle (\Delta N)^2 \rangle_{TV}$ are identical, while equation (2.8) for $\langle (\Delta x)^2 \rangle$ can be reduced to

$$\langle (\Delta x)^2 \rangle = \frac{kT}{N\left(\dfrac{\partial \mu_A}{\partial x} - \dfrac{\partial \mu_B}{\partial x}\right)_{pT}}, \tag{2.10}$$

since

$$\left(\frac{\partial \mu_A}{\partial x} - \frac{\partial \mu_B}{\partial x}\right)_{pT} = \left(\frac{\partial \mu_A}{\partial x} - \frac{\partial \mu_B}{\partial x}\right)_{\rho T} + \frac{1}{\rho^2}\left(\frac{\partial p}{\partial x}\right)_{\rho T}\left(\frac{\partial \rho}{\partial x}\right)_{pT}. \tag{2.11}$$

Again, (2.7), (2.8) and (2.11) give

$$\langle (\Delta N)^2 \rangle = NkT\left(\frac{\partial \rho}{\partial p}\right)_{xT} + V^2\left(\frac{\partial \rho}{\partial x}\right)_{pT}^2 \langle (\Delta x)^2 \rangle, \tag{2.12}$$

showing how the effect of concentration fluctuations is added to the compressibility term which forms the well-known simple fluid limit. Furthermore, in the case of an ideal gas mixture, it is easily shown that

$$\langle (\Delta N)^2 \rangle = \langle N \rangle, \quad \langle (\Delta x)^2 \rangle = x_A x_B / \langle N \rangle \quad \text{and} \quad \langle \Delta N \Delta x \rangle = 0.$$

The above equations permit discussion of the condensation or evaporation, and the phase separation, of a binary mixture. The Jacobian in the denominator of equation (2.5) can be rewritten as

$$\frac{1}{N^2}\left(\frac{\partial p}{\partial \rho}\right)_{xT}\left(\frac{\partial \mu_A}{\partial x} - \frac{\partial \mu_B}{\partial x}\right)_{pT}. \tag{2.13}$$

Its vanishing implies large fluctuations—infinite in the present theory, but finite if higher order terms in the above Taylor expansions are included (see Hirschfelder, Curtiss and Bird 1954). In the case of phase separation

$$\left(\frac{\partial \mu_A}{\partial x} - \frac{\partial \mu_B}{\partial x}\right)_{pT} = 0, \qquad \left(\frac{\partial p}{\partial \rho}\right)_{xT} > 0, \tag{2.14}$$

implying that $\langle (\Delta x)^2 \rangle$ becomes infinite by virtue of (2.10). At the same time (2.12) shows that $\langle (\Delta N)^2 \rangle$ becomes infinite because of the overwhelming contribution of the second term, the effect of the first term being negligible. We use this conclusion in § 6.

Condensation, on the other hand, is not in general accompanied by the vanishing of $(\partial p/\partial \rho)_{xT}$ as in simple fluid theory, nor is it relevant to our main theme. A full discussion would require reference to diagrams of concentration against temperature and pressure: see, for example, Prigogine and Defay (1954), who discuss the temperature at which $(\partial p/\partial \rho)_{xT}$ actually vanishes.

3. Fluctuations, Molecular Pair- and Radial-Distribution Functions

We divide the volume V into a large number of cells of volume $d\mathbf{r}_i$ accommodating only one molecule, their centres denoted by the vectors $\mathbf{r}_i$, and consider the delta functions $\Delta_A(\mathbf{r}_i)$ and $\Delta_B(\mathbf{r}_i)$, either of which is unity if the mass-centre of an A or B system lies within the ith cell, and zero otherwise. Clearly

$$\Delta_A(\mathbf{r}_i)\Delta_A(\mathbf{r}_i) = \Delta_A(\mathbf{r}_i), \qquad \Delta_B(\mathbf{r}_i)\Delta_B(\mathbf{r}_i) = \Delta_B(\mathbf{r}_i), \tag{3.1}$$

whereas

$$\Delta_A(\mathbf{r}_i)\Delta_B(\mathbf{r}_i) = 0. \tag{3.2}$$

Then, by definition of a mean value,

$$\langle \Delta_A(\mathbf{r}_i) \rangle = \langle N_i^A \rangle = \rho_A d\mathbf{r}_i, \tag{3.3}$$

where $\langle N_i^A \rangle$ is the mean number of A systems in the ith cell, and

$$\langle \Delta_A(\mathbf{r}_i)\Delta_A(\mathbf{r}_j) \rangle = \langle N_i^A N_j^A \rangle = n_{AA}(r_{ij})d\mathbf{r}_i d\mathbf{r}_j, \tag{3.4}$$

where $n_{AA}(r_{ij})$ is the molecular distribution function for a pair of A systems, and similarly for B systems. Also

$$\langle \Delta_A(\mathbf{r}_i)\Delta_B(\mathbf{r}_j) \rangle = \langle N_i^A N_j^B \rangle = n_{AB}(r_{ij})d\mathbf{r}_i d\mathbf{r}_j \tag{3.5}$$

independent of the order of the suffices A and B. It then follows from (3.3) that if $\Delta N_i^A = N_i^A - \langle N_i^A \rangle$ then

$$\langle (\Delta N_i^A)^2 \rangle = \langle (\Delta_A(\mathbf{r}_i) - \rho_A d\mathbf{r}_i)^2 \rangle = \rho_A d\mathbf{r}_i(1 - \rho_A d\mathbf{r}_i). \tag{3.6}$$

Similarly

$$\langle (\Delta N_i^B)^2 \rangle = \rho_B d\mathbf{r}_i(1 - \rho_B d\mathbf{r}_i), \tag{3.7}$$

and by (3.2),

$$\langle \Delta N_i^A \Delta N_i^B \rangle = -\rho_A \rho_B d\mathbf{r}_i^2. \tag{3.8}$$

Also, from (3.4) and (3.5),

$$\langle \Delta N_j^x \Delta N_k^y \rangle = \langle N_j^x N_k^y \rangle - \langle N_j^x \rangle \langle N_k^y \rangle = (n_{xy}(r_{jk}) - \rho_x \rho_y) d\mathbf{r}_j d\mathbf{r}_k. \tag{3.9}$$

Here, and subsequently, $x, y = A, B$ for brevity.
Now

$$\langle(\Delta N)^2\rangle = \sum_i \langle(\Delta N_i^A)^2\rangle + \sum_{j\neq k}\sum \langle \Delta N_j^A \Delta N_k^A\rangle + \sum_i \langle(\Delta N_i^B)^2\rangle$$
$$+ \sum_{j\neq k}\sum \langle \Delta N_j^B \Delta N_k^B\rangle + 2\sum_i \langle \Delta N_i^A \Delta N_i^B\rangle + 2\sum_{j\neq k}\sum \langle \Delta N_j^A \Delta N_k^B\rangle : \quad (3.10)$$

therefore, using (3.6)–(3.10), introducing radial distribution functions $g_{xy}(r) = n_{xy}(r)/\rho_x\rho_y$, and replacing summation by integration, we obtain

$$\frac{\langle(\Delta N)^2\rangle}{\langle N\rangle} = 1 + \rho\int_0^\infty \{x_A^2(g_{AA}(r)-1) + x_B^2(g_{BB}(r)-1)$$
$$+ 2x_A x_B(g_{AB}(r)-1)\} d\mathbf{r}, \quad (3.11)$$

where

$$x = x_A = N_A/N \quad \text{and} \quad x_B = 1 - x = N_B/N.$$

In forming the analogous equations for $\langle(\Delta x)^2\rangle$ and $\langle \Delta N \Delta x\rangle$, we use

$$\Delta x = \frac{\Delta N_A}{\langle N\rangle} - \frac{\langle N_A\rangle}{\langle N\rangle^2}\Delta N, \quad (3.12)$$

where

$$\Delta N = \Delta N_A + \Delta N_B, \qquad \Delta N_A = \sum_i \Delta N_i^A \quad \text{and} \quad \Delta N_B = \sum_j \Delta N_j^B. \quad (3.13)$$

It then follows, from (3.6)–(3.9), that

$$\langle(\Delta x_A)^2\rangle = \frac{x_A x_B}{\langle N\rangle}\left[1 + \rho x_A x_B \int_0^\infty \{g_{AA}(r) - 2g_{AB}(r) + g_{BB}(r)\} d\mathbf{r}\right] \quad (3.14)$$

and

$$\langle \Delta N \Delta x_A\rangle = \rho x_A x_B \int_0^\infty \{x_A(g_{AA}(r)-1) - x_B(g_{BB}(r)-1)$$
$$+ (x_B - x_A)(g_{AB}(r)-1)\} d\mathbf{r}. \quad (3.15)$$

4. Direct and Indirect Correlation Functions

The direct correlation function $f(r)$ for a simple fluid has been defined by Rosenfeld (1951). Denoting by $\langle \Delta N_k^{(j)}\rangle$ the mean number of systems in the kth cell when all the other cells, denoted collectively by the running index $j \neq k$, have fixed fluctuations ΔN_j from their mean numbers $\langle N_j\rangle$, he suggests that $\langle \Delta N_k^{(j)}\rangle$ will depend linearly on the fluctuations in the other cells, and on their distances away, and consequently writes

$$\langle \Delta N_k^{(j)}\rangle = \sum_{j\neq k} \Delta N_j f(r_{jk}) d\mathbf{r}_k. \quad (4.1)$$

The extension to binary mixtures is, we suggest, in an obvious notation,

$$\langle \Delta N_k^{A(j)} \rangle = \sum_{j \neq k} \Delta N_j^A f_{AA}(r_{jk}) d\mathbf{r}_k + \sum \Delta N_j^B f_{BA}(r_{jk}) d\mathbf{r}_k, \tag{4.2}$$

$$\langle \Delta N_k^{B(j)} \rangle = \sum_{j \neq k} \Delta N_j^B f_{BB}(r_{jk}) d\mathbf{r}_k + \sum \Delta N_j^A f_{AB}(r_{jk}) d\mathbf{r}_k, \tag{4.3}$$

where the second summation is in each case unrestricted in order to take account of the effect upon $\langle \Delta N_k^{B(j)} \rangle$ of ΔN_k^A, and upon $\langle \Delta N_k^{A(j)} \rangle$ of ΔN_k^B. Here $f_{AA}(r)$, $f_{BB}(r)$, $f_{AB}(r)$ and $f_{BA}(r)$ are the four direct correlation functions for a binary mixture.

Multiplying equation (4.2) by ΔN_i^A and taking mean values, we have

$$\langle \Delta N_i^A \Delta N_k^A \rangle = \sum_{j \neq ki} \langle \Delta N_i^A \Delta N_j^A \rangle f_{AA}(r_{jk}) d\mathbf{r}_k + \sum_{j \neq i} \langle \Delta N_i^A \Delta N_j^B \rangle f_{BA}(r_{jk}) d\mathbf{r}_k$$
$$+ \langle (\Delta N_i^A)^2 \rangle f_{AA}(r_{ik}) d\mathbf{r}_k + \langle \Delta N_i^A \Delta N_i^B \rangle f_{BA}(r_{ik}) d\mathbf{r}_k. \tag{4.4}$$

Replacing the mean values by the appropriate expressions from equations (3.6)–(3.9) and defining indirect correlation functions $h_{xy}(r)$ by

$$h_{xy}(r) = g_{xy}(r) - 1 \tag{4.5}$$

with

$$\lim_{r \to 0} h_{xy}(r) = -1, \tag{4.6}$$

we find that

$$\rho_A^2 h_{AA}(r_{ik}) = \sum_{j \neq ki} \rho_A^2 h_{AA}(r_{ij}) f_{AA}(r_{jk}) d\mathbf{r}_j + \sum_{j \neq i} \rho_A \rho_B h_{AB}(r_{ij}) f_{BA}(r_{jk}) d\mathbf{r}_j$$
$$+ \rho_A (1 - \rho_A d\mathbf{r}_i) f_{AA}(r_{ik}) - \rho_A \rho_B d\mathbf{r}_i f_{BA}(r_{ik}). \tag{4.7}$$

The restrictions on the first two summations can be waived by incorporating the last two terms, using (4.6) and assuming

$$\lim_{r \to 0} f_{AA}(r) d\mathbf{r} = 0. \tag{4.8}$$

On replacing summation by integration, an integral equation is obtained, namely

$$\rho_A^2 h_{AA}(r) = \rho_A f_{AA}(r) + \rho_A^2 \int_0^\infty h_{AA}(r^1) f_{AA}(\mathbf{r}^1 - \mathbf{r}) d\mathbf{r}^1$$
$$+ \rho_A \rho_B \int_0^\infty h_{AB}(r^1) f_{BA}(\mathbf{r}^1 - \mathbf{r}) d\mathbf{r}^1. \tag{4.9}$$

Three more integral equations follow on multiplying (4.2) by ΔN_i^B, and (4.3) by ΔN_i^A and ΔN_i^B, and repeating the above procedure. They can all be solved by introducing the Fourier transforms

$$\left.\begin{aligned} sH_{xy}(s) &= (2\pi)^{-\frac{1}{2}}\int_{-\infty}^{+\infty} rh_{xy}(r)\sin rs\,dr, \\ sF_{xy}(s) &= (2\pi)^{-\frac{1}{2}}\int_{-\infty}^{+\infty} rf_{xy}(r)\sin rs\,dr. \end{aligned}\right\} \tag{4.10}$$

With the abbreviations c_A, $c_B=(2\pi)^{\frac{3}{2}}\rho_A$, $(2\pi)^{\frac{3}{2}}\rho_B$, we finally obtain

$$\rho_A H_{AA}(s) = F_{AA}(s) + c_A H_{AA}(s)F_{AA}(s) + c_B H_{AB}(s)F_{BA}(s), \tag{4.11}$$

$$\rho_B H_{AB}(s) = F_{AB}(s) + c_A H_{AA}(s)F_{AB}(s) + c_B H_{AB}(s)F_{BB}(s), \tag{4.12}$$

$$\rho_A H_{AB}(s) = F_{BA}(s) + c_B H_{BB}(s)F_{BA}(s) + c_A H_{AB}(s)F_{AA}(s), \tag{4.13}$$

$$\rho_B H_{BB}(s) = F_{BB}(s) + c_B H_{BB}(s)F_{BB}(s) + c_A H_{AB}(s)F_{AB}(s), \tag{4.14}$$

from which we can express $F_{xy}(s)$ or, ultimately, $f_{xy}(r)$ in terms of $H_{xy}(s)$ or $g_{xy}(r)$, and conversely, and prove the following determinantal equation:

$$\left|\begin{matrix} 1-(2\pi)^{\frac{3}{2}}F_{AA}(s) & (2\pi)^{\frac{3}{2}}F_{BA}(s) \\ (2\pi)^{\frac{3}{2}}F_{AB}(s) & 1-(2\pi)^{\frac{3}{2}}F_{BB}(s) \end{matrix}\right\| \begin{matrix} 1+(2\pi)^{\frac{3}{2}}\rho_A H_{AA}(s) & (2\pi)^{\frac{3}{2}}\rho_B H_{AB}(s) \\ (2\pi)^{\frac{3}{2}}\rho_A H_{AB}(s) & 1+(2\pi)^{\frac{3}{2}}\rho_B H_{BB}(s) \end{matrix}\right| = 1. \tag{4.15}$$

This is analogous to the simple fluid result of Rushbrooke and Scoins (1953), namely

$$\{1-(2\pi)^{\frac{3}{2}}F(s)\}\{1+(2\pi)^{\frac{3}{2}}\rho H(s)\} = 1, \tag{4.16}$$

and suggests the form which a generalization to multi-component mixtures is likely to take.

The solutions of equations (4.11)–(4.14) also satisfy

$$x_A F_{AB}(s) = x_B F_{BA}(s), \tag{4.17}$$

and can be expressed in a very simple manner in terms of the Born-Green theory as extended to binary mixtures by Fournet (1951) and Rushbrooke (1952), who gave appropriate solutions for the radial distribution functions.

We find eventually that

$$\rho_x f_{xy}(r) = \rho_x \rho_y \epsilon_{xy}[e^{-\phi_{xy}(r)/kT} - 1] \tag{4.18}$$

analogous to the simple fluid case discussed by Rushbrooke and Scoins (1953). Here, the $\phi_{xy}(r)$ are the intermolecular pair potential energies, the ϵ_{xy} are the adjustable parameters of the Born-Green theory, and $\epsilon_{AB}=\epsilon_{BA}$. Finally, Rushbrooke (1952) has shown that this Born-Green type of approximation responds to the vanishing of the Jacobian of (2.13) and hence to the phenomena of phase separation and condensation.

5. Consideration of the Main Equations

We equate expressions (3.11), (3.14) and (3.15) for $\langle(\Delta N)^2\rangle$, $\langle(\Delta x)^2\rangle$ and $\langle\Delta N\Delta x\rangle$ in terms of radial distribution functions to the corresponding equations (2.7), (2.10) and (2.9) in terms of thermodynamic functions, because the former involve integration throughout that fixed volume held constant in the latter, thermodynamic, expressions. Then, using (2.12) to eliminate $\left(\frac{\partial\mu_A}{\partial x}-\frac{\partial\mu_B}{\partial x}\right)_{pT}$, and solving for $(\partial p/\partial\rho)_{xT}$, $(\partial p/\partial x)_{\rho T}$ and $(\partial\rho/\partial x)_{pT}$, we simplify the results by making use of an obvious notation for permanants (+) and determinants (−). We obtain

$$kT\left(\frac{\partial\rho}{\partial p}\right)_{xT}=\frac{{}^{-}\begin{vmatrix}1+x_A\rho\int_0^\infty(g_{AA}(r)-1)d\mathbf{r} & x_B\rho\int_0^\infty(g_{AB}(r)-1)d\mathbf{r}\\ x_A\rho\int_0^\infty(g_{AB}(r)-1)d\mathbf{r} & 1+x_B\rho\int_0^\infty(g_{BB}(r)-1)d\mathbf{r}\end{vmatrix}}{{}^{+}\begin{vmatrix}1+x_B\rho\int_0^\infty(g_{AA}(r)-g_{AB}(r))d\mathbf{r} & x_B\\ 1+x_A\rho\int_0^\infty(g_{BB}(r)-g_{AB}(r))d\mathbf{r} & x_A\end{vmatrix}} \tag{5.1}$$

and

$$-\frac{1}{\rho kT}\left(\frac{\partial p}{\partial x}\right)_{\rho T}=\frac{{}^{-}\begin{vmatrix}\rho\int_0^\infty(g_{AA}(r)-g_{AB}(r))d\mathbf{r} & x_B\\ \rho\int_0^\infty(g_{BB}(r)-g_{AB}(r))d\mathbf{r} & x_A\end{vmatrix}}{{}^{-}\begin{vmatrix}1+x_A\rho\int_0^\infty(g_{AA}(r)-1)d\mathbf{r} & x_B\rho\int_0^\infty(g_{AB}(r)-1)d\mathbf{r}\\ x_A\rho\int_0^\infty(g_{AB}(r)-1)d\mathbf{r} & 1+x_B\rho\int_0^\infty(g_{BB}(r)-1)d\mathbf{r}\end{vmatrix}}, \tag{5.2}$$

$(\partial\rho/\partial x)_{pT}$ being obtainable from

$$\left(\frac{\partial p}{\partial x}\right)_{\rho T}\left(\frac{\partial x}{\partial\rho}\right)_{pT}\left(\frac{\partial\rho}{\partial p}\right)_{xT}=-1. \tag{5.3}$$

Equations (5.1) and (5.2) can be expressed in terms of the direct correlation functions of equations (4.11)–(4.14), using the well-known relation between a space integral and a Fourier transform. We thus obtain

$$\frac{1}{kT}\left(\frac{\partial p}{\partial\rho}\right)_{xT}=1-\int_0^\infty\{x_Af_{AA}(r)+x_Bf_{BB}(r)+x_Af_{AB}(r)+x_Bf_{BA}(r)\}d\mathbf{r}, \tag{5.4}$$

analogous to the simple fluid result

$$\frac{1}{kT}\left(\frac{\partial p}{\partial\rho}\right)_T=1-\int_0^\infty f(r)d\mathbf{r} \tag{5.5}$$

of Rushbrooke and Scoins (1953), and also

$$\frac{1}{\rho kT}\left(\frac{\partial p}{\partial x_A}\right)_{\rho T}=\int_0^{\infty}(f_{BB}(r)-f_{AA}(r)-f_{AB}(r)+f_{BA}(r))d\mathbf{r}. \tag{5.6}$$

The importance of the direct correlation function in binary mixture theory is shown by the simplicity of these results.

If now we substitute into (5.4) and (5.6) the appropriate solutions (4.18) for the direct correlation functions in terms of Born-Green theory, making use of the equations given by Rushbrooke (1952) for the parameters ϵ_{xy}, we find that

$$\frac{1}{kT}\left(\frac{\partial p}{\partial \rho}\right)_{xT}=1+2\rho\{x_A^2B_{AA}+2x_Ax_BB_{AB}+x_B^2B_{BB}\}$$
$$+3\rho^2\{x_A^3C_{AAA}+3x_A^2x_BC_{AAB}+3x_Ax_B^2C_{ABB}+x_B^3C_{BBB}\} \tag{5.7}$$

and

$$\frac{1}{\rho kT}\left(\frac{\partial p}{\partial x_A}\right)_{\rho T}=2\{x_AB_{AA}+(x_B-x_A)B_{AB}-x_BB_{BB}\}$$
$$+3\rho^2\{x_A^2C_{AAA}+x_A(2x_B-x_A)C_{AAB}-x_B(2x_A-x_B)C_{ABB}-x_B^2C_{BBB}\}. \tag{5.8}$$

The B_{xy} and the C_{xyz} are the well-known second and third virial coefficients, and it is easily verified by straightforward differentiation that (5.7) and (5.8) are consistent with the truncated virial expansion

$$p=\rho kT[1+\rho\{x_A^2B_{AA}+2x_Ax_BB_{AB}+x_B^2B_{BB}\}$$
$$+\rho^2\{x_A^3C_{AAA}+3x_A^2x_BC_{AAB}+3x_Ax_B^2C_{ABB}+x_B^3C_{BBB}\}+\ldots]. \tag{5.9}$$

This consistency represents an approximate solution to the formal problem of demonstrating that substitution into equations (5.1) and (5.2) of the cluster integral expansions of the radial distribution functions given by Rushbrooke (1952) is consistent with the complete virial expansion for a binary mixture.

6. X-ray Scattering by a Binary Mixture

It is well known (see, for example, de Boer 1949) that at distance R from the scatterer, the intensity $I(s)$ of radiation of wavelength λ scattered through angle θ from the direction of a primary beam of intensity I_0 by a simple fluid of N systems each of scattering power $f(s)$ is given by

$$I(s)=I_0Nf^2(s)\left(\frac{e^2}{mc^2}\right)^2\frac{1+\cos^2\theta}{2R^2}\left\{1+4\pi\rho\int_0^{\infty}r^2(g(r)-1)\frac{\sin rs}{rs}dr\right\}, \tag{6.1}$$

where $s = 4\pi\lambda^{-1} \sin^{\theta}/2$ and e^2/mc^2 is the so-called classical radius r_0 of the electron. It can easily be shown that the corresponding formula for a binary mixture of systems of scattering powers $f_A(s)$ and $f_B(s)$ is

$$I(s) = I_0 N r_0^2 \frac{1+\cos^2\theta}{2R^2} \Big\{ x_A f_A^2(s) + x_B f_B^2(s)$$

$$+ 4\pi\rho \int_0^\infty r^2 \{ x_A^2 f_A^2(s)(g_{AA}(r) - 1) + x_B^2 f_B^2(s)(g_{BB}(r) - 1)$$

$$+ 2x_A x_B f_A(s) f_B(s)(g_{AB}(r) - 1)\} \frac{\sin rs}{rs} dr \Big\}. \tag{6.2}$$

From equation (6.1) $g(r)$ can be determined by a suitable Fourier inversion of the experimental results. Unfortunately, this cannot be done directly from (6.2) because there are now two scattering powers dependent upon s, and three radial distribution functions. However, equation (6.2) can be rewritten in the form

$$I(s) = I_0 N r_0^2 \frac{1+\cos^2\theta}{2R^2} \Big\{ (x_A f_A + x_B f_B)^2 \Big[1 + \rho \int_0^\infty \Big\{ x_A^2 (g_{AA} - 1)$$

$$+ x_B^2 (g_{BB} - 1) + 2x_A x_B (g_{AB} - 1) \Big\} \frac{\sin rs}{rs} d\mathbf{r} \Big]$$

$$+ 2(f_A - f_B)(x_A f_A + x_B f_B) \Big[\rho x_A x_B \int_0^\infty \Big\{ x_A (g_{AA} - 1) - x_B (g_{BB} - 1)$$

$$+ (x_B - x_A)(g_{AB} - 1) \Big\} \frac{\sin rs}{rs} d\mathbf{r} \Big]$$

$$+ (f_A - f_B)^2 x_A x_B \Big[1 + \rho x_A x_B \int_0^\infty \Big\{ g_{AA} + g_{BB} - 2g_{AB} \Big\} \frac{\sin rs}{rs} d\mathbf{r} \Big] \Big\}, \tag{6.3}$$

where the three terms in square brackets differ from equations (3.11) (3.15) and (3.14) for $\langle(\Delta N)^2\rangle/\langle N\rangle$, $\langle \Delta N \Delta \mathbf{x}\rangle$ and $\langle(\Delta x)^2\rangle$, respectively, only by the presence in the integrands of the Fourier factor $(\sin rs)/rs$. We can therefore rewrite (6.3) in the form of a summation over fluctuations $\Delta\rho_i$ and Δx_i in number density and concentration in the volume cells of § 3, obtaining

$$I(s) = I_0 r_0^2 \frac{1+\cos^2\theta}{2R^2} \cdot V^2 \cdot \sum_j \sum_k \Big\{ (x_A f_A + x_B f_B)^2 \langle \Delta\rho_j \Delta\rho_k \rangle$$

$$+ 2\rho (f_A - f_B)(x_A f_A + x_B f_B) \langle \Delta\rho_j \Delta x_k \rangle$$

$$+ \rho^2 (f_A - f_B)^2 \langle \Delta x_j \Delta x_k \rangle \Big\} \frac{\sin sr_{jk}}{sr_{jk}}, \tag{6.4}$$

where

$$\Delta x_i = \frac{\Delta N_i^A}{\langle N \rangle} - \frac{\langle N_A \rangle}{\langle N \rangle^2} \Delta N_i \tag{6.5}$$

analogous to (3.12). In the case of X-ray scattering these volume cells must accommodate only one molecule, because the wavelength of the radiation is comparable with the intermolecular separation. In the case of critical opalescence under visible irradiation, however, the wavelength is so long that the volume cells can be large enough to possess thermodynamic functions. Furthermore, near the temperature of phase separation, $\langle (\Delta x)^2 \rangle$ becomes so large that it overwhelms the contribution to $\langle (\Delta N)^2 \rangle$ of the term $NkT(\partial\rho/\partial p)_{xT}$ in equation (2.12), which together with (2.9) gives, therefore, for hese larger volume cells

$$\langle (\Delta\rho_i)^2 \rangle \doteqdot \left(\frac{\partial\rho}{\partial x}\right)_{pT}^2 \langle (\Delta x_i)^2 \rangle \qquad \langle \Delta\rho_i \Delta x_i \rangle = \left(\frac{\partial\rho}{\partial x}\right)_{pT} \langle (\Delta x_i)^2 \rangle. \tag{6.6}$$

This enables us to rewrite (6.4) in terms of fluctuations in concentration alone, for which (3.14) is still valid, because the range of intermolecular correlation is now of the same order as the linear dimensions of the container. The scattered intensity I_{crit} in this context reduces therefore to

$$I_{\text{crit}} = I_0 r_0^2 \frac{1+\cos^2\theta}{2R^2} \frac{N}{\rho^2} \left\{ (x_A f_A(s) + x_B f_B(s)) \left(\frac{\partial\rho}{\partial x}\right)_{pT} + \rho(f_A(s) - f_B(s)) \right\}^2 \left[x_A x_B \left(1 + \rho x_A x_B \int_0^\infty \left\{ g_{AA}(r) + g_{BB}(r) - 2g_{AB}(r) \right\} \frac{\sin rs}{rs} d\mathrm{r} \right) \right], \tag{6.7}$$

from which a concentration correlation function

$$g_x(r) = g_{AA}(r) + g_{BB}(r) - 2g_{AB}(r) \tag{6.8}$$

can be obtained by Fourier inversion of experimental observations of critical opalescence.

Now, Furth and Williams (1954) defined a function $g_c(r)$ purely by analogy with simple fluid theory, namely

$$I_{\text{crit}} \propto \int_0^\infty g_c(r) \frac{\sin rs}{rs} d\mathrm{r}, \tag{6.9}$$

so that no account was taken of the difference between the scattering powers of the A and B systems. The experimental results of Quantie (1954) cannot, however, be reinterpreted in terms of the $g_x(r)$ of equation (6.8) because of the lack of information about $f_A(s)$, $f_B(s)$ and $(\partial\rho/\partial x)_{pT}$. We are therefore unable to comment on their division of binary mixtures into two classes, one of which exhibits a region of "negative correlation" which they attribute to the effect of diffusion, although their idea of "concentration waves" appears to owe much to an incorrect assumption that large fluctuations in concentration are unaccompanied by the fluctuations in number predicted by equation (2.12).

Acknowledgments

One of us (F. J. P.) is indebted to the Ministry of Education for a maintenance grant under the Further Education and Training Scheme for ex-Servicemen, without which this research would not have been possible.

References to Literature

de Boer, J., 1949. *Reps. Progr. Phys.*, **12**, 305.

Fournet, G., 1951. *J. Phys. Rad.*, **12**, 292.

Furth, R., and Williams, C. L., 1954. *Proc. Roy. Soc. Lond.*, A, **224**, 104.

Green, H. S., 1952. *Molecular Theory of Fluids.* North-Holland Publishing Co., Amsterdam.

Hirschfelder, J. O., Curtiss, C. F., and Bird, R. B., 1954. *Molecular Theory of Liquids and Gases,* Wiley, New York.

Kirkwood, J. G., and Goldberg, R. J., 1950. *J. Chem. Phys.*, **18**, 54.

Klein, M. J., and Tisza, L., 1949. *Phys. Rev.*, **76**, 1861.

Landau, L., and Lifshitz, E., 1938. *Statistical Physics.* Oxford University Press.

Mayer, J. E., 1939. *J. Phys. Chem.*, **43**, 71.

Munster, A., 1953. *Z. Phys.*, **136**, 179.

Ornstein, L. S., and Zernike, F., 1914 *Proc. Acad. Sci. Amst.*, **17**, 795.

Prigogine, I., and Defay, R., 1954. *Chemical Thermodynamics.* Longmans, Green, London.

Quantie, C., 1954. *Proc. Roy. Soc. Lond.*, A, **224**, 90.

Rosenfeld, L., 1951. *Theory of Electrons.* North-Holland Publishing Co., Amsterdam.

Rushbrooke, G. S., 1952. *Phil. Mag.*, **43**, 1276.

Rushbrooke, G. S., and Scoins, H. I., 1953. *Proc. Roy. Soc. Lond.*, A, **216**, 203.

Stockmayer, W. H., 1950. *J. Chem. Phys.*, **18**, 58.

Zernike, F., 1916. *Proc. Acad. Sci. Amst.*, **17**, 793.

(*Issued separately May* 22, 1957)

On the Statistical Mechanical Theory of Solutions*

R. M. Mazo†
Enrico Fermi Institute for Nuclear Studies
University of Chicago, Chicago, IL

A perturbation expansion of the excess free energy of a binary mixture based on the theory of composition fluctuations in an open system is presented. The theory is applied to classical mixtures and quantum mechanical isotope mixtures, and comparisons with conformal solution theory and Chester's theory of isotope mixtures are made. The advantage of the present theory is that it contains in summed form, in each order of perturbation, terms which in the direct expansion of the partition function would be considered to be higher order.

INTRODUCTION

In recent years there has been a great interest in what may be called perturbation theories of solutions.[1–8] These are theories which attempt to express the properties of a solution in terms of those of some suitable pure substance plus correction terms. The theories are usually couched in terms of a Taylor's series development of the partition function with respect to some appropriate parameters.

There are also several exact theories of solutions,[9–11] which have, until now, been mainly useful for providing power series expansions in the concentrations of the various solutes. The object of this paper is to demonstrate that one of these exact theories, that of Kirkwood and Buff,[10] is a particularly convenient starting point for a perturbation expansion which has the advantage of including, in a given order of perturbation, some of the terms which would be considered as higher order in the conventional treatments.

In this paper we develop the theory for both classical solutions and quantum mechanical isotope mixtures, to the lowest nonvanishing order of approximation in each case. This is the first order in the classical case, and second order for the isotope mixtures. Part I contains the general theory, and Parts II and III the specializations to the two cases considered. We restrict ourselves to binary mixtures, and assume that all internal partition functions are independent of the state of aggregation.

* This research supported by the U. S. Atomic Energy Commission.

† Present address: Gates and Crellin Laboratories, California Institute of Technology, Pasadena, California.

1 H. C. Longuet-Higgins, Proc. Roy. Soc. (London) **A205**, 247 (1951).
2 W. B. Brown and H. C. Longuet-Higgins, Proc. Roy. Soc. (London) **A209**, 416 (1951).
3 W. B. Brown, Proc. Roy. Soc. (London) **A240**, 561 (1957).
4 W. B. Brown, Phil. Trans. Roy. Soc. (London) **A250**, 175 (1957).
5 R. L. Scott, J. Chem. Phys. **25**, 193 (1956).
6 Prigogine, Bellemans, and Englert-Chwoles, J. Chem. Phys. **24**, 518 (1956).
7 Salsburg, Wojtowicz, and Kirkwood, J. Chem. Phys. **26**, 1533 (1957).
8 Wojtowicz, Salsburg, and Kirkwood, J. Chem. Phys. **27**, 505 (1957).
9 W. G. McMillan and J. E. Mayer, J. Chem. Phys. **13**, 276 (1945).
10 J. G. Kirkwood and F. P. Buff, J. Chem. Phys. **19**, 774 (1951).
11 T. Hill, J. Am. Chem. Soc. **74**, 4885 (1957).

I. GENERAL THEORY

The main result of Kirkwood and Buff which we shall use is Eq. (22) of their paper, which expresses the change of chemical potential of one component of a binary mixture with composition. If we denote the ingredients of the mixture by a and b the equation reads

$$\frac{1}{kT}\left(\frac{\partial\mu_b}{\partial x_b}\right)_{T,P}=\frac{1}{x_b}+\frac{\rho x_a(2G_{ab}-G_{aa}-G_{bb})}{1+\rho x_a x_b(G_{aa}+G_{bb}-2G_{ab})}, \tag{1}$$

x_a is the mole fraction of a $[N_a/(N_a+N_b)]$, ρ is the number density $(N_a+N_b)/V$, and the $G_{\alpha\beta}$ are concentration fluctuations.

$$G_{\alpha\beta}=V\left[\frac{\langle n_\alpha n_\beta\rangle-N_\alpha N_\beta}{N_\alpha N_\beta}-\frac{\delta_{\alpha\beta}}{N_\alpha}\right]; \tag{2}$$

the $\langle\ \ \rangle$ denotes averaging with respect to the grand canonical ensemble,

$$\langle n_\alpha n_\beta\rangle=e^{-PV/kT}\sum_{n_b,n_a}\frac{z_a^{n_a}z_b^{n_b}}{n_a!\,n_b!}n_\alpha n_\beta Q_{n_a+n_b}, \tag{3}$$

where $Q_{n_a+n_b}$ is the configurational partition function for n_a+n_b molecules, and z_α is the fugacity of component α. We have written $N_\alpha=\langle n_\alpha\rangle$.

Clearly $(kT)^{-1}(\partial\mu_a/\partial x_b)_{T,P}$ can be found from Eq. (1) by the application of the Gibbs-Duhem equation. The derivatives of the chemical potentials may be integrated at constant pressure and temperature, and formal expressions for the chemical potentials themselves obtained.

$$\mu_a(T,P,x)=\mu_a^0(T,P)+kT\ln x_a-kT\int_1^{x_a}\frac{\rho(1-y)\Gamma dy}{1+\rho y(1-y)\Gamma}, \tag{4}$$

where we have written

$$\Gamma=G_{aa}+G_{bb}-2G_{ab},$$

ρ and Γ are, of course, functions of composition.

From Eq. (4) one immediately finds that the excess Gibbs function is

$$g^E=kT\left[x_a\int_0^{x_b}\frac{\rho\Gamma y dy}{1+\rho\Gamma y(1-y)}+x_b\int_0^{x_a}\frac{\rho\Gamma y dy}{1+\rho\Gamma y(1-y)}\right]. \tag{5}$$

It should be clear from the appearance of Boltzmann's constant, k, instead of the usual gas constant, R, that all our thermodynamic functions are reckoned per molecule, not per mole. So far our development has been formally exact, but the concentration dependence of Γ is unknown, and so we cannot even begin to evaluate the integrals. This is where the perturbation expansion enters.

Suppose that in the grand partition function

$$e^{PV/kT}=\sum\frac{z_a^{n_a}z_b^{n_b}}{n_a!\,n_b!}Q_{n_a+n_b}(\lambda), \tag{6}$$

we assume that the configurational partition function $Q_{n_a+n_b}$ depends on a parameter, or set of parameters, which we shall denote by the single letter, λ, in such a way that for $\lambda=0$, $Q_{n_a+n_b}$ is the configurational partition function of some pure system, while for some particular $\lambda\neq0$, $Q_{n_a+n_b}$ is the configurational partition function for the actual system of interest.‡ If we write a Taylor series expansion for Γ,

$$\Gamma=\sum_{l=0}^{\infty}\Gamma^{(l)}\lambda^l/l!$$

$$\Gamma^{(l)}=(\partial^l\Gamma/\partial\lambda^l)_{T,P,N_a,N_b}|_{\lambda=0}, \tag{7}$$

we can explicitly compute the first nonvanishing term and use this in Eq. (5) for the excess Gibbs function.§ If there is more than one parameter, λ, Eq. (7) should be understood to be a symbolic expression of the Taylor expansion of a function of several variables.

It is clear that $\Gamma^{(0)}$ vanishes identically. Qualitatively this is because the reference solution is ideal. To prove this mathematically we note that

$$\Gamma=\left(\frac{z_a^2}{N_a^2}\frac{\partial^2}{\partial z_a^2}+\frac{z_b^2}{N_b^2}\frac{\partial^2}{\partial z_b^2}-\frac{2z_az_b}{N_aN_b}\frac{\partial^2}{\partial z_a\partial z_b}\right)\frac{PV}{kT}, \tag{8}$$

where the differentiations are at constant T, V, λ. When $\lambda=0$, $Q_{n_a+n_b}$ depends only on the sum n_a+n_b, and not on n_a or n_b individually. From

$$N_\alpha/z_\alpha=[(\partial/\partial z_\alpha)(PV/kT)]_{T,V,z_\beta,\lambda} \tag{9}$$

we find that $N_a/z_a=N_b/z_b$ at $\lambda=0$. Therefore

$$\Gamma^{(0)}=\frac{z_a^2}{N_a^2}\left(\frac{\partial}{\partial z_a}-\frac{\partial}{\partial z_b}\right)^2\frac{PV}{kT}=\frac{z_a^2}{N_a^2}\left(\frac{\partial}{\partial z_a}-\frac{\partial}{\partial z_b}\right)\left(\frac{N_a}{z_a}-\frac{N_b}{z_b}\right)=0. \tag{10}$$

The $\Gamma^{(n)}$ are defined as derivatives at constant T, P, N_a, N_b, but the structure of the grand partition function makes it more convenient to work at constant T, V, z_a, z_b. We therefore write

$$(\partial\Gamma/\partial\lambda)_{T,P,N_a,N_b}=(\partial\Gamma/\partial V)_{T,V,z,\lambda}(\partial V/\partial\lambda)_{T,P,N}+(\partial\Gamma/\partial\ln z_a)_{T,V,z_b,\lambda}(\partial\ln z_a/\partial\lambda)_{T,P,N}+(\partial\Gamma/\partial\ln z_b)_{T,V,z,\lambda}(\partial\ln z_b/\partial\lambda)_{T,P,N}+(\partial\Gamma/\partial\lambda)_{T,V,z}. \tag{11}$$

‡ This method of expansion was suggested by work of L. Nosanow and J. E. Mayer (unpublished).

§ Mr. W. B. Brown has very kindly pointed out to the author that for a critical mixing phase $\partial\mu_b/\partial x_b=0$ and hence Γ is infinite. The following development, therefore, cannot be used near critical mixing, nor used to predict critical mixing. This is a considerable restriction on the theory.

When $\lambda=0$ the first three terms vanish, since $\Gamma\equiv 0$. Thus

$$\Gamma^{(1)}=(\partial\Gamma/\partial\lambda)_{T,V,z}|_{\lambda=0}$$

$$=V\left[\frac{1}{N_a^2}\frac{\partial\langle n_a(n_a-1)\rangle}{\partial\lambda}+\frac{1}{N_b^2}\frac{\partial\langle n_b(n_b-1)\rangle}{\partial\lambda}-\frac{2}{N_aN_b}\frac{\partial\langle n_an_b\rangle}{\partial\lambda}\right]_{\lambda=0}$$

$$-2V\left[\frac{\partial\ln N_a}{\partial\lambda}\frac{\langle n_a(n_a-1)\rangle}{N_a^2}+\frac{\partial\ln N_b}{\partial\lambda}\frac{\langle n_b(n_b-1)\rangle}{N_b^2}-\frac{\langle n_an_b\rangle}{N_aN_b}\left(\frac{\partial\ln N_a}{\partial\lambda}+\frac{\partial\ln N_b}{\partial\lambda}\right)\right]_{\lambda=0}. \quad (12)$$

The second bracket vanishes at $\lambda=0$ for the same reason that Γ vanishes, so that

$$\Gamma^{(1)}=V\left[\frac{1}{N_a^2}\frac{\partial\langle n_a(n_a-1)\rangle}{\partial\lambda}+\frac{1}{N_b^2}\frac{\partial\langle n_b(n_b-1)\rangle}{\partial\lambda}-\frac{2}{N_aN_b}\frac{\partial\langle n_an_b\rangle}{\partial\lambda}\right]_{\lambda=0}. \quad (13)$$

In order to write $\Gamma^{(2)}$, which we shall need for the quantum mechanical problem, we note that

$$(\partial\Gamma/\partial V)_{T,z,\lambda}\equiv 0$$

for all λ. This may be proved by actually taking the volume derivative of Eq. (8), but it is obvious that Γ is an intensive quantity, and so cannot change with volume when the other intensive variables characterizing the system are held constant. Therefore

$$(\partial^2\Gamma/\partial\lambda^2)_{T,P,N}=(\partial\Gamma/\partial\ln z_a)_{T,V,z_b,\lambda}(\partial^2\ln z_a/\partial\lambda^2)_{T,P,N}$$
$$+(\partial\Gamma/\partial\ln z_b)_{T,V,z_a,\lambda}(\partial^2\ln z_b/\partial\lambda^2)_{T,P,N}$$
$$+(\partial\ln z_a/\partial\lambda)_{T,P,N}[\partial/\partial\lambda(\partial\Gamma/\partial\ln z_a)_{T,V,z_b,\lambda}]_{T,P,N}$$
$$+(\partial\ln z_b/\partial\lambda)_{T,P,N}[\partial/\partial\lambda(\partial\Gamma/\partial\ln z_b)_{T,V,z_a,\lambda}]_{T,P,N}$$
$$+[\partial/\partial\lambda(\partial\Gamma/\partial\lambda)_{T,V,z}]_{T,P,N}. \quad (14)$$

When $\lambda=0$ this becomes

$$\Gamma^{(2)}=2(\partial\ln z_a/\partial\lambda)_{T,P,N}(\partial^2\Gamma/\partial\lambda\partial\ln z_a)_{T,V,z}$$
$$+2(\partial\ln z_b/\partial\lambda)_{T,P,N}(\partial^2\Gamma/\partial\lambda\partial\ln z_b)_{T,V,z}$$
$$+(\partial^2\Gamma/\partial\lambda^2)_{T,V,z}\quad(\lambda=0). \quad (15)$$

Higher coefficients may be computed in the same manner, but become progressively more complicated, and have not been investigated. What remains, for present purposes, is to evaluate $\Gamma^{(1)}$ and $\Gamma^{(2)}$ for appropriate types of λ dependence.

II. CLASSICAL MIXTURES

In classical statistical mechanics, translational contributions to the partition function lead to additive contributions to the thermodynamic functions, and so cancel in the calculation of thermodynamic excess functions. The perturbation which gives rise to the excess functions is the difference in forces between the molecules of the several components. If the actual potential energy of a system of $\mathbf{n}(=n_a+n_b)$ particles is $U_\mathbf{n}$, we define a fictitious potential, $\tilde{U}_\mathbf{n}$, by

$$\tilde{U}_\mathbf{n}(\lambda)=U_n^0+\lambda(\tilde{U}_\mathbf{n}-U_n^0)=U_n^0+\lambda\Delta U_\mathbf{n}, \quad (16)$$

where U_n^0 is a potential which is independent of n_a and n_b separately, but only depends on the total number of particles. Clearly,

$$\tilde{U}_\mathbf{n}(\lambda=0)=U_n^0;\qquad \tilde{U}_\mathbf{n}(\lambda=1)=U_\mathbf{n}. \quad (17)$$

In classical statistics, $Q_\mathbf{n}$ of Eq. (6) is just

$$Q_\mathbf{n}(\lambda)=\int\exp(-\tilde{U}_\mathbf{n}/kT)d\{\mathbf{n}\}, \quad (18)$$

where $d\{\mathbf{n}\}$ means $\prod_{i=1}^{n}d\mathbf{r}_i$. Our task is now to compute the λ derivatives of Eq. (13) and evaluate the power series expansion, Eq. (7) at $\lambda=1$. This can be done as follows: taking $\partial\langle n_an_b\rangle/\partial\lambda$ as an example

$$\left.\frac{\partial\langle n_an_b\rangle}{\partial\lambda}\right|_{\lambda=0}$$
$$=\frac{\partial}{\partial\lambda}\left[e^{-PV/kT}\sum\frac{z_a^{n_a}}{(n_a-1)!}\frac{z_b^{n_b}}{(n_b-1)!}\int\exp-(\tilde{U}_\mathbf{n}/kTd\{\mathbf{n}\}\right]$$
$$=-\langle n_an_b\rangle^0[(\partial/\partial\lambda)(PV/kT)]_{\lambda=0}-(kT)^{-1}e^{-PV/kT}$$
$$\sum\frac{z_a^{n_a}}{(n_a-1)!}\frac{z_b^{n_b}}{(n_b-1)!}\int\Delta U_\mathbf{n}\exp(-U_n^0/kT)d\{\mathbf{n}\}$$
$$=-\langle n_an_b\rangle^0[(\partial/\partial\lambda)(PV/kT)]_{\lambda=0}-(kT)^{-1}\langle n_an_b\Delta U_\mathbf{n}\rangle^0, \quad (19)$$

where the superscript zero indicates that the average is to be taken with respect to the potential of the reference solution. The term in $\partial(PV/kT)/\partial\lambda$ is common to each component of $\Gamma^{(1)}$, and cancels when the sum is formed at $\lambda=0$. We therefore have

$$\Gamma^{(1)}=-\frac{V}{kT}\left[\frac{\langle n_a(n_a-1)\Delta U_\mathbf{n}\rangle^0}{N_a^2}+\frac{\langle n_b(n_b-1)\Delta U_\mathbf{n}\rangle^0}{N_b^2}-\frac{2\langle n_an_b\Delta U_\mathbf{n}\rangle^0}{N_aN_b}\right]. \quad (20)$$

We now specialize further, and assume that both $U_\mathbf{n}$ and U_n^0 are the sum of pair potentials

$$U_\mathbf{n}=\sum_{\alpha,\beta}\sum_{i\neq j}u_{\alpha\beta}(|\mathbf{r}_i^\alpha-\mathbf{r}_j^\beta|)\quad \alpha,\beta=a,b$$
$$U_n^0=\sum_{i\neq j}u^0(|\mathbf{r}_i-\mathbf{r}_j|) \quad (21)$$

in which case, for example

$$\langle n_an_b\Delta U_\mathbf{n}\rangle^0=\tfrac{1}{2}\langle n_a^2(n_a-1)\Delta u_{aa}\rangle^0+\langle n_a^2n_b^2\Delta u_{ab}\rangle^0$$
$$+\tfrac{1}{2}\langle n_a^2(n_b-1)n_a\Delta u_{bb}\rangle^0$$
$$\Delta u_{\alpha\beta}=u_{\alpha\beta}(\gamma)-u^0(\gamma). \quad (22)$$

When the terms in Eq. (22) are expressed in terms of molecular distribution functions

$$F_{\mathbf{n}}{}^0=e^{-PV/kT}\frac{z_a{}^{n_a}}{\rho_a{}^{n_a}}\frac{z_b{}^{n_b}}{\rho_b{}^{n_b}}$$

$$\sum_m \frac{z_a{}^{m_a}}{m_a!}\frac{z_b{}^{m_b}}{m_b!}\int \exp(-U_{m+2}{}^0/kT)d\{\mathbf{m}\} \qquad (23)$$

and added to the corresponding representations of the first two terms of Eq. (20), a great deal of favorable cancellation occurs and we are left with

$$\Gamma^{(1)}=-(kT)^{-1}\int[u_{aa}+u_{bb}-2u_{ab})F_2{}^0dv, \qquad (24)$$

where $F_2{}^{(0)}$ is the pair correlation function|| for the reference system. Such cancellation was anticipated because of the symmetry of Γ, and this was the reason we chose to start with Eq. (1) instead of one of the other equations of reference 10. For example, if one attempts to calculate the partial molecular volume from Eq. (17) of reference 10, integrals involving $F_3{}^{(0)}$ occur, and this is a function about which we have very little knowledge. However, the partial molecular volume may be gotten from the chemical potential, Eq. (4), in the usual manner.

The second-order terms of the theory, $\Gamma^{(2)}$, are quite complex, involving distribution functions up to order 4. We shall not exhibit them here, confining our interest for the present to the lowest nonvanishing order of approximation. The complexity of the second-order terms is a feature common to all perturbation theories thus far proposed.

The expression (25) is not obviously related to any simple property of the reference fluid. However, if we assume that the solution is conformal in the sense of Longuet-Higgins,[1]

$$u_{\alpha\beta}=\epsilon_{\alpha\beta}f(r/\sigma_{\alpha\beta}), \qquad (25)$$

and, furthermore, assume that all the $\sigma_{\alpha\beta}$ are equal (molecules of "equal size)", then $\Gamma^{(1)}$ becomes

$$\Gamma^{(1)}=-(2d_{ab}/kTV\rho^2)\Delta E_v, \qquad (26)$$

where ΔE_v is the energy of vaporization of the reference fluid to the ideal gas state, and

$$d_{ab}=(2\epsilon_{ab}-\epsilon_{aa}-\epsilon_{bb})/\epsilon_{aa} \qquad (27)$$

if we choose pure a as the reference component. If we now regard $\rho\Gamma^{(1)}x_ax_b$ as small compared to unity and neglect it in the denominators of Eq. (5), the integrals of Eq. (5) may easily be evaluated, yielding

$$g^E=-d_{ab}(\Delta E_v/N)x_ax_b, \qquad (28)$$

which is just the result of Longuet-Higgins' first-order conformal solution theory.[1] The same result may be obtained without assuming equal sizes by further expanding $\Gamma^{(1)}$ in a power series in the deviations of the potential parameters from those of the reference system, keeping only first-order terms, and assuming $\sigma_{ab}=(\sigma_{aa}+\sigma_{bb})/2$, as did Longuet-Higgins. If we merely set the denominator equal to unity, without making any assumption about the potentials, we retrieve the first-order term of Nosanow and Mayer.

Thus, our device of writing the derivative of the chemical potential in the form of a quotient, and expanding the numerator and denominator separately in Taylor's series, has implicitly summed some of the higher order terms which occur in the straightforward expansion of the partition function. When we do not neglect these terms arising from the denominators; (a) the excess Gibbs function is no longer strictly proportional to the energy of vaporization and (b) is no longer parabolic in the mole fraction. To the order of approximation to which we are working, however, the excess Gibbs function is still a symmetrical function of the mole fractions.

Returning now, for illustrative purposes, to the assumption that $u_{\alpha\beta}=\epsilon_{\alpha\beta}f(r)$, we may find the excess volume and excess entropy by differentiation of g^E with respect to pressure and temperature respectively. We find

$$g^E=-2d_{ab}\frac{\Delta E_v}{N}\left[x_a\int_0^{x_b}\frac{ydy}{1-2d_{ab}(\Delta E_v/RT)y(1-y)}+x_b\int_0^{x_a}\frac{ydy}{1-2d_{ab}(\Delta E_v/RT)y(1-y)}\right], \qquad (29a)$$

$$v^E=\frac{-2d_{ab}}{N}\left(\frac{\partial\Delta E_v}{\partial P}\right)_T\left[x_a\int_0^{x_b}\frac{ydy}{[1-2d_{ab}(\Delta E_v/RT)y(1-y)]^2}+x_b\int_0^{x_a}\frac{ydy}{[1-2d_{ab}(\Delta E_v/RT)y(1-y)]^2}\right], \qquad (29b)$$

$$s^E=\frac{2d_{ab}}{N}\left[x_a\int_0^{x_b}\frac{y[(\partial\Delta E_v/\partial T)-2d_{ab}y(1-y)(\Delta E_v)^2/RT^2]}{[1-2d_{ab}(\Delta E_v/RT)y(1-y)]^2}+x_b\int_0^{x_a}\frac{y[(\partial\Delta E_v/\partial T)-2d_{ab}y(1-y)(\Delta E_v)^2/RT^2]}{[1-2d_{ab}(\Delta E_v/RT)y(1-y)]^2}\right]. \qquad (29c)$$

By subtraction

$$h^E=\frac{-2d_{ab}}{N}\left(T\frac{\partial\Delta E_v}{\partial T}-\Delta E_v\right)\times\left[x_a\int_0^{x_b}\frac{ydy}{[1-2d_{ab}(\Delta E_v/RT)y(1-y)]^2}+x_b\int_0^{x_a}\frac{ydy}{[1-2d_{ab}(\Delta E_v/RT)y(1-y)]^2}\right]. \qquad (29d)$$

Thus the four excess functions are not proportional. Only the pair v^E, h^E retains this property of conformal solution theory.

|| Denoted by g in reference 10. We use here the notation of J. E. Mayer.

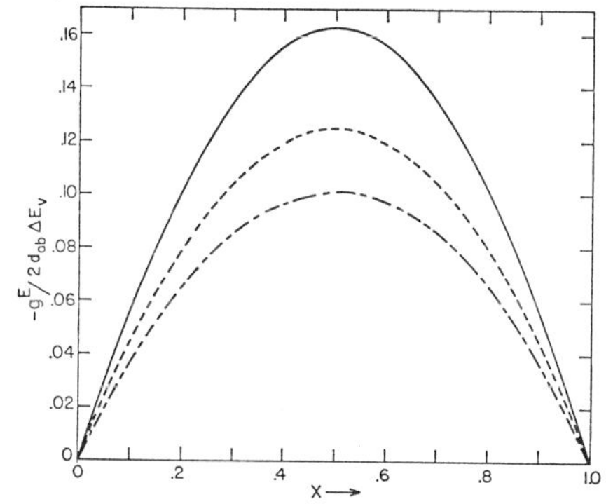

FIG. 1. Excess Gibb function *vs* mole fraction. ——— present work, $2d_{12}\Delta E_v/RT=1.1$; —·—·— present work, $2d_{12}\Delta E_v/RT=-1.1$. — — — conformal solution theory.

In order for $(\partial\mu_b/\partial\chi_b)_{T,P}$ to remain finite when $x_b>0$, $\rho x_a x_b \Gamma \geq -1$. For our present approximation this means $d_{ab}\leq 2RT/\Delta E_v \simeq 0.2$. For values of d_{ab} in the neighborhood of this upper limit we do not expect first-order terms to be sufficient, in any case. Therefore this limitation on d_{ab} plays no essential role in the theory.

The full integrals, Eq. (29), not neglecting the denominators, may be found in tables.[12] We have computed g^E for two values of d_{ab}, ± 0.05, and for $\Delta E_v/RT=11$ (Trouton's rule). The results are shown, together with the prediction of conformal solution theory, in Fig. 1. It is clear that retention of the denominator can make an appreciable contribution to the excess Gibbs function. The contribution will be even greater for the other excess functions because the denominator is squared. For example, the excess enthalpy has been calculated from Eq. (29d) for $d_{ab}=0.05$, and was found to be 70% greater than that predicted by conformal solution theory at $x_a=0.5$, while g^E is only 30% greater.

There are now two questions which we may ask. How does the present theory compare with experiment, and what light does it throw on conformal solution theory. Regarding the first point, the theory with retention of only first-order terms cannot be expected to give good agreement with experiment for many materials, for all excess functions are proportional to d_{ab}, and Waelbroeck[13] has shown that at least for benzene-cyclohexane mixtures, d_{ab} is much too small for the theory to be able to account for the experimental deviations from ideality. The situation may possibly be improved if we actually compute $\Gamma^{(1)}$ from Eq. (24), instead of assuming equal sizes, as we did for simplicity. Secondly, all the excess functions are symmetrical functions of the mole fractions, and many binary systems are known for which this is not the case. Thirdly, the ratio of v^E to h^E is just the same as for first-order conformal solutions, and so this ratio must be positive, in contradiction to experiment on certain systems. We do not believe that perturbation theories of the present type (expansion with respect to some potential parameter) will be as useful for quantitative numerical work as the theories based on order-disorder concepts.[4,7,8] Their main use, we believe, is as guides in the formulation of ideas and models of solutions to serve as "boundary conditions" so to speak.¶

This brings us to the second question. In conformal solution theory d_{ab} is often treated as an adjustable parameter, because of the lack of knowledge of what its value should really be. That this is consistent point of view is usually inferred from the near equality of d_{ab} determined by the experimental values of different excess quantities. According to our theory, although the d_{ab} obtained from v^E and h^E ought to be the same, the d_{ab} obtained from g^E and h^E, for example, should differ. By this we mean that the computed curves in Fig. 1 are very nearly parabolic, and thus one can attempt to fit them by conformal solution theory with an effective d_{ab}. When d_{ab} is negative, the effective value should be less than the actual value, and vice versa for $d_{ab}>0$. Similarly if one attempts to fit both g^E and h^E by parabolas the effective values of d_{ab} will differ for the two excess functions, and, in fact

$$d_{ab}^{\text{eff}}(h^E)>d_{ab}^{\text{eff}}(g^E),$$

when $d_{ab}(\text{actual})>0$, and vice versa, when $d_{ab}<0$. Because of the inadequacy of first-order theory, this conclusion should perhaps not be taken too seriously. It is difficult to calculate d_{ab} by comparison of theory and experiment according to our development, because this parameter does not enter the theory in a linear fashion as it does in conformal solution theory. For this reason we have merely tried to see how our development might modify conformal solution theory, to which it is closely related.

The theory has, in addition, some methodological interest. It gives directly the Gibbs function rather than the Helmholtz function, and furthermore, in each order of perturbation, contains in summed form, via the denominator, some of the "higher order" terms. It will be interesting to extend the theory by making the potential U_n^0 concentration dependent, in the manner of Brown[4] and Kirkwood *et al.*[7,8] Our development does not apply to this case, since we have treated $Q_{n_a+n_b}$ as constant when taking fugacity derivatives.

III. ISOTOPE MIXTURES

For classical systems, deviations from ideality are caused solely by differences in intermolecular forces.

[12] H. B. Dwight, *Tables of Integrals and Other Mathematical Data* (The Macmillan Company, New York, 1947), numbers 160.01, 160.11, 160.12.

[13] F. Waelbroeck, J. Chem. Phys. **23**, 749 (1955); J. chim. phys. **54**, 710 (1957).

¶ This point of view has been put forward by Brown[3] in some detail.

For quantum mechanical systems, there are also effects due to mass differences and due to symmetry which remain even when all forces are identical. A perturbation theory which expands about a pure reference system cannot treat symmetry effects. The reason for this has been discussed by Chester.[14] However, we can treat the mass effect, which may be expected to be dominant, for example, for hydrogen isotopes in the liquid range.

Chester,[14] and Prigogine *et al.*[15] have discussed isotope mixtures at low temperatures from the point of view of the Helmholtz function. Our formulation is also applicable, and gives essentially the results of Chester, corrected for the fact that we mix at constant pressure, while he mixes at constant volume, and with the new feature of the typical denominator which arises from our method of expansion. We calculate only the excess Gibbs function to second order in the perturbation, making use of some of the intermediate steps in Chester's analysis.

The quantum mechanical system is characterized by a Hamiltonian

$$H_{\mathbf{n}}=\sum_{i=1}^{n_a}\frac{P_i^2}{2m_a}+\sum_{j=1}^{n_b}\frac{P_j^2}{2m_b}+U_n, \tag{30}$$

which we may write as

$$H_{\mathbf{n}}=\sum_{i=1}^{n_a+n_b}k_i+U_n+\lambda_a\sum_{j=1}^{n_a}k_j+\lambda_b\sum_{l=1}^{n_b}k_l$$

$$k_i=P_i^2/2m_0;\qquad \lambda_a=(m_0-m_a)/m_a;$$

$$\lambda_b=(m_0-m_b)/m_b. \tag{31}$$

Thus we now have two perturbation parameters, which are no longer to be given the value unity at the end of the calculation, but are determined by the isotopic masses and the mass of the chosen reference material, m_0. The perturbation parameters, λ_a, λ_b, enter the grand partition function through the fact that

$$Q_{\mathbf{n}}=Tr\exp(-H_{\mathbf{n}}/kT). \tag{32}$$

The trace is to be taken with unsymmeterized wave functions (Boltzmann statistics).

Now the meaning of z_α must be changed. In the classical case $\mu_\alpha=\mu_\alpha{}^*+kT\ln z_\alpha$, where $\mu_\alpha{}^*$ is the chemical potential of α in the ideal gas state. For the way we have written the quantum mechanical grand partition function, z_α must be defined by $\mu_\alpha=\mu_\alpha{}^++kT\ln z_\alpha$, where $\mu_\alpha{}^+$ is the chemical potential associated with the internal state of species α. This arises because one can no longer simply split off the translational part of the partition function. This reinterpretation of z_α does not affect our general formalism.

[14] G. V. Chester, Phys. Rev. **100**, 446 (1955).
[15] I. Prigogine, *The Molecular Theory of Solutions* (North-Holland Publishing Company, Amsterdam, 1957), Chap. XIX.

For a binary isotope mixture there are two perturbation parameters, and so the first-order perturbation must be written

$$\sum_{i=a,b}\lambda_i\left(\frac{\partial\Gamma}{\partial\lambda_i}\right)_{T,z,V,\lambda=0}$$

$$=V\frac{z_a^2}{N_a^2}\left(\frac{\partial}{\partial z_a}-\frac{\partial}{\partial z_b}\right)^2\sum_i\lambda_i\left(\frac{\partial}{\partial\lambda_i}\frac{PV}{kT}\right)_{\lambda=0}. \tag{33}$$

Using the formulas of Chester[14] for $\partial Q_{\mathbf{n}}/\partial\lambda_i$ the summation over i in Eq. (33) may be written

$$e^{-PV/kT}\sum_n\frac{z_a^{n_a}}{n_a!}\frac{z_b^{n_b}}{n_b!}Q_n\{k_\alpha\}_n\sum_i\lambda_i n_i$$

$$=e^{-PV/kT}\sum_n\frac{z_a^{n_a}}{n_a!}\frac{z_b^{n_b}}{n_b!}Q_{n+1}\{k_\alpha\}_{n+1}(\lambda_a z_a+\lambda_b z_b), \tag{34}$$

where $\{k_\alpha\}_n$ is the mean kinetic energy of a particle in the reference fluid, evaluated by the canonical ensemble for n particles. The first application of

$$(\partial/\partial z_a-\partial/\partial z_b)$$

to Eq. (34) yields

$$e^{-PV/kT}(\lambda_a-\lambda_b)\sum\frac{z_a^{n_a}}{n_a!}\frac{z_b^{n_b}}{n_b!}Q_{n+1}\{k_\alpha\}_{n+1},$$

as the only nonvanishing term. A second application of the same operator yields zero. We must therefore investigate the second-order terms, which are

$$\frac{1}{2!}\sum_{i,j}\lambda_i\lambda_j\left[\sum_{\alpha=a,b}\left(\frac{\partial\ln z_\alpha}{\partial\lambda_i}\right)_{T,P,N}\left(\frac{\partial^2\Gamma}{\partial\lambda_j\partial z_\alpha}+\frac{\partial^2\Gamma}{\partial\lambda_i\partial\lambda_j}\right]_{\lambda=0}. \tag{35}$$

The first set of terms vanishes, since

$$\sum_j\left(\frac{\partial^2\Gamma}{\partial\lambda_j\partial z_\alpha}\right)_{\lambda=0}\lambda_j=\frac{\partial}{\partial z_\alpha}\sum_j\left(\frac{\partial\Gamma}{\partial\lambda_j}\right)_{\lambda=0}\lambda_j=0 \tag{36}$$

by our results on the first-order terms. It is rather long and tedious to write out the remaining terms of Eq. (35) in terms of averages, but one can show that the only nonvanishing contributions are

$$G_0\left[\sum_i\lambda_i\left(\frac{\partial\ln N_a}{\partial\lambda_i}-\frac{\partial\ln N_b}{\partial\lambda_i}\right)\right]^2$$

$$+\frac{V}{2}\frac{z_a^2}{N_a^2}\left(\frac{\partial}{\partial z_a}-\frac{\partial}{\partial z_b}\right)^2\sum_{i,j}\frac{\partial^2}{\partial\lambda_i\partial\lambda_j}\frac{PV}{kT}, \tag{37}$$

where G_0 is the density fluctuation of the reference system, and is given by

$$G_0=\kappa^0kT-1/\rho^0, \tag{38}$$

κ being the compressibility, and the superscript zero, denoting the reference solution. The first term of expression (37) is

$$\frac{G^0}{(kT)^2}(\lambda_a-\lambda_b)^2\bar{k}^2, \tag{39}$$

where

$$\bar{k}=\frac{\langle n_a\{k_\alpha\}_n\rangle^0}{N_a}=\frac{\langle n_b\{k_\alpha\}_n\rangle^0}{N_b}. \tag{40}$$

The second term of (37) is

$$\frac{N}{\rho^0(kT)^2}(\lambda_a-\lambda_b)^2\left[\frac{\langle n_a n_b\{k_\alpha, k_\beta\}_n\rangle^0}{N_a N_b}-\bar{k}^2\right], \tag{41}$$

where

$\{k_\alpha, k_\beta\}_n$

$$Q_n^{-1}\sum_{l_1,l_2}(k_\alpha)_{l_1,l_2}(k_\beta)_{l_1,l_2}$$

$$\times[\exp(-\beta E_{l_1})-\exp(-\beta E_{l_2})]/\beta(E_{l_2}-E_{l_1})$$

[reference 14, Eq. (2.12)].

Adding (39) and (41) we get for the second-order terms

$$\Gamma^{(2)}=\frac{(\lambda_a-\lambda_b)^2}{(kT)^2}\left[G_0\bar{k}^2+\frac{N}{\rho^0}\left(\frac{\langle n_a n_b\{k_\alpha, k_\beta\}_n\rangle^0}{N_a N_b}-\bar{k}^2\right]. \tag{42}$$

The derivation of (39) and (41) from (37) is rather long, but follows the same line of argument used in the proof that the first-order terms vanish. We therefore do not reproduce it here.

We may substitute expression (42) directly in Eq. (5) to find the excess Gibbs function. Since Γ, to our order of approximation, is independent of composition all averages being taken with respect to the reference system, the integrals may be evaluated just as in the classical case. Now, however, the constant parameters of the integrand can no longer be interpreted as *thermodynamic* properties of the reference system. If we neglect Γ in the denominator of the integrand we recover the results of Chester, with two differences. (a) All averages are defined with respect to the grand canonical ensemble, not the canonical ensemble which Chester used. (b) There is an additional term, proportional to G_0. This difference arises because Chester's mixing process is carried out at constant volume, while ours is carried out at constant pressure.

We have slurred over one point here by claiming that the integrands of Eq. (5) contained only concentration independent parameters. This is not strictly so, since ρ in Eq. (5) is the actual number density existing at concentration y, $1-y$. To a first approximation however, we may take $\rho=\rho^0$, since $\rho-\rho^0$ is at least of order λ_i, and introduction of the correction terms would give some terms of the third order in λ, which we do not wish to consider at present.

Since the parameters of the quantum mechanical theory are not related to measurable properties of the reference component, one needs accurate measurements of g^E over a wide concentration range in order to test the theory. To the best of the author's knowledge, such data are not available for liquids to which the present theory should be applicable (hydrogen or perhaps helium isotope mixtures). The concentration range and the accuracy need be great in order to find a difference between our and Chester's theories since the deviation of g^E from a parabolic curve must be detected, and such deviation is likely to be small. If Γ is small enough, our development is consistent with Chester's, and all the conclusions of his paper will follow. If Γ in all the conclusions of his paper will follow. If Γ in the denominator of Eq. (5) cannot be neglected, then Chester's arguments about corresponding states must be modified.

ACKNOWLEDGMENTS

The author would like to thank Professor J. E. Mayer for several helpful discussions, and Mr. L. Nosanow for the use of his manuscript on the perturbation theory of solutions. The idea for this research arose from lectures on the molecular theory of solutions which the author, together with colleagues, gave at the University of Chicago. He would like to thank the auditors of these lectures for their patient attention and friendly criticism.

Thermodynamic Properties of Mixtures from Fluctuation Solution Theory

John P. O'Connell
Department of Chemical Engineering
University of Virginia, Charlottesville, VA

ABSTRACT
Fluctuation Solution Theory develops exact relations between integrals of molecular correlation functions and concentration derivatives of pressure and chemical potential. These properties can be usefully correlated, particularly in strongly nonideal liquid mixtures, yielding unique paths to solution thermodynamic property values. This review describes the basic relations for normal and "reactive" components in terms of integrals of the total correlation function and the centers direct correlation function. Obtaining the more desirable properties of density and activity through integration of model equations is shown. Comments on the behavior of these integrals are made for a wide variety of pure and mixed liquids including those with normal, polar, aqueous, dissolved supercritical and electrolyte species. Finally, successful correlations are demonstrated for the above systems.

INTRODUCTION

Fluctuation Solution Theory is important not only for understanding the molecular processes of mixtures, it also provides general statistical mechanical relationships for certain thermodynamic properties that can be utilized for physical property correlations of mixtures with arbitrary numbers of components. In particular, concentration derivatives of the chemical potential and pressure are given in terms of molecular correlation functions without any assumption about the form of the intermolecular potential energy. This paper describes these expressions and the author's experience in using them for modeling equations of thermodynamic properties of strongly nonideal systems. The emphasis will be on the knowledge and concepts that may not have been fully described in the available literature rather than be a synopsis of the published articles.

BASIC EXPRESSIONS FOR MOLECULAR COMPONENTS

Total Correlation Functions

The earliest work on fluctuation theory was by Kirkwood and Buff (1951) who related integrals of the statistical mechanical total correlation function to the macroscopic manifestations of molecular concentration fluctuations in the grand canonical ensemble. They gave fundamental expressions that connected the matrix of integrals of the total correlation function (TCFI), $\mathbf{H}$, to the statistical fluctuations of the elements of the vector of numbers of molecules of each component, $\mathbf{N}$, and to the variation of the number of molecules of component i, N_i, with a change of the chemical potential of component j, μ_j, at fixed temperature, T, total volume, V^T, and other chemical potentials, $\mu_{k \neq j}$, where i and j are an arbitrary pair[1]

[1]Matrix notation allows much more compact writing and computer symbolic programming. For readers who find it unfamiliar or confusing, the most effective way to follow developments such as this is to write out completely the two- or three-component cases.

$$[\mathbf{A}]_{ij} \equiv \frac{RT}{N}\left(\frac{\partial N_i}{\partial \mu_j}\right)_{T,V^T,\mu_{k\neq j}} \tag{1.1}$$

$$= \left([(\mathbf{NN})^T]_{ij} - N_iN_j\right)/N^2 \tag{1.2}$$

$$= \left(\delta_{ij}N_i + N_iN_j[\mathbf{H}]_{ij}/N\right)/N \tag{1.3}$$

Here the total correlation function is defined as the difference between the pair correlation function (or radial distribution function) at a particular distance from a particle center of mass and the average solution value of unity

$$h_{ij} \equiv g_{ij} - 1 \tag{1.4}$$

the TCFI is the dimensionless spatial integral

$$[\mathbf{H}]_{ij} \equiv H_{ij} \equiv \rho \int h_{ij}d\mathbf{r} \equiv \rho G_{ij} \tag{1.5}$$

where ρ is the molar density, N/V^T, δ_{ij} is the Kronecker delta and G_{ij} is the fluctuation integral. All matrices given above are symmetric.

The effect of a change of composition made while all but one of the chemical potentials remains constant - similar to changes in McMillan-Mayer solution theory (1945) - are reciprocals of the elements of the $\mathbf{A}$ matrix that yield the derivative

$$\frac{N}{RT}\left(\frac{\partial \mu_i}{\partial N_j}\right)_{T,V^T,\mu_{k\neq i}} = 1/[\mathbf{A}]_{ij} \tag{1.6}$$

$$= N/\left(\delta_{ij}N_i + N_iN_j[\mathbf{H}]_{ij}/N\right) \tag{1.7}$$

A more desirable quantity is the change of chemical potential of any component i with the number of molecules of any other component j at fixed T, V^T, and other mole numbers, $N_{k\neq j}$, (the canonical ensemble variables) which is related to elements of the inverse matrix of $\mathbf{A}$

$$\frac{N}{RT}\left(\frac{\partial \mu_i}{\partial N_j}\right)_{T,V^T,N_{k\neq i}} = [\mathbf{A}^{-1}]_{ij} \tag{1.8}$$

To obtain the derivatives that include some of the $\mathbf{N}$ and μ fixed as in multicomponent McMillan-Mayer systems requires complex expressions of inverses of Minors of $\mathbf{A}$. This has not yet been developed in general terms although Friedman and Ramanathan (1970) gave some expressions for a special case of pairs of common-ion strong electrolytes.

All of these equations are expressed in terms of the total number of molecules which can be converted to moles with Avogadro's number, N_A. Expressions in terms of intensive variables involve dividing $\mathbf{N}$ by V^T to get concentrations, $\rho_i \equiv N_i/V^T$. The result for equation (1.8) is

$$\frac{\rho}{RT}\left(\frac{\partial \mu_i}{\partial \rho_j}\right)_{T,\rho_{k\neq j}} = [\mathbf{A}^{-1}]_{ij} \tag{1.9}$$

A set of independent variables in terms of V^T or ρ is not the most convenient for many purposes, so a transformation must be done to obtain the Lewis-Randall set T, P, $\mathbf{N}$ or $\mathbf{x}$. This is achieved in two steps which not only yield chemical potential derivatives with these variables, but also generate a new relation which connects system density and pressure. Thus, the formalism provides equations for both chemical potential (or activity coefficient) and density from Lewis-Randall variables. The final results for binary systems and for the case of one solvent ($0 < x_1 \leq 1$) and multiple solutes ($x_i < 1$) were given by Kirkwood and Buff (1951) who transformed the $\mathbf{A}$ matrix for this case.

O'Connell (1971) gave matrix expressions for the general case with all solution compositions and introduced full matrix notation. The final equations for TCFI will not be repeated here because of their complexity and the fact that they have not been used by the author. However the procedure will be illustrated.

First, there is the transformation

$$\frac{N}{RT}\left(\frac{\partial\mu_i}{\partial N_j}\right)_{T,VT,N_{k\neq j}} = \frac{N}{RT}\left(\frac{\partial\mu_i}{\partial N_j}\right)_{T,P,N_{k\neq j}} + \frac{\rho\overline{V}_i\overline{V}_j}{\kappa_T RT} \tag{1.10}$$

where $\overline{V}_i$ is the partial molar volume of component i,

$$\overline{V}_i \equiv N_A\left(\frac{\partial VT}{\partial N_i}\right)_{T,P,N_{k\neq i}} \tag{1.11}$$

and κ_T is the isothermal compressibility of the mixture

$$\kappa_T \equiv -\frac{1}{VT}\left(\frac{\partial VT}{\partial P}\right)_{T,\mathbf{N}} = \frac{1}{\rho}\left(\frac{\partial \rho}{\partial P}\right)_{T,\mathbf{x}} \tag{1.12}$$

Then, the Gibbs-Duhem equation can be used on equation (1.10) to obtain the (complicated) expression for the partial derivative in terms of TCFI. Further, vector operations on equation (1.10) yield equations for the partial molar volume and solution isothermal compressibility in terms of the TCFI. Finally, proper manipulation of the equations gives the concentration derivative of the osmotic pressure.

All of these equations are for *derivatives* of the chemical potential, total volume and osmotic pressure. Further steps are needed to obtain the desired properties. In particular, integration of sets of these partial derivatives from a suitable reference state to the solution state is done (O'Connell, 1981), as described below. This feature is both the strength and the challenge of fluctuation solution thermodynamics; it yields activity coefficients as well as densities of mixtures of arbitrary numbers of components in terms of molecular quantities amenable to modeling, but it is formulated in terms of complex expressions and requires reference state information.

Direct Correlation Functions

In addition to the above formulation in terms of the TCFI, the Ornstein-Zernike Equation (1914) has been used to express the same thermodynamic quantities in term of integrals of the direct correlation function. The results are expressed much more simply than in terms of the TCFI and there are no divergences at fluid critical points. The integrated form of the multicomponent O-Z Equation is

$$[\mathbf{H}]_{ij} = [\mathbf{C}]_{ij} + [\mathbf{HXC}]_{ij} \tag{1.13}$$

where $[\mathbf{C}]_{ij}$, the direct correlation function integral (DCFI) is the dimensionless spatial integral of the direct correlation function (dcf)

$$[\mathbf{C}]_{ij} \equiv C_{ij} \equiv \rho\int c_{ij}d\mathbf{r} \tag{1.14}$$

Note that here C_{ij} is dimensionless whereas elsewhere in this volume the symbol C_{ij} refers to the dimensional form without the density ρ. Also, $\mathbf{X}$ is a diagonal matrix of the mole fractions

$$[\mathbf{X}]_{ij} \equiv \delta_{ij}x_i \tag{1.15}$$

The expressions that arise for the quantities of Equations (1.9) to (1.12) are much simpler in terms of the DCFI. The basic expression relating **A**, **H**, and **C** arises from combination of Equations (1.8), (1.10) and (1.13).

$$[\mathbf{A}]^{-1} = [\mathbf{X} + \mathbf{XHX}]^{-1} = \mathbf{X}^{-1} - \mathbf{C} \tag{1.16}$$

This then leads to

$$\frac{\rho}{RT}\left(\frac{\partial\mu_i}{\partial\rho_j}\right)_{T,\rho_{k\neq j}} = \frac{\delta_{ij}}{x_i} - C_{ij} \tag{1.17}$$

or, in terms of the mole fraction activity coefficient

$$\ln\gamma_i \equiv (\mu_i - \mu_i^o)/RT - \ln x_i \tag{1.18}$$

$$\rho\left(\frac{\partial\ln\gamma_i}{\partial\rho_j}\right)_{T,\rho_{k\neq j}} = 1 - C_{ij} \tag{1.19}$$

The connection between the canonical and the Lewis-Randall variables for activity coefficients is the same as for chemical potentials in Equation (1.10) so

$$N\left(\frac{\partial\ln\gamma_i}{\partial N_j}\right)_{T,VT,N_{k\neq j}} = N\left(\frac{\partial\ln\gamma_i}{\partial N_j}\right)_{T,P,N_{k\neq j}} + \frac{\rho\overline{V}_i\overline{V}_j}{\kappa_T RT} \tag{1.20}$$

which leads to the activity coefficient derivative

$$\frac{N}{\rho\kappa_T RT}\left(\frac{\partial\ln\gamma_i}{\partial N_j}\right)_{T,P,N_{k\neq j}} = $$

$$= \sum_k\sum_l x_k x_l[(1-C_{ij})(1-C_{kl}) - (1-C_{ik})(1-C_{jl})] \tag{1.21}$$

the partial molar volume

$$\frac{\overline{V}_i}{\kappa_T RT} \equiv \left(\frac{\partial P/RT}{\partial\rho_i}\right)_{T,\rho_{k\neq j}} \tag{1.22}$$

$$= \sum_j x_j(1-C_{ij}) \tag{1.23}$$

and the reduced bulk modulus

$$\frac{1}{\rho\kappa_T RT} \equiv \left(\frac{\partial P/RT}{\partial\rho}\right)_{T,\mathbf{x}} \tag{1.24}$$

$$= \sum_i\sum_j x_i x_j(1-C_{ij}) \tag{1.25}$$

For a component $i \neq 1$ in a single solvent 1, the osmotic pressure, Π, is related to composition by

$$\frac{N_1}{\rho RT}\left(\frac{\partial\Pi}{\partial N_i}\right)_{T,\mu_1,N_1,N_{j\neq i}} = x_1\sum_{j>1}\frac{N_j}{kT}\frac{\partial\mu_j}{\partial N_i}\Bigg)_{T,N_1,N_{k\neq i}} \tag{1.26}$$

$$= \left[\frac{-N_1}{RT}\left(\frac{\partial\mu_1}{\partial N_i}\right)_{T,P,N_1,N_{k\neq i}}\right]/\rho\overline{V}_1 \tag{1.27}$$

$$= x_1\left[C_{1i} + \sum_j x_j(1-C_{ij})\right] \qquad i \neq 1 \tag{1.28}$$

or

$$\sum_{j>1} x_j\left(N\frac{\partial\ln\gamma_j}{\partial N_i}\right)_{T,\mu_1,N_1,N_{k\neq i}} = \sum_{j>1} x_j\left[(1-C_{ij}) - (1-C_{1i})\right] \tag{1.29}$$

The general expressions for multicomponent solvents are more complex.

Comments on Mathematical Forms

The mathematical form of the relations given above has several significant aspects. First, the number of terms in equation (1.21) are many fewer than might be supposed; for a binary mixture there is only one term for each i-j pair, e.g., for $i = j = 1$ only the term for $k = l = 2$ exists. Second, for pure components, there is only one equation, (1.24), for the compressibility. Thus, the expressions do not treat pure component free energies such as are needed to find saturation pressures.

Third, for a binary mixture, there are three DCFI, C_{11}, C_{12}, and C_{22}. Having values of the binary activity coefficient derivative, the partial molar volume for one of the components, the isothermal compressibility and the density allows values of all three DCFI to be obtained by substitution in Equations (1.21), (1.23) and (1.25).

$$(1 - C_{11}) = \frac{\rho \overline{V}_1^2}{\kappa_T RT} + x_2 \left(\frac{\partial ln \gamma_1}{\partial x_1}\right)_{T,P} \tag{1.30}$$
$$(1 - C_{12}) = \frac{\rho \overline{V}_1 \overline{V}_2}{\kappa_T RT} - x_1 \left(\frac{\partial ln \gamma_1}{\partial x_1}\right)_{T,P} \tag{1.31}$$
$$(1 - C_{22}) = \frac{\rho \overline{V}_2^2}{\kappa_T RT} + x_1 \left(\frac{\partial ln \gamma_2}{\partial x_2}\right)_{T,P} \tag{1.32}$$

The TCFI can be calculated from the DCFI by integration of the Ornstein-Zernike Equation (1914) and matrix inversion,

$$H_{11} = [(1 - x_2 C_{22})/D_c - 1]/x_1 \tag{1.33}$$
$$H_{12} = C_{12}/D_c \tag{1.34}$$
$$H_{22} = [(1 - x_1 C_{11})/D_c - 1]/x_2 \tag{1.35}$$

where

$$D_c = (1 - x_1 C_{11})(1 - x_2 C_{22}) - x_1 x_2 C_{12}^2 \tag{1.36}$$

The quantity D_c is the central quantity of the phase stability matrix. Only when D_c is positive can a single phase be thermodynamically stable. Systems for which D_c changes sign with composition will have partial miscibility. As discussed below, calculated values of D_c may change sign even if the system does not show immiscibility. This usually arises from inappropriate equations to fit activity coefficient data.

There is an ideal solution contribution to the C_{ij} when the activity coefficient derivatives disappear, the partial molar volumes are the inverse of the pure molar densities and the solution molar volume and isothermal compressibility are linear functions of mole fraction. Then, the DCFI vary inversely with mole fraction. For example,

$$(1 - C_{11})^{IS} = \frac{V_1^{o2}}{\left(x_1 V_1^o \kappa_{T1}^o + x_2 V_2^o \kappa_{T2}^o\right) RT} \tag{1.37}$$

Only in strongly nonideal solutions do the second terms of equations (1.30) to (1.32) significantly exceed the first terms. However, the nonidealities in both terms cause deviations from ideal solution behavior. Their behavior and modeling will be noted below.

Fourth, examination (Campanella, 1984) of various activity coefficient models with temperature and mole fraction as the independent variables suggests why "local composition" models are so successful for extending binary parameter expressions to multicomponent systems. For example, the binary Margules relations for $\ln \gamma_i$ are much different than those from the Van Laar Equation because the latter has a denominator involving composition. Further, it is not obvious how to extend the Van Laar Model to multicomponent systems. The NRTL and Wilson Equations for excess Gibbs Energy and activity coefficient (Prausnitz, et al., 1986) appear much different from each other. However, at the derivative level, a pattern begins to emerge where the expressions for activity coefficients of component i from all the different models involve ratios of polynomials and the principle mathematical variation is captured in them by terms such as

$$\sum_j \frac{x_j A_{ij}}{(\sum_k x_k A_{kj})} \tag{1.38}$$

Perhaps simple empirical models can be developed by examining the mathematical form of the derivative.

Fifth, the most desirable properties are the *integrals* of these equations. To actually accomplish this requires that the model for the DCFI be given in terms of the independent variables T and ρ, as in traditional equations of state in the van der Waals form. Here, given a model for the C_{ij} as a function of (T, ρ), integration of the set of equations (1.22) for all i uses the desired independent variables $(T, P, \mathbf{x})$ to obtain ρ from the reference state properties $T, P_o, \rho_o, \mathbf{x}_o$.

$$\frac{(P-P_o)}{RT} = \sum_i \sum_j \int_{x_{oj}\rho_o}^{x_j\rho} \rho_j \left.\frac{(1-C_{ij})}{\rho}\right|_{T,\rho_{k\neq j}} d\rho_j \tag{1.39}$$

Then, integration is performed to obtain the activity coefficient using the set of equations (1.19)

$$\ln \gamma_i = \sum_j \int_{x_{oj}\rho}^{x_j\rho} \left.\frac{(1-C_{ij})}{\rho}\right|_{T,\rho_{k\neq j}} d\rho_j \tag{1.40}$$

Although the forms of equations (1.39) and (1.40) may imply some complexity in the integrations, Mathias and O'Connell (1981) show how this can be easily accomplished.

Finally, only for pure components (where there is only one DCFI) and for strongly nonideal solutions (where the unlike $1 - C_{ij}$ are significantly different from the arithmetic or geometric mean of the like $1 - C_{ii}$ and $1 - C_{jj}$) will correlations have any chance of success. Thus, as shown by Wooley (1987) (see below) and Hamad, et al., (1989) even aqueous alcohol solutions show relatively small deviations from these mean values leading to a "small difference of large numbers" problem in equation (1.21) which cannot be avoided by using equation (1.19) because it reappears in equation (1.40). This means that only the cases of supercritical and solid substances sparingly soluble in liquids, salts and perhaps for partially miscible components can estimated values of the various DCFI be used for accurate predictions. As a result, this author's work has focussed on the pressure effects on pure and mixed liquids, and on supercritical and ionic systems.

Comments on Molecular Implications

From the microscopic point of view, these results have some subtle implications. When spherical, the variation of c_{ij} ranges from negative (but not infinite) at zero separation, through zero at about the zero of the pair potential energy, then positive becoming zero at large separations. Figure 1 shows the variation of the dimensionless pair potential, u/ϵ_{ij}, h_{ij} and c_{ij} for a fluid at liquid-like densities. In addition to the simplicity of equations (1.19), (1.23) and (1.25) compared to the results for the TCFI, the fact that the dcf is short-ranged and has only one extremum suggests that this development may be of greater utility. This was the basic rationale for the author's developments described below.

When a DCFI is less than zero, the (repulsive) excluded volume forces dominate, while if it is greater than zero, attractive and/or long range forces dominate. The former is true for normal pure liquids where the range of C is from -10 to -1000. There are three cases where the latter occurs. The first is in pure fluids near the critical point where $C \approx 1$. The second is in dilute gases at lower reduced temperatures since $C = -2B\rho + ...$ with B (≤ 0) the second virial coefficient. The third case is for pairs of salt components at low concentrations where the Debye-Hückel Limiting Law gives $C_{ij}^{DHLL} = S_\gamma \rho / 4I^{\frac{1}{2}} + \ldots > 0$ where S_γ is related to the solvent dielectric constant and density and I is the ionic strength (see below). As

I approaches zero, all DCFI between two salts of the same charge type diverge in the same way.

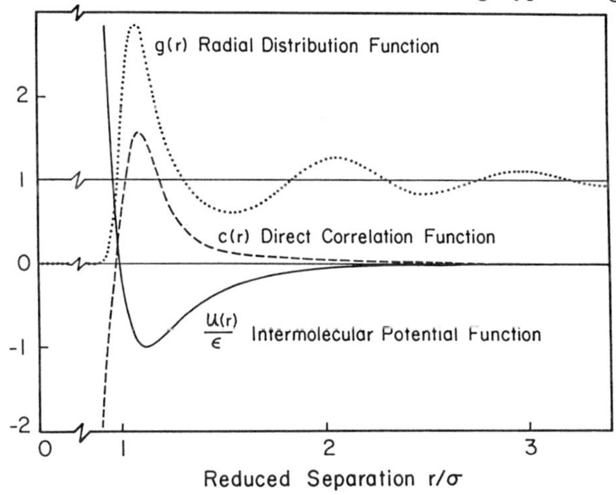

Figure 1. Molecular Interaction Functions for Spherical Systems.

Equation (1.13) assumes that the last term (associated with indirect correlations) can be integrated in such a way that the TCFI and DCFI can be separated. The discussion in Chapter 5 of Gray and Gubbins (1984) indicates that for molecular fluids, this either requires that a strong angular condition be obeyed or that the integrals may not be directly connected to microscopic angle-averaged or perturbation correlation functions. In the present work, the DCFI is treated as a macroscopic quantity in the same way that the partition function is in cell or lattice theory models (Reed and Gubbins, 1973). For nearly spherical systems, this lack of connection is not of concern and, in fact, the correlational success obtained for wide varieties of molecular fluids when equation (1.25) is used may imply that natural systems follow this formulation. On the other hand, use of microscopic models directly in the integral of equations (1.14) may not be appropriate for some polar and nonspherical molecular components.

The use of higher order dcfs allows connections between various DCFI (which are for pairs of components) via recurrence relations (O'Connell, 1971; Perry and O'Connell, 1984). The basic equation connecting pair DCFI's to triplet DCFI's for spherically symmetric systems is

$$\rho^2 \left.\frac{\partial C_{ij}/\rho}{\partial \rho_k}\right)_{T,\rho_{l\neq k}} = \rho^2 \left.\frac{\partial C_{ik}/\rho}{\partial \rho_j}\right)_{T,\rho_{l\neq j}} \tag{1.41}$$

$$= C_{ijk}$$

$$\equiv \rho^2 \int\int c^{(3)}_{ijk} d\vec{r}_j d\vec{r}_k \tag{1.42}$$

Hamad, et al. (1989) give a similar relation involving P and x_i. The functional Taylor's expansion method (Percus, 1964) shows that for pure components, the CHNC approximation requires that $C^{CHNC}_{iii} = 0$, whereas the Percus-Yevick approximation gives $C^{PY}_{iii} = -C^2_{ii}$.

BASIC EXPRESSIONS FOR SPECIES FROM REACTIVE COMPONENTS

The above relations are rigorously true for atomic and molecular components, but there are other situations of interest in fluids. One for which the fluctuation thermodynamics has been explored in some detail involves cases where the components used to make up the thermodynamic system can react (dissociate, associate, solvate) to form new species. Examples of such reactive components include electrolytes that ionize, organic acids that dimerize and absorbing solvents for acid gases which solvate. While the molecular, or *component* treatment is still correct, a more desirable goal is to model TCFI and/or DCFI between the *species* which are actually, or conceptually, present. This would lead to correlations with parameters from, for example, ions, rather than salts. Then, multicomponent expressions would require minimal or no information beyond single- or binary-component systems. Thus, for example, completely reacting systems, e.g., strong electrolytes, would have no "salt" parameters nor equilibrium constants. (For partially reacting systems, both would be needed.) Whether the result would be the same as for other "chemical theories" where the situation often involves too many parameters to fit to the system data, or whether the species quantities can be obtained from other systems is still to be investigated.

The fact that the species arise from components via well-defined reactions leads to a certain "connectedness" that appears in the correlation function integrals. This is separate from the long-ranged behavior of electrostatic forces and of site-site developments. Perry et al. (Perry and O'Connell, 1984; Perry, et al., 1988) developed the general expressions for fluctuation thermodynamic properties in multicomponent systems. Their complexity precludes repeating all of them here. However, the case of completely reacting components is given so that the results can be implemented for strong electrolyte models (see below).

The essential equations are again of the vector and matrix form. For a set of n_0 components whose numbers of moles are given by a vector $\mathbf{N_0}$ with n_0 elements, the reactions will yield a set of n species whose numbers of moles are given by a vector $\mathbf{N}$ with n elements. The reactions (specifically the stoichiometric coefficients) lead to a nxn_0 projection matrix identified as $\mathbf{W}$. Thus

$$\mathbf{N} = \mathbf{W}\mathbf{N_0} \tag{2.1}$$

While Perry and O'Connell (1984) give general expressions, the most interesting result for the present purposes with electrolyte solutions is that for complete dissociation (Perry, et al., 1988). For two components i and j which dissociate into $\nu_{i\alpha}$ and $\nu_{j\beta}$ species, the desired component j concentration derivative of the component i activity coefficient is not only given by the molecular DCFI of equation (1.19) above, C_{0ij}, it can also be expressed in terms of DCFIs between $\alpha\beta$ pairs of the species from the reactions. With this notation (subscript $_0$ for components), the result is

$$\rho\left(\frac{\partial \ln \gamma_i}{\partial \rho_{0j}}\right)_{T,\rho_{0k\neq j}} = 1 - C_{0ij} \tag{2.2}$$

where

$$\rho \quad \equiv N/V^T \tag{2.2}$$

$$N \quad \equiv \sum_i \sum_\alpha \nu_{i\alpha} N_{0i} \tag{2.3}$$

$$\rho_0 \quad \equiv N_0/V^T \tag{2.4}$$

$$N_0 \quad \equiv \sum_i N_{0i} \tag{2.5}$$

$$\nu_i \quad \equiv \sum_\alpha \nu_{i\alpha} \tag{2.6}$$

$$C_{0ij} = \sum_\alpha \sum_\beta \nu_{i\alpha}\nu_{j\beta} C_{\alpha\beta}/\nu_\alpha \nu_\beta \tag{2.7}$$

In essence, the development shows that for complete reactions, a direct substitution of the proper collection [equations (2.7)] of species DCFI can be made for the component DCFI to obtain the fluctuation thermodynamic properties of such solutions. As shown below, this leads to a particularly simple development for strong electrolytes. Unfortunately, the general case of partial reactions is much more complicated since the general formulation involves derivatives of the equilibrium extent of reaction which must be connected to the equilibrium constant. This case has not been addressed in any detail.

CORRELATIONS AND RESULTS FOR VARIOUS CASES

Pure Liquids

Apparently the first useful correlation of fluctuation thermodynamic properties was that of Brelvi and O'Connell (1972, 1975ab) who described a one-parameter form of corresponding states for the DCFI with a characteristic volume to scale the density of a wide variety of pure fluids. This led to a simple and reliable correlation for the change of density of liquids when pressurized. Given the saturated molar volume at ambient conditions, the density at a higher pressure could be predicted to within 1% if the temperature were less than the critical. This work was taken a step further by Huang and O'Connell (1987) with an extremely accurate and reliable 3-parameter corresponding states correlation over the same liquid range. In this case the characteristic parameters were in volume, V^*, temperature, T^*, and DCFI, C^*.

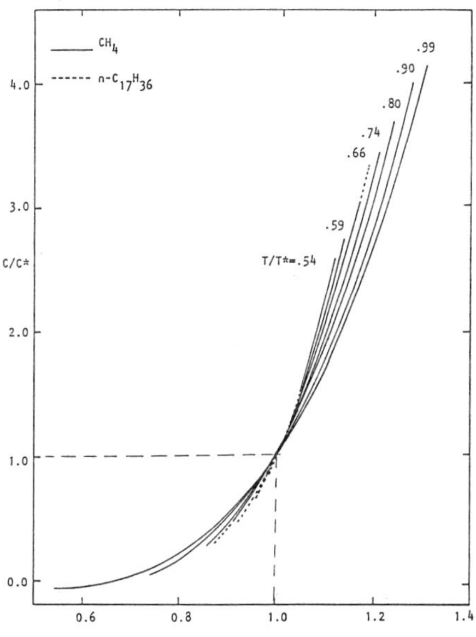

Figure 2. Corresponding States Behavior of Reduced Bulk Modulus of Liquids.

Figure 2 shows the behavior of the scaled reduced bulk modulus, C/C^* over wide ranges of density. There are two significant features. The first is the weak temperature dependence. There is even a density where no variation with T occurs: at lower densities the DCFI becomes less negative with temperature while at high densities the opposite occurs. Second, there is a strong density dependence. The correlation consists of characteristic DCFI and volume values (where the temperature dependence disappears) and a characteristic temperature. Table 1 shows results for reduced density, ρV^*, and temperature, T/T^*, using the correlation for some of the nearly 300 substances and mixtures analyzed by Huang (1986). The predicted densities including those up to very high pressures are generally within 1%. Predictions can be made with estimates of V^* and T^* (but with fitted values of V^* because of the great sensitivity to

Table 1: CHARACTERISTIC PARAMETERS AND COMPARISONS WITH LIQUID COMPRESSION DATA

From Y-H. Huang (1986).

Substance	$-C^*$	$V^*, \frac{cm^3}{mol}$	T^*, K	ρV^*	T/T^*	AAE,%
CH_4	14.7	38.72	191	0.7-1.3	0.5-0.9	0.1
$n-C_4H_{10}$	32.4	90.61	397	0.8-1.3	0.5 0.9	0.5
$C_6H_5NO_2$	48.3	109.18	586	1.0-1.1	0.5-0.6	0.3
$C_6H_5NH_2$	48.3	96.84	576	1.0-1.1	0.5-0.6	0.2
CCl_2F_2	37.3	75.97	373	0.8-1.1	0.7-0.9	1.5
Acetic Acid	32.8	55.78	494	0.9-1.1	0.6-0.7	0.3
Ethylene Carbonate	68.3	65.28	646	0.9-1.1	0.5-0.6	1.4
$Si(-OCH_4)_4$	52.8	122.75	402	0.8-1.2	0.7-0.9	1.1
$AgNO_3$	63.3	44.56	942	1.0-1.1	0.6-0.8	0.5
$(n-C_6H_{13})_4NBF_4$	231.9	484.14	809	0.9-1.1	0.5-0.6	0.1
Silicone Oil (Bayer M100)	312.6	1.0522^+	555	0.9-1.1	0.5-0.6	0.4
Poly (n-butyl-methacrylate)	288.7	1.0522^+	962	1.0-1.2	0.4-0.5	0.4

$^+ cm^3/gm$

it) from group contributions. The mixture results for both the Brelvi and Huang methods are commensurate with their accuracies for pure components. With estimates of liquid heat capacities, these correlations can also be used to obtain speeds of sound in liquids.

The thermodynamic implications of the temperature-independent DCFI and the nature of the slope of the isotherms on pressure - density plots were discussed by Huang and O'Connell (1987). In particular, an indication of the balance between attractive and repulsive forces can be found at the crossover state. Also, Mathias (1978), Campanella (1984), Telotte (1985) and Huang (1986, 1987) investigated whether current models based on either equations of state or molecular theory yield the same behavior as shown by these simple correlations. The summary conclusion is no. For all liquids, the density variation is weaker at lower densities than for rigid bodies. (At the critical point, a real fluid has a DCFI value of 1 whereas rigid bodies always have the DCFI ≤ 0. This difference dominates much of the liquid range.) The variation is about the same at very high densities as given by spherical and nonspherical rigid-body equations, particularly the Weeks-Chandler-Andersen(WCA) reference fluid (see, e.g., Reed and Gubbins, 1973) which has a temperature and density dependent hard sphere diameter[2]. However, at normal liquid densities, no linear density term, such as a generalized van der Waals equation would give, is adequate to make the correction.[3] This suggests that perturbation theories based on rigid bodies must require elaborate corrections to achieve good descriptions of TCFI and DCFI. Finally, even those cubic equations of state which give reliable saturation densities give poor compressibilities; their form does not agree with the correlation. This indicates that their accuracy with vapor-liquid equilibria, whether with

[2]Huang and O'Connell(1987) imply that the real and rigid body variations are not similar in any density range. In fact, depending on the connection of the reduced densities, behavior at high densities can appear to be similar, particularly for polyatomic substances.

[3]Mathias and O'Connell (1981) use such an equation, but it is not intended to yield highly accurate density changes with compression.

Table 2: PARTIAL MOLAR VOLUMES AT INFINITE DILUTION OF GASES IN LIQUIDS

From S. W. Brelvi (1972).

Solute (1)	Solvent(2)	T, K	$V_2, \frac{cm^3}{mol}$	$\overline{V}^{\infty}_{1,Calc}, \frac{cm^3}{mol}$	$\overline{V}^{\infty}_{1,Exp}, \frac{cm^3}{mol}$
CH_4	$n-C_7H_{16}$	300	148.6	56.6	55.4
H_2	C_6H_6	298	89.0	31.1	35.0
		403	102.2	57.4	58.3
			101.2	53.9	53.7
			100.0	51.4	49.7
		433	107.4	66.9	72.1
			105.7	63.0	64.0
			104.4	56.7	57.5
		473	116.1	92.9	104.
			113.3	81.6	85.6
			111.1	69.0	72.5
CH_3Cl	C_6H_5Cl	298	100.9	52.9	52.8
N_2	CH_3OH	273	39.6	48.6	47.0
		298	40.6	54.0	52.0
		323	41.8	60.7	54.0
CO_2	H_2O	273	18.01	32.4	32.0
		298	18.06	32.8	33.0
		323	18.22	34.0	33.0

good or poor saturation densities, does not occur because their isotherms are correct. They apparently have compensating errors.

Partial Molar Volumes at Infinite Dilution

The second quantity be correlated by DCFI models was the partial molar volume at infinite dilution for gases in liquids (Brelvi and O'Connell, 1972) and for salts in water (Cooney and O'Connell, 1987) over wide ranges of conditions and substances. The form of the correlation was to write equation (1.22) at infinite dilution and express the unlike DCFI, C_{ij}^{∞}, in terms of reduced density. Brelvi and O'Connell's correlation (1972) also involved only the characteristic volumes of the gas and the solvent, but was reasonably accurate for many types of substances over wide ranges of solvent density (and temperature). Table 2 shows some of the comparisons with data.

Cooney and O'Connell's examination (1987) of aqueous solution salt-water (unlike) DCFI found that C_{ij}^{∞} could be described by a 2-parameter correlation in terms of water density and temperature to cover the range from water at its critical point all the way to 100MPa and room temperature. One of the parameters seemed to depend only on the charge type of the salt. Table 3 shows some of the comparisons with data for various salts. Preliminary work indicates that the general behavior of this correlation is preserved by NaCl at supercritical temperatures as well.

The forms of the above correlations also have implications for solute i partial molar volumes at the solvent j critical point (where κ_j goes to ∞ so $1-C_{jj}$ goes to zero). Brelvi's correlation indicates that at the solvent critical point C_{ij} for supercritical i remains positive and finite. This is consistent with the observation that at the solvent critical point, the gas partial molar volume approaches $+\infty$. At the same time, Cooney's work shows that C_{ij} is positive and independent of temperature below about 2.8 times the water critical density, becoming nearly

Table 3: PARTIAL MOLAR VOLUMES OF SALTS IN WATER

From W.R. Cooney and J.P. O'Connell (1987)

$$\frac{\overline{V}_S^\infty}{\kappa_W RT} = a\,(1.0 - 7.0\text{x}10^{-9} exp\,[4500/T])\,exp\,(106.1\rho_W) + b$$

Salt	a	b	$\rho_W V_W^c$	T, K	Err in $\overline{V}_S^\infty$, %
$BaCl_2$	0.1707	-35.16	2.7-3.1	323-473	0.9
HCl	0.0857	-14.15	2.7-3.1	298-473	0.3
NH_4ClO_4	0.2167	-17.82	2.7-3.1	323-473	0.6
K_2SO_4	0.1887	-34.55	2.7-3.1	323-473	0.7
$NaCl$	0.0790	-12.72	1.9-3.5	273-623	2.0
$MgCl_2$	0.1395	-35.50	2.7-3.1	323-473	0.8

independent of density at about twice the critical. Thus, the salt partial molar volume goes to $-\infty$ at the water critical. These observations can be summarized as

$$\begin{aligned} \frac{\overline{V}_i^\infty}{\kappa_T RT} &\equiv \lim_{x_j \to 1} \frac{\overline{V}_i}{\kappa_T RT} \\ &= \lim_{x_j \to 1} (1 - C_{ij}) \\ &\equiv (1 - C_{ij}^\infty) \end{aligned} \tag{3.1}$$

$$\rho_j^{Pure}\overline{V}_i^\infty = \frac{1 - C_{ij}^\infty}{1 - C_{jj}^{Pure}} \tag{3.2}$$

and

$$\lim_{j \to critical} \begin{cases} \kappa_j & = \lim_{j \to critical} \frac{1}{\rho_j^c RT\left(1 - C_{jj}^{Pure}\right)} \\ & = +\infty \\ (1 - C_{jj}^{Pure}) & = 0 \\ \rho_j^{Pure}\overline{V}_i^\infty & = +\infty \; for \; T_i^c < T_j^c \\ \rho_j^{Pure}\overline{V}_i^\infty & = -\infty \; for \; T_i^c > T_j^c \end{cases} \tag{3.3}$$

Of course, the variation of κ near the critical follows nonclassical, or critical-scaling, behavior (see, e.g., Rowlinson and Swinton, 1983), implying that $1 - C_{jj}^{Pure}$ is nonclassical near its zero value. It is not known if $1 - C_{ij}^\infty$ is classical or nonclassical at the critical of component j.

Nonelectrolyte Liquid Solutions

As mentioned above, there is not much likelihood of using estimated values of DCFI to obtain activity coefficients in miscible nonelectrolyte solutions because of the form of equation (1.21). However, this sensitivity can be exploited to test models for activities by taking their derivatives to arrive at solution DCFI and comparing them with data. This has been done a few times as indicated Hamad, et al.(1989) and also by Brelvi (1972) and Wooley (1987). In the work of Wooley, equations (1.30) - (1.32) were used with data on activity coefficients, partial molar volumes and isothermal compressibilities.

Brelvi (1972) found DCFI for both subcritical and supercritical systems from available data. Wooley and O'Connell (1990) have completed more extensive evaluations of the TCFI and DCFI for 26 subcritical binary nonelectrolyte systems ranging from nonpolar hydrocarbon and globular systems to strongly nonideal polar and associating systems to aqueous systems. The most indicative property for classifying systems appears to be

$$\Delta C \equiv (1 - C_{11})(1 - C_{22}) - (1 - C_{12})^2 \tag{3.4}$$

as found in the binary form of equation (1.21). The four characteristic variations of ΔC are shown in Figure 3 as $1/\Delta C$ versus x_1. Some values are small everywhere, as cyclopentane and cyclooctane, while others are large and significantly curved, particularly for alcohol-hydrocarbon systems. Interestingly, a 2-parameter model assuming linearity of this function was quite successful in describing activity coefficients for many of the complex systems studied (Wooley, 1987). Wooley found that the C_{ij} results were sensitive to the form of activity coefficient correlation used. There were cases where the Wilson and NRTL Equations gave results that were close enough for the "true" results to probably be very close to the values given. However, for others, the NRTL Model yielded only partially miscible systems, the Wilson Model gave variations with strong extrema and only the Modified Margules Relation (Abbott and Van Ness, 1975) gave apparently reliable results.

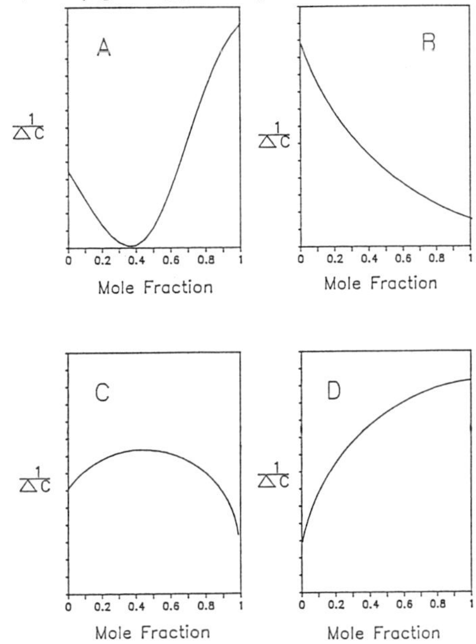

Figure 3. The Four Characteristic Variations of the DCFI Difference of Equation (3.4) in Nonelectrolyte Liquid Solutions. $\Delta C \equiv (1 - C_{11})(1 - C_{22}) - (1 - C_{12})^2$

Table 4 shows representative values of infinite dilution DCFI of liquids with supercritical components (Brelvi and O'Connell, 1972). The unlike integrals are clearly not equal to the geometric or arithmetic means of the like integrals. This indicates the solution is strongly nonideal so correlations of DCFI may be used. In every case, C_{12}^{∞} is less negative than the mean of C_{11}^{∞} and C_{22}^{Pure}. This probably arises from the (negative) direct correlations usually associated with excluded volume effects being greater between the solute molecules than the unlike interactions would suggest. This is similar to gas phase second virial coefficients for supercritical (1) compared to subcritical (2) components which can have $0 \simeq B_{11} > B_{12} >> B_{22}$.

Brelvi and O'Connell (1972), Mathias and O'Connell (1981) and Campanella, et al. (1987) demonstrate application of simple models for the DCFI which describe liquids with dissolved supercritical components. The present discussion is taken from the latter two references.

The objective of the correlation is to obtain expressions for fugacities of liquid components which could be used for vapor-liquid equilibria calculations. For supercritical components,

Table 4: INFINITE DILUTION DCFI FOR BINARY LIQUIDS WITH ONE SUPERCRITICAL COMPONENT

From S.W. Brelvi (1972).

Gas(1)	Solvent(2)	T, K	C_{22}	C_{12}	C_{11}
N_2	C_6H_6	298	-37.13	-19.02	-8.51
N_2	CCl_4	298	-37.84	-19.73	-9.16
H_2	C_6H_6	298	-37.13	-15.05	-3.28
H_2	$n-C_8H_{18}$	298	-54.04	-15.45	-2.80
		418	-14.50	-4.69	-0.68
CH_4	C_3H_8	310	-3.91	-4.59	-4.08

the composition range is mostly dilute, so the Henry's Law reference state is often chosen. In addition, the pressure is usually elevated so that the isothermal fugacity depends on both composition and pressure. As illustrated by Prausnitz, et al. (1986), a traditional way of writing this expression is

$$\ln \hat{f}_i(T,P,\mathbf{x}) = \ln\left[x_i H_i(T)\right] + \ln \gamma_i^*(T,\mathbf{x}) + \int_{P_i^o}^{P} \left.\frac{\overline{V}_i(T,p,\mathbf{x})}{RT}\right|_{T,\mathbf{x}} dp \tag{3.5}$$

where the dependence of the quantities on solution state, $T, P, \mathbf{x}$, are shown here explicitly. These include Henry's Constant, H_i, at the reference pressure, P_i^o, the partial molar volume at the solution state, $\overline{V}_i$ and the unsymmetric convention activity coefficient, γ_i^*, with [4]

$$\lim_{x_i \to 0} \ln \gamma_i * = 0 \tag{3.6}$$

At the low concentrations usually encountered, it is common to use the Krichevsky - Ilinskaya (KI) approximations[5]

$$\ln \gamma_i^{*KI}(T,\mathbf{x}) \simeq -A_{ii}(T)\left[2x_i - x_i^2\right] \tag{3.7}$$

$$P_i^0 \equiv P_j^{Sat} \tag{3.8}$$

$$\int_{P_i^o}^{P} \left.\frac{\overline{V}_i(T,p,\mathbf{x})}{RT}\right|_{T,\mathbf{x}} dp \simeq \overline{V}_{ij}^{\infty}(T)\left[\frac{(P-P_j^{Sat})}{RT}\right] \tag{3.9}$$

The composition effect [equation (3.7)] and the pressure effect [equation (3.9)] are separable and of opposite sign. The partial molar volume of the latter correction is connected to dilatometric experiments where $\overline{V}_{ij}^{\infty}(T)$ can be measured. Experimentally, $\ln \hat{f}_i(T,P,\mathbf{x}) - \ln x_i$ can be obtained from $\ln(y_i \hat{\phi}_i P/x_i)$ [with $\hat{\phi}_i(T,P,\mathbf{y})$ the vapor phase fugacity coefficient]. If this is

[4]Note that this identification of reference state is not complete for a mixed solvent since specifying only one mole fraction is not enough for systems with more than two components. This leaves the user with the choice of specifying more mole fractions for the reference state or using this specification and having Henry's Constant depend on solvent composition. While the latter is often taken because γ_i^* is can be very close to unity at low pressures, this merely transfers modeling of the composition effect on solute fugacity to H. The alternative is described below.

[5]The subscript on the parameter A_{ii} of the unsymmetric convention Porter Model [equation (3.7)] has been chosen to reinforce the fact that composition deviations from Henry's Law occur because the solute - solute interaction is different from the solvent - solute interaction which is accounted for in the ideal solution state. If all interactions were the same, Henry's constant would equal the fugacity of pure component i as a real liquid at the system T and P_i^o rather than as a hypothetical pure liquid (Van Ness and Abbott, 1982). Deviations from Raoult's Law occur for the same reason, but the parameter in the symmetric convention Porter model (Prausnitz, et al., 1986) is written A_{ij} because the deviations are due to solute - solvent interactions being different from the ideal solution state where all interactions are the same.

Table 5: VAPOR-LIQUID EQUILIBRIA FOR BINARY SOLUTIONS WITH ONE SUPERCRITICAL COMPONENT

From E.A. Campanella, P.M. Mathias and J.P. O'Connell, (1987).

Gas(1)	Solvent(2)	T, K	x_1 Max	x_1^{Err}, RMS	x_1^{Err} Max
H_2	$n-C_8H_{18}$	463-533	0.23	0.0006	0.0014
H_2	CH_3OH	294-413	0.08	0.0012	0.0034
N_2	NH_3	273-373	0.10	0.0009	0.0028
CH_4	$m-CH_3C_6H_4OH$	463-663	0.21	0.0087	0.0303
H_2	Quinoline	462-702	0.21	0.0010	0.0022
H_2	Tetralin	463-662	0.28	0.0029+	0.0081+
H_2	$CH_2(C_6H_5)_2$	463-702	0.31	0.0044+	0.0113+
H_2	"CLPP" Coal Oil	463-543	0.19	0.0026+	0.0037+

+ Predicted Using H_2 - Quinoline Henry's Constant

plotted versus $P - P_j^{Sat}$, (the Krichevsky - Kasarnovsky plot) the result is nearly linear even if the slope gives an apparent value of $\overline{V}_{ij}^{\infty}/RT$ that is different from the volumetric value (See, e.g., Orentlicher and Prausnitz, 1964).[6]

Campanella, et al (1987), building on the work of Mathias and O'Connell (1981), described the volumes and activities for a large variety of liquids with supercritical components while avoiding estimating the two terms of opposite sign. Their fundamental equation for the fugacity is

$$\ln \hat{f}_i(T, P, \mathbf{x}) = \ln [x_i H_{iR}(T)] + \ln \gamma_i^+(T, P, \mathbf{x}) \tag{3.10}$$

where the reference state for the activity coefficient is carefully defined

$$\lim_{\substack{x_R \to 1 \\ P \to P_R^{Sat}}} \gamma_i^+(T, P, \mathbf{x}) = 1 \tag{3.11}$$

with component R being a reference component which is a liquid at the system temperature. All solutions use this state for their reference composition and pressure. The model was based on a two-parameter corresponding states form of C_{ij} from van der Waals equation and involved a single binary parameter for each solvent/solute pair of components. The key to the correlation's success arises uniquely from fluctuation solution theory by integrating Equation (1.38) to find the solution density and Equation (1.39) to find the activity coefficient γ_i^+ from a pure solvent reference state. This worked extremely well for many different systems as indicated in Table 5. One appealing aspect of the work was that the densities and deviations from Henry's Law in a single solvent j were quite insensitive to the value of the model's only binary parameter. However, the accuracy did depend strongly on the infinite dilution fugacities, i.e., Henry's Constant, H_{ij}.

An accurate and convenient way to obtain H_{ij} is with a particular limit for the integration

[6]In fact, the phase equilibrium experiments will give close to linear variations whether the independent variable is $2x_i - x_i^2$ (where the slope would apparently be A_{ii}), or is $P - P^{Sat}$ (where the slope would apparently be $\overline{V}_{ij}^{\infty}/RT$.) Thus, the corrections to Henry's Law are not either term alone and the actual values should be in between those from the two forms.

Table 6: SOLUBILITY OF GASES IN MIXED SOLVENTS

From E.A. Campanella, P.M. Mathias and J.P. O'Connell, (1987).

Gas(1)	Solvent (2)	Solvent(3)	T, K	No. Data	H_{1m}^{Err}, %
N_2	Chloroform	Acetone	298	9	1
CO	Toluene	Aniline	298	5	4
C_2H_4	Methanol	Acetone	298	9	4
$n-C_4H_{10}$	Acetone	n-Butanol	298	12	5
Ar	Acetic Acid	Water	293-323	40	4
O_2	Methanol	Water	273-313	30	6
CO_2	n-Propanol	Water	283-313	46	7

of Equation (1.39) for supercritical component i. Taking the limits to be infinite dilution in solvent j and infinite dilution in solvent R connects the two Henry's Constants H_{ij} and H_{iR} via the limiting activity coefficient.

$$\gamma_i^{+\infty} \equiv \lim_{x_j \to 1} \gamma_i^+ \tag{3.12}$$

$$= \frac{H_{ij}}{H_{iR}} \tag{3.13}$$

$$= \int_{\rho_R^{Pure}}^{\rho_j^{Pure}} \frac{1-C_{ij}}{\rho}\bigg|_{T,\rho_{k \neq j}} d\rho_j \tag{3.14}$$

Then, the activity of a supercritical component at low concentrations in a solvent can be predicted from Henry's Constant in the Reference solvent, the densities of both pure solvents and a good model for the C_{ij}. The definition of "good" is to have an accurate enough value of the binary constant to obtain Henry's Constant, H_{ij}, in the desired solvent j from equation (3.13). It was found that a temperature independent quantity was often adequate for temperature variations of more than 100 K.

This same technique has been used successfully for Henry's constants in mixed solvents, H_{im}. In this case, the results can be sensitive to values of the mixed solvent density, particularly in aqueous solutions. Table 6 gives some of the mixed solvent results of Campanella, et al. (1987). The accuracy is uniquely high for aqueous systems and uses only binary information to predict ternary and higher order systems.

At the present time, models for vapor-liquid equilibria are best written in terms of full equations of state. This technique avoids the inevitable discontinuities of fugacities in the critical region when the liquid phase value uses activity coefficients and the vapor phase uses fugacity coefficients obtained from different models. Their limitation is the difficulty of finding mixing rules for their parameters that allow predictions, particularly in the liquid phase. The correlation of Campanella, et al.(1987) is primarily intended to guide mixing rule formulation and provide insight into liquid behavior, though the work showed that the model can also be useful with the gas phase virial or other equations of state for systems well removed from their critical lines.

Strong Electrolyte Solutions

A final area where the author has investigated application of Fluctuation Solution Theory is strong electrolytes. As mentioned above, the formalism for species of reactive components encompasses all systems including partially reacting substances, but these have not been addressed because the basic modeling for charged species has not been completed.

There is an additional complexity associated with ionic species because of the long-range electrostatic forces. While the dcfs diverge because they vary as r^{-1} at long distances, the reactive component treatment shows how charge neutrality requires that all of these effects cancel out for thermodynamic purposes. Thus, there is a convergent dcf

$$c_{ij}^0 \equiv c_{ij} + \frac{z_i z_j e^2}{kTr} \tag{3.15}$$

which is integrated to obtain a convergent DCFI, C_{ij}^0. It is this quantity that is used in the formulae given above.

The infinite dilution limits of the various C_{ij}^0 are well determined. The limiting solvent - solvent DCFI, $1 - C_{11}^{Pure}$, is the pure solvent bulk modulus, $\rho/\kappa_{T1}^{Pure}RT$. The solvent - salt DCFI, $1 - C_{12}^{\infty}$, is $\overline{V}_{02}^{\infty}(T)/\nu_2\kappa_{T1}^{Pure}RT$. The salt - salt DCFI, $C_{22}^{0\infty}$, must obey the Debye-Huckel Limiting Law

$$C_{22}^{0DHLL} = \lim_{x_2 \to 0} C_{22} = \lim_{x_2 \to 0}\left[\sum_i \sum_j \nu_{i2}\nu_{j2} z_i^2 z_j^2 S_{\gamma 1}\rho I^{-1/2}/\nu_2^2\right] \tag{3.16}$$

where the sums are over all ions in salt 2 [they arise from Equation (2.7)], the ionic charges are z_i, the molar ionic strength is

$$I \equiv \sum_{ions} \nu_{i2} z_i^2 N_{02}/V^T \tag{3.17}$$

$$\equiv \omega_2 \rho x_{02} \tag{3.18}$$

with $x_{02} \equiv N_{02}/N$ and the solvent properties are in

$$S_{\gamma 1} \equiv \left[\frac{2\pi e^6 N_A}{\epsilon_1^3 (kT)^3}\right]^{1/2} \tag{3.19}$$

with e the electronic charge and ϵ_1 the solvent dielectric constant. Obviously, C_{22} goes to ∞ at zero salt concentration. It should also be noted that the development also leads naturally to the higher order activity coefficient term for unsymmetric salts given by Friedman (1962).

The theoretical work reviewed by Blum (1980) and Stell, et al. (1981) as well as used by Krienke and Weigl (1988), indicates that the dcf can be split into contributions from long-range (ionic and multipolar electrostatic) and short-range forces. Although Stell et al. explicitly expand the electrostatic effects for ions and dipoles in a series in the inverse Debye length

$$\kappa \equiv 2^{3/2} e \left[\frac{\pi N_A}{\epsilon_1 kT}\right] I^{1/2} \tag{3.20}$$

the series converges only for the dcf and C_{22}^0 but every term in the series contributes to C_{12} and C_{11}. Thus, a thermodynamic approach must be taken to obtain the correct macroscopic result for these DCFI. Use of Equation (1.41) leads to the Debye-Huckel Limiting Law electrostatic terms for C_{12} and C_{11}. The results are

$$C_{12}^{DHLL} = -\frac{3}{4}\left.\frac{\partial \ln \epsilon_1}{\partial \ln \rho_1^{Pure}}\right|_T \frac{\omega_2 e^2 \kappa}{\epsilon_1 kT} \tag{3.21}$$

and

$$C_{11}^{DHLL} = -\left(2\left.\frac{\partial^2 \ln \epsilon_1}{\partial\left(\rho_1^{Pure}\right)^2}\right|_T - 3\left[\left.\frac{\partial \ln \epsilon_1}{\partial \rho_1^{Pure}}\right|_T\right]^2\right)\frac{\kappa^3 \rho}{16\pi} \tag{3.22}$$

These lead to the expressions used by Redlich and Meyer (1964) and Gucker (1933) for the electrostatic corrections to the salt partial molar volume and solution compressibility, respectively. As shown by Perry, et al. (1988), Extended Debye-Huckel (EXTDH) expressions can also be derived. For example,

$$\nu_2 \ln \gamma_{\pm 2}^{EXTDH} = -C_{22}^{EXTDH} = -\frac{S_{\gamma 1} I^{1/2}}{\omega_2} \sum_i \sum_j \frac{\nu_{i2}\nu_{j2} z_i^2 z_j^2}{1+a_{ij}\kappa} + \cdots \tag{3.23}$$

where a_{ij} is the diameter for the interaction of primitive-ions. This may be the only rigorous development that leads to true ionic sizes in the EXTDH model rather than the usual single ion diameter (see, e.g., Friedman, 1962). Similar corrections are given for C_{12}^{EXTDH} and C_{11}^{EXTDH}.

To establish full models, the approach has been to examine the behavior of the DCFI for strong salts (2) in water (1) using available experimental data at 298.15K. Equations (1.30) - (1.32) were used by Marshall (1989) to arrive at Figures 4 through 6 for $1 - C_{11}$, $1 - C_{12}$ and $1 - C_{22}^0$ for the set of four salts made up of the ions Na^+, Li^+, Cl^-, Br^- from very low concentrations to saturation. These were chosen because extensive data exist, they show significant differences in density variation with composition, the lithium salts have very high solubilities and the match of salt to ion data should lead to models for ionic quantities.

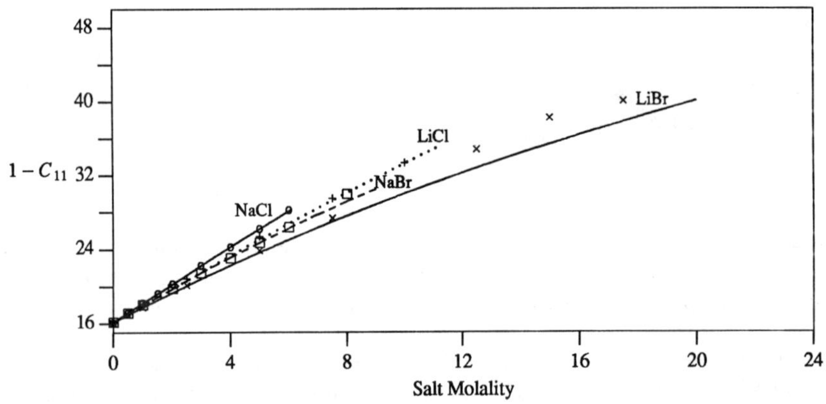

Figure 4. Water - Water DCFI in Aqueous Salt Solutions at 298K.

The variation of $1 - C_{11}$ in Figure 4 is relatively simple and shows an ordering consistent with the ion sizes. The variation of $1 - C_{12}$ in Figure 5 is somewhat more complex.

Finally, Figure 6 shows that the variation of $1 - C_{22}^0$ is the most complicated dependence on composition, particularly at low salt concentrations where C_{22}^0 is positive. This arises from the electrostatic interactions between the ions, though the Debye-Huckel Limiting Law overestimates this effect at finite contributions as shown by the minimum in Figure 7 for a "convergent" salt-salt DCFI where the Limiting Law contribution is removed.

$$1 - C_{22}^{Convergent} \equiv 1 - C_{22}^0 - (1 - C_{22}^{DHLL}) \tag{3.24}$$

At finite salt concentrations, the values of the C_{12} are always much less negative than any mean of the C_{11} and C_{22}^0 even where the electrostatic effects on C_{22}^0 are unimportant. This is likely to be due to hydration effects of the ions with the water. The work of Tanger and Pitzer (1988) demonstrates how this might be taken into account.

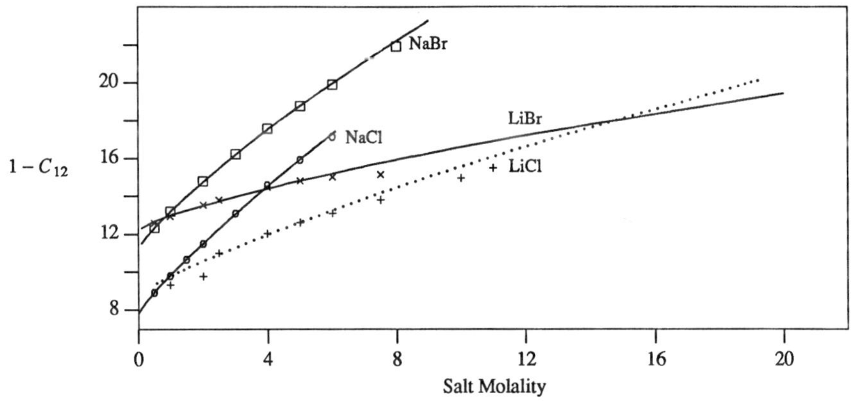

Figure 5. Water - Salt DCFI in Aqueous Salt Solutions at 298K.

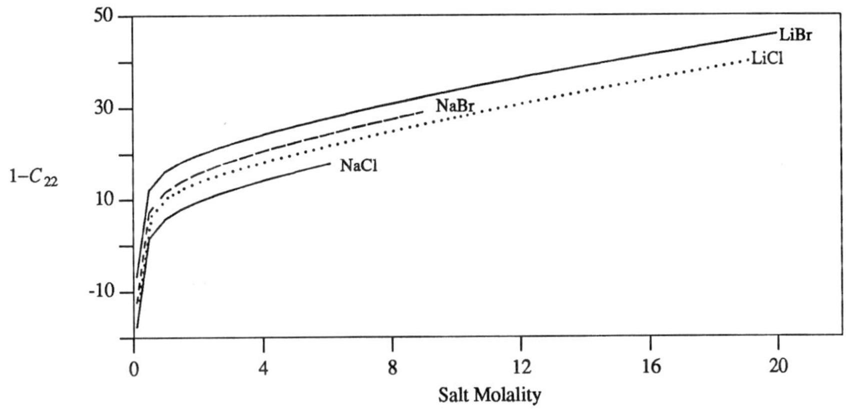

Figure 6. Salt - Salt DCFI in Aqueous Salt Solutions at 298K.

Since most salts have limited solubility, the initial models for the DCFI were of the form

$$C_{ij} \simeq C_{ij}^{LR} + C_{ij}^{V} + C_{ij}^{MM} \tag{3.25}$$

where the terms for long-range contributions (LR) are from the above development, the short-range contributions, C_{ij}^{V}, are from an osmotic virial series and there is a contribution, C_{ij}^{MM}, to correct the LR terms from their McMillan-Mayer origin to the desired canonical thermodynamic variables as described by Friedman (1972) and Pailthorpe, et al.(1984).

If the McMillan-Mayer conversion is only for the LR term (and thus quite small), the virial coefficients are in the canonical system. If the conversion is for all contributions to the correlations, as suggested by Pailthorpe, et al., they are osmotic virial coefficients of the sort described by Rushbrooke and Scoins (1953). The latter should have fundamental ionic additivity with simple rules (such as arithmetic or geometric means) for the unlike coefficients.

In either case, the form of the C_{ij}^V is

$$C_{ij}^V = \rho \sum_\alpha \sum_\beta \nu_{i\alpha}\nu_{j\beta}\left[F_{\alpha\beta} + \rho\sum_\gamma x_{0\gamma}\Phi_{\alpha\beta\gamma}\right] \tag{3.26}$$

where the $F_{\alpha\beta}$ and $\Phi_{\alpha\beta\gamma}$ depend only on the ions, α, β and γ, temperature and solvent.

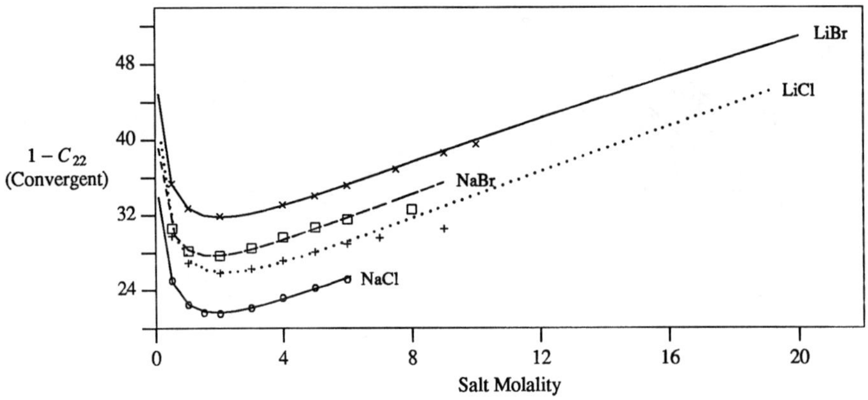

Figure 7. "Convergent" Salt-Salt DCFI in Aqueous Salt Solutions at 298K.
$1 - C_{22}^{Convergent} \equiv (1 - C_{22}^0) - (1 - C_{22}^{DHLL})$

For a single-salt (α and $\beta = +, -$), single-solvent (1) system, there are a total of six $F_{\alpha\beta}$ and ten $\Phi_{\alpha\beta\gamma}$ though those with $\alpha = \beta = \gamma = 1$ can be found directly from pure solvent densities as a function of pressure, and those with γ and either α or β equal unity can be found from the salt partial molar volume at infinite dilution (which is the sum of its ionic values) and its pressure dependence. If mixing rules are adopted for the rest of the ion-ion parameters, there are four unknowns that remain, F_{++}, F_{--}, Φ_{+++} and Φ_{---}. These should be ionic quantities independent of the salt they originated from, so they should be able to be found from the set of data described above.

It is also possible to treat the virial coefficients as salt parameters.

$$C_{ij}^V = \rho\left[F_{ij} + \rho\sum_k x_{0k}\Phi_{ijk}\right] \qquad i, j, k = 1, 2 \tag{3.27}$$

Then, values of F_{11} and Φ_{111} are found from pure solvent data and F_{12} and Φ_{112} are found from the salt partial molar volume at infinite dillution. This leaves three more parameters to be fitted to data for each salt, F_{22}, Φ_{122} and Φ_{222}. The results are not very sensitive to the hard-sphere diameters in equation (3.23), a_{ij}, so estimates by Marcus (1983) or Pauling radii can be used additively.

Initial attempts at modeling have shown mixed success, though there remain a number of options. Figure 8 shows the 298.15K densities obtained by using ionically additive values for the virial coefficients in equation (3.26) and hard spheres in the EXTDH term of equation (3.23). The calculated values agree to within 2% of the data and these predictions are substantially the same as the values found when the virial coefficients were fitted as salt parameters in equation (3.27). This confirms the insensitivity of calculated densities to model details.

Calculated values of the activity coefficients are not as accurate. It is possible to fit the data to a high degree of accuracy up to about 5 molar when the three salt parameters of (3.27) are fitted. The major source of error seems to be in the EXTDH term which forces the virial terms to assume values that describe the low concentration data accurately (±.002) but then cannot describe the higher concentrations. This is consistent with the experience of Pailthorpe, et al. (1984). The virial parameter values obtained from this fitting do not show ionic additivity. The results are considerably outside the experimental error if attempts are made to use ionic parameters in (3.26).

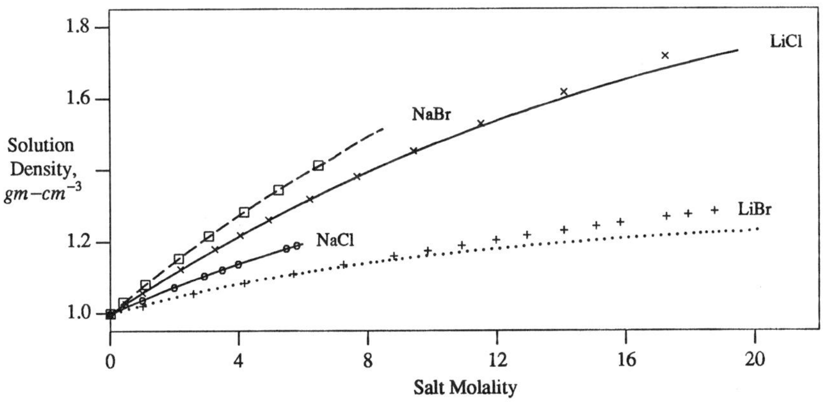

Figure 8. Densities of Aqueous Salt Solutions at 298K.

At this stage, examination of the EXTDH expression is appropriate and the use of osmotic virial terms would more likely lead to ionic additivity. Also, the treatment of weak electrolytes and mixed solvents is yet to be undertaken.

CONCLUSIONS

The unique form of Fluctuation Solution Theory that involves correlation function integrals as the basic quantity for derivatives of the properties of interest has led to a number of useful correlations of liquid thermodynamic properties based on macroscopic parameterizations. Using pure component properties at ambient pressures, results have been obtained for densities of pure and mixed compressed liquids, partial molar volumes and solution densities, and activity coefficients of supercritical and ionizing substances in liquids. Much remains to be done in treatments of this type as well as to connect the empirical success described here to molecular theories for the correlation functions.

ACKNOWLEDGEMENTS

The more than twenty years of research work described herein have been principally supported (in chronological order) by the Donors of the Petroleum Research Fund, the National Science Foundation and the Office of Basic Energy Sciences of the Department of Energy. The author is grateful for this assistance as well as to the students identified above whose efforts uncovered and elaborated upon these theories and models and to many supportive colleagues around the world.

REFERENCES

Abbott, M. M., and H. C. Van Ness 1975. Vapor-liquid equilibrium. part III. Data reduction with precise expressions for G^E. AIChE Jour. 21: 62.

Blum, L., 1980. In Theoretical Chemistry Advances and Perspectives, Vol. 5, ed. H. Eyring and D. Henderson, 1. New York: Academic Press.

Brelvi, S. W., 1972. Applications of the direct correlation function solution theory to the thermodynamic properties of fluids and fluid mixtures. PhD Dissertation: University of Florida.

Brelvi, S. W., and J. P. O'Connell 1972. Corresponding states correlation for liquid compressibility and partial molar volumes of gases at infinite dilution in liquids. AIChE Jour. 18: 1239.

Brelvi, S. W., and J. P. O'Connell 1975a. A generalized isothermal equation of state for dense liquids. AIChE Jour. 21: 171.

Brelvi, S. W., and J. P. O'Connell 1975b. Generalized prediction of isothermal compressibilities and an isothermal equation of state for liquid mixtures. AIChE Jour. 21:1024.

Campanella, E. A., 1984. Some applications of fluctuation solution theory to multiphase systems of nonelectrolytes. PhD Dissertation: University of Florida.

Campanella, E. A., P. M. Mathias and J. P. O'Connell 1987. Equilibrium properties of liquids containing supercritical substances.AIChE Jour. 33: 2057.

Cooney, W. R., and J. P. O'Connell 1987. Correlation of partial molar volumes at infinite dilution of salts in water. Chem. Eng. Comm. 56: 341.

Friedman, H. L., 1962. Ionic Solution Theory, New York: Wiley - Interscience.

Friedman, H. L., 1972. Lewis-Randall to McMillan-Mayer conversion for the thermodynamic excess functions of solutions, I, II, III. J. Soln. Chem. 1: 387, 413, 419.

Friedman, H. L., and P. S. Ramanathan 1970. Theory of mixed electrolyte solutions and applications in a model for aqueous lithium chloride - cesium chloride. J. Phys. Chem. 74: 3756.

Gray, C. G., and K. E. Gubbins 1984. Theory of Molecular Fluids, Vol 1, Oxford: Clarendon Press.

Gucker, F. T., 1933. The apparent molal heat capacity, volume and compressibility of electrolytes. Chem. Rev. 13: 111.

Hamad, E. Z., G. A. Mansoori, E. Matteoli and L. Lepori 1989. Relations among concentration fluctuation integrals in mixtures (theory and experiments). Z. Physik. Chem. Neue Folge 162: 27.

Huang, Y. H., 1986. Thermodynamic properties of compressed liquids and liquid mixtures from fluctuation theory. PhD Dissertation: University of Florida.

Huang, Y. H., and J. P. O'Connell 1987. Corresponding states correlation for the volumetric properties of compressed liquids and liquid mixtures. Fluid Phase Equil. 37: 75.

Kirkwood, J. G., and F. P. Buff 1951. The statistical mechanical theory of solutions. J. Chem. Phys. 19: 774.

Krienke, H., and G. Weigl 1988. Ionic Interactions in a polarizable dipolar hard-sphere solvent. Ann. Physik 45: 313.

Marcus, Y. J., 1983. Ionic radii in aqueous solutions. J. Soln. Chem. 12: 271.

Marshall, K. A., 1989. A liquid equation of state for aqueous strong electrolytes. PhD Dissertation: University of Florida.

Mathias, P. M. 1978. Thermodynamic properties of high-pressure liquid mixtures containing supercritical components. PhD Dissertation: University of Florida.

Mathias, P. M., and J. P. O'Connell 1981. Molecular thermodynamics of solutions containing supercritical components. Chem. Eng. Sci. 36: 1123.

McMillan Jr., W. G., and J. E. Mayer 1945. The statistical mechanics of multicomponent solutions. J. Chem. Phys. 13: 276.

O'Connell, J. P., 1971. Thermodynamic properties of solutions based on correlation functions. Mol. Phys. 20: 27.

O'Connell, J. P., 1981. Thermodynamic properties of solutions and the theory of fluctuations. Fluid Phase Equil. 6: 21.

Orentlicher, M., and J. M. Prausnitz 1964. Thermodynammics of hydrogen solubility in cryogenic solvents at high pressures. Chem. Eng. Sci. 19: 775.

Ornstein, L. S., and F. Zernike 1914. Fortuitous density changes and opalescence at the critical point of an element. Verslag Akad.Wetenschappen 23: 582.

Pailthorpe, B. A., D. A. Mitchell and B. W. Ninham 1984. Ion-solvent interactions and the activity coefficients of real electrolyte solutions. J. Chem. Soc. Faraday Trans. II 80: 115.

Percus, J. K. 1964. In Classical Fluids, ed. H.L. Frisch and J.L. Lebowitz, II-33. New York: Benjamin.

Perry, R. L., H. Cabezas, Jr. and J. P. O'Connell 1988. Fluctuation thermodynamic properties of strong electrolyte solutions. Mol. Phys. 63: 189

Perry, R. L., and J. P. O'Connell 1984. Fluctuation thermodynamic properties of reactive components from species correlation function integrals. Mol. Phys. 52: 137.

Prausnitz, J. M., R. N. Lichtenthaler and E. G. Azevedo 1986. Molecular Thermodynamics of Fluid Phase Equilibria, 2nd Ed., Englewood Cliffs, NJ: Prentice-Hall.

Redlich, O., and D. M. Meyer 1964. The molal volumes of electrolytes. Chem. Rev. 64: 221.

Reed III, T. M., and K. E. Gubbins 1973. Applied Molecular Theory. New York: McGraw-Hill.

Rowlinson, J. S., and F. L. Swinton 1983. Liquids and Liquid Mixtures, 3rd Ed., London: Butterworths Scientific.

Rushbrooke, G. S., and H. I. Scoins 1953. The theory of fluids. Proc. Roy. Soc. A. 216: 203.

Stell, G., G. N. Patey and J. S. Høye 1981. Dielectric constants of fluid models: statistical mechanical theory and its quantitative implementation. Adv. Chem. Phys. 48: 183.

Tanger IV, J. C., and K. S. Pitzer 1989. Calculation of the thermodynamic properties of aqueous electrolytes to 1000° C. and 5000 bar from a semicontinuum model for ion hydration. J. Phys. Chem. 93: 4941.

Telotte, J. C., 1985. Thermodynamic properties of multicomponent mixtures from the solution of groups approach to direct correlation function solution theory. PhD Dissertation: University of Florida.

Van Ness, H. C., and M. M. Abbott 1982. Classical Thermodynamics of Nonelectrolyte Solutions. New York: McGraw-Hill.

Wooley, R. J., 1987. Fluctuation thermodynamic properties of nonelectrolyte liquid mixtures. PhD Dissertation: University of Florida.

Wooley, R. J., and J. P. O'Connell 1990. A database of fluctuation thermodynamic properties and molecular correlation function integrals for a variety of binary liquids. Fluid Phase Equilibria, in press.

Structure and Properties of Supercritical Fluid Mixtures from Kirkwood-Buff Fluctuation Theory and Integral Equation Methods

Henry D. Cochran
Chemical Technology Div.
Oak Ridge National Lab., Oak Ridge, TN 37831

Lloyd L. Lee and Daivd M. Pfund
Dept. of Chemical Engineering
University of Oklahoma, Norman, OK 73019

ABSTRACT

Large density fluctuations near critical points are associated with divergence in pure fluid isothermal compressiblity and solute partial molar volume, resulting in striking solubility behavior. Supercritical solution behavior has been studied within the framework of Kirkwood-Buff fluctuation theory. In addition to the usual Kirkwood-Buff equations for the partial molar volume and isothermal compressibility, expressions may be obtained for the number of excess solvent molecules about a solute molecule, solubility of a pure incompressible solid (as a function of pressure or density), and variation of fugacity with composition for dilute solutions. The large, negative partial molar volume and the dramatic increase in solubility occur with increases in the long-range structure of the fluid as shown by the correlation functions obtained from integral equation theories. Fluctuation theory provides a connection between the microscopic structure and the thermodynamic properties of the fluid and therefore offers an explanation of supercritical behavior. *A priori* calculations based on distribution functions from accurate integral equation theories yield quantitative predictions for structural and thermodynamic properties of model supercritical solutions. Reasonable approximations lead to simple algebraic expressions for solubility and other properties that show excellent fit to experimental data.

INTRODUCTION

For over a century it has been known (Hannay and Hogarth 1879, 1880) that low volatility solids exhibit markedly enhanced solubility in dense supercritical gases compared with the solubility that would be predicted for an ideal gas mixture; enhancement of solubility by several orders of magnitude is not uncommon. Furthermore, solubility may be caused to vary in a controlled fashion over a hundred to a thousandfold by small changes in pressure or temperature near the critical point (CP). The transport properties of solutions in supercritical solvents typically fall between those of gaseous mixtures and liquid solutions; thus, compared with liquid solvents, supercritical solvents may exhibit improved mass transfer and heat transfer performance. Several common gases, *e.g.*, carbon dioxide and ethylene, have convenient critical temperatures and are common supercritical solvents. These solvents are relatively benign in their impacts on human health and on the environment compared with many common liquid solvents. And they may be recovered and recycled by a simple expansion, leaving the nonvolatile solute in a solvent-free state.

Because of these characteristics, supercritical solvents have been used recently in a number of technological applications. The first commercial application of supercritical fluid extraction, the removal and recovery of caffein from coffee (Zosel 1974, Rozelius 1974), is typical of many applications to natural products. Fluid phase processing of heat-sensitive, nonvolatile materials is possible without contact by toxic solvents. Other similar applications include extraction of hops (Vitzthum, *et al.*, 1978), fragrances (Caragay 1981), fish oil (Suzuki, *et al.*, 1986), *etc*. The use of supercritical solvents as media for chromatographic separations (Rosset 1988) offers the solvent power of liquid chromatography, the universal-sensitive detection of gas chromatography, and clean solute recovery not characteristic of either liquid or gas chromatography. Several commercial supercritical fluid chromatography systems are now available. Ethylene is polymerized in supercritical solutions (Ehrlich and Kurpen 1963), and other reactions in supercritical solvents (Subramaniam and McHugh 1986) may have commercial interest. Supercritcal carbon dioxide is used in enhanced petroleum recovery (Stalkup 1983). Supercritical fluids are used for recovering heavy hydrocarbons from distillation residues (Leonard 1981) and for regenerating spent sorbents (Modell 1977). Numerous other applications are being explored.

Supercritical solutions present several challenges for theoretical modeling. Fluid properties change drastically close to and become nonclassical very near CPs. Furthermore, the conditions of interest and importance span a wide density range. Perturbation theories, well-suited to liquid-like densities, may become inaccurate near and below the critical density (Jonah *et al.*, 1983). The virial equation of state, on the other hand, is applicable only to about half the critical density (Prausnitz *et al.*, 1986). Supercritical solutions frequently involve large, heavy solute molecules in a solvent of small, light molecules. The large differences in molecular size and interaction energy are challenging to the combining rules for estimating parameters of the unlike molecular interaction from the pure component parameters (Kreglewski 1984) and to the mixing rules that are applied in technological applications of equations of state (Haselow *et al.*, 1986). In addition, the solutions are typically quite dilute; this leads to poor statistics for the solute-solvent and, particularly, for the solute-solute distribution functions in molecular simulations (Petsche and Debenedetti 1988). Correlation of density fluctuations becomes long-ranged and fluctuations become large near a CP; these factors present difficulties for both molecular simulations (Petsche and Debenedetti 1988) and integral equations (Cochran *et al.* 1988a, Cochran and Lee 1989). We have found that fluctuation theory, particularly within the Kirkwood-Buff formalism, provides an appropriate and useful approach to theoretical understanding and practical modeling of supercritical solutions.

Through fluctuation theory, one can obtain bulk fluid properties from knowledge of the microscopic fluid structure. In addition, it has recently become clear, both from experimental evidence and theoretical analyses (Eckert *et al.*, 1983, 1986, Yonker, *et al.*, 1986, Johnston and Haynes, 1987, Johnston, *et al.*, 1988, 1989, Cochran *et al.*, 1988a, 1988b, Kim and Johnston 1987a, 1987b, Kajimoto *et al.*, 1988, Debenedetti 1987, and Debenedetti and Kumar 1986) that solvation structure changes rapidly near a CP. Fluctuation theory provides a connection between the microscopic structure and the thermodynamic properties of the fluid and therefore offers an explanation of supercritical behavior. Eckert and co-workers (Eckert *et al.*, 1983, 1986) inferred the collapse of *ca.* 100 solvent molecules about a solute molecule from the large, negative partial molar volume of solutes in supercritical solutions. Likewise, Kim and Johnston (Kim and Johnston 1987a, 1987b) determined a local solvent density surrounding a solute molecule more than 50% greater than the bulk solvent density from the frequency shift in the UV absorption maximum for phenol blue dissolved in supercritical ethylene. Debenedetti (Debenedetti 1987, Debenedetti and Kumar 1986) calculated that a solute molecule is solvated to form clusters of *ca.* 100 solvent molecules at CP using the fluctuation theory of Kirkwood and Buff (Kirkwood and Buff 1951) to interpret experimental partial volume data (for naphthalene dissolved in supercritical CO_2, for example (Eckert *et al.*, 1983, 1986)). Additional experimental evidence for the clustering phenomenon is provided by several recent studies (Brennecke and Eckert 1988, 1989, Collins, *et al.*, 1988). A related phenomenon was noted by Brennecke and Eckert (Brennecke and Eckert 1988, 1989) in the fluorescence spectra of pyrene dissolved in supercritical carbon dioxide and ethylene. The presence of a peak ascribed to the excimer-dimer suggested to them very strong solute-solute interactions in the supercritical solution even at bulk pyrene mole fraction 10^{-5}.

Improved understanding of the clustering phenomena at a molecular level can come from molecular distribution functions, especially the solvent-solute correlation function which could be obtained from neutron scattering experiments, computer simulation, or integral equation theories. To date, no scattering experiments of supercritical solutions have been carried out for this purpose. Computer simulation, discussed elsewhere (Petsche and Debenedetti 1988), can give exact results for the short-range structure. The molecular dynamics results also indicate the lifetime of the solvent cluster about a solute molecule. For dilute solute species, the structure (particularly the long-range structure near a CP) is more easily calculated by integral equations. The extent of solute-solvent aggregation can be calculated from the solution of these integral equations using fluctuation theory. Furthermore, integral equations can shed light on the solute-solute aggregation. It should be recognized that results from such mean-field-like calculations cannot be accurate very close to the CP where behavior is non-classical. Like most cases of practical interest, our calculations do not approach the CP so closely and the integral equation methods are appropriate.

Through the formalism of fluctuation theory, the bulk properties of supercritical solutions are related to the pair correlation functions by the Kirkwood fluctuation integrals, $G_{\alpha\beta}$.

$$G_{\alpha\beta} = G_{\beta\alpha} = \int_0^\infty dr \left\{ \left[g_{\alpha\beta}(r) - 1 \right] 4\pi r^2 \right\}, \tag{1}$$

Although the correlation range tends to infinity at a CP, the magnitude of the pair correlation function, $g_{\alpha\beta}(r)$, remains always of order unity. Thus, Münster (1965) shows that G_{AA} and G_{BB} diverge at a CP, but that

$$\lim_{T \to T_c} \left\{ \frac{G_{AB}^2}{G_{AA} G_{BB}} \right\} = 1. \tag{2}$$

Thus, G_{AB} also tends to infinity at a CP, and the correlation functions and bulk properties are describable by a single correlation function (Münster 1965). Divergence of the $G_{\alpha\beta}$ may, in fact, be taken as an indicator of criticality (Münster 1965, Hoye and Stell 1985, McGuigan and Monson 1987). Many supercritical solutions of experimental or technological importance are quite dilute; we shall address below the implications of the asymptotic condition of infinite dilution in supercritical solutions. By way of introduction let us simply state Debenedetti's results (Debenedetti 1987)

$$\lim_{x_B \to 0} \left\{ G_{AA} + G_{BB} - 2G_{AB} \right\} = G_{AA}^0 + G_{BB}^0 - 2G_{AB}^0 = \text{finite except at a CP.} \tag{3}$$

$$\lim_{x_B \to 0} \left\{ G_{BB} - G_{AB} \right\} = G_{BB}^0 - G_{AB}^0 = \text{finite except at a CP.} \tag{4}$$

where here and after the superscript zero indicates the value of the quantity at the limit of infinite solute dilution. These limiting expressions (Equation (3) and (4)) were assumed but not proved by Kirkwood and Buff (Kirkwood and Buff 1951). Fluctuation theory then presents a rich and appropriate formalism within which to explore the structure and properties of supercritical solutions.

We show that the Kirkwood-Buff (KB) formalism could be exploited for formulating solubility relations that give quantitative information on supercritical phase compositions. A quantitative understanding of the fluctuation integrals is obtained on the molecular level by evaluating the correlation functions using integral equations for model fluid mixtures. In the next section, we review the KB equations for thermodynamic properties of supercritical solutions. Familiar (Kirkwood and Buff 1951) expressions are summarized for the partial molar volume and isothermal compressibility. Then we review the derivations of expressions for the solubility of an incompressible solid, for the size of the solvent-solute cluster, and for derivatives of the solubility. In the following section, we present results of integral equation calculations from distribution function theory. These results include the variations of the fluctuation integrals for supercritical solutions, giving insights into solvent-solute clustering and solute-solute aggregation, mechanical properties, and chemical potential. Comparison with experimental results is made to illustrate the physical picture. Finally, we review how approximate models may be developed for practical use in the expressions from fluctuation theory for properties of dilute supercritical solutions.

KIRKWOOD-BUFF EQUATIONS

Isothermal Compressibility, Partial Molar Volume, and Cluster Size

The familiar expressions for partial molar volume and isothermal compressibility from KB solution theory (Kirkwood and Buff 1951) are quite useful in treating supercritical solutions and will be called upon later. For a c-component system we follow the usual definitions.

$$\mathbf{A \cdot B = I} \tag{5}$$

where

$$A_{\alpha\beta} \equiv \frac{V}{kT}\left[\frac{\partial \mu_\alpha}{\partial \langle N_\beta \rangle}\right]_{T,V,N_{\gamma \neq \beta}} = \frac{1}{kT}\left[\frac{\partial \mu_\alpha}{\partial \rho_\beta}\right]_{T,\rho_{\gamma \neq \beta}}, \tag{6}$$

$$B_{\alpha\beta} \equiv \frac{kT}{V}\left[\frac{\partial \langle N_\alpha \rangle}{\partial \mu_\beta}\right]_{T,V,\mu_{\gamma \neq \beta}} = kT\left[\frac{\partial \rho_\alpha}{\partial \mu_\beta}\right]_{T,\mu_{\gamma \neq \beta}} = \rho_\alpha \rho_\beta G_{\alpha\beta} + \rho_\alpha \delta_{\alpha\beta}, \tag{7}$$

I is the unit diagonal matrix, and $\delta_{\alpha\beta}$ is the Kronecker delta. Then, for a binary mixture of A and B

$$\rho \bar{v}_B = \frac{1 + \rho x_A (G_{AA} - G_{AB})}{1 + \rho x_A x_B \Delta} \tag{8}$$

and

$$\rho kT \kappa_T = \frac{1 + \rho x_A G_{AA} + \rho x_B G_{BB} + \rho^2 x_A x_B (G_{AA} G_{BB} - G_{AB}^2)}{1 + \rho x_A x_B \Delta} \tag{9}$$

where

$$\Delta = G_{AA} + G_{BB} - 2G_{AB}. \tag{10}$$

For a binary solution of B at infinite dilution in A, Equations (8) and (9) simplify to

$$\rho_A^0 \bar{v}_B^0 = 1 + \rho_A^0 (G_{AA}^0 - G_{AB}^0) \tag{11}$$

and

$$\rho_A^0 kT \kappa_{TA}^0 = 1 + \rho_A^0 G_{AA}^0. \tag{12}$$

Figure 1 shows the partial molar volume of naphthalene dissolved in supercritical carbon dioxide (Eckert *et al.*, 1983)); on the same plot is the isothermal compressibility of pure carbon dioxide at the same conditions (Huang *et al.*, 1985). Large, negative values of the solute partial molar volume at conditions near the CP of the pure solvent are characteristic of dilute supercritical solutions. Direct connection to molecular fluctuations is made when $B_{\alpha\beta}$ and $G_{\alpha\beta}$ are expressed in terms of the fluctuations in number of molecules in a volume V; thus,

$$B_{\alpha\beta} = kT\left[\frac{\partial \rho_\alpha}{\partial \mu_\beta}\right]_{T,\mu_{\gamma \neq \beta}} = \frac{\langle N_\alpha N_\beta \rangle - \langle N_\alpha \rangle \langle N_\beta \rangle}{V} \tag{13}$$

and

$$G_{\alpha\beta} = V \frac{\langle N_\alpha N_\beta \rangle - \langle N_\alpha \rangle \langle N_\beta \rangle - \delta_{\alpha\beta} \langle N_\beta \rangle}{\langle N_\alpha \rangle \langle N_\beta \rangle}. \tag{14}$$

For dilute solutions, the size of the cluster of solvent molecules A surrounding a dissolved solute molecule B is related to the solute partial molar volume and the solvent isothermal compressibility (Kim and Johnston 1987a, Debenedetti 1987):

$$\Gamma_{AB} = \rho_A^0 kT \kappa_{TA}^0 - \rho_A^0 \overline{v}_B^0 \tag{15}$$

where the cluster size is defined as the excess number of A molecules surrounding a B molecule compared with the bulk value

$$\Gamma_{AB} \equiv \rho_A^0 G_{AB}^0 = \rho_A^0 \int_0^\infty dr \left\{ \left[g_{AB}^0(r) - 1 \right] 4\pi r^2 \right\}. \tag{16}$$

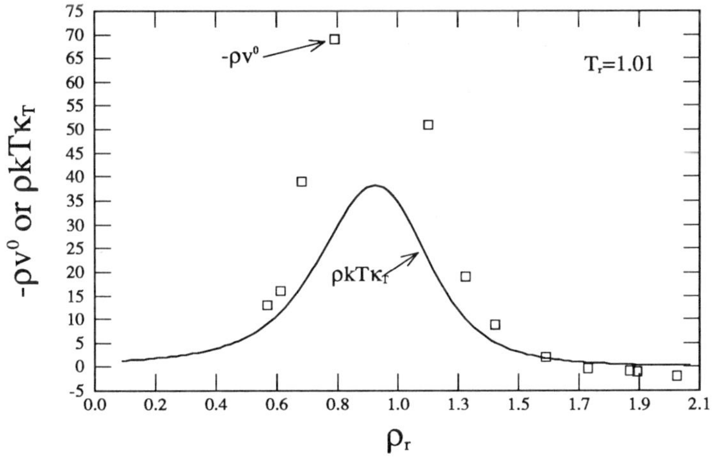

Figure 1. Dimensionless partial molar volume (Eckert *et al.* 1983) and dimensionless isothermal compressibility (Huang *et al.* 1985) of naphthalene dissolved in carbon dioxide *vs*. reduced density.

Solubility

The KB formulas could be employed in the development of solubility expressions in a supercritical mixture. For consideration of solubility (Cochran *et al.*, 1987, 1988a, Pfund *et al.*, 1988) of any solute i in a supercritical solvent j, we begin by expressing the number density of i as a function of the temperature and the chemical potential of each species in that phase:

$$\rho_i = \rho_i(T, \mu_1, \ldots, \mu_c). \tag{17}$$

Then the total differential of the solute density is given by

$$d\rho_i = \left[\frac{\partial \rho_i}{\partial T}\right]_{\mu_k} dT + \sum_{k=1}^{c} \left[\frac{\partial \rho_i}{\partial \mu_k}\right]_{T,\mu_{m \neq k}} d\mu_k. \tag{18}$$

Noticing that the partial derivatives on the righthand side of Equation (18) are given directly by Equation (7), we find for isothermal conditions

$$kT d\rho_i = \rho_i G_{ij} dP + \sum_{k \neq j}^{c} \left[\rho_i \rho_k (G_{ik} - G_{ij}) + \delta_{ik} \rho_i \right] d\mu_k \tag{19}$$

where we have made use of the isothermal Gibbs-Duhem equation

$$dP = \sum_{k=1}^{c} \rho_k d\mu_k. \tag{20}$$

Equation (19) is a general expression for the variation of solute density in an isothermal, homogeneous, fluid phase. For the case of equilibrium of a supercritical solution with a second phase ′ of fixed composition

$$d\mu_k = d\mu_k' = \overline{v}_k' dP \tag{21}$$

and

$$\frac{kTd\rho_i}{\rho_i} = \left\{ (G_{ij} + \overline{v}_i') + \sum_{k \neq j}^{c} \left[\rho_k \overline{v}_k' (G_{ik} - G_{ij}) \right] \right\} dP. \tag{22}$$

The first term on the righthand side of Equation (22) contains the solute-solvent fluctuation integral G_{ij} and the partial molar volume of the solute in the second phase $\overline{v}_i'$. The second term contains the difference between the solute-solute fluctuation integrals G_{ik} and the solute-solvent fluctuation integral G_{ij}. It is from this term that any observed effects must arise on the solubility of solute i because of the concentration of solute i, the presence of other solutes k, or the presence of added cosolvent k.

In the common case that a solvent j is at equilibrium with an incompressible solid phase of pure solute i, then

$$\frac{kTd\rho_i}{\rho_i} = \left\{ G_{ij} + v_i^s + \rho_i v_i^s (G_{ii} - G_{ij}) \right\} dP. \tag{23}$$

where the superscript s indicates the solid phase. At infinite dilution of the solute i, Equation (23) becomes

$$\frac{kTd\rho_i}{\rho_i} = \left\{ G_{ij}^0 + v_i^s \right\} dP. \tag{24}$$

This expression will be used further to develop some useful, approximate models in a later section.

The slope of the solubility curve, $(dx/dP)_T$, becomes large near the CP. Gitterman and Procaccia (Gitterman and Procaccia 1983) examined the slope for the case of a pure, incompressible, solid B at equilibrium with a supercritical solvent A:

$$\left[\frac{\partial x_B}{\partial P} \right]_{T,eq} = \frac{(v_B^s - \overline{v}_B)}{\left[\frac{\partial \mu_B}{\partial x_B} \right]_{T,P}}. \tag{25}$$

In terms of the KB formulas, Equation (25) can be expressed as

$$\left[\frac{\partial x_B}{\partial P} \right]_{T,eq} = \frac{v_B^s - \dfrac{1 + \rho_A (G_{AA} - G_{AB})}{(\rho_A + \rho_B + \rho_A \rho_B \Delta)}}{kT \left[\dfrac{1}{x_B} - \dfrac{\rho_A \Delta}{(1 + \rho_A x_B \Delta)} \right]} \tag{26}$$

Fugacity Coefficient

Debenedetti and Kumar (Debenedetti and Kumar 1986) observed that for dilute binary solutions (in particular, typical supercritical solutions), the natural logarithm of the solute fugacity coefficient divided by its value at infinite dilution is directly proportional to the solute mole fraction; thus

$$\ln\left(\frac{\phi_B}{\phi_B^0}\right) = -Kx_B. \tag{27}$$

The proportionality coefficient, K, was related by Debenedetti and Kumar to partial molar volume and to partial molar enthalpy. We have shown (Cochran and Lee 1987, 1988) that the proportionality coefficient can be related to the Kirkwood fluctuation integrals by

$$K = \left[G_{AA}^0 + G_{BB}^0 - 2G_{AB}^0\right]\rho_A^0 = \Delta^0\rho_A^0, \tag{28}$$

and that the approximate expressions developed in a later section for use in Equation (24) can be used for estimates of the proportionality coefficient.

CALCULATIONS FROM DISTRIBUTION FUNCTION THEORY

To obtain fluctuation integrals for use in equations in the previous section and to investigate the microstructure of supercritical solutions, we have studied (Cochran *et al.*, 1988b, Cochran and Lee 1989) Lennard-Jones (LJ) mixtures that simulate the carbon dioxide-naphthalene system using integral equation theory. Since real CO_2 and $C_{10}H_8$ are far from LJ particles, the LJ fluid is used only as a model, the rationale being that the interesting properties of supercritical solutions are common for all fluids exhibiting critical behavior, regardless of their specific intermolecular forces. So, by studying simple models we can bring out the the important features without the complicating details.

To calculate the pair correlation functions at various conditions characteristic of supercritical solutions, we used the accurate Hybrid Mean Spherical Approximation (HMSA) integral equation of Zerah and Hansen (Zerah and Hansen 1986). The HMSA equation is an interpolation between the soft mean spherical approximation at small separation distances and the hypernetted chain closure at large separations. The interpolating parameter was chosen to achieve consistency between the isothermal compressibility from the compressibility equation and that from the virial equation. The calculations would have been impractical without the use of an efficient and robust algorithm for solution of the Ornstein-Zernike equation. We employed the approach proposed by Labik, *et al.* (Labik *et al.*, 1985). Details of the calculations are given in the original references (Cochran *et al.*, 1988b, Cochran and Lee 1989).

The Reference Hypernetted Chain (RHNC) closure (Lado 1973) was used in independent, corroborative calculations for the same mixtures. Although quantitative values of distribution functions from the two theories differed slightly, the general features were the same; the RHNC theory gives somewhat higher peaks, but they occur at the same intermolecular separation. In closely related work (Pfund *et al.*, 1989a, 1989b, Lee 1988, Lee and Shing 1989) we have explored methods of determining chemical potential in solutions from molecular distribution functions.

We used the following Lennard-Jones parameters for the CO_2 (A) — $C_{10}H_8$ (B) mixture:

molecular pair		ε/k (K)	σ (A)
CO_2—CO_2	(A—A)	225.3	3.794
$C_{10}H_8$—$C_{10}H_8$	(B—B)	554.4	6.199
CO_2—$C_{10}H_8$	(A—B)	353.4	4.997

The integral equation results we show are entirely from the HMSA equation and are for very dilute, $x_B=10^{-9}$, solutions; we found the results to be unaffected for solute mole fractions from 10^{-9} to 10^{-4}. Dilute solutions are characteristic of many experimentally-studied supercritical solutions.

The pair correlation functions $g_{\alpha\beta}$ *vs.* reduced separation distance, $r^*=r/\sigma_{AA}$, are shown in Figure 2 for a typical, dilute supercritical solution using the LJ parameters given above and at $T^*=kT/\varepsilon_{AA}=1.37$ (T=308.4K), $\rho^*=\rho\sigma_{AA}^3=0.40$ ($\rho=0.00732\text{A}^{-3}$), and $x_B=10^{-9}$; the calculated pressure for this state is 8.7 MPa (86 atm). In the solvent-solute pair correlation function, $g_{AB}(r^*)$, one can clearly see the clustering of solvent molecules about a dissolved solute molecule, with first, second, and even third solvation shells. The long-rangedness of $g_{BB}(r^*)$ is also evident. At large separation distances, $g_{BB}(r^*)$ remains greater than one; therefore, there is a long-range structure of solute molecules about solute molecules.

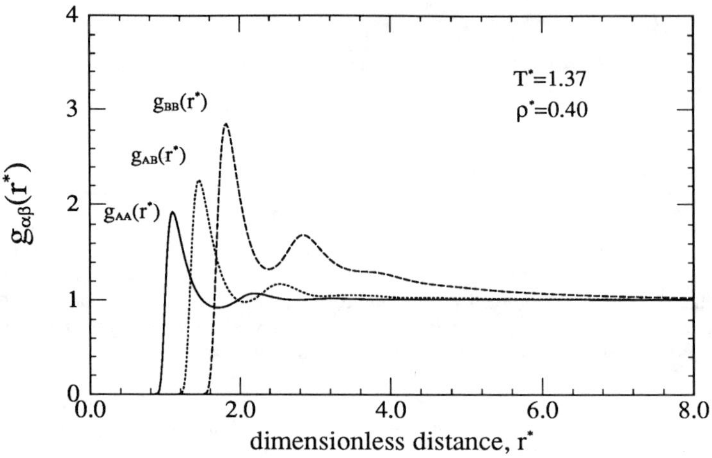

Figure 2. Pair correlation functions at $T^* = 1.37$ and $\rho^* = 0.40$ For the LJ fluid the CP occurs at about $T^* = 1.35$ and $\rho^* = 0.35$. Reprinted from Cochran and Lee 1989 with permission. © 1989 American Chemical Society.

Figures 3a and 3b show how g_{AB} varies with density. We see liquid-like structure with first, second, third, ... maxima/minima oscillating about 1.0 at the highest density (ρ^*=0.6). Using Equation (14), we calculated the size of the solvent-solute cluster to be about −1 solvent molecule for this state. Thus, at this state the presence of one solute molecule excludes about one solvent molecule. The peak maxima are lower and the peak minima are shallower at intermediate density (ρ^*=0.5). Still the long-range maxima and minima oscillate about 1.0, and for this state the size of the solvent-solute cluster was calculated to be about +2. Thus, for intermediate density each solute molecule is surrounded by about two excess solvent molecules compared with the bulk average concentration. At near-critical density (ρ^*=0.4) the oscillations in g_{AB} subside into a persistent long-range correlation, exhibiting a long-range tail that is greater than 1.0. (The CP of the pure LJ solvent is at about T^*=1.35 and ρ^*=0.35). See particularly Figure 3b which clearly shows this long-range structure. The solvent-solute cluster for this state was calculated to be +14.65 solvent molecules. At states farther from the CP the solute molecule excluded solvent molecules, but near the CP there is a large buildup of longer-range local solvent density around a solute molecule.

To examine the radial structure of the solute-solvent cluster we have plotted in Figure 4 the number of excess solvent molecules within a sphere of radius $L(=r/\sigma_{AA})$ around a solute molecule, $\rho G_{AB}(L)$, *vs.* L for the same states as Figure 3. For the state closest to the CP (ρ^*=0.40) it can be seen that the solvent-solute cluster has grown to almost 15 excess solvent molecules. It can be estimated from Figure 4 that the first solvation sphere contributes about 4 excess solvent molecules to the cluster. The near-critical cluster consists mainly of a diffuse, longer-range buildup of local solvent density about the solute which extends 10-14 solvent diameters from the solute. Molecular dynamics calculations (Petsche and Debenedetti 1988) suggest that the cluster maintains its structural integrity for significant periods of time although the solvent molecules constituting the cluster change identity more rapidly. These clusters are characteristic of solvation structure near the CP and are responsible for the striking behavior of thermodynamic properties near the CP.

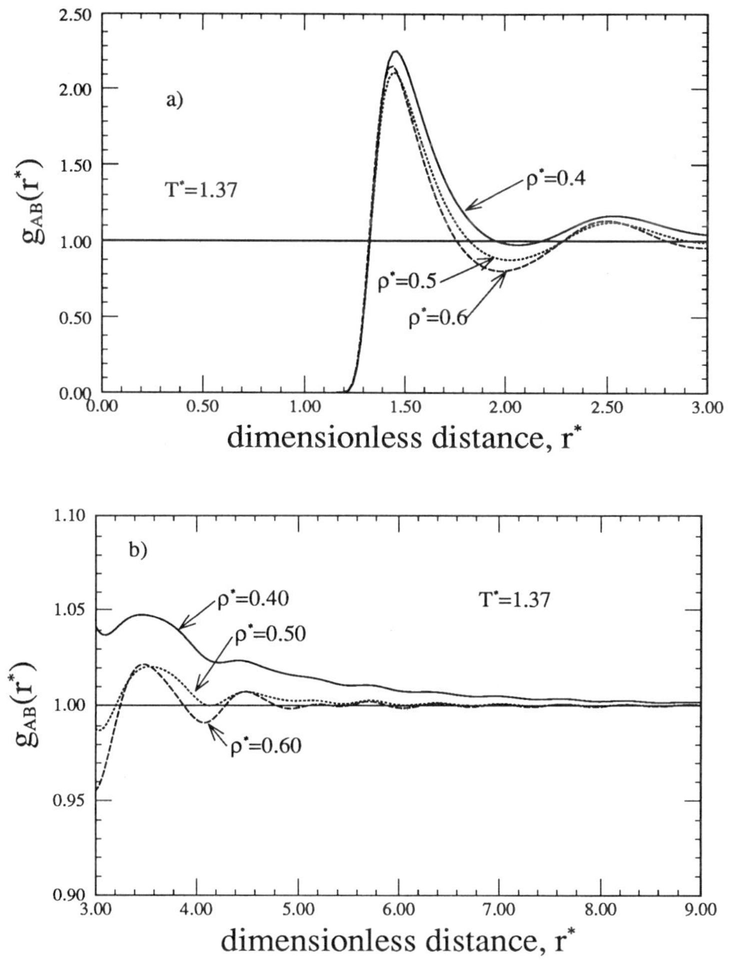

Figure 3. Solute-solvent pair correlation functions at T^* = 1.37 and ρ^* = 0.40, 0.50, and 0.60. a) Short-range structure. b) Long-range structure. Reprinted from Cochran and Lee 1989 with permission. © 1989 American Chemical Society.

The effect of temperature on the size of the solvent-solute cluster *vs.* density is shown in Figure 5 at T^*=1.37, 1.41, and 1.46 (35, 45, and 55 °C). The size of the cluster grows rapidly as density increases from zero toward the CP, and the effect is stronger as the temperature approaches the CP. The rapid decrease in the cluster size as density is increased away from the CP is also evident. At temperatures higher above the CP the cluster is smaller and the structure is shorter in range. These calculations support the notion of a large solvent-solute cluster near the CP as suggested from spectroscopic results (Kim and Johnston 1987a, 1987b). The estimation (Eckert *et al.*, 1986, Debenedetti 1987) that the cluster approaches 100 solvent molecules based on the partial molar volume data for naphthalene in supercritical carbon dioxide at this temperature (Eckert *et al.* 1983, 1986) also appears entirely consistent with the predictions from the integral equation theory.

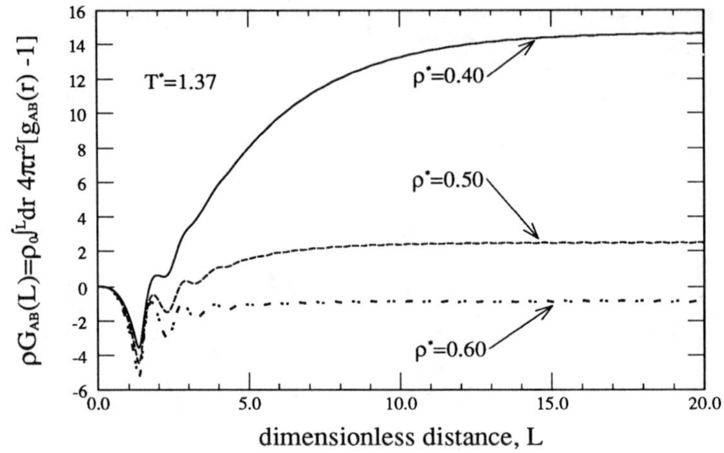

Figure 4. Number of excess solvent molecules within a sphere of radius L around a solute molecule. Same conditions as Figure 2.

$$\rho_A G_{AB}(L) = \rho_A \int_0^L dr\ 4\pi r^2 \left[g_{AB}(r) - 1\right]$$

This graph shows the contributions to the fluctuation integral from the successive neighborhoods. Reprinted from Cochran and Lee 1989 with permission. © 1989 American Chemical Society.

Pair correlation functions become long-ranged near the CP and convergence difficulties can be experienced. As detailed in the original works, the results presented here are for fully converged states; states closer to the CP could not be calculated by this method. As a practical matter, the range of the calculations discussed ($\leq 40.96\sigma_{AA}$) limits convergence at $T^*=1.37$ to densities below 0.15 and densities above 0.4. This range exceeded that of previous, converged calculations of G_{ij} for liquid mixtures by Kojima, *et al.*, (Kojima, *et al.*, 1984). The Percus-Yevick approximation for a pure fluid (McGuigan and Monson 1987, 1989) does not experience as severe difficulties.

The partial molar volume of naphthalene in supercritical carbon dioxide predicted by theory is compared with data (Eckert *et al.* 1983, 1986) in Figure 6. The degree of agreement between prediction and experiment is judged to be very satisfactory since neither naphthalene nor carbon dioxide would be expected to be described quantitatively by the LJ potential. Furthermore, no binary interaction parameter was used with the Lorentz-Berthelot estimates for the CO_2—$C_{10}H_8$ interaction.

Figures 7, 8, and 9 show the solvent-solvent, solute-solvent, and solute-solute fluctuation integral *vs.* density at three temperatures. The qualitative similarity of the variation of the three fluctuation integrals is evident. We noted in Figure 2 the qualitative similarity of the pair correlation functions for one state. Such similarity in correlation functions and fluctuation integrals is characteristic of these dilute, near-critical LJ mixtures.

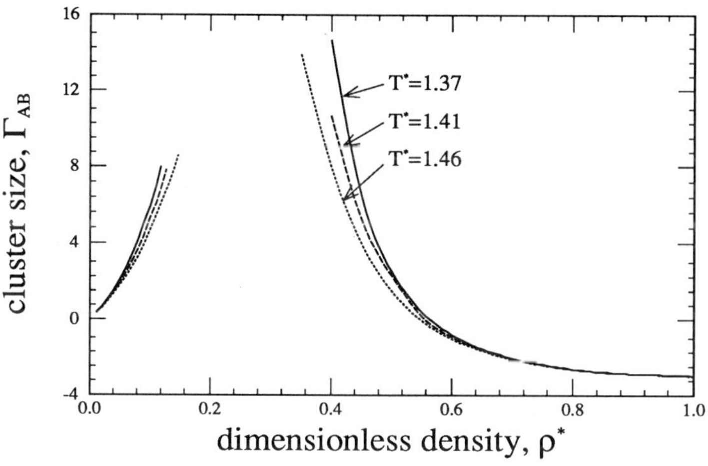

Figure 5. Size of solvent-solute cluster *vs.* density at $T^* = 1.37$, 1.41, and 1.46.

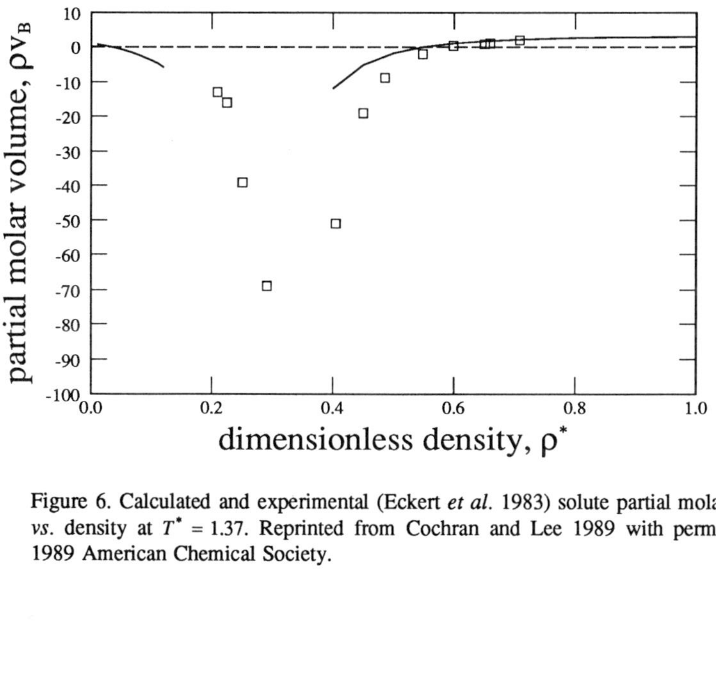

Figure 6. Calculated and experimental (Eckert *et al.* 1983) solute partial molar volume *vs.* density at $T^* = 1.37$. Reprinted from Cochran and Lee 1989 with permission. © 1989 American Chemical Society.

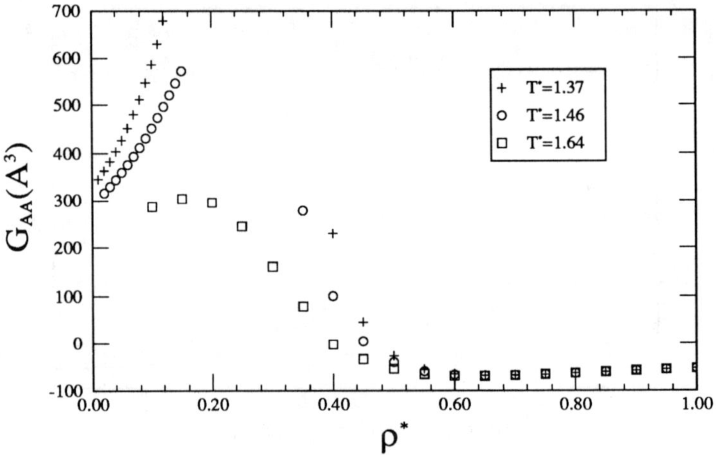

Figure 7. Solvent-solvent fluctuation integral *vs.* density for three temperatures.

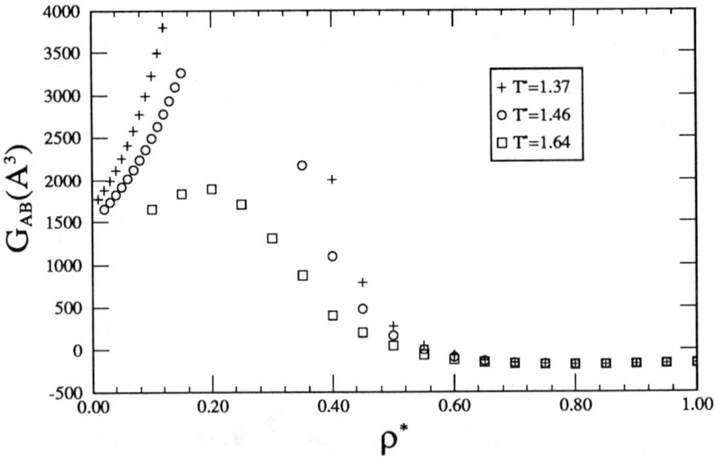

Figure 8. Solute-solvent fluctuation integral *vs.* density for three temperatures.

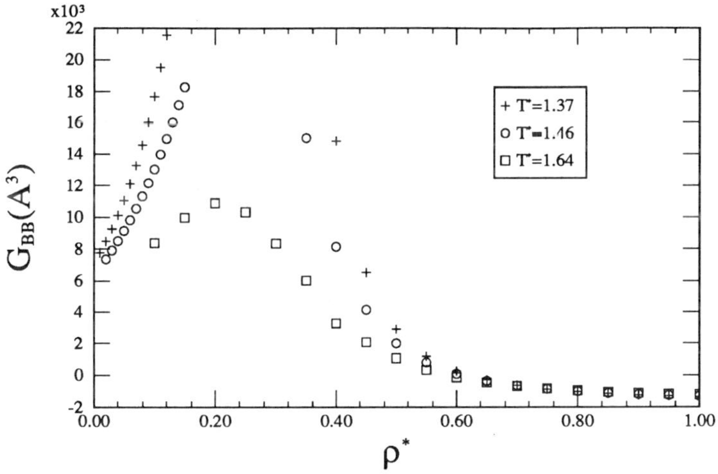

Figure 9. Solute-solute fluctuation integral *vs.* density for three temperatures.

We found (Cochran *et al.,* 1988b) that moderate changes in the magnitude of the solute size and energy parameters resulted in only small, quantitative changes in both the solute-solvent and solute-solute fluctuation integrals; see Figures 10 and 11. McGuigan and Monson (McGuigan and Monson 1987, 1989) have used the KB formalism in the Percus-Yevick approximation to locate the spinodal curve for various LJ mixtures and to determine the critical locus as an extremum in the $P—T$ projection of the spinodal curve. They have reported the effects of molecular interaction energy, size, and shape on the critical locus and on the partial molar volume of an infinitely dilute solute near the CP.

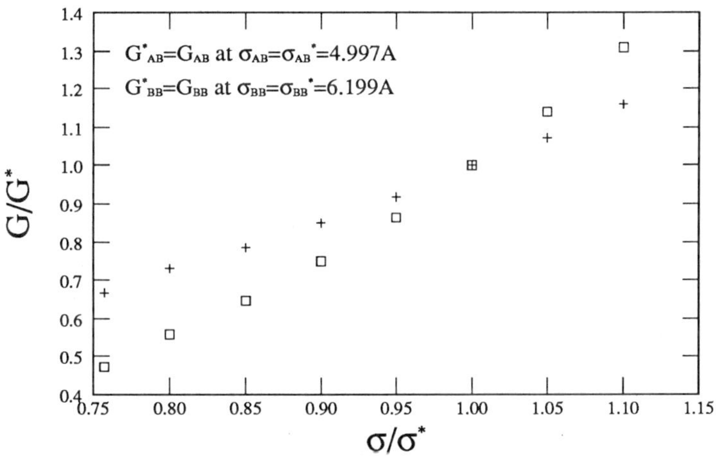

Figure 10. Relative change in fluctuation integrals with solute size parameter at $T^* = 1.37$ and $\rho^* = 0.40$. Crosses: G_{AB}/G^*_{AB} *vs.* $\sigma_{AB}/\sigma^*_{AB}$. Squares: G_{BB}/G^*_{BB} *vs.* $\sigma_{BB}/\sigma^*_{BB}$.

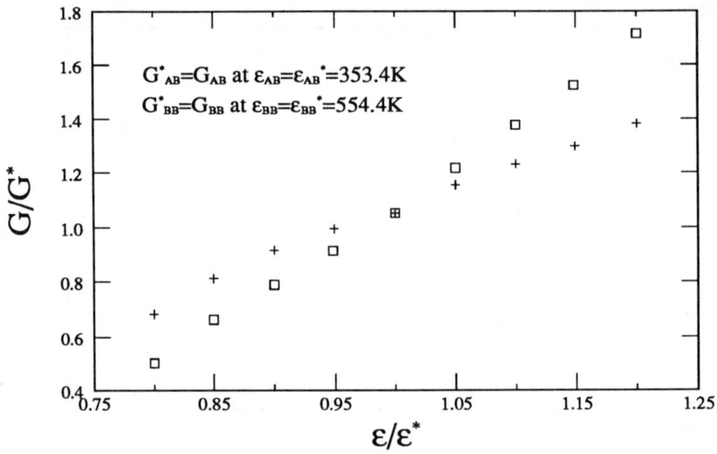

Figure 11. Relative change in fluctuation integrals with solute energy parameter at $T^* = 1.37$ and $\rho^* = 0.40$. Crosses: G_{AB}/G^*_{AB} *vs.* $\varepsilon_{AB}/\varepsilon^*_{AB}$. Squares: G_{BB}/G^*_{BB} *vs.* $\varepsilon_{BB}/\varepsilon^*_{BB}$.

One might ask whether these results from integral equation theory may shed any light on the critical exponents. Substantial evidence (*e.g.*, Watts 1968, 1969, Baxter 1968, Henderson and Murphy 1972, Guerrero *et al.*, 1975, Foiles and Ashcroft 1981, Fishman 1981) not reviewed here indicates that the approximate integral equation theories such as those used in the work described here do not show quantitatively accurate critical scaling. The Percus-Yevick approximation exhibits mean-field (classical) scaling behavior (Watts 1968, Baxter 1968, Henderson and Murphy 1972, Guerrero *et al.*, 1975). Although the HMSA equation has not been investigated in this regard, the HNC equation, which constitutes the long-range part of the HMSA equation, does not exhibit a physically realistic critical point (Watts 1969, Guerrero *et al.*, 1975, Foiles and Ashcroft 1981).

In order to explore the possibility of solute-solute aggregation near the CP we show the solute-solute pair correlation functions in Figures 12a and 12b at the same conditions as the solute-solvent functions in Figures 3a and 3b. The aggregation of the solute molecules is qualitatively similar to but quantitatively stronger than the solvation structure shown in Figure 3. At present we do not have a useful definition for the solute-solute cluster size (corresponding to Equation (14) for the solute-solvent cluster, for example). A quantitative interpretation of the fluorescence spectra of Brennecke and Eckert (Brennecke and Eckert 1988, 1989) in terms of the local density of solute molecules surrounding a given solute molecule has not been given for the band attributed to a solute-solute excimer-dimer; nevertheless, their qualitative interpretation suggesting a significant solute-solute aggregation near the CP appears to be supported by these theoretical results.

In other, related work, the question of solute-solute aggregation has been explored further (Wu, *et al.*, 1989) using the RHNC integral equation theory (Lado 1973). New methods for determining the chemical potential, using distribution functions from integral equation theory, have also been developed recently (Pfund *et al.*, 1989a, 1989b, Lee 1988, Lee and Shing 1989).

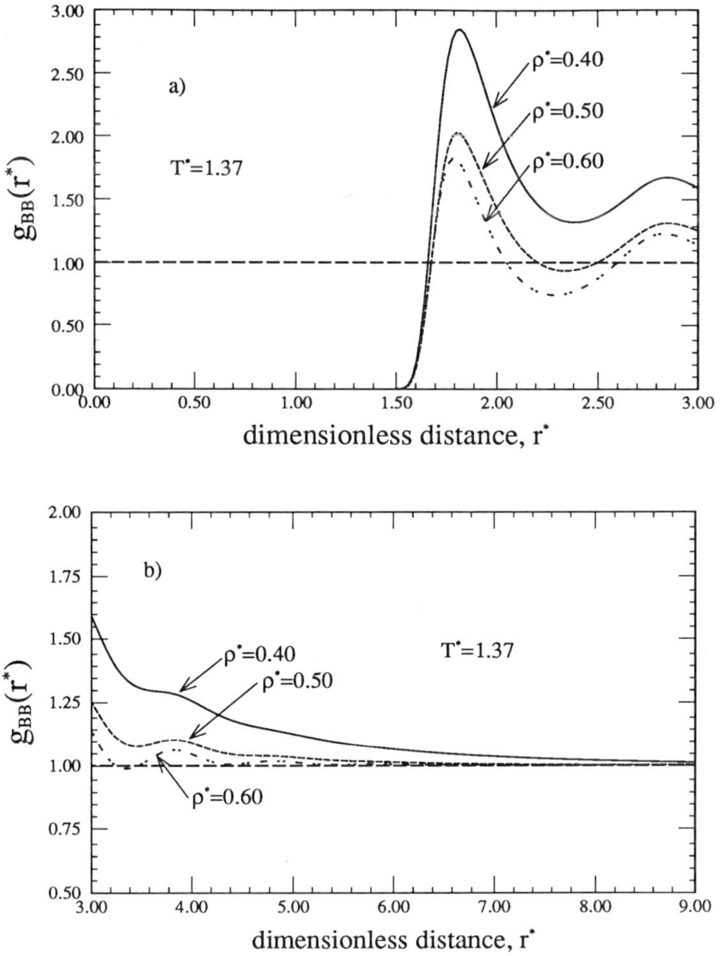

Figure 12. Solute-solute pair correlation functions. Same conditions as Figure 3. a) Short-range structure. b) Long-range structure. The curve at $\rho^* = 0.4$ shows a double peak with shallow minimum indicating solute-solute aggregation. Reprinted from Cochran and Lee 1989 with permission. © 1989 American Chemical Society.

APPROXIMATE MODELS FOR G_{ij}

For the dilute supercritical solutions of LJ molecules considered in the previous section, we have observed that the solvent-solvent, solute-solvent, and solute-solute pair correlation functions and fluctuation integrals are qualitatively similar. This suggests the possibility of developing simple but useful models by suitable approximations. The basic strategy has been to assume that the qualitative behavior of the fluctuation integrals at the infinite dilution limit is reflected by that of the pure solvent and then to use molecular solution approximations for quantitative adjustment. The solvent-solvent fluctuation integral for the pure solvent may be calculated from an appropriate equation of state for the pure solvent:

$$G_{jj}^0 = (kT\rho_j^0\kappa_{Tj}^0 - 1)/\rho_j^0 = \left\{kT\left[\frac{\partial\rho_j}{\partial P}\right]_T^0 - 1\right\}/\rho_j^0. \tag{29}$$

The hope was that relatively crude approximations might lead to useful models because of the information about fluid structure which is captured by an accurate equation of state for the pure solvent. For the common supercritical solvents, *e.g.*, CO_2 and C_2H_4, very accurate equations of state are available (Huang, *et al.*, 1985, Jahangiri, *et al.*, 1986).

The simple models that have been developed (Cochran *et al.*, 1987, 1988a, Pfund *et al.*, 1988) may be useful for semi-quantitative prediction (*i.e.*, without benefit of binary data) and have yielded very accurate fits to data. We shall describe three models.

The first model accounts for an excluded volume for each species and assumes that the infinite dilution fluctuation integrals scale with a parameter α_{ij}. The excluded volumes V_{ij} and V_{jj} and the parameter α_{ij} may be fitted or estimated from pure molecular parameters ($\alpha_{ij} \approx \varepsilon_{ij}\sigma_{ij}^3/\varepsilon_{jj}\sigma_{jj}^3$):

$$G_{ij}^0 + V_{ij} \approx \alpha_{ij}(G_{jj}^0 + V_{jj}). \tag{30}$$

We shall call this model the excluded volume (EV) model. When the EV model is inserted into Equation (24), a simple, algebraic expression for the solubility results:

$$\ln(\rho_i kT/p_i^{sat}) \approx -\alpha_{ij}\ln(f_j^0/\rho_j^0 kT) + [v_i^s - V_{ij} + \alpha_{ij}V_{jj}][P - p_i^{sat}]/(kT). \tag{31}$$

where p_i^{sat} is the equilibrium vapor pressure of pure i and f_j^0 is the fugacity of pure j.

The second model includes the volume of the solvation sphere, the scaling parameter α_{ij}, and a local composition term of the form of the NRTL model (Renon and Prausnitz 1968). The volume of the solvation sphere, V_{ij}, the scaling parameter α_{ij}, and the parameters β_0 and β_1 might be estimated or fitted:

$$G_{ij}^0 + V_{ij} \approx \alpha_{ij}\left[G_{jj}^0 + \frac{V_{ij}}{\alpha_{ij}}\exp(\beta_1 + \beta_0/T)\right] \tag{32}$$

We shall call this model the local composition (LC) model. When substituted into Equation (24), the LC model also integrates to a simple algebraic expression:

$$\ln(\rho_i kT/p_i^{sat}) \approx -\alpha_{ij}\ln(f_j^0/\rho_j^0 kT) + \left\{v_i^s - V_{ij}\left[1 + \exp\left[\beta_1 - \beta_0/T\right]\right]\right\}\left\{P - p_i^{sat}\right\}/(kT). \tag{33}$$

The third model we shall describe uses the Carnahan-Starling (Carnahan and Starling 1969) equation for hard spheres in a hard sphere expansion conformal solution model (Mansoori and Leland 1972):

$$G_{ij}^0 \approx \frac{\pi\sigma_{ij}^3}{6}\frac{\left[\frac{\varepsilon_{ij}}{\varepsilon_{jj}} - 1\right]\left[8 - 2y\right]}{(1 + 4y + 4y^2 - 4y^3 + y^4)} + \alpha_{ij}G_{jj}^0. \tag{34}$$

We shall call this model the hard sphere expansion conformal solution (CS) model. The CS model must be evaluated numerically. The σ and ε parameters are the usual molecular size and interaction energy, or they may be treated as adjustable parameters in fitting data; y is the usual hard sphere packing fraction based on the pure solvent diameter $y = \frac{\pi}{6}\rho\sigma_{jj}^3$.

The EV and LC models (Equations (30) and (32)) have been tested rather thoroughly (Pfund *et al.*, 1988) for solubility using Equation (24), but the CS model (Equation (34)) has been subjected to only limited testing and has not been previously published. The EV, LC, or CS models may be substituted into the expressions above to yield models for the partial molar volume, solute-solvent cluster size, solubility of a pure incompressible solid, Gitterman-Procaccia slope of the solubility curve, and Debenedetti-Kumar proportionality coefficient.

Tables 1, 2, and 3 (Pfund *et al.*, 1988) compare the fit of the EV model, the LC model, and the Peng-Robinson (PR) equation of state with van der Waals 1-fluid mixing rules (Peng and Robinson 1976) for a number of solutes in supercritical carbon dioxide and ethylene respectively.

Table 1. Regression Results — Excluded Volume Model

System	Number of points	V_{12} cm^3/gmol value ± s	ε_{12}/k Kelvin value ± s	error in calculated solubility <|%ε|>	$\sum\varepsilon^2$
Carbon Dioxide Systems					
Acridene	28	294.4±0.4	530.18±0.04	30.59	0.103
Benzoic acid	55	300.0±3.7	512.38±0.41	19.72	0.631
2,3 dimethyl naphthalene	15	419.1±5.1	356.39±0.38	11.67	0.809
2,6 dimethyl naphthalene	15	371.2±3.6	385.94±0.31	9.80	0.283
Fluorene	30	387.1±210.4	401.73±16.47	15.77	0.371
Hexamethyl benzene	27	409.6±139.5	363.56±9.81	41.59	0.187
Naphthalene	68	309.3±8.6	411.38±0.84	20.27	2401.84
1,4 naphtho quinone	18	350.5±2.1	459.17±0.19	7.92	0.168
Phenanthrene	36	496.4±1.6	343.68±0.11	23.37	0.132
Pyrene	22	450.5±86.4	402.89±6.19	21.28	0.0053
Triphenyl methane	22	596.2±161.4	330.10±9.04	72.01	0.087
Ethylene Systems					
Benzoic acid	15	413.4±0.2	511.31±0.23	16.12	0.181
2,3 dimethyl naphthalene	18	256.8±13.5	738.06±1.92	15.08	21.334
2,6 dimethyl naphthalene	18	345.4±8.8	565.79±0.93	12.98	1.736
Fluorene	24	518.5±573.5	445.37±38.24	22.49	4.048
Hexamethyl benzene	24	577.2±271.4	398.66±15.97	17.17	1.225
Naphthalene	84	379.9±55.1	480.77±5.02	23.99	38.648
Phenanthrene	45	425.3±2.8	526.98±0.26	19.58	0.234
Pyrene	15	623.7±140.0	438.09±8.25	13.73	0.0254
		Average absolute percent error: 22.58			

Both the EV and LC models outperform the PR model. The EV model in Equation (24) fails to account adequately for the variation of solubility with temperature when the excluded volume V_{ij} and the scaling parameter αij are treated as temperature-independent constants; see Figure 13, for example. With parameters fitted at each temperature, however, the EV model in Equation (24) gives a truly excellent fit to the solubility; see Figure 14, for example. With temperature-independent parameters, the LC model in Equation (24) produces a far more satisfactory fit to the temperature variation of solubility than the EV model (see Figure 15), but the LC model of the temperature variation does not perform so well as the EV model

Table 2. Regression Results — Local Composition Model

System	V_{12} cm³/gmol value ± *s*	β_1 Kelvin value ± *s*	β_0 (-) value ± *s*	error in calculated solubility <|%ε|>	$\sum\varepsilon^2$
Carbon Dioxide Systems					
Acridene	367.1±1.4	-902.4±0.7	2.848±0.048	17.98	0.183
Benzoic acid	328.8±0.6	-285.0±0.5	1.318±0.035	21.06	0.299
2,3 dimethyl naphthalene	306.4±1.2	-71.6±1.6	0.430±0.117	11.35	0.864
2,6 dimethyl naphthalene	297.7±0.6	-122.9±0.4	0.650±0.029	5.24	0.157
Fluorene	319.1±17.9	-238.3±19.3	1.014±1.345	8.25	0.025
Hexamethyl benzene	320.9±24.4	-367.7±48.9	1.265±3.432	35.90	0.079
Naphthalene	270.9±0.4	-1104.4±1.2	3.736±0.088	10.58	1.401
1,4 naphtho quinone	328.0±0.4	-38.3±0.1	0.550±0.006	7.46	0.108
Phenanthrene	346.7±1.2	-71.1±0.9	0.335±0.067	24.79	0.455
Pyrene	345.7±8.4	-216.2±7.9	0.985±0.560	16.94	0.00051
Triphenyl methane	385.4±44.2	-144.8±60.1	0.577±4.466	70.98	0.077
Ethylene Systems					
Benzoic acid	455.4±1.	-169.0±0.6	1.063±0.043	12.19	0.146
2,3 dimethyl naphthalene	414.5±2.7	-453.4±1.5	2.078±0.106	12.79	2.497
2,6 dimethyl naphthalene	422.7±1.2	-187.6±0.8	1.170±0.055	13.27	0.493
Fluorene	484.4±90.0	-131.2±46.6	0.868±3.357	17.88	0.632
Hexamethyl benzene	495.3±49.9	-104.9±23.6	0.723±1.821	11.39	0.214
Naphthalene	403.2±106.6	-436.4±61.7	1.899±4.600	17.04	7.085
Phenanthrene	446.0±1.4	+20.4±0.4	0.543±0.029	19.49	0.236
Pyrene	536.4±28.4	-59.1±10.6	0.667±0.746	14.06	0.0062

Average absolute percent error: 18.4

Adapted from Pfund *et al.* 1988 with permission. © Elsevier Science Publishers.

with temperature-dependent parameters. Figures 16 and 17 illustrate the prediction of partial molar volume using the EV and LC models in Equation (15) with the temperature-dependent parameters in the EV model and temperature-independent parameters in the LC model from fitting solubility data. The CS model has not really been tested for fitting solubility data, but preliminary calculations indicate it may be superior to both the EV and LC models. For example, the CS model with temperature-independent parameters gives an excellent fit to solubility data for one system as shown in Figure 18. The EV, LC, and CS models have not been used for calculation of other solution properties (except the solute partial molar volume) to date. The success with these simple models for solubility and partial molar volume, however, suggests that further study may be warranted.

CONCLUDING REMARKS

Fluctuation theory, particularly within the KB formalism, has proved to be an appropriate and useful approach to the study of the structure and properties of supercritical fluid mixtures. The large density fluctuations near CPs are associated with divergence in pure fluid isothermal compressibility and solute partial molar volume, leading to striking solubility behavior and other effects. In addition to the familiar KB equations for the partial molar volume and isothermal compressibility, we have reviewed expressions for the size of a solvent cluster around a solute molecule, degree of solute aggregation, solubility of a pure, incompressible solid (as a function of pressure or density), and variation of fugacity with composition for dilute solutions. The negative partial molar volume and solubility enhancement phenomena occur with increases in the long-range order of the fluid as shown by the correlation functions obtained from integral equation theories. Fluctuation theory provides a connection between the microscopic structure and

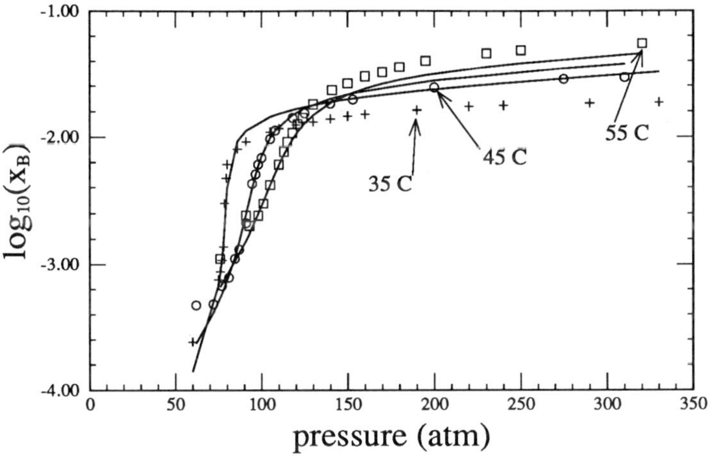

Figure 13. The correlation of naphthalene solubility in carbon dioxide from the EV model with temperature-independent parameters. Data are from Tsekhanskaya *et al.* 1964.

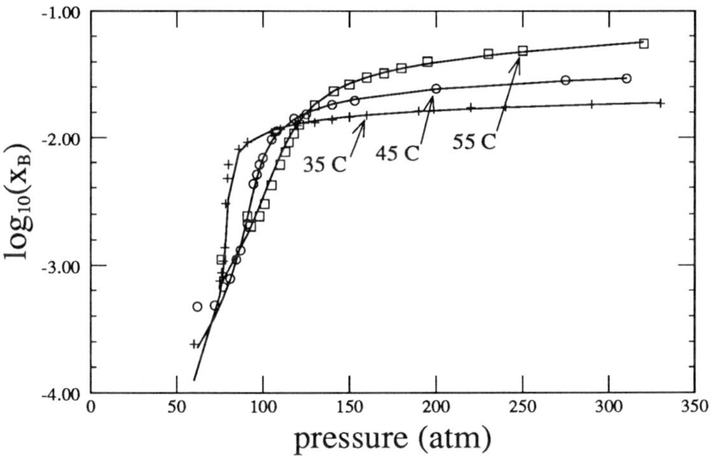

Figure 14. The correlation of naphthalene solubility in carbon dioxide from the EV model with temperature-dependent parameters. Data are from Tsekhanskaya *et al.* 1964.

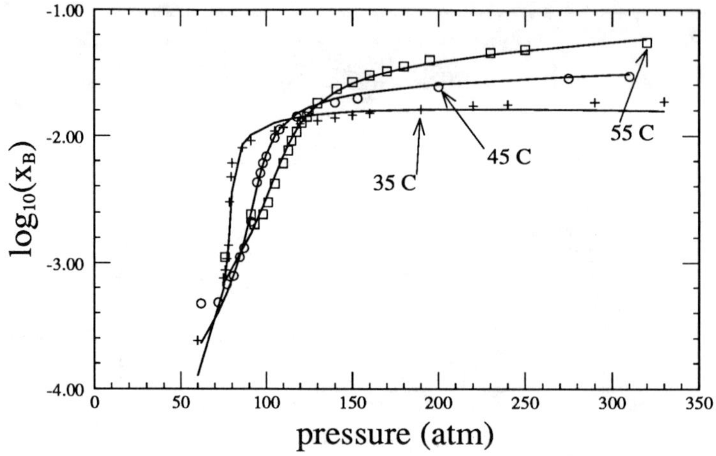

Figure 15. The correlation of naphthalene solubility in carbon dioxide from the LC model with temperature-independent parameters. Data are from Tsekhanskaya *et al.* 1964.

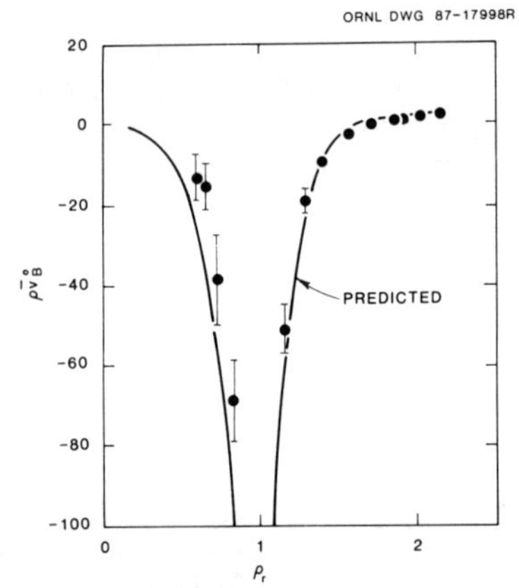

Figure 16. Partial molar volume of naphthalene in carbon dioxide predicted by the EV model with temperature-dependent parameters. Data are from Eckert *et al.* 1983.

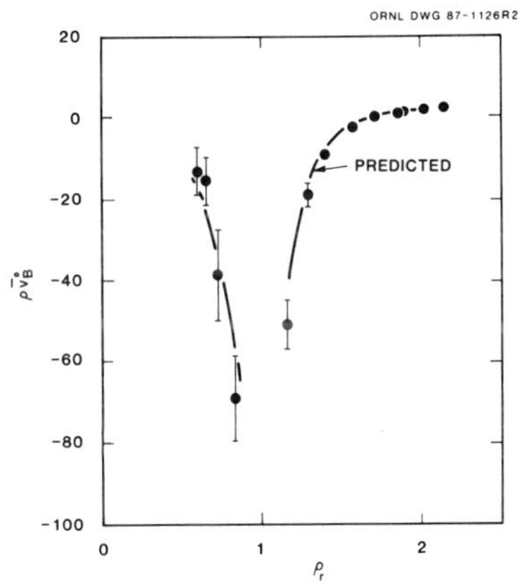

Figure 17. Partial molar volume of naphthalene in carbon dioxide predicted by the LC model with temperature-independent parameters. Data are from Eckert *et al.* 1983.

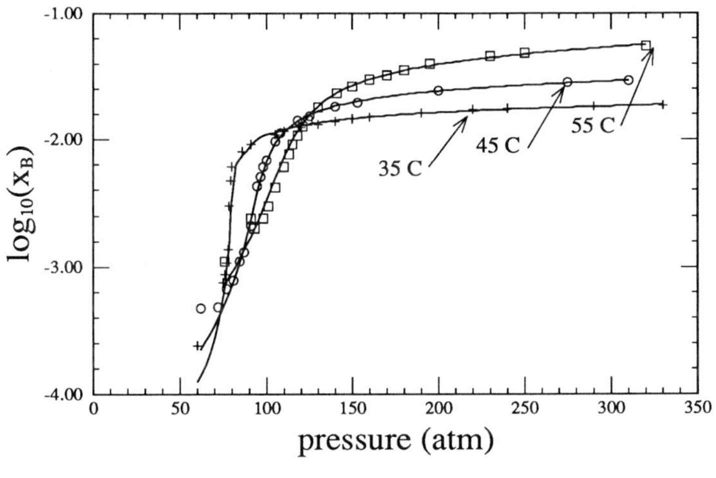

Figure 18. The correlation of naphthalene solubility in carbon dioxide from the CS model with temperature-independent parameters. Data are from Tsekhanskaya *et al.* 1964.

Table 3. Regression Results — Peng-Robinson Model

System	Number of points	Interaction parameter k_{12} value ± s	error in calculated solubility <|%ε|>	$\sum\varepsilon^2$
Carbon Dioxide Systems				
Acridene	28	0.0996±0.0009	19.11	0.530
Benzoic acid	55	0.0108±0.0014	27.15	1.396
2,3 dimethyl naphthalene	15	0.1059±0.0015	20.70	1.367
2,6 dimethyl naphthalene	15	0.1017±0.0006	9.30	0.167
Fluorene	30	0.0838±0.0448	11.48	0.103
Hexamethyl benzene	27	0.0939±0.0849	22.02	0.276
Naphthalene	68	0.0983±0.0013	27.19	13.114
1,4 naphthoquinone	18	0.0556±0.0010	31.28	0.731
Phenanthrene	36	0.1396±0.0010	35.21	0.356
Pyrene	22	0.1242±0.0498	54.88	0.013
Triphenyl methane	22	0.1050±0.0604	28.03	0.096
Ethylene Systems				
Benzoic acid	15	-0.0613±0.0013	18.93	0.305
2,3 dimethyl naphthalene	18	0.0157±0.0013	17.43	6.053
2,6 dimethyl naphthalene	18	0.0182±0.0007	9.07	0.641
Fluorene	24	0.0046±0.1040	19.57	1.548
Hexamethyl benzene	24	0.0181±0.0908	20.49	0.632
Naphthalene	84	0.0164±0.1889	38.41	34.627
Phenanthrene	45	0.0512±0.0006	24.34	0.238
Pyrene	15	0.0501±0.0718	22.34	0.078
	Average absolute percent error: 26.42			

Adapted from Pfund *et al.* 1988 with permission. © Elsevier Science Publishers.

the thermodynamic properties of the fluid and therefore offers an explanation of supercritical behavior. *A priori* calculations based on distribution functions from accurate integral equation theories yield quantitative predictions for structural and thermodynamic properties of supercritical solutions. Reasonable approximations lead to simple algebraic expressions for solubility and other properties that show excellent fit to experimental data.

ACKNOWLEDGMENT

This work has been supported in part by the Division of Chemical Sciences, Office of Basic Energy Sciences, U. S. Department of Energy under contract DE-AC-05-840R21400 with Martin Marietta Energy Systems, Inc. and subcontract with the University of Oklahoma. The authors thank Professors Eckert, Debenedetti, Johnston, and Monson for providing recent results in advance of publication.

REFERENCES

Baxter, R. J. 1968. Percus-Yevick Equation for Hard Spheres with Surface Adhesion. J. Chem. Phys. 49: 2770-2774.

Brennecke, J. F. and C. A. Eckert. 1988. Molecular Interactions from Fluorescence Spectroscopy. Proc. Int. Symp. on Supercritical Fluids, Nice, M. Perrut, Ed.: 263-270.

Brennecke, J. F. and C. A. Eckert. 1989. Fluorescence Spectroscopy Studies of Intermolecular Interactions in Supercritical Fluids. Supercritical Fluid Science and Technology, K. P. Johnston and J. M. L. Penninger, Eds., ACS Symp. Ser. 406: 14-26.

Caragay, A. B. 1981. Supercritical Fluids for Extraction of Flavors and Fragrances from Natural Products. Perfumer and Flavorist B6: 43.

Carnahan, N. F. and K. E. Starling. 1969. Equation of State for Nonattracting Rigid Spheres. J. Chem. Phys. 51: 635-638.

Cochran, H. D., L. L. Lee, and D. M. Pfund. 1987. Application of the Kirkwood-Buff Theory of Solutions to Dilute Supercritical Solutions. Fluid Phase Equilib. 34: 219-234.

Cochran, H. D. and L. L. Lee. 1987. General Behavior of Dilute Binary Solutions. AIChE J. 33: 1391-1393.

Cochran, H. D. and L. L. Lee. 1988. General Behavior of Dilute Binary Solutions: Erratum. AIChE J. 34: 170.

Cochran, H. D., D. M. Pfund, and L. L. Lee. 1988a. Theoretical Models of Thermodynamic Properties of Supercritical Solutions. Sep. Sci. and Tech. 23: 2031-2047.

Cochran, H. D., D. M. Pfund, and L. L. Lee. 1988b. Study of Fluctuations in Supercritical Solutions by an Integral Equation Method. Proc. Int. Symp. on Supercritical Fluids, Nice, M.Perrut, Ed.: 245-252.

Cochran, H. D. and L. L. Lee. 1989. Solvation Structure in Supercritical Fluid Mixtures Based on Molecular Distribution Functions. Supercritical Fluid Science and Technology, K. P. Johnston and J. M. L. Penninger, Eds., ACS Symp. Ser. 406: 27-38.

Collins, N. A., P. G. Debenedetti, and S. Sundaresan. 1988. Disproportionation of Toluene over ZSM-5 under Near-Critical Conditions. AIChE J. 34: 1211-1214.

Debenedetti, P. G. and S. K. Kumar. 1986. Infinite Dilution Fugacity Coefficients and the General Behavior of Dilute Binary Systems. AIChE J. 32: 1253-1261.

Debenedetti, P. G. 1987. Clustering in Dilute, Binary Supercritical Mixtures: A Fluctuation Analysis. Chem. Eng. Sci. 42: 2203-2212.

Eckert, C. A., D. H. Ziger, K. P. Johnston, and T. K. Ellison. 1983. The Use of Partial Molar Volume Data to Evaluate Equations of State for Supercritical Fluid Mixtures. Fluid Phase Equilib. 14: 167-175.

Eckert, C. A., D. H. Ziger, K. P. Johnston, and S. Kim. 1986. Solute Partial Molar Volumes in Supercritical Fluids. J. Phys. Chem. 90: 2738-2746.

Ehrlich, P. and J. J. Kurpen. 1963. Phase Equilibria of Polymer-Solvent Systems at High Pressures Near their Critical Loci. J. Polym. Sci. 1: 3217-3229.

Fishman, S. 1981. Fourier Space Analysis of the Yvon-Born-Green Equation in the Critical Region. Physica 109A: 382-402.

Foiles, S. M. and N. W. Ashcroft. 1981. Absence of Power-Law Behavior of the Hypernetted Chain Equation. Phys. Rev. A 24: 424-428.

Gitterman, M. and I. Procaccia. 1983. Quantitative Theory of Solubility in Supercritical Fluids. J. Chem. Phys. 78: 2648-2654.

Guerrero, M. I., G. Saville, and J. S. Rowlinson. 1975. Solutions of the Percus-Yevick and Hyper-Netted Chain Equations near the Gas-Liquid Critical Point. Mol. Phys. 29: 1941-1944.

Hannay, J. B. and J. Hogarth. 1879. On the Solubility of Solids in Gases. Proc. Roy. Soc. London Ser. A 29: 324-326.

Hannay, J. B. On the Solubility of Solids in Gases II. Proc. Roy. Soc. London Ser. A 30: 484-489.

Haselow, J. S., S. J. Han, R. A. Greenkorn, and K. C. Chao. 1986. Equation of State for Supercritical Extraction. Equations of State Theories and Applications ACS Symp. Ser. 300: 156-178.

Henderson, D. and R. D. Murphy. 1972. Critical Point in the Percus-Yevick Theory. Phys. Rev. A 6: 1224-1232.

Hoye, J. S. and G. Stell. 1985. Toward a Liquid State Theory with Accurate Critical Region Behavior. Int. J. Thermophys. 6: 561-571.

Huang, F. H., M. H. Li, L. L. Lee, K. E. Starling, and F. T. H. Chung. 1985. An Accurate Equation of State for Carbon Dioxide. J. Chem. Eng. Japan 18: 490-496.

Jahangiri, M., R. T. Jacobsen, R. B. Stewart, and R. D. McCarty. 1986. A Thermodynamic Property Formulation for Ethylene from the Freezing Line to 450 K at Pressures to 260 MPa. Int. J. Thermophys. 7: 491-501.

Johnston, K. P. and C. Haynes. 1987. Extreme Solvent Effects on Reaction Rate Constants at Supercritical Fluid Conditions. AIChE J. 33: 2017-2026.

Johnston, K. P., W. Flarsheim, B. Hrnjez, A. Mehta, M. Fox, and A. Bard. 1988. Solvent Effect on Chemical Reactions at Supercritical Fluid Conditions. Proc. Int. Symp. on Supercritical Fluids, Nice, M. Perrut, Ed.: 907-915.

Johnston, K. P., S. Kim, and J. Combes. 1989. Spectroscopic Determination of Solvent Strength and Structure in Supercritical Fluid Mixtures. Supercritical Fluid Science and Technology, K. P. Johnston and J. M. L. Penninger, Eds., ACS Symp. Ser. 406: 52-70.

Jonah, D. A., K. S. Shing, V. Venkatasubramanian, and K. E. Gubbins. 1983. Molecular Thermodynamics of Dilute Solutions in Supercritical Solvents. Chemical Engineering at Supercritical Conditions Ann Arbor Science: 221-244.

Kajimoto, O., M. Futakami, T. Kobayashi, K. Yamasak. 1988. Charge-Transfer-State Formation in Supercritical Fluid: (N,N-Dimethylamino)benzonitrile in CF_3H. J. Phys. Chem. 92: 1347-1352.

Kim, S. and K. P. Johnston. 1987a. Molecular Interactions in Dilute Supercritical Fluid Solutions I&EC Res. 26: 1206-1213.

Kim, S. and K. P. Johnston. 1987b. Clustering in Supercritical Fluid Mixtures. AIChE J. 33: 1603-1611.

Kirkwood, J. G. and F. P. Buff. 1951. The Statistical Theory of Solutions I. J. Chem. Phys. 19: 774-777.

Kojima, K., T. Kato, and H. Nomura. 1984. Estimation of the Parameters in the Kirkwood-Buff Theory of Solution using Percus-Yevick Fluid Mixtures. J. Solution Chem. 13: 151-165.

Kreglewski, A. 1984. Equilibrium Properties of Fluids and Fluid Mixtures Texas A&M University Press: 76.

Labik, S., A. Malijevsky, and P. Vonka. 1985. A Rapidly Convergent Method of Solving the OZ Equation. Mol. Phys. 56: 709-715.

Lado, F. 1973. Perturbation Correction for the Free Energy and Structure of Simple Fluids. Phys. Rev. A 87: 2548-2552.

Lee, L. L. 1988. Application of the Zero-Separation Theorems of Molecular Correlation Functions to the Chemical Potentials in Supercritical Mixtures. Proc. Int. Symp. on Supercritical Fluids , Nice, M. Perrut, Ed.: 207-215.

Lee, L. L. and K. S. Shing 1989. A Test Particle Approach to the Zero Value Theorems of Molecular Correlation Functions. J. Chem. Phys. 91: 477-488.

Leonard, R. E. 1981. Supercritical Process for Producing Deasphalted, Demineralized, and Deresined Oils. U. S. Patent 4,290,880.

McGuigan, D. B. and P. A. Monson. 1987. Vapor-Liquid Critical Lines Predicted by Distribution Function Theories Mol. Phys. 62: 3-18.

McGuigan, D. B. and P. A. Monson. 1989. Analysis of Infinite Dilution Partial Molar Volumes Using a Distribution Function Theory. Fluid Phase Equilib.: submitted.

Mansoori, G. A. and T. W. Leland. 1972. Statistical Thermodynamics of Mixtures. Trans. Far. Soc. II 68: 320-344.

Modell, M. 1977. Process Using a Supercritical Fluid for Regenerating Synthetic Organic Polymeric Adsorbents and Waste Water Treatment Embodying the Same. U. S. Patent 4,061,566.

Münster, A. 1965. Critical Fluctuations. Fluctuation Phenomena in Solids, R. E. Burgess, Ed., Academic Press: 179.

Peng, D.-Y. and D. B. Robinson. 1976. A New Two Constant Equation of State. I&EC Fund. 15: 59-64.

Petsche, I. B. and P. G. Debenedetti. 1989. Solute-Solvent Interactions in Infinitely Dilute Supercritical Mixtures: A Molecular Dynamics Investigation. J. Chem. Phys. 91: 7075-7084.

Pfund, D. M., L. L. Lee, and H. D. Cochran. 1988. Application of the Kirkwood-Buff Theory of Solutions to Dilute Supercritical Solutions II. The Excluded Volume and Local Composition Models. Fluid Phase Equilib. 39: 161-192.

Pfund, D. M., L. L. Lee, and H. D. Cochran. 1989a. Chemical Potential Prediction in Realistic Fluid Models with Scaled Particle Theory Int. J. Thermophys. 11: 73-86.

Pfund, D. M., H. D. Cochran, and L. L. Lee. 1989b. Chemical Potentials from Integral Equations. AIChE Annual Meeting, San Francisco, paper 42e.

Prausnitz, J. M., R. N. Lichtenthaler, and E. E. Avezedo. 1986. Molecular Thermodynamics of Fluid Phase Equilibria, 2nd Ed., Prentice-Hall: 171.

Renon, H. and J. M. Prausnitz. 1968. Local Compositions in Thermodynamic Excess Functions for Liquid Mixtures. AIChE J. 14: 135-144.

Rosset, R. 1988. New Results in Supercritical Fluid Chromatography. Proc. Int. Symp. on Supercritical Fluids, Nice, M. Perrut, Ed.: 357-359.

Rozelius, W., O. Vitzthum, and P. Hubert. 1974. Method for the Production of Caffein-Free Extract. U.S. Patent 3,843,824.

Stalkup, F. I. 1983. Status of Miscible Displacement. J. Pet. Tech. 35: 815.

Subramaniam, B. and M. A. McHugh. 1986. Reactions in Supercritical Fluids — A Review. I&EC Proc. Des. and Dev. 25: 1-12.

Suzuki, Y., M. Shimazu, K. Arai, and S. Saito. 1986. Supercritical Extraction of Fish Oil and Fractionation of Methyl Esters with CO_2. AIChE Annual Meeting, Miami.

Tsekhanskaya, Y. V., M. B. Iomtov, and E. V. Mushkina. 1964. Solubility of Naphthalene in Ethylene and Carbon Dioxide under Pressure. Russ. J. Phys. Chem. 38: 1173-1176.

Vitzthum, O., P. Hubert, and W. Sirtl. 1978. Production of Hops Extracts. U. S. Patent 4,104,409.

Watts, R. O. 1968. Percus-Yevick Equation Applied to a Lennard-Jones Fluid. J. Chem. Phys. 48: 50-55.

Watts, R. O. 1969. Hypernetted-Chain Approximation Applied to a Modified Lennard-Jones Fluid J. Chem. Phys. 50: 1358-1365.

Wu, R.-S., L. L. Lee, and H. D. Cochran. 1989. The Structure of Dilute Supercritical Solutions: Clustering of Solvent and Solute Molecules and the Thermodynamic Effects. I&EC Res.: submitted.

Yonker, C. R., S. L. Frye, D. R. Kalkwarf, and R. D. Smith. 1986. Characterization of Supercritical Fluid Solvents Using Solvatochromic Shifts. J. Phys. Chem. 90: 3022-3026.

Zerah, G. and J. P. Hansen. 1986. Self-Consistent Integral Equations for Fluid Pair Distribution Functions: Another Attempt J. Chem. Phys. 84: 2336-2343.

Zosel, K. 1974. Process for Recovering Caffein. U. S. Patent 3,806,619.

New Fluctuation Theory Approaches for Calculation of Total, Partial Molar Properties, and Phase Equilibria of Mixtures

E. Z. Hamad and G. A. Mansoori
Department of chemical Engineering
University of Illinois, Chicago, IL 60680

ABSTRACT

This paper consists of the review of the contributions of the authors and their collaborators on the development of new fluctuation theory approaches for calculation of total, partial molar properties, and phase equilibria of mixtures. These contributions consist of three parts:

(i) Development of new mixing rules based on the fluctuation and conformal solution theories. The resulting mixing rules are shown to be applicable for calculation of properties of molecular fluids with varying size and interaction energy differences.

(ii) Development of a new fluctuation theory technique for derivation of analytic expressions for chemical potentials and activity coefficients of components of mixtures. This new technique is based upon the newly developed exact relations among the mixture direct correlation function integrals and the closures for unlike-interaction direct correlation function integrals. It is shown that vapor-liquid and liquid-liquid phase equilibria of mixtures can be predicted accurately by this technique. It is also demonstrated that this technique is capable of predicting phase splitting in liquid mixtures quite accurately.

(iii) Introduction of the concept of local particle numbers and compositions with respect to fluctuation integrals. This has also resulted in a rigorous definition of the radius of the sphere of influence of local compositions which can be utilized in order to specify this radius at different thermodynamic state conditions.

One major advantage of these new fluctuation theory approaches is the fact that it is now possible to produce analytic expressions for total and partial molar properties of mixtures at, both, infinite dilution and finite concentrations. Also because of their independence from the nature of intermolecular interaction potentials in solutions these approaches are applicable for mixtures consisting of complex molecules with or without intermolecular polar and association forces.

INTRODUCTION

One of the interesting problems in statistical mechanics is the prediction of mixture properties from those of the pure components comprising the mixture. There exist a number of techniques to do this. The conformal solution theory (Brown 1957; Massih and Mansoori 1983) has offered a practical and rather accurate approach to this problem. In this theory it is assumed there exists a hypothetical pure fluid that has the same properties as those of the mixture at the same density and temperature. This hypothetical pure fluid is described by the same equation of state as the pure fluids which form the mixture. The parameters of the hypothetical pure fluid are related to the parameters of the components of the mixture, composition, and possibly temperature and density by what are usually called the mixing rules. Many mixing rules have been suggested in the literature. Some are based on empirical grounds, and others are derived by making certain assumptions about the mixture radial distribution functions (Mansoori and Ely 1985; Mansoori 1986; Mansoori and Leland 1972). The latter approach will be used here to derive mixing rules using the fluctuation theory of mixtures.

By joining the concepts of the conformal solution theory and the fluctuation theory of mixtures it is possible to derive mixing rules which can be then used for mixture property calculations (Hamad et al, 1987). Two sets of calculations using these mixing rules are performed and reported here: In the first set an equimolar mixture of hard-spheres is studied. The deviations of the Helmholtz free energy, the compressibility factor and the isothermal compressibility from the exact hard-sphere mixture properties are calculated as a function of reduced density at three different molecular size ratios. In the second set of calculations the mixing rules are used to derive the Henry's constant for a binary mixture of Lennard-Jones fluid. The Henry's constant is calculated for: (i) The same energies of interaction ε_{ij}, but at different size ratios; (ii) The same molecular sizes σ_{ij}, but at different energies of interaction. The calculations are compared with the results of the van der Waals mixing rules and the computer simulation data.

Definition of activity coefficients provides a convenient way of describing the non-idealities encountered in solution thermodynamics (Guggenheim 1952; Chao and Greenkorn 1975; Kreglewski 1984). The activity coefficient of component i, γ_i, in a mixture is defined by the following relation

$$\ln (x_i \gamma_i) = (\mu_i - \mu_{io})/kT \qquad (1)$$

where μ_i is the chemical potential of component i in the mixture, μ_{io} is the standard state chemical potential of i, k is the Boltzmann constant and T is the absolute temperature. There exist very limited theoretical understanding through which one can derive expressions for activity coefficients using the principles of statistical mechanics (Haile and Mansoori 1983). As a result, most of the available analytic expressions for activity coefficients are empirical or semi-empirical which are based on classical thermodynamic correlations of excess properties.

Lattice theories of statistical mechanics were successful in providing us with analytic expressions for activity coefficients (Guggenheim 1952)`. There exist numerous group-contribution expressions for activity coefficients which are based on the lattice theories of statistical mechanics. Due to the high degree of combinatorial approximations which are used in the lattice theories, their resulting activity coefficient expressions have met with limited success.

Another statistical mechanical route for the development of analytic expressions of activity coefficients is the use of the Gibbs, or NPT, ensemble (Mansoori 1980). The Gibbs ensemble, joined with the conformal solution theory of statistical mechanics, have provided us with the formalism in systematically deriving analytic expressions of activity coefficients. The success of this technique is presently limited to simple molecular fluid mixture for which the intermolecular potential energy functions are available. In the case of polar mixtures and mixtures possessing hydrogen-bonding species the route through the Gibbs ensemble is subject to further investigation.

The third statistical mechanical route to activity coefficients is the fluctuation solution theory of statistical mechanics (Kirkwood and Buff 1951). By utilizing this theory, activity coefficients in solutions are obtained from the relations between the chemical potentials, μ_i, and the fluctuation integrals, G_{ij}'s. According to this theory the expression for the chemical potential of component i, μ_i, in a binary mixture takes the following form:

$$x_i\,[\partial\mu_i/\partial x_i]_{P,T} = kT/[1+x_1x_2\rho\,(G_{11}+G_{22}-2G_{12})] \tag{2}$$

where

$$G_{ij} = \int_0^\infty [g_{ij}(r)-1]4\pi r^2 dr\,, \tag{3}$$

x_i is the mole fraction of component i, P is the pressure, ρ is the density and $g_{ij}(r)$ is the radial distribution function of the pair of molecules i and j. Expression for $[\partial\mu_i/\partial x_i]_{P,T}$ are also written in terms of direct correlation integrals

$$x_i\,[\partial\mu_i/\partial x_i]_{P,T} = kT\;\frac{1 - x_1\rho C_{11} - x_2\rho C_{22} + x_1x_2\rho^2(C_{11}C_{22}-C_{12}^2)}{1-\sum\sum x_i x_j \rho\, C_{ij}}\,, \tag{4}$$

where

$$C_{ij} = \int_0^\infty c_{ij}(r)4\pi r^2 dr \tag{5}$$

and the direct correlation function, $c_{ij}(r)$, is defined with respect to the radial distribution function, $g_{ij}(r)$, in an n-component mixture by

$$c_{ij}(r_{12}) = [g_{ij}(r_{12})-1] - \sum_{k=1}^{n} x_k \rho \int_V c_{ik}(r_{13})[g_{jk}(r_{23})-1]dr_3. \tag{6}$$

Finally, the activity coefficient, γ_i, of component i is obtained by integrating Eqs. (2) or (4) with respect to the mole fraction. Equation (1) then becomes

$$\ln \gamma_i = \int_{x_i}^{1} (1-x_i)\rho\,(G_{11}+G_{22}-2G_{12})\,[1+x_1x_2\rho(G_{11}+G_{22}-2G_{12})]^{-1}\,dx_i \tag{7}$$

This equation, in principle, can be used to calculate the activity coefficients provided information about G_{ij} integrals are available. One major advantage of this formalism is the fact that there is no need for information about the intermolecular potential energy functions of species of the mixture as it is the case for other mixture theories. However, the lack of sufficient knowledge about G_{ij} (or C_{ij}) integrals and their relationships has limited the utility of this theory. For mixtures of species with model intermolecular potential energy functions it has been possible to calculate G_{ij} (or C_{ij}) integrals and as a result the activity coefficients from statistical mechanics. Eq. (7) has had also some utility for complex solutions (mixtures consisting of highly asymmetric, polar, and hydrogen-bonding species) only at infinite dilutions.

The fourth statistical mechanical route to activity coefficients is a new fluctuation theory introduced recently (Hamad and Mansoori 1989) that provides us with a strong tool for developing analytic expressions for activity coefficients of multi-component mixtures using Eq. (7) and a number of newly discovered characteristics of direct correlation function integrals in mixtures. This theory is shown to be applicable to complex multicomponent mixtures at finite concentrations as well as solutions at infinite dilutions. It has also been applied for phase equilibrium (vapor-liquid and liquid-liquid equilibrium) of mixtures with success.

There has been a great deal of interest during the past few decades in the concept of local composition and macroscopic thermodynamic properties of mixtures (Wilson 1964; Renon and Prausnitz 1968; Prausnitz 1969; McDermott and Ashton 1977; Nakanishi, et al. 1979, 1982, 1983; Mollerup 1981; Kemeny and Rassmussen 1981; Whiting and Prausnitz 1982; Fischer 1983; Mathias and Copeman 1983; Hu et al. 1983; Lee et al. 1983; Gierycz and Nakanashi 1984; Hoheisel and Kohler 1984; see also Rubio et al's chapter in this book). These analyses have been based on lattice theories of fluids, distribution functions, computer simulation studies, and empirical interpretation of local composition. In a more recent report it is demonstrated that the fluctuation theory can be used to interpret the concept of local compositions from statistical mechanical point of view (Mansoori and Ely 1985). In this work it is also shown how common liquid theory approximations such as the van der Waals one-fluid theory can lead to different local composition expressions. The statistical mechanical expressions, dealing with the incorporation of the local composition concept in a theoretically rigorous manner are introduced using explicit relation for isothermal compressibility in terms of the local particle numbers. The isothermal compressibility of a mixture is given by (Kirkwood and Buff 1951)

$$\kappa_T = |\underline{\underline{B}}| \,/\, \{\rho^2 kT \sum_i \sum_j x_i\, x_j\, |\underline{\underline{B}}|_{ij}\} \tag{8}$$

where $\underline{\underline{B}}$ is a n x n matrix with its elements $B_{ij}=\rho x_i[\delta_{ij}+x_j\rho G_{ij}]$ and $|\underline{\underline{B}}|_{ij}$ is the cofactor of of the element B_{ij} in the determinant $|\underline{\underline{B}}|$, δ_{ij} is the Kronecker delta, and G_{ij} is the fluctuation integral as defined by Eq. (3)

Different phenomenological and statistical mechanical approximations to local compositions are presented and compared and expressions for the partial molar volume and chemical potential of a mixture component in terms of local particle numbers are derived (Mansoori and Ely 1985). In the present state of the art review the part of the paper of Mansoori and Ely which is related to the fluctuation theory interpretation of the local composition concept will be discussed.

FLUCTUATION THEORY OF MIXING RULES

As it is shown in the first paper in this monograph the total thermodynamic properties of fluids are related to the radial distribution function by the energy, virial and compressibility equations of statistical mechanics. The statistical mechanical equation for the internal energy and virial equations require the assumption of a form for the intermolecular potential energy function. This is not the case with the compressibility equation, which is independent of this assumption. The statistical mechanical compressibility equation for pure fluids and mixtures are also reported in the first paper of this monograph.

To derive mixing rules (Mansoori and Ely 1985; Hamad, et al. 1987) from energy and virial equations we equate the internal energy and compressibility factor of the mixture to the same properties of the hypothetical pure fluid. Then, we scale the resulting equations using the respective size parameters to get the following set of two equations (Mansoori 1986):

$$\sigma_x^3 \varepsilon_x \int_0^\infty \varepsilon_x(r/\sigma_x) g_x(r/\sigma_x) 4\pi (r/\sigma_x)^2 d(r/\sigma_x)$$

$$= \sum_{i=1}^{n}\sum_{j=1}^{n} \sigma_{ij}^3 \varepsilon_{ij} x_i\, x_j \int_0^\infty \phi_{ij}(r/\sigma_{ij}) g_{ij}(r/\sigma_{ij}) 4\pi (r/\sigma_{ij})^2 d(r/\sigma_{ij}) \tag{9}$$

and

$$\sigma^3_x\varepsilon_x\int_0^\infty \phi'_x(r/\sigma_x)g_x(r/\sigma_x)4\pi(r/\sigma_x)^3 d(r/\sigma_x)$$

$$=\sum_{i=1}^{n}\sum_{j=1}^{n}\sigma_{ij}^3\varepsilon_{ij}x_i x_j\int_0^\infty \phi'_{ij}(r/\sigma_{ij})g_{ij}(r/\sigma_{ij})4\pi(r/\sigma_{ij})^3 d(r/\sigma_{ij}) \tag{10}$$

In the above two equations the subscript x refers to the hypothetical pure fluid, and $\phi=u/\varepsilon$ where u is the pair intermolecular potential and $-\varepsilon$ is the value of this potential at its minimum. At this stage let us assume conformality of the intermolecular potential energy functions of species of the mixture and the hypothetical pure fluid in the following form

$$u_{ij} = (\varepsilon_{ij}/\varepsilon_{oo})\, u_{oo}(r\sigma_{ij}/\sigma_{oo}) \tag{11}$$

where the subscript oo indicates a pure reference fluid. In addition to assuming conformal potentials, we assume conformality of the radial distribution functions, g_{ij}, of the species of the mixture and the hypothetical pure fluid. According to the latter assumption, the scaled radial distribution functions in the mixture and the hypothetical pure fluid are all identical (Mansoori 1986), i. e.

$$g_{11}(r/\sigma_{11}) = g_{22}(r/\sigma_{22}) =...= g_{ij}(r/\sigma_{ij}) =...= g_{xx}(r/\sigma_{xx}) \tag{12}$$

The above two assumptions will make the integrals in the two sides of Eq.'s (9) and (10) equal, and hence, they will cancel from both sides of these equations. Finally Eq.'s (9) and (10) will reduce to the following single equation

$$\sigma_x^3\varepsilon_x = \sum_{i=1}^{n}\sum_{j=1}^{n}\sigma^3_{ij}\varepsilon_{ij}x_i x_j \tag{13}$$

In the above derivations it is assumed that the molecular interaction potential function has two parameters, ε and σ. Equation (13) gives one relation between the parameters of the hypothetical pure fluid (σ_x & ε_x) and the parameters (σ_{ij} & ε_{ij}) and compositions (x_i) in the mixture. To define σ_x and ε_x uniquely we need another relation between (σ_x & ε_x) and (σ_{ij}, ε_{ij} & x_i). This second expression can be obtained by using the conformal solution approximation for g_{ij} in the fluctuation integral G_{ij} which can be written in the following form

$$G_{ij} = \sigma_{ij}^3\int_0^\infty [g_{ij}(r^*)-1]4\pi\, r^{*2}dr^* = (\sigma_{ij}/\sigma_x)^3\, G_x \tag{14}$$

where $r^*=r/\sigma_{ij}$. Let us also scale the hypothetical pure fluid isothermal compressibility equation in the following form

$$\rho kT\kappa_{Tx} = 1 + \rho\sigma_x^3\int_0^\infty [g(r^*)-1]4\pi r^{*2}dr^* = 1 + \rho(\sigma_x/\sigma_{ij})^3 \; G_{ij} \tag{15}$$

This leads to the relation:

$$G_{ij} = \Delta_x(\sigma_{ij}/\sigma_x)^3 \;/\rho \tag{16}$$

where

$$\Delta_x = \rho kT/\kappa_{Tx} - 1. \tag{17}$$

By replacing G_{ij} as defined by Eq. (16) in the expression for the mixture isothermal compressibility as given by Eq. (8) ands derived in the first chapter of this monograph we will derive the following expression

$$\rho\, kT\kappa_{Tx} = \; |\underline{\underline{B}}| \;/\; \rho \textstyle\sum_i \sum_j x_i\; x_j\; |\underline{\underline{B}}|_{ij} \tag{18}$$

where the elements of the determinant $|\underline{\underline{B}}|$ are given by

$$B_{ij} = \rho\, x_i\; [\,\delta_{ij} + \; x_j \Delta_x\, (\sigma_{ij}/\sigma_x)^3\,] \tag{19}$$

Equation (18), which is the second needed relation between (σ_x & ε_x) and (σ_{ij}, ε_{ij} & x_i; i,j=1,2,...,n) is an implicit equation between σ_x, which is also dependent on temperature and density, and σ_{ij} (i,j=1,2,...,n). Equations (13) and (18) form the necessary set of mixing rules to relate σ_x and ε_x to the properties of the mixture. Note that Eq. (13) has the same form as the first (attractive) van der Waals mixing rule (Mansoori and Ely 1985; Mansoori 1986). However, the difference between the two mixing rules comes from Eq.(18) which is also density and temperature dependent.

It is important to realize that since the mixing rule reported as Eq. (18) is temperature and density dependent and it is derived from the compressibility equation it should only be used to calculate the mixture isothermal compressibility κ_T. Any other thermodynamic property can then be calculated from the appropriate thermodynamic relation between that property and κ_T. Note that such limitations will not exist for mixing rules which are independent of density and temperature.

As an example, let us write Eq's (13) and (18) for a two component mixture. In this case Eq.(13) will take the following form

$$\sigma_x^3\varepsilon_x = x_1^2\sigma_{11}^3\varepsilon_{11}+2x_1x_2\sigma_{12}^3\varepsilon_{12}+x_2^2\sigma_{22}^3\varepsilon_{22} \tag{20}$$

The determinant $|\underline{\underline{B}}|$ in Eq.(18) reduces to the following 2x2 determinant

$$|\underline{\underline{B}}| = \rho \begin{vmatrix} x_1+x_1^2\Delta_x(\sigma_{11}/\sigma_x)^3 & x_1x_2\Delta_x(\sigma_{12}/\sigma_x)^3 \\ x_1x_2\Delta_x(\sigma_{12}/\sigma_x)^3 & x_2+x_2^2\Delta_x(\sigma_{22}/\sigma_x)^3 \end{vmatrix} \tag{21}$$

Substituting Eq.(18) in Eq.(21) and solving for σ_x^3 we get

$$\sigma_x^3 = (1/2)\Big[\sigma_{x\,(vdw)}^3 -(\sigma_{11}^3-2\sigma_{12}^3+\sigma_{22}^3)x_1x_2\Delta_x + \{[\sigma_{x\,(vdw)}^3 -(\sigma_{11}^3-2\sigma_{12}^3+\sigma_{22}^3)x_1x_2\Delta_x]^2+4(\sigma_{11}^3\sigma_{22}^3-\sigma_{12}^3)x_1x_2\Delta_x\}^{1/2}\Big] \tag{22}$$

where $\sigma_{x\,(vdw)}^3$ is the second (respulsive) van der Waals mixing rule for molecular diameters in a binary mixture

$$\sigma_{x\,(vdw)}^3 = x_1^2\sigma_{11}^3+2x_1x_2\sigma_{12}^3+x_2^2\sigma_{22}^3 \tag{23}$$

Equations (20) and (22) constitute the fluctuation theory mixing rules for binary mixtures while Eq.'s (20) and (23) constitute the van der Waals mixing rules for binary mixtures. These sets of mixing rules will be used later to calculate properties of hard-sphere mixtures and Lennard-Jones mixtures.

CHEMICAL POTENTIALS AND ACTIVITY COEFFICIENTS

A: <u>THE CASE OF P, T, & x AS THE INDEPENDENT VARIABLES.</u>

In deriving chemical potentials and activity coefficients from the fluctuation theory of mixtures when P, T, & x (pressure, temperature, and composition) are the independent variables, we consider a single-phase multi-component mixture consisting of n components. Chemical potential, μ_i, of every species in this mixture is generally a function of n+1 independent variables, i.e.

$$\mu_i = \mu_i\,(T, P, x_1, x_2, \ldots, x_{n-1})$$

Using this expression one can write the following expressions between the mixed second derivatives of the chemical potential (Hamad and Mansoori 1989):

$$\partial[\partial\mu_i/\partial x_j]/\partial P = \partial[\partial\mu_i/\partial P]\partial x_j = [\partial\overline{v}_i/\partial x_j]_{P,T} \qquad i,j=1,\ldots,n \tag{24}$$

$$\partial[\partial\mu_i/\partial x_j]/\partial x_k = \partial[\partial\mu_i/\partial x_k]/\partial x_j \qquad i,j,k=1,\ldots,n \text{ and } j\neq k \tag{25}$$

In the above equations, $\partial\mu_i/\partial x_j$ and partial molar volume, $\overline{v}_i$, are related to the direct correlation function integrals by (Pearson and Rushbrooke 1957; O'Connell 1981)

$$N_i(\partial\mu_i/\partial N_j)_{T,P,N_{k\neq j}} = kT\Big[\delta_{ij}-x_i\Big\{1+\rho C_{ij}-\rho\sum_{k=1}^{n}x_k(C_{ik}+C_{jk})+\rho^2\sum_{k=1}^{n}\sum_{\ell=1}^{n}x_kx_\ell(C_{ik}C_{j\ell}-C_{ij}C_{k\ell})\Big\}$$

$$/\{1-\rho\sum_{k=1}^{n}\sum_{\ell=1}^{n}x_k x_\ell C_{k\ell}\}]\tag{26}$$

and

$$\overline{v}_i = (\rho^{-1})(1-\rho\sum_{j=1}^{n}x_j C_{ij})/(1-\rho\sum_{j=1}^{n}\sum_{k=1}^{n}x_j x_k C_{jk}).\tag{27}$$

By replacing Eq.s (26) and (27) in Eq.s (24) and (25) we will derive a number of expressions between C_{ij}'s of the mixture.

In an n-component mixture there exist n(n+1)/2 direct correlation function integrals which are considered as unknowns in Eq.s (24) and (25). Note that there are n(n-1) relations of the type of Eq. (24) and n(n-1)(n-2)/2 relations of the type of Eq. (25). Of all these equations only n(n+1)/2-1 of them are independent. Since there are n(n+1)/2 unknowns then one additional relation will be needed to determine all the unknowns regardless of the number of components involved. The resulting equations for direct correlation function integrals consist of a set of non-linear partial differential equations. General analytic solution of this set of equations for multi-component mixtures does not seem feasible at the present time. Numerical solution of these equations is involved and lengthy and may require the use of multi-variable finite difference techniques. In what follows we report the analytic solution of these equations for mixtures at low to moderate pressures.

A1: Analytic Solutions for Low to Moderate Pressures: For liquids at low to moderate pressures it may be assumed that the activity coefficient is not dependent on pressure. In mathematical form using Eq. (1) this means

$$[\partial \ln\gamma_i/\partial P]_{x_i,T} = [\partial(\mu_i - \mu_{i0})/\partial P]_{x_i,T}/kT = 0$$

As a result of this approximation Eq. (24) reduces to $[\partial\overline{v}_i/\partial x_j]_{P,T}=0$, or $\overline{v}_i=v_i$, where v_i is the molar volume of pure component i. Then Eq. (27) can be written in the following form:

$$\rho\sum_{j=1}^{n}x_j C_{ij} = 1-v_i\rho(1-\rho\sum_{i=1}^{n}\sum_{j=1}^{n}x_i x_j C_{ij}) \qquad i=1,\ldots,n\tag{28}$$

where v_1 is the molar volume of pure component i. By considering that

$$\rho(1-\sum\sum x_i x_j\rho C_{ij}) = (kT\kappa_T)^{-1}\tag{29}$$

Eq. (28) then becomes

$$\sum_{j=1}^{n} x_j \rho C_{ij} = 1 - v_i / (kT\kappa_T) \qquad i=1,\ldots,n \tag{30}$$

where the mixture isothermal compressibility is now given by

$$\kappa_T = -\sum x_i (\partial v_i / \partial p) \Big/ \sum x_i v_i \tag{31}$$

The set of relations in Eq. (25) do not simplify upon making the assumption of low pressure. To overcome this difficulty and to provide the one additional relation needed, the following closure relation is assumed to be valid for all the cross direct correlation function integrals (C_{ij}, $i \neq j$) (Hamad 1988; Hamad, et al. 1989):

$$C_{ij} = \alpha_{ji} C_{ii} + \alpha_{ij} C_{jj} \qquad i,j = 1,\ldots,n \text{ and } i \neq j \tag{32}$$

where α_{ji} and α_{ij} are binary interaction parameters ($\alpha_{ji} \neq \alpha_{ij}$). In an earlier publication (Hamad, et al. 1989) it is shown that this form of the closure expression is quite satisfactory in representing the cross direct correlation function integrals of varieties of non-polar, polar, and hydrogen-bonding binary mixtures. Substituting Eq. (32) in Eq. (30) gives the following set of linear equations:

$$\begin{array}{l} \left(\sum x_j \alpha_{j1}\right) \rho C_{11} + x_2 \alpha_{12} \rho C_{22} + \ldots + x_n \alpha_{1n} \rho C_{nn} = 1 - v_1/(kT\kappa_T) \\ x_1 \alpha_{21} \rho C_{11} + \left(\sum x_j \alpha_{j2}\right) \rho C_{22} + \ldots + x_n \alpha_{2n} \rho C_{nn} = 1 - v_2/(kT\kappa_T) \\ \cdots\cdots\cdots\cdots\cdots\cdots\cdots\cdots\cdots \\ \cdots\cdots\cdots\cdots\cdots\cdots\cdots\cdots\cdots \\ \cdots\cdots\cdots\cdots\cdots\cdots\cdots\cdots\cdots \\ x_1 \alpha_{n1} \rho C_{11} + x_2 \alpha_{n2} \rho C_{22} + \ldots + \left(\sum x_j \alpha_{jn}\right) \rho C_{nn} = 1 - v_n/(kT\kappa_T) \end{array} \tag{33}$$

where summations run from j=1->n, and $\alpha_{ii}=1$. The above relations represent a set of n linear equations in the unknowns ρC_{ii}. To solve for ρC_{ii} it is more convenient to use matrix notations:

$$\underline{\underline{A}} \cdot \underline{C} = \underline{b} \tag{34}$$

where $\underline{\underline{A}}$ is a matrix with elements $A_{ii} = \sum x_\ell \alpha_{\ell i}$, $A_{ij} = x_j \alpha_{ij}$ and $\underline{b}$ and $\underline{C}$ are vectors with elements $b_i = 1 - v_i / kT\kappa_T$ and $C_i = \rho C_{ii}$, respectively. The solution of Eq. (34), is:

$$\rho C_{kk} = \sum_{i=1}^{n} |\underline{\underline{A}}|_{ik} b_i \Big/ |\underline{\underline{A}}| \tag{35}$$

where $|\underline{\underline{A}}|_{ik}$ is the cofactor of the element A_{ik} in the determinant $|\underline{\underline{A}}|$.

By using the direct correlation function integrals calculated above one can derive the expression for the activity coefficients of a species in a multi-component mixture in the following form

$$N(\partial \ln\gamma_k/\partial N_k) = 1-\rho C_{kk}+v_k^2\,[kT\Sigma x_i\,(\partial v_i/\partial p)_T]^{-1} \tag{36}$$

As an example of application of this theory, it will be used for derivation of activity coefficients and vapor-liquid and liquid-liquid equilibria calculation of binary mixtures.

B: <u>THE CASE OF T & ρ_i (i=1,2,...,n) AS THE INDEPENDENT VARIABLES</u>

In deriving chemical potentials and activity coefficients from the fluctuation theory of mixtures when T & ρ_i (i=1,2,...,n) (temperature and component densities) are the independent variables we consider a system which is described by the following set of independent variables $\{T, \rho_1, \rho_2, \ldots, \rho_n\}$, where T is the absolute temperature, $\rho_i = N_i/V$, N_i is the number of molecules i in the system, V is the system volume, and n is the number of components in the mixture. In the grand canonical ensemble the following expression for $\partial\mu_i/\partial\rho_j$ is derived (Pearson and Rushbrooke 1957; O'Connell 1971):

$$[\partial\mu_i/\partial\rho_j]_{T,\rho_{k\neq j}} = kT\,\{\,\delta_{ij}/\rho_i - C_{ij}\,\} \tag{37}$$

where $\rho_{k\neq j}$ stands for the set $(\rho_1, \rho_2, \ldots, \rho_n)$ excluding ρ_j. This equation is actually equivalent to Eq. (4) except for the change of variables from $\{P, T, \& x_i\,(i=1,2,...,n-1)\}$ to $\{T \& \rho_i\,(i=1,2,...,n)\}$. Exact relations can be derived among the direct correlation integrals, C_{ij}, by equating the mixed second derivatives of the chemical potentials. For a binary mixture with T, ρ_1, ρ_2 as the independent variables one has:

$$(\partial C_{12}/\partial\rho_1)_{T,\rho_2} = (\partial C_{11}/\partial\rho_2)_{T,\rho_1} \tag{38}$$

and

$$(\partial C_{12}/\partial\rho_2)_{T,\rho_1} = (\partial C_{22}/\partial\rho_1)_{T,\rho_2} \tag{39}$$

The above two equations are exact, but one needs a closure relation among C_{11}, C_{22}, and C_{12} to solve for the three integrals. Variety of closure expressions for C_{12} may be considered. In what follows we use simple geometric mean and weighted arithmetic mean closures in order to solve analytically the above set of equations for chemical potentials.

B1: <u>Geometric Mean Closure</u>: In this case we derive analytic solution for chemical potentials using the simple geometric mean closure, $C_{12}=(C_{11}C_{22})^{1/2}$. By substituting this expression for C_{12} in Eq.s (38) and (39) and using Eq. (37), we can derive the following expression for, μ_{1r}, the residual chemical potential of component 1 in a binary mixture

$$(\partial\mu_{1r}/\partial\rho_2)^2(\partial^2\mu_{1r}/\partial\rho_1^2) - 2(\partial\mu_{1r}/\partial\rho_1)(\partial\mu_{1r}/\partial\rho_2)(\partial^2\mu_{1r}/\partial\rho_1\partial\rho_2) + (\partial\mu_{1r}/\partial\rho_1)^2(\partial^2\mu_{1r}/\partial\rho_2^2) = 0 \tag{40}$$

This nonlinear partial differential equation can be solved analytically utilizing its symmetry with respect to ρ_1 & ρ_2 variables. First, we write this differential equation in the following form

$$(\partial\mu_{1r}/\partial\rho_1)(\partial R/\partial\rho_2) - (\partial\mu_{1r}/\partial\rho_2)(\partial R/\partial\rho_1) = 0 \tag{41}$$

where $R=(\partial\mu_{1r}/\partial\rho_2)/(\partial\mu_{1r}/\partial\rho_1)$. Integration of Eq. (41) results in

$$R = (\partial\mu_{1r}/\partial\rho_2)/(\partial\mu_{1r}/\partial\rho_1) = h(\mu_{1r})$$

or

$$(\partial\mu_{1r}/\partial\rho_2) = h(\mu_{1r})(\partial\mu_{1r}/\partial\rho_1) \tag{42}$$

where h is an unknown function to be determined from the pure component properties. Now, integrating Eq. (42) will produce

$$\mu_{1r} = F[\rho_2 h(\mu_{1r}) + \rho_1] \tag{43}$$

where F is also an unknown function which can be determined from the following condition

$$\mu_{1r}(\rho_2=0) = \mu_{1r,p} = kT f_1(\rho) \tag{44}$$

In this equation subscript (p) indicates a pure component property. From equation (44) we can conclude $F(t)=kTf_1(t)$ and as a result Eq. (43) will become

$$\mu_{1r}/kT = f_1[\rho_2 h(\mu_{1r}) + \rho_1] \tag{45}$$

In order to determine the functional form of h one needs first to find the chemical potential of component 2. Using the following thermodynamic relation

$$\partial\mu_{1r}/\partial\rho_2 = \partial\mu_{2r}/\partial\rho_1 \tag{46}$$

one gets

$$(\partial\mu_{2r}/\partial\rho_1) = h(\mu_{1r})\,(\partial\mu_{1r}/\partial\rho_1) = \partial H/\partial\rho_1 \qquad (47)$$

where H is defined by

$$H(\mu_{1r}) = \int h(\mu_{1r})d\mu_{1r}.$$

Integrating Eq. (47) with respect to ρ_1 gives

$$\mu_{2r} = H(\mu_{1r}) + K(\rho_1) \qquad (48)$$

The function $K(\rho_1)$ can be shown to be zero considering assumption of $C_{12}=(C_{11}C_{22})^{1/2}$ and the fact that μ_{1r} and μ_{2r} vanish at zero total density. As a result Eq. (48) becomes

$$\mu_{2r} = H(\mu_{1r}) \qquad (49)$$

At $\rho_1=0$ the chemical potential of component 2 in pure state, $\mu_{2r}(\rho_1=0)=\mu_{2r,p}=kTf_2(\rho)$, is now used to determine the function $H(\rho_{1r})$. From Eq.s (47), (48) and (49) one can show

$$h = dH/dt = f_1^{-1}(t)/f_2^{-1}(t) \qquad (50)$$

where f_1^{-1} is the inverse function so that $f(f^{-1}(t))=t$. Using the thermodynamic relation

$$\rho(\partial\mu_{ir,p}/\partial\rho) = (\partial p_{ir,p}/\partial\rho), \qquad (51)$$

the following relation is finally obtained

$$p_{2r}[f_2^{-1}(t)] = p_{1r}[f_1^{-1}(H)] \qquad (52)$$

where $p_{ir}[\rho]$ is the residual pressure of pure component i. Joining Equations (45) and (52) will give us the chemical potential according to the simple geometric mean closure, $C_{12}=(C_{11}C_{22})^{1/2}$.

B2: The Arithmetic Mean Closure: In this case the weighted arithmetic mean closure, $C_{12}=\alpha_{21}C_{11}+\alpha_{12}C_{22}$, can be used to derive analytic expressions for the chemical potentials of components in a mixture. When this closure is combined with Eq.s (37) and (38), and (39) the following linear partial differential equation for μ_{1r} is obtained:

$$\alpha_{21}(\partial^2\mu_{1r}/\partial\rho_1{}^2) - (\partial^2\mu_{1r}/\partial\rho_1\partial\rho_2) + \alpha_{12}(\partial^2\mu_{1r}/\partial\rho_2{}^2) = 0 \tag{53}$$

The solution of this differential equation depends on the quantity $(1\text{-}4\alpha_{21}\alpha_{12})$. When $(1\text{-}4\alpha_{21}\alpha_{12})$ is positive the differential equation is hyperbolic and the solution is obtained in terms of a power series expansion. When $(1\text{-}4\alpha_{21}\alpha_{12})$ is zero, Eq. (22) is parabolic with the following solution

$$\mu_{1r} = F(2\alpha_{21}\rho_1+\rho_2) + \rho_2\, H(2\alpha_{21}\rho_1+\rho_2) \tag{54}$$

where F and H are undetermined functions. The choice of $\alpha_{12}= (1/4)\alpha_{21}$ in $C_{12}=\alpha_{21}C_{11}+\alpha_{12}C_{22}$ has then the advantage of producing a closed-form solution.

As in the case of geometric mean closure, the pure component chemical potentials are used to determine the unknown functions F and H. The chemical potential of component 1 in a binary mixture then takes the following form

$$\mu_{1r} = a_{1r,p}(\rho_1+2\alpha_{21}\rho_2) + kT\left\{x_1Z_{1r,p}(\rho_1+2\alpha_{21}\rho_2)+x_2Z_{2r,p}(\alpha_1/2\alpha_{21}+\rho_2)\right\}/(x_1+2\alpha_{21}x_2) \tag{55}$$

where $a_{ir,p}$ and $Z_{ir,p}$ are the residual molar Helmholtz free energy and residual compressibility factor of pure component i, respectively. The mixture compressibility factor then takes the form:

$$Z = x_1Z_{1,p}(\rho_1+2\alpha_{21}\rho_2)+x_2Z_{2,p}(\rho_1/2\alpha_{21} +\rho_2) \tag{56}$$

The above expression reduces to the ideal mixture case when α_{21}=1/2, which is equivalent to assuming the validity of the simple (unweighted) arithmetic mean closure for C_{12}. It is interesting to note that according to Equation (56) all the mixture virial coefficients, obtained by expanding this equations in powers of density have the correct dependences on composition (quadratic for B, cubic for C, and so on).

FLUCTUATION THEORY ASPECT OF LOCAL COMPOSITIONS

According to statistical mechanics, the local structure or environment experienced by a molecule in a mixture is reflected in the n-body correlation functions which are essentially the probabilities of observing a certain configuration of n molecules. Of particular interest is the radial distribution function $g_{ij}(r,[\rho_k],T,[\sigma_{kl}],[\varepsilon_{kl}])$ which is proportional to the probability of observing a molecule of type i at a distance r from a central molecule of type j in the mixture. If one assumes spherically symmetric molecules, the radial distribution function depends upon the temperature T as well as the set of component number-densities $[\rho_k]$ and potential energy size $[\sigma_{kl}]$ and energy $[\varepsilon_{kl}]$ parameters for all pairs in the mixture.

Given these functions it is possible to calculate local particle numbers, i.e., the number of molecules of type i contained within a sphere of radius R_{ij} centered around a molecule of type j via the relation (Nakanishi 1982, 1983)

$$n_{ij}(R_{ij}) = [(N_i-\delta_{ij})/V]\int_0^{R_{ij}} g_{ij}(r, [\rho_k], T, [\sigma_{kl}], [\varepsilon_{kl}])4\pi r^2 dr \tag{57}$$

where g_{ij} is the radial distribution function for the ij pair, V is the system volume and N_i is the total number of molecules of type i in the mixture and δ_{ij} is the Kronecker delta function. Normally N_i is large compared with the δ_{ij} in Eq. (57) so that $N_i - \delta_{ij} \sim N_i$.

As mentioned above, Eq. (57) is strictly only valid for mixtures of molecules whose intermolecular potentials are spherically symmetric. Extension of these definitions to angle-dependent radial distribution functions is straightforward. According to some previous investigations (Lebowitz and Percus 1983; Massih and Mansoori 1983) one can effectively represent the properties of polar molecules through angle averaging techniques which provide an effective spherical potential. Consequently, the local composition about a central j molecule is defined in terms of Eq. (57), viz.

$$x_{ij}(R_{ij}) = n_{ij}(R_{ij}) \Big/ \sum_{k=1}^{n} n_{kj}(R_{ij}) \tag{58}$$

where n is the number of components in the mixture.

Exact relationships between the x_{ij} and x_{ji} as well as relationships between the $[x_{ij}]$ and the bulk mole fractions $[x_i]$ are not well known due to the fact that the composition dependence of the radial distribution functions in mixtures is not well understood. It is possible, however, to derive several relations between the local and bulk mole fractions and local particle numbers.

Generally R_{ij} is a function of r and T and does not have to have the same value for each binary pair. If, however, $R_{ij}=R_{ji}$ or if R_{ij} is a universal constant, then since $g_{ij} = g_{ji}$, we find

$$n_{ij}(R_{ij})/n_{ji}(R_{ij}) = x_i/x_j \tag{59}$$

$$n_{ij}(R_{ij})/n_{jj}(R_{jj}) = x_{ij}/x_{jj} \tag{60}$$

and combining (59) and (60) we find

$$n_{ii}(R_{ii})/n_{jj}(R_{jj}) = (x_{ii}/x_{jj})(x_{ij}/x_{ji})(x_j/x_i) \tag{61}$$

Finally, differentiating Eq. (57) and rearranging terms one obtains

$$g_{ij}(R_{ij}) = [V/(4\pi N_i)]\,(R_{ij}^2)\,[(dn_{ij})/(dR_{ij})] \tag{62}$$

Equation (62) can be used to express the concept of local compositions with respect to fluctuation integrals (Mansoori and Ely 1985). Isothermal compressibility $\kappa_T=(1/\rho)(\partial\rho/\partial P)_T$, of a multicomponent mixture with respect to the mixture radial distribution functions is expressed by Eq. (8). To express the isothermal compressibility, Eq. (8), with respect to the local particle numbers we write Eq. (3) in the following form (Mansoori and Ely 1985)

$$G_{ij} = \int_0^{R_{ij}} (g_{ij} - 1)4\pi r^2 dr + \int_{R_{ij}}^{\infty} (g_{ij} - 1)4\pi r^2 dr$$

Substituting Eq. (62) in the first integral and subsequent integration yields

$$G_{ij} = 1/(\rho x_i)\,[n_{ij}(R_{ij}) - \langle n_i(R_{ij})\rangle] + \int_0^{\infty} R_{ij}(g_{ij} - 1)4\pi r^2 dr \tag{63}$$

where $\langle n_i(R_{ij})\rangle = 4\pi r x_i R_{ij}^{\,3}/3$ is the average number of molecules of type i in a spherical volume of radius R_{ij}. R_{ij}, the radius of the sphere of influence of local compositions, is chosen such that the integral remaining in Eq. (63) vanishes (Mansoori and Ely 1985). The details of this choice are given in the concluding section. We find, therefore, that

$$G_{ij} = 1/(\rho x_i)\,\Delta n_{ij}(R_{ij}) \tag{64}$$

where $\Delta n_{ij}(R_{ij})=n_{ij}(R_{ij})-\langle n_{ij}(R_{ij})\rangle$. Note that for a pure fluid using Eq. (64) in Eq. (8) results in

$$\rho k T \kappa_T = 1 + \rho G = 1 + \Delta n_x(R)$$

where the subscript x denotes a pure fluid. Substituting Eq. (64) into $B_{ij}=\rho x_i[\delta_{ij}+x_j\rho G_{ij}]$ one finds

$$B_{ij} = \rho x_i[\delta_{ij} + (x_j/x_i)\Delta n_{ij}(R_{ij})] \tag{65}$$

Equations (8) and (65) together express the isothermal compressibility of a mixture with respect to the local particular numbers.

Other thermodynamic properties of interest to fluid phase equilibria calculations, which can also be expressed rigorously with respect to the local particle numbers, are the partial molar volumes and chemical potentials of the mixture components. According to the fluctuation theory (see Chapter One) the partial molar volume of component i of a n-component mixture is given by the following equation

$$\overline{v}_i = (1/\rho)\sum_{j=1}^{n} x_j |\underline{\underline{B}}|_{ij} \Big/ \left(\sum_{j=1}^{n}\sum_{k=1}^{n} x_j x_k |\underline{\underline{B}}|_{jk}\right) \tag{66}$$

where $\underline{\underline{B}}$ is a n x n matrix with its elements $B_{ij}=\rho x_i[\delta_{ij}+x_j\rho G_{ij}]$ and $|\underline{\underline{B}}|_{ij}$ is the cofactor of of the element B_{ik} in the determinant $|\underline{\underline{B}}|$, δ_{ij} is the Kronecker delta, and G_{ij} is the fluctuation integral as defined by Eq. (3). Likewise, the chemical potential of a component of the mixture can be expressed through the fluctuation theory of solutions as discussed in Chapter One in the following form

$$(\partial\mu_i/\partial N_j)_{T,P,N_{k\neq j}} = [(kT)/(\rho x_i x_j)]\,[(|\underline{\underline{\Delta}}|_{ij})/|\underline{\underline{\Delta}}|] \tag{67}$$

In this equation $|\underline{\underline{\Delta}}|$ symbolizes the determinant of the n x n matrix $\underline{\underline{\Delta}}$ elements Δ_{ij} given by

$$\Delta_{ij} = (\delta_{ij}/\rho x_i) + (1/\rho x_i) + G_{ij} + G_{11} + G_{1i} + G_{1j} \tag{68}$$

and $|\underline{\underline{\Delta}}|_{ij}$ is the cofactor of term Δ_{ij}. In terms of the local particle numbers we can write

$$\Delta_{ij} = (1/\rho x_i)\left\{\delta_{ij} + \Delta n_{ij}(R_{ij}) + (x_i/x_1)[1+\Delta n_{11}(R_{11}) - \Delta n_{1i}(R_{1i}) - \Delta n_{1j}(R_{1j})]\right\} \tag{69}$$

By replacing different local composition approximations in Eq. (69) one may solve Eq.s (66) and (69) simultaneously for the chemical potential of components in the mixture.

CALCULATIONS AND RESULTS

A: TEST OF MIXING RULES FOR HARD-SPHERE MIXTURES.

For a hard-sphere fluid mixture, Equation (9), which is a result of the energy equation, will vanish. On the other hand, Equation (10), which is a result of the virial equation, will reduce to the van der Waals (repulsive) mixing rule for σ as given by Eq.(23). Thus, we have two mixing rules for σ; one is derived from the virial equation and the other is derived from the compressibility equation. For comparison both of these mixing rules will be used to calculate mixture properties. In such calculations one needs also a pure fluid equation of state which can be joined with the mixing rules. For a pure fluid of hard-spheres the Carnahan-Starling equation of state is used here (Carnahan and Starling 1969)

$$p/kT\rho = (1+y+y^2-y^3)\big/(1-y)^3 \tag{70}$$

where $y=\pi\rho\sigma^3/6$. The quantities $(\partial p/\partial\rho)_T$ and hence Δ_x appearing in Eq. (22) can be derived from the above equation as the following:

$$(\partial p/\partial \rho)_T = kT(1+4y+4y^2-4y^3+y^4)/(1-y)^4 \tag{71}$$

$$\Delta_x = (-8y+2y^2)/(1+4y+4y^2-4y^3+y^4) \tag{72}$$

For a binary mixture, σ_x at a given ρ, T and x_i can be calculated by solving Eqs (22) and (72) simultaneously. The isothermal compressibility, κ_T, of the mixture can be calculated directly from Eq. (71) after obtaining σ_x. However, the compressibility factor Z, and the Helmholtz free energy A, of the mixture have to be calculated from the following thermodynamic relations, respectively

$$Z = p/kT\rho = 1 + [1/(kT\rho)]\int_0^{\rho}\left[[\partial p/\partial \rho]_{\sigma=\sigma_x} - kT\right]d\rho \tag{73}$$

$$A - A^\circ = \int_0^{\rho}(p-kT\rho)d\rho/\rho^2 \tag{74}$$

where A° is the ideal gas Helmholtz free energy. Equations (73) and (74) can be integrated numerically to obtain A and Z, respectively. The percentage deviations of A, Z and κ_T, as calculated using the above equations, from those predicted by the exact hard-sphere mixture equation of state, are plotted versus the mixture reduced density in Figures 1-3 for equimolar binary hard-sphere mixtures.

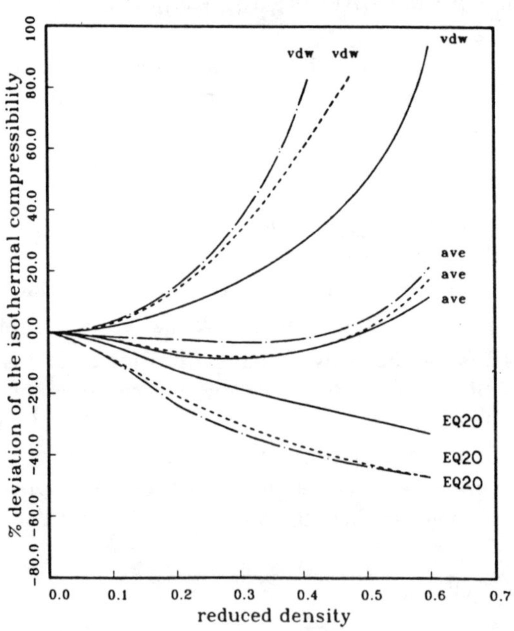

Figure 1. The percentage deviation of the isothermal compressibility of three hard-sphere mixtures from the hard-sphere equation of state (Mansoori et al., 1971) as predicted by: the van der Waals mixing rule for hard spheres, Eqs (13) & (23), (vdw); the fluctuation mixing rule for hard spheres, Eqs. (13) & (20, 22); and Eqs. (13) & (76) (ave). The curves represent: —— $\sigma_{22}/\sigma_{11}=2$, - - - - $\sigma_{22}/\sigma_{11}=4$, —·—·— $\sigma_{22}/\sigma_{11}=6$.

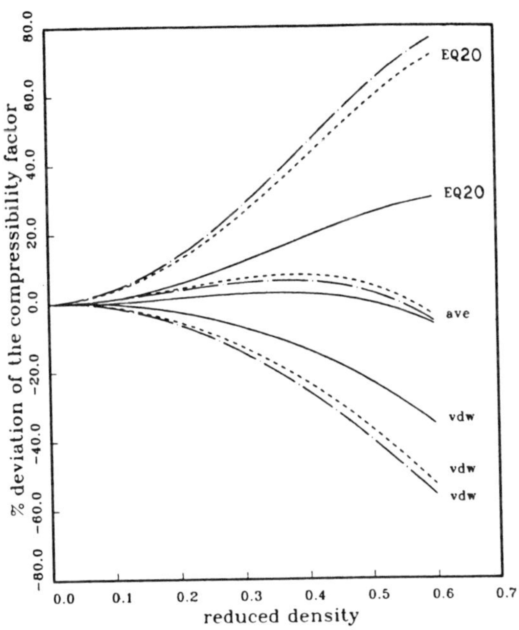

Figure 2. The percentage deviation of the compressibility factor of three hard-sphere mixtures from the hard-sphere equation of state (Mansoori, et al. 1971) as predicted by the van der Waals mixing rule for hard spheres and the fluctuation mixing rule for hard spheres. See Figure 1 for notations.

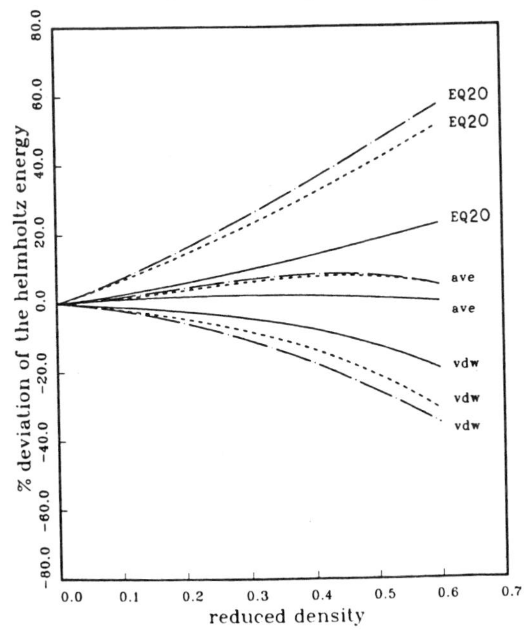

Figure 3. The percentage deviation of the Helmholtz free energy of three hard-sphere mixtures from the hard-sphere equation of state (Mansoori, et al. 1971) as predicted by the van der Waals mixing rule for hard spheres and the fluctuation mixing rule for hard spheres. See Figure 1 for notations.

The equimolar mixture reduced density, ξ, reported in Figures 1-3 is defined as (Mansoori et al. 1971)

$$\xi = \pi\rho(\sigma_{11}^3+\sigma_{22}^3)/12 \tag{75}$$

Calculations reported in Figures 1-3 were performed for three different size ratios of σ_{22}/σ_{11}=2, 4, 6 (Hamad, et al. 1987). According to these figures, as the size ratio increases the deviations from the exact mixture results increase. This reflects the approximate nature of the mixing rules. For comparison, similar calculations were repeated using the van der Waals mixing rule as given by Eq. (23). These results are also reported in Figures 1-3. Deviations based on the van der Waals mixing rule are always in the opposite direction of the deviations based on the mixing rule resulting from the fluctuation theory. The reason for the opposite deviations resulting from the two mixing rules lies in the fact that at the same density the van der Waals mixing rule predicts a lower value for σ_x than the fluctuation mixing rule. The exact value of σ_x lies somewhere in between the two calculations. The thermodynamic stability condition $[\partial p/\partial\rho]_T>0$ implies that pressure is a monotonic increasing function of the reduced density $\rho\sigma_x^3$. Therefore, one may conclude that at higher value of σ_x the pressure will be higher. Similar arguments can be made for the Helmholtz free energy and the isothermal compressibility.

From mathematical point of view it can be also concluded that the value of σ_x calculated from Eq.(22) is larger than the calculation by the van der Waals mixing rule, Eq.(23): According to Eq.(15) since $\int[g-1]r^{*2}dr \le 0$ then $\Delta_x \le 0$. Also since $(\sigma_{11}^3-2\sigma_{12}^3+\sigma_{22}^3) \ge 0$ this implies that $\sigma_x^3{}_{[Eq(22)]} \ge \sigma^3_{vdw}-(\sigma_{11}^3-2\sigma_{12}^3+\sigma_{22}^3)x_1x_2\Delta_x$ when $(\sigma_{11}^3\sigma_{22}^3-\sigma_{12}^6) < 0$ or, since $\Delta_x\le 0$, $\sigma_x^3{}_{[Eq(22)]} \ge \sigma^3_{vdw}$. The latter inequality also holds when $\Delta_x > 0$, but small compared to $(\sigma_{11}^3-2\sigma_{12}^3+\sigma_{22}^3)$, and since it can be shown that $(\sigma_{11}^3-2\sigma_{12}^3+\sigma_{22}^3)-(\sigma_{11}^3\sigma_{22}^3-\sigma_{12}^6) \ge 0$, then we conclude that $\sigma_x^3{}_{[Eq(22)]} \ge \sigma_{vdw}$. These observations suggest that if one uses an average of the two mixing rules then the result will be better than either of the two. For simplicity the arithmetic average of the two mixing rules will be used which is:

$$\sigma_x^3 = (3/4)\,\sigma_{x\,(vdw)}^3 - (1/4)\Big[(\sigma_{11}^3-2\sigma_{12}^3+\sigma_{22}^3)x_1x_2\Delta_x+\big\{[\sigma_{x\,(vdw)}^3-(\sigma_{11}^3 - 2\sigma_{12}^3+\sigma_{22}^3)x_1x_2\Delta_x]^2+4(\sigma_{11}^3\sigma_{22}^3-\sigma_{12}^6)x_1x_2\Delta_x\big\}^{1/2}\Big] \tag{76}$$

The Helmholtz free energy, the compressibility factor, and the isothermal compressibility can be calculated using the procedure discussed above. The results for the average mixing rule are shown in Figures 1-3. It is clear that Eq.(76) gives a better prediction than either Eq.(22) or the van der Waals mixing rule as reported by Eq. (23).

B: <u>HENRY'S CONSTANTS OF LENNARD-JONES FLUID MIXTURES</u>

The Henry's constant can be calculated from the fluctuation mixing rules, Eqs (13) and (22), and the van der Waals mixing rules, Eqs (13) and (23), for a mixture of Lennard-Jones (LJ) fluids. For the pure LJ fluid the equation of state proposed by Nicolas et al. (1979) will be used. To derive the Henry's constant we use the following relation

$$\ln(H_i/\rho kT) = \mu^{\infty}_{ir}/kT \tag{77}$$

where H_i is the Henry's constant for component i and μ^{∞}_{ir} is the residual chemical potential of component i in the mixture at infinite dilution. The chemical potential can be calculated from the relation

$$\mu_{ir}/kT = \int_0^{\rho} [\partial p/\partial N_i (N/kT) - \rho] d\rho/\rho^2 \tag{78}$$

To use Eqs (13) and (22) one has to write Eq.(78) in terms of the derivative $(\partial p/\partial \rho)_T$, i. e.

$$\mu_{ir}/kT = \int_0^{\rho} [(N/kT)\,\partial\{\int_0^{\rho} (\partial p/\partial \rho)_{T(\sigma=\sigma_x,\, \varepsilon=\varepsilon_x)} d\rho\}/\partial N_i - \rho] d\rho/\rho^2 \tag{79}$$

$$\mu_{ir}/kT = (N/kT)\int_0^{\rho}\Big[\int_0^{\rho}\{\partial\varepsilon_x/\partial N_i\,\partial p^*/\partial\rho + \varepsilon_x(\rho\,\partial\sigma_x/\partial N_i + \sigma_x\rho/N)\,\partial^2 p^*/\partial\rho^2$$
$$- T^*\partial^2 p^*/\partial\rho\partial T^*\,\partial\varepsilon_x/\partial N_i\}d\rho + (1/N)\int_0^{\rho}\varepsilon_x\,\partial p^*/\partial\rho\, d\rho - \rho\Big] d\rho/\rho^2 \tag{80}$$

To derive the expression for the Henry's constant from Eq. (80) one needs $\partial\sigma_x^3/\partial N_i$ and $\partial\varepsilon_x/\partial N_i$. For binary mixtures, in the limit of $x_1 \rightarrow 0$, $\partial\sigma_x^3/\partial N_1$ and $\partial\varepsilon_x/\partial N_1$ will be in the following forms

$$N[\partial\sigma_x^3/\partial N_1]_{x_1=0} = 2\sigma_{12}^3 - 2\sigma_{22}^3 + A\Delta_{22}/\sigma_{22}^3 - C\Delta_{22} \tag{81}$$

$$N[\partial\varepsilon_x/\partial N_1]_{x_1=0} = 2\varepsilon_{12}(\sigma_{12}/\sigma_{22})^3 - 2\varepsilon_{22} - (\varepsilon_{22}/\sigma_{22}^3)N\partial\sigma_x^3/\partial N_1 \tag{82}$$

In addition to the above conditions one has:

Limit of ε_x (as $x_1 \rightarrow 0$) = ε_{22} and limit of σ_x (as $x_1 \rightarrow 0$) = σ_{22}.

Using the above relations in Eq. (80) and after a lengthy, but straight forward, manipulation one obtains:

$$\mu_{1r}^{\infty}/kT = \mu_{2r}/kT+2h(f-1)u_{2r}/kT$$
$$+2(h-1)(Z_2-1)+(h-1)^2\{Z_2-1-u_{2r}/kT-F(T,\rho)\} \quad (83)$$

where μ_{2r}, u_{2r} and Z_2 are the residual chemical potential, the residual internal energy and the compressibility factor of pure component 2 at the same temperature and density of the mixture, and

$$f \equiv \varepsilon_{12}/\varepsilon_{22} \quad (84)$$

$$h \equiv (\sigma_{12}/\sigma_{22})^3 \quad (85)$$

$$F(T,\rho) = \int_0^{\rho}\left[\int_0^{\rho}(\rho\partial^2 p/\partial\rho^2+T\partial^2 p/\partial T\partial\rho)/(\partial p/\partial\rho)d\rho-\rho\right]d\rho/\rho^2 \quad (86)$$

For the van der Waals mixing rules the expression for the residual chemical potential is given by

$$\mu_{1r}^{\infty}/kT = \mu_{2r}/kT+2h(f-1)u_{2r}/kT+2(h-1)(Z_2-1) \quad (87)$$

Note that the difference between this expression and Eq. (83) is the last term of Eq. (83), the quadratic term in (h-1), which is missing from Eq. (87). In Figures 4 and 5, Eq.s (83) and (87) are compared to the simulation data for different values of f and h when $\rho\sigma_{22}^3$=0.7 and kT/ε_{22}=1.2. At these conditions Eq.s (83) and (87) become:

$$\mu_{1r}^{\infty}/kT = -1.986-7.946h(f-1)-0.485(h-1)+12.03(h-1)^2 \quad (88)$$

$$\mu_{1r}^{\infty}/kT = -1.986-7.946h(f-1)-0.485(h-1) \quad (89)$$

Figure 4 shows the simulation data and Eqs (88) and (89) for h =1, i.e. for mixture of molecules with the same sizes, but different energies of interaction. In this case fluctuation mixing rules, Eq.s (13&22), and the van der Waals mixing rules, Eq.s (13&23), predict the same results for μ_{1r}^{∞}, which turns out to be a straight line. The agreement is generally good except for the small region near f=0.

Figure 5 shows the results for f=1, i.e. the energies of interaction are the same for the two fluids, but the molecules have different sizes. In this case the agreement is only good in the region near h=1. While the van der Waals mixing rules predict a linear behavior, fluctuation mixing rules predict the correct qualitative behavior of the simulation data. The curved shape is due to the quadratic, $(h-1)^2$, term in Eq. (83) which is missing from the van der Waals expression, Eq. (87). This term can be thought of as a correction to the van der Waals expression. However, this correction is too large, and is making the prediction to go to the opposite side of the simulation data. It is worth mentioning that this simulation data (where the size ratio is different) is not very accurate as pointed out by Shing, et al. (1988), but it should give an idea about the qualitative behavior of μ_{1r}^{∞}.

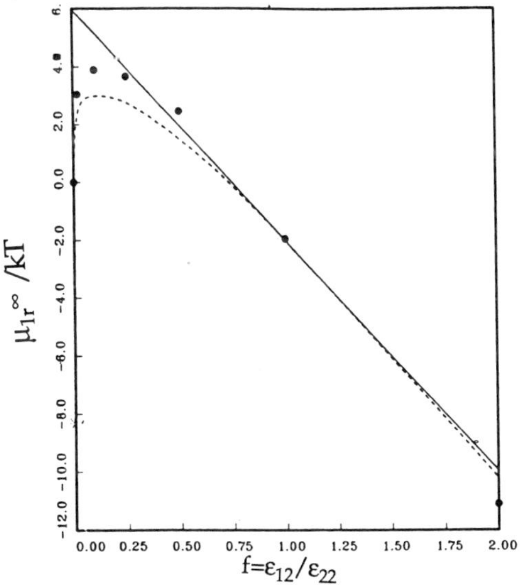

Figure 4. Variation of μ_{1r}^{∞} /kT with f=$\varepsilon_{12}/\varepsilon_{22}$. The points represent the simulation data (Shing 1982, Shing et al. 1988) and the curves represent the following : — van der Waals mixing rules, Eqs. (13 & 23), and - - - - the average mixing rules, Eqs. (13 & 76).

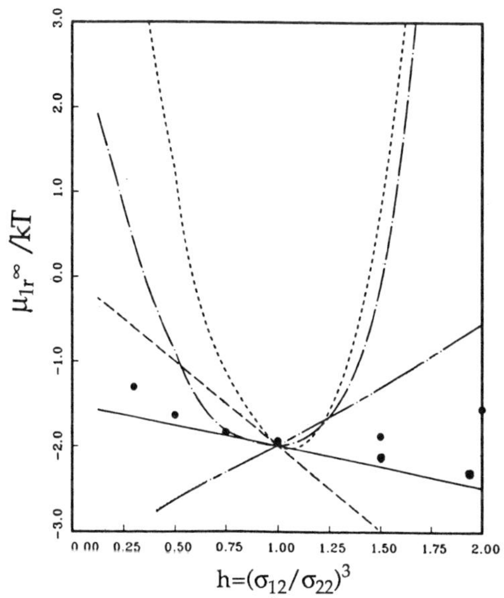

Figure 5. Variation of μ_{1r}^{∞} /kT with h=$(\sigma_{12}/\sigma_{22})^3$. The points represents the simulation data (Shing 1982, Shing et al. 1988) and the curves represent the following : — van der Waals, Eqs. (13 & 23); the fluctuation mixing rules - - - - Eqs (13 & 22); -·-·- the average mixing rules, Eqs. (13 & 76); ...... Eq. (97) with σ_x calculated from Eq. (23); —·-·— Eq. (97) with σ_x calculated from Eq. (22).

As observed in the case of hard-sphere mixture the deviations of the two mixing rules are on opposite sides from the exact values. The average mixing rule, Eq.(76), gives the following expression for μ_{1r}^{∞}

$$\mu_{1r}^{\infty}/kT = \mu_{2r}/kT+2h(f-1.25)u_{2r}/kT$$
$$+2(1.25h-1)(Z_2-1)+(h-1)^2\{Z_2-1-u_{2r}/kT-F(T,\rho)\} \quad (90)$$

This expression gives a better prediction of the actual value when h is far from unity which corresponds to the ideal mixture. The reason for this can be seen from the derivative of σ with respect to N_1 for the arithmetic-average mixing rule. This derivative is given by

$$N[\partial\sigma_x^3/\partial N_1]_{x_1=0} = (5/2)\sigma_{12}^3-2\sigma_{22}^3+(1/2)(A/\sigma_{22}^3-C)\Delta_{22} \quad (91)$$

This expression predicts a non-zero value of the derivative at equal sizes ($\sigma_{11}=\sigma_{22}$). This is the cause of the incorrect behavior near h=1. However, since the ideal mixture behavior is known ($\mu_{1r}^{\infty}/kT=\mu_{2r}/kT$), one can subtract the additional terms which show up in Eq.(90) when we put h=f=1 (the extra terms are (1/2) [Z-1- u_{2r}/kT]). The resulting expression is then:

$$\mu_{1r}^{\infty}/kT = \mu_{2r}/kT+2[h(f-1.25)+0.25]u_{2r}/kT+2.5(h-1)(Z_2-1)$$
$$+ (h-1)^2\{Z_2-1-u_{2r}/kT-F(T,\rho)\} \quad (92)$$

This equation is plotted in Figures 4 and 5. It is clear that this equation is better than Eq. (82), but still there is a need for improvement at small and large values of h.

Henry's constant can also be calculated directly (without going through mixing rules) from the fluctuation theory. The derivative of the chemical potential of component one, in a binary mixture, with respect to composition is given by:

$$(1/kT)(\partial\mu_1/\partial x_1)_{T,p} = 1/x_1-\{x_2\rho(G_{11}+G_{22}-2G_{12})\}/\{1+x_1x_2\rho(G_{11}+G_2-2G_{12})\} \quad (93)$$

For a component of an ideal gas mixture

$$(1/kT)(\partial\mu_1^0/\partial x_1)_{T,p} = 1/x_1 \quad (94)$$

From Eqs (93) and (94) one obtains the relation for the residual chemical potential

$$(\partial\mu_{1r}/\partial x_1)_{T,p} = -kT\{x_{2r}\rho\,(G_{11}+G_{22}-2G_{12})\}/\{1+x_1x_2\rho\,(G_{11}+G_{22}-2G_{12})\} \quad (95)$$

Integrating Eq. (95) at constant T and p from x_1=0 to x_1=1 one obtains:

$$\mu_{1r}^{\infty} = \mu_{1r}^{pure} + kT\int_0^1 \{x_2\rho(G_{11}+G_{22}-2G_{12})\} / \{1+x_1x_2\rho(G_{11}+G_{22}-2G_{12})\}dx_1 \quad (96)$$

The conformal solution approximation for the radial distribution function, Eq. (12), can be used to integrate the above equation. Substituting Eq.(16) in Eq. (96) one gets

$$\mu_{1r}^{\infty} = \mu_{1r}^{pure} + (\sigma_{11}^3 - 2\sigma_{12}^3 + \sigma_{22}^3)kT$$

$$\times\int_0^1 x_2\Delta_x / (\sigma_x^3 + (\sigma_{11}^3 - 2\sigma_{12}^3 + \sigma_{22}^3)x_1x_2\Delta_x)dx_1 \quad (97)$$

In the case of equal molecular sizes, $(\sigma_{11}^3-2\sigma_{12}^3+\sigma_{22}^3)=0$, and the integral in Eq.(97) drops. On the other hand, if the molecular sizes are different, the integral has to be evaluated numerically by using an appropriate equation of state to calculate Δ_x. The term σ_x^3 can be calculated from either Eq.(22) or Eq. (23), which gives almost the same result as shown in Figure 5. The curves were terminated at h=0.409 which corresponds to the saturation condition at $T^*=1.2$. Beyond this point the Nicolas et. al. LJ equation of state predicts a vapor phase.

Equation (97) predicts the correct qualitative dependence of μ_{1r}^{∞} on f as seen in figure 4. On the other hand, it predicts almost a linear dependence of μ_{1r}^{∞} on h which is not like the simulation data as reported in Figure 5. In all cases studied the prediction around h=f=1 is accurate, which is expected because of the ideal behavior of the mixtures at these values of f and h.

C: PHASE EQUILIBRIUM IN BINARY COMPLEX MIXTURES

For binary liquid mixtures at low to moderate pressures the following expressions for the direct correlation function integrals were derived

$$\rho C_{11} = \frac{W(1-\alpha_{12}v_2/v_1)x_2^2 + Wx_1(\alpha_{12}x_1+x_2)v_1/v_2 + (1-\alpha_{12})x_2 + \alpha_{12}x_1}{\alpha_{12}x_1^2 + x_1x_2 + \alpha_{21}x_2^2} \quad (98)$$

$$\rho C_{22} = \frac{W(1-\alpha_{21}v_1/v_2)x_1^2 + Wx_2(\alpha_{21}x_2+x_1)v_2/v_1 + (1-\alpha_{21})x_1 + \alpha_{21}x_2}{\alpha_{12}x_1^2 + x_1x_2 + \alpha_{21}x_2^2} \quad (99)$$

$$\rho C_{12} = \frac{W(\alpha_{12}x_1^2 + \alpha_{21}x_1x_2v_1/v_2 + \alpha_{12}x_1x_2v_2/v_1 + \alpha_{21}x_2^2) + \alpha_{12}x_1 + \alpha_{21}x_2}{\alpha_{12}x_1^2 + x_1x_2 + \alpha_{21}x_2^2} \quad (100)$$

In these expressions α_{21} and α_{12} are the binary parameters, and

$$W = -\,[kT(x_1\kappa_{T1}/v_2+x_2\kappa_{T2}/v_1)]^{-1} \qquad (101)$$

where κ_{Ti} is the isothermal compressibility of the pure component i.

Given the direct correlation function integrals, activity coefficients can be obtained by substituting the C_{ij}'s in Eq. (36), which for binary mixtures can also be written as

$$\ln \gamma_1 = -\int_0^{x_2}(\rho C_{12}-W-1)dx_2/x_1 \qquad (102)$$

Substituting for ρC_{12} from Eq. (100) and performing the integration, analytical expression for activity coefficient results

$$\ln \gamma_1 = -L_1\tau \ln[x_1+(L_2/L_1)x_2]+(1/2)(L_1\tau-1)\ln\left[a/\alpha_{12}x_2^{\,2}+(\alpha_{12}^{\,-1}-2)x_2+1\right]$$
$$+(1/2)[(2\alpha_{12}L_2-L_1)\tau+1-2\tau_{12}]/q\ \ln\left[(1+x_2/r^-)/(1+x_2/r^+)\right] \qquad (103)$$

where $L_1 = kT\kappa_{T1}/v_2$, $L_2=kT\kappa_{T2}/v_1$, $a=\alpha_{21}+\alpha_{12}-1$, $q=(1-4\alpha_{21}\alpha_{12})^{1/2}$,
$r^{\pm}=(1/2)(1-2\alpha_{12}\pm q)/a$

and

$$\tau = (\alpha_{21}v_1/v_2+\alpha_{12}v_2/v_1-1)/(\alpha_{21}L_1^{\,2}+\alpha_{12}L_2^{\,2}-L_1L_2). \qquad (104)$$

The expression for $\ln\gamma_2$ is obtained by inter-changing subscripts 1 and 2 in Eq. (103) and in the expression for $r^{\pm}$. The quantities a, q and τ remain unchanged under this operation. To test the activity coefficient expressions, vapor-liquid and liquid-liquid equilibria calculations of a number of binary systems are here carried out.

D: <u>APPLICATIONS FOR VAPOR-LIQUID PHASE EQUILIBRIA CALCULATIONS</u>

To test the possibility of vapor-liquid equilibrium (VLE) prediction by the new activity coefficient expressions a number of complex systems are chosen at low pressures. The vapor phase is assumed to be an ideal gas and the vapor pressure of the pure components, $p_i^{\,o}$ is represented by the Antoine equation. Two kinds of pure liquid property data are needed in the activity coefficient equation: They are the specific volume and isothermal compressibility data. The VLE data are fitted to the activity coefficient expressions by minimizing the following objective function

$$S = \Sigma(p-p_{exp})^2/S_p+\Sigma(y-y_{exp})^2/S_y \qquad (105)$$

where S_p and S_y are the sum of squares resulting from minimizing $\Sigma(p-p_{exp})^2$ and $\Sigma(y-y_{exp})^2$ individually. The results of calculations are shown in Table 1.

Table 1. Parameters of the present activity coefficient model and comparisons with Wilson correlation (Hamad and Mansoori 1989).

System	T/°C	α_{21}	α_{12}	Eq.(103) S_p	Eq.(103) S_y	Wilson S_p	Wilson S_y
Methanol+Water	60.	0.3869	0.3562	2.0	0.10	1.0	0.07
Ethanol+Water	40.	0.2900	0.3561	1.4	0.09	1.1	0.06
Ethanol+Water	55.	0.2920	0.3799	0.7	0.14	0.4	0.13
Ethanol+Water	70.	0.2942	0.4048	1.6	0.11	0.6	0.13
1-Propanol+Water	30.	0.2392	0.2722	.07	0.46	.09	0.64
Ethyl Acetate+ Ethanol	40.	0.5105	0.2989	.03	0.04	0.02	0.04
Ethanol+Benzene	25.	0.1028	0.6393	.009	0.008	.004	0.01
Ethanol+Benzene	40.	0.09741	0.6458	0.2	0.07	0.1	0.05
Ethanol+Benzene	55.	0.09688	0.6498	0.9	0.13	0.5	0.13
Acetone+ Chloroform	25.	0.2907	0.6404	0.04	0.24	0.04	0.24

In this table, $S_p = 10^{-2}\,\Sigma(p-p_{exp})^2$, and $S_y = 10^2\,\Sigma(y-y_{exp})^2$.
Experimental data taken from VLE Data Collection (Dechema).

In this table comparisons are also made with the results of Wilson (1964) correlation which is selected because it gave the best over all fit to the large number of mixtures studied in VLE calculations. Table 1 also includes parameters α_{21} and α_{12}, and optimized values of S_p and S_y. Figures 6-8 show the vapor-liquid equilibrium composition (x-y) diagrams for the systems methanol+water, 1-propanol+water and ethanol+benzene.

According to Table 1 and Figures 6-8 it can be concluded that the present theory can predict the vapor-liquid equilibria of the systems studied as good as the Wilson (1964) equation.

E: APPLICATIONS FOR LIQUID-LIQUID PHASE EQUILIBRIA CALCULATIONS

Phase splitting in liquid mixtures occurs when two phases have a lower total Gibbs free energy than one phase at constant temperature and pressure. A necessary condition for stability of a two-component homogeneous mixture is (Van Ness and Abbott 1982):

$$d\mu_i / dx_i > 0 \qquad i=1,2 \qquad (106)$$

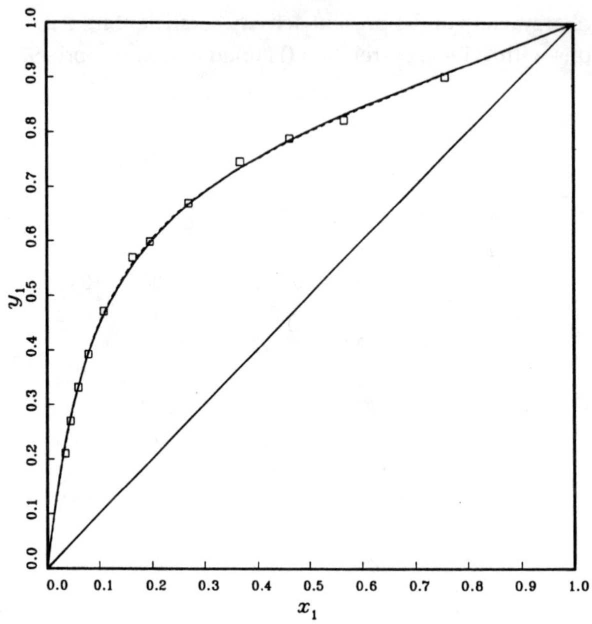

Figure 6. The vapor-liquid equilibrium composition (x-y) diagram for methanol (1) + water (2) at 60 C. The squares are the experimental data, the solid curve is the present theory and the dashed curve is Wilson correlation.

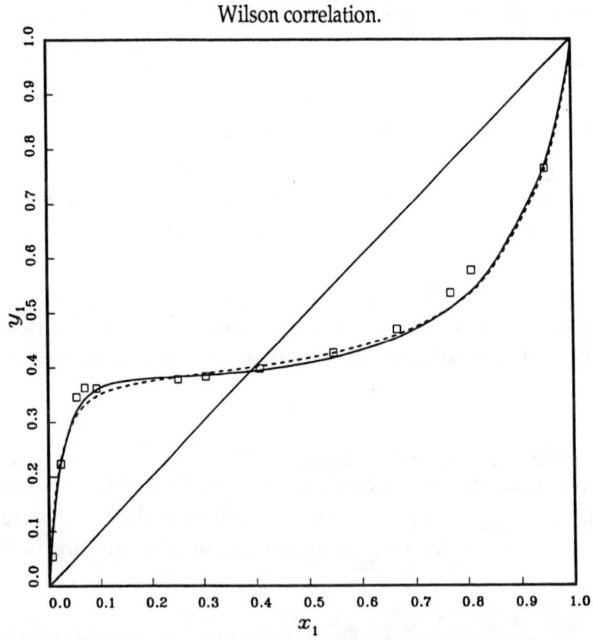

Figure 7. The vapor-liquid equilibrium composition (x-y) diagram for 1-propanol (1) + water (2) at 30 C. The squares are the experimental data, the solid curve is the present theory and the dashed curve is Wilson correlation.

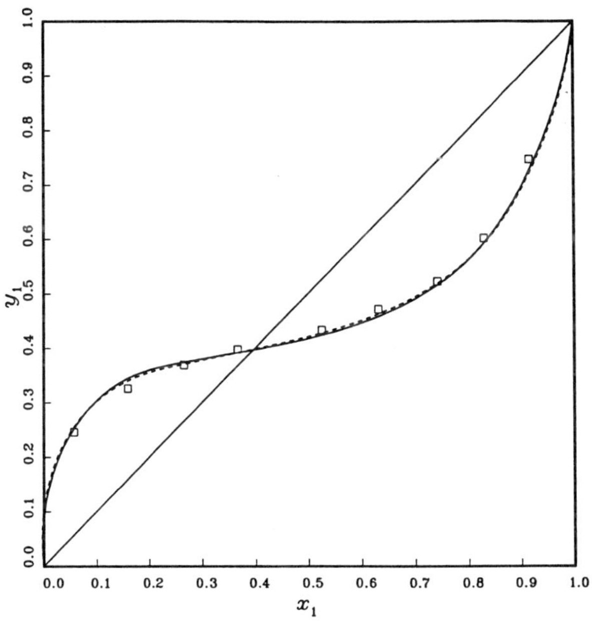

Figure 8. The vapor-liquid equilibrium composition (x-y) diagram for ethanol (1) + benzene (2) at 55 C. The squares are the experimental data, the solid curve is the present theory and the dashed curve is Wilson correlation.

In terms of the activity coefficients, Eq. (106) becomes:

$$d(\ln\gamma_i)/dx_i + 1/x_i > 0 \tag{107}$$

Substituting for $\ln \gamma_i$ in this equation from Eq. (103) gives

$$(\alpha_{12}x_1+\alpha_{21}x_2)(L_1x_1+L_2x_2)- x_1x_2(\alpha_{21}v_1/v_2+\alpha_{12}v_2/v_1 -1) > 0 \tag{108}$$

provided parameters α_{21} and α_{12} are positive. For liquid-liquid equilibria (LLE) the left-hand side of inequality (108) should be negative. Considering the magnitudes of the different quantities one concludes that the present activity coefficient model is capable of predicting LLE. The Wilson (1964) correlation, on the other hand, is known to be unable to predict LLE, because it always satisfies inequality (107).

The present activity coefficient theory is tested versus experimental LLE data. For the variation of the isothermal compressibilities with temperature, which is needed in this calculation, a three-parameter equation is used (Brostow and Maynadier

1979).

Temperature and pure component volume dependence of parameters α_{21} and α_{12} for a number of organic binary systems exhibiting LLE are represented by the following expressions (Hamad and Mansoori 1989):

$$\alpha_{21} = (v_2/v_1)\exp(\alpha_{o21}/RT) \tag{109}$$

$$\alpha_{12} = \alpha_{o12}(v_1/v_2)\ , \tag{110}$$

in which α_{o21} and α_{o12} are constants (independent of temperature and composition). The experimental LLE data is fitted to the activity coefficient equation with Eqs. (109) and (110) for α_{21} and α_{12} by minimizing the difference between the activities in the two liquid phases:

$$S = \sum\{1-x_1^{(1)}\gamma_1^{(1)}/[x_1^{(2)}\gamma_1^{(2)}]\}^2 + \sum\{1-x_2^{(1)}\gamma_2^{(1)}/[x_2^{(2)}\gamma_2^{(2)}]\}^2 \tag{111}$$

where superscripts (1) and (2) are for phases 1 and 2, respectively.

Table 2 shows values of parameters α_{o21} and α_{o12} and optimum S for a number of binary systems.

The variation of the compositions of two LLE systems (methanol+heptane and formic acid+benzene) with temperature is shown in Figures 9 and 10.

Table 2. LLE Parameters of the present activity coefficient model for different binary mixtures.

System	α_{o21}	α_{o12}	$SX10^3$
Methanol+Heptane	-1070.9	1.0243	0.12
Methanol+Cyclohexane	-980.29	1.0191	0.21
Formic Acid+Benzene	-748.15	0.97983	1.7
Phenol+Octane	-1077.0	1.0031	7.3
Methanol+Carbondisulfide	-797.20	1.0186	0.45
Nitrobenzene+Hexane	-975.97	1.0027	2.0
1,3-Dihydroxybenzene+ Benzene	-937.02	1.0032	5.5

In this table: $S = \Sigma\{1-x_1^{(1)}\gamma_1^{(1)}/[x_1^{(2)}\gamma_1^{(2)}]\}^2 + \Sigma\{1-x_2^{(1)}\gamma_2^{(1)}/[x_2^{(2)}\gamma_2^{(2)}]\}^2$. α_{o21} dim.: J/mol.

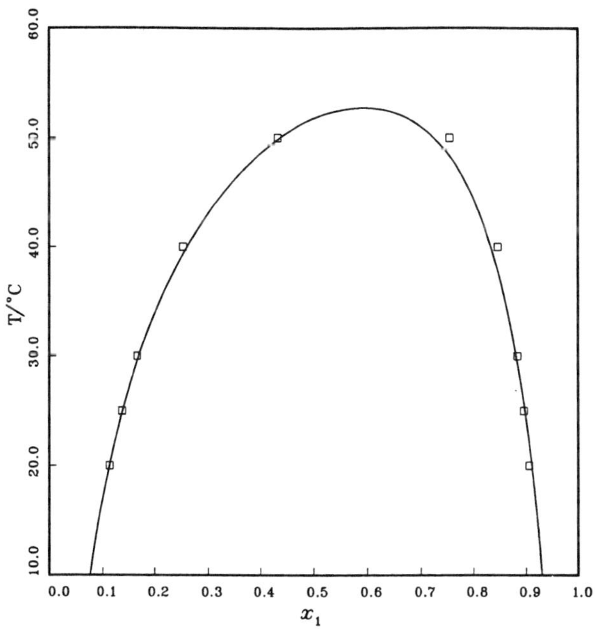

Figure 9. The composition versus temperature (x-T) diagram for liquid-liquid equilibrium of methanol (1) + heptane (2). The squares are the experimental data and the solid curve is the present theory.

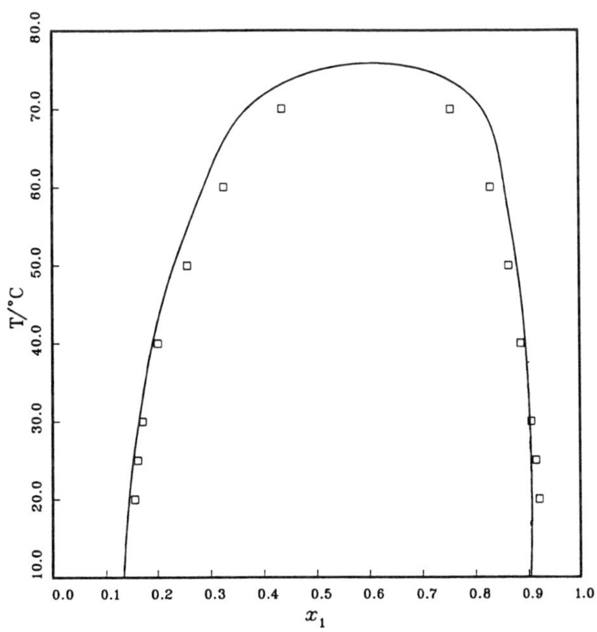

Figure 10. The composition versus temperature (x-T) diagram for liquid-liquid equilibrium of formic acid (1) + benzene (2). The squares are the experimental data and the solid curve is the present theory.

According to these figures the agreement with the data is satisfactory. The largest deviation of theory from experimental data occurs near the upper critical solution temperature.

Overall, the present technique is capable of formulating analytic expressions for activity coefficients in mixtures. For this purpose it is necessary to define closure expressions for the cross direct correlation function integrals. Application of the activity coefficient expressions resulting from this technique for vapor-liquid and liquid-liquid equilibria calculations has been as successful as the other activity coefficient expressions available.

F: THE RADIUS OF THE SPHERE OF INFLUENCE OF LOCAL COMPOSITIONS

In the basic definition of the local particle numbers, Eq. (57), R_{ij}, the radius of the sphere of influence of local compositions must be defined. In the formulation which is provided above, no numerical value of this radius was specified for the sake of generality of the results. In the calculation of partial molar properties one has to take into account the numerical value of this radius. To assign a value of the radius of the sphere of influence of local compositions, one must consider the basic approximation which was made in deriving Eq. (64) from Eq. (63). According to this approximation the radius of the sphere of effectiveness of local composition can be defined by the following equation

$$\int_{R_{ij}}^{\infty}(g_{ij}(r) - 1)4\pi r^2 dr = 0 \tag{112}$$

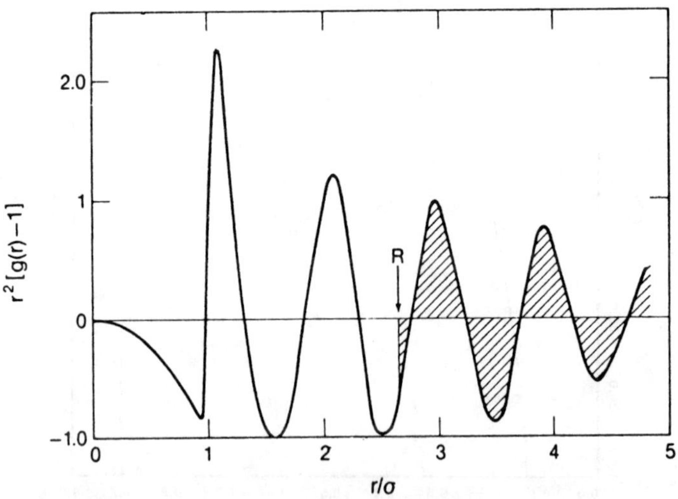

Figure 11. Schematic diagram of the radial dependence of the isothermal compressibility integrand $r^2[g(r)-1]$ in a dense fluid. The approximate radius of influence of local composition (R) is shown as the point where the oscilations of g(r) about unity cancel each other.

This expression indicates that R_{ij} should be defined such that the dashed areas in Figure 11 above and below the horizontal axis are equal to each other.

Knowing that g_{ij} is generally a function of temperature, density, and composition it becomes clear that R_{ij} will also have to be a function of temperature, density, and composition. Since g_{ij} varies significantly with density it can be expected that R_{ij} will be a strong function of density decreasing as the density increases.

It is certainly possible that Eq. (112) can have more than one solution and one must, therefore, impose some secondary constraint to obtain the correct value. For dense soft-spheres it is shown that [Hanley and Evans (1981)] the statistical mechanical energy and virial equations reproduce the "experimental" values when the upper limit of integration is changed from infinity to roughly 2.5 molecular diameters. In particular, Figure 6 of Hanley and Evans (1981) shows that properties such as the internal energy and pressure are completely determined by interactions of relatively short range. It would seem safe, therefore, to choose the solution to Eq. (112) which is in the vicinity of 2-3 molecular diameters. Nonetheless, it is clear that in order to have a more quantitative knowledge about R_{ij} one has to use accurate radial distribution function data, preferably, using computer simulation values. A second problem is that the intermolecular potential is usually truncated at 2.5 molecular diameters in simulation studies. Thus, it is difficult to obtain accurate g_{ij} information which one can use to study Eq. (112). Recently, molecular dynamics computer simulation data for radial distribution functions of binary Lennard-Jones mixtures with a size ratio of two and enegy ratios of 0.5 to 3.0 have been published (Huber and Ely 1989). This data may be used for the study of the radius of influence of local compositions.

It has become possible to introduce the concept of local particle numbers and compositions with respect to fluctuation integrals. Incorporation of the local composition concept in the fluctuation theory has produced a direct method for calculating partial molar properties from different molecular theories of solutions. The method does not require the use of mixing rules or any other approximation about the nature of the mixture under consideration. Finally, this study has also resulted in a rigorous equation, Eq. (112), for the definition of the radius of the sphere of influence of local compositions which can be utilized in order to specify this radius at different thermodynamic state conditions.

CONCLUSIONS

The techniques presented here have allowed us to perform calculations of total, partial molar, and phase equilibria of mixtures using the concept of fluctuation theory. There are several points to be noted about the advantages of these techniques versus the existing state-of-the-art of thermodynamic calculations of mixtures and fluctuation theory: (i) The present techniques allow us to perform thermodynamic calculations for the whole range of mixture compositions and not just at the infinite dilution and high concentration limits as it had been the case for most of the other fluctuation theory techniques. (ii) The present techniques are applicable for mixtures consisting of species with large differences in molecular size, shape, and energetics. They are specifically useful for polar and associating molecular fluids for which, generally, no accurate intermolecular potential energy functions are available, (iii) With the application of the weighted arithmetic mean closure for the cross direct correlation function integrals it has become possible to derive analytic expressions for activity coefficients in complex mixtures. the resulting activity coefficient expressions allow us to perform vapor-liquid and liquid-liquid equilibria computations of mixtures. Accuracy of these calculations are as good as the best available phase equilibria computational techniques for mixtures.

ACKNOWLEDGEMENT: This research is supported by the Chemical Sciences Division of the Office of Basic Energy Sciences, the U.S. Department of Energy, Grant # DE-FG02-84ER13229.

REFERENCES

Brostow, W. and P. Maynadier. 1979. Temperature dependence of the isothermal compressibility of liquids. High Temp. Sci. : 11, 7-21.

Brown, W.B. 1957a. The second-order theory of conformal solutions. Proc. R. Soc. London, Ser A 240: 561-580.

Brown, W.B. 1957b. The statistical thermodynamics of mixtures of Lennard-Jones molecules. I: Random mixtures. Philos. Trans. R. Soc. London, Ser A 250: 175-220.

Brown, W.B. 1957c. The statistical thermodynamics of mixtures of Lennard-Jones molecules. II: Deviations from random mixing. Philos. Trans. R. Soc. London, Ser A 250: 221-246.

Carnahan, N.F. and K.E. Starling. 1969. Rigid sphere equation of state. J. Chem. Phys. 51: 635-638.

Chao, K.C. and R.A. Greenkorn. 1975. Thermodynamics of Fluids. New York: Marcel-Dekker.

Fischer, J. 1983. Remarks on molecular approaches to liquid mixtures. Fluid Phase Equilibria 10: 1-7.

Gierycz, P. and K. Nakanishi. 1984. Local composition in binary mixtures of Lennard-Jones fluids with differing sizes of components. Fluid Phase Equilib. 16: 255-273.

Guggenheim, E.A. 1952. Mixtures. Oxford: Clarendon Press.

Haile, J.M. and G.A. Mansoori. eds. 1983. Molecular-Based Study of Fluids. Washington D.C: American Chemical Society.

Hamad, E.Z. 1988. Statistical Mechanical Modeling of Complex Mixtures: Distribution Function Approach. PhD Thesis, University of Illinois, Chicago, IL.

Hamad, E.Z., G.A. Mansoori, and J.F. Ely. 1987. Conformality of the Kirkwood-Buff solution theory of statistical mechanics. J. Chem. Phys. 86: 1478- 1484.

Hamad, E.Z., G.A. Mansoori, E. Matteoli and L. Lepori. 1989. Relations between mixture fluctuation integrals. Z. Phys. Chem. Neue Folge. 162, 27-45.

Hamad, E.Z. and G.A. Mansoori. 1989. A fluctuation solution theory of activity coefficients (Phase equilibria in associating molecular solutions). J. Chem. Phys. (to appear).

Hanley, H.J.M. and D.J. Evans. 1981. Behavior of a nonconformal mixture via computer simulation. Int. J. Thermophys. 2: 1-19,

Hoheisel, C. and F. Kohler. 1984. Local composition in liquid mixtures. Fluid Phase Equilib. 16: 13-24.

Hu, Y., E.G. Azevedo, and J.M. Prausnitz. 1983. The molecular basis for local compositions in liquid mixture models. Fluid Phase Equilib. 13: 351-360.

Huber, M.L. and J.F. Ely. 1989. Properties of Lennard-Jones mixtures at various temperatures and energy ratios with a size ratio of two. NIST Technical Note 1331: U.S. Government Printing Office, Washington, D.C.

Kemeny, S. and P. Rasmussen. 1981. A derivation of local composition expressions from partition functions. Fluid Phase Equilib. 7: 197-203.

Kirkwood, J.G. and F.P. Buff. 1951. The statistical mechanical theory of solutions:I. J. Chem. Phys. 19: 774-777.

Kreglewski, A. 1984. Equilibrium Properties of Fluids and Fluid Mixtures. College Station, TX: Texas A&M University Press.

Lebowitz, J.L. and J.K. Percus. 1983. Spherization of nonspherical interactions. J. Chem. Phys. 79: 443-444.

Lee, L.L., T.H. Chung and K.E. Starling. 1983. A molecular theory for the thermodynamic behavior of polar mixtures. I. The statistical-mechanical local composition model. Fluid Phase Equilib. 12: 105-124.

Mansoori, G.A., N.F. Carnahan, K.E. Starling and T.W. Leland. 1971. Equilibrium thermodynamic properties of mixtures of hard spheres. J. Chem. Phys. 54: 1523-1525.

Mansoori, G.A. 1986. Molecular basis of activity coefficient (isobaric-isothermal ensemble approach). Fluid-Phase Equilib. 4: 61-69.

Mansoori, G.A. and J.F. Ely. 1985. Statistical Mechanical theory of local compositions. Fluid Phase Equilibria 22: 253-275.

Mansoori, G. A. (1986). Mixing rules for cubic equations of state. ACS Symposium Series 300, Washington, DC: American Chemical Society.

Mansoori, G.A. and T.W. Leland, Jr. 1972. Statistical thermodynamics of mixtures (a new version for the theory of conformal solutions). J.Chem. Soc. Farad. Trans. II 68: 320-344.

Massih, A.R. and G.A Mansoori. 1983. Conformal-solution theory of polar fluids: The statistical mechanical basis of shape factors. Fluid Phase Equilib. 10: 57-72.

Mathias, P.M. and T.W. Copeman. 1983. Extension of the Peng-Robinson equation of state to complex mixtures: evaluation of the various forms of the local composition concept. Fluid Phase Equilib. 13: 91-108.

McDermott, C. and N. Ashton. 1977. Note on the definition of local composition. Fluid Phase Equilib. 1: 33-35.

Mollerup, J. 1981. A note on excess Gibbs energy models, equations of state and the local composition concept. Fluid Phase Equilib. 7: 121-138.

Nakanishi, K. and K. Toukulio. 1979. Molecular dynamics studies of Lennard-Jones liquid mixtures. J. Chem. Phys. 70: 5848-5850.

Nakanishi, K., S. Okazaki, K. Ikari, T. Higuchi and H. Tanaka. 1982. Free energy of mixing, phase stability, and local composition in Lennard-Jones liquid mixtures. J. Chem. Phys. 76: 629-636.

Nakanishi, K. and H. Tanaka. 1983. Molecular dynamic studies on the local composition in Lennard-Jones liquid mixtures of nonspherical molecules. Fluid Phase Equilib. 13: 371-380.

Nicolas, J.J., K.E. Gubbins, W.B. Street and D.J. Tildesley. 1979. Equation of state for Lennard-Jones fluid. Molec. Phys. 37: 1429-1454.

O'Connell, J.P. 1971. Thermodynamic properties of solutions based on correlation functions. Molec. Phys. 20: 27-33.

Pearson F.J.,and G. S. Rushbrooke. 1957. On the theory od binary fluid mixtures. Proc. Roy. Soc. Edinburgh, Sec A, A64, 305-317.

Prausnitz, J.M. 1969. Molecular Thermodynamics of Fluid Phase Equilibria. Englewood Cliffs, New Jersey: Prentice-Hall.

Renon, H. and J.M. Prausnitz. 1968. Local compositions in thermodynamics excess functions for liquid mixtures. AIChE J. 14: 135-144.

Shing, K.S. and K.E. Gubbins. 1982. The chemical potential in dense fluids and fluid mixtures via computer simulation. Molec. Phys. 46: 1109-1128.

Shing, K.S., K.E. Gubbins and K. Lucas. 1988. Henry constants in non-ideal fluid mixtures: computer simulation and theory. Mol. Phys. 65: 1235-1252.

Van Ness, H.C. and M.M. Abbott. 1982. Classical Thermodynamics of Nonelectrolyte Solutions. New York: McGraw-Hill Book Company.

Vapor-Liquid Equilibrium Data Collection (DECHEMA, Frankfurt, 1977).

Whiting, W.B. and J.M. Prausnitz. 1982. Equations of state for strongly nonideal fluid mixtures: application of local compositions toward density-dependent mixing rules. Fluid Phase Equilib. 9: 119-147.

Wilson, G.M. 1964. Vapor-liquid equilibrium. XI: A new expression for the excess free energy of mixing. J. Am. Chem. Soc. 86: 127-130.

An Alternative Approach to Modeling Excess Gibbs Free Energy in Terms of Kirkwood-Buff Integrals

Ariel Augusto Chialvo
Department of Chemical Engineering
Princeton University, Princeton, NJ 08544

ABSTRACT

The Kirkwood-Buff formalism is used to obtain rigorous thermodynamic constraints for activity coefficients and their composition slopes at infinite dilution. Nine patterns of nonideality are studied in terms of particular linear combination of K-B integrals $\Delta(TPx) = G_{11} + G_{22} - 2G_{12}$. The analysis of the infinite dilution behavior of $\Delta(x)_{TP}$ allows predicting, by only knowing the infinite dilution activity coefficients and the corresponding composition slopes, (a) whether a liquid binary mixture shows positive, negative or positive-negative deviation from a reference ideal solution, (b) whether the composition dependence of the mixture's activity coefficients presents extrema, and (c) whether the liquid mixture exhibits one- or two-phases.

Based on the previous analysis, we propose three simple algebraical representations for the composition dependence of Δ. Then we generate three new sets of correlations for the activity coefficients of binary mixtures, by introducing the modeled $\Delta(x)_{TP}$ into the K-B equations. The new correlations involve two important features: (a) they model a combination of K-B integrals - $\Delta(x)_{TP}$ - i.e., the second composition derivative of the excess Gibbs free energy instead of the excess Gibbs free energy itself, and (b) for a given composition dependence of $\rho\Delta(x)_{TP}$, the resulting correlations for the activity coefficients take different algebraical forms depending on whether they deal with positive or negative deviation from the ideal solution.

Test of the new correlations on highly nonideal binary mixtures suggest that a simple (crude) composition dependence for $\Delta(x)_{TP}$ suffices to obtain excellent agreement with experimental data, without recourse to statistical adjustment of data.

1. INTRODUCTION

The description of fluid phase equilibrium in chemical engineering is usually accomplished by correlations of either PVT equations of state or of activity coefficients

(Prausnitz et al ,1967; 1986) , i.e., fitting a model to experimental data to describe (interpolate) how mixture properties change with state conditions and composition. Correlations are typically based on uncertain assumptions that are incorporated into the model as guesses of unknown functional relationships. Although assumptions might be well-founded conjectures, usually they are incorporated ad hoc and justified a posteriori. The justifications often take the form of demonstrating some dependence between model parameters and molecular quantities; however, such correspondence does not necessarily imply causal connection. A typical example is given by the recent interpretation of Wilson's parameters (Wilson ,1964) as exponentials of the differences of potentials of mean force in the mixture (Lee et al , 1983).

Any serious development of equations for correlating fluid-phase equilibrium needs: (a) a rigorous statistical mechanical basis, (b) the identification of the dominant source of nonideality (Shukla et al.1988), and (c) the connection between excess properties and particular equilibrium behaviour. To accomplish this task here, we first take advantage of an exact statistical mechanical formalism, the Kirkwood-Buff (K-B) (Kirkwood and Buff,1951), to analyse the relations between phase equilibrium and microscopic structure (integrals over spatial pair distribution functions $g_{ij}(r_{ij})$) , and then use the gained insights to propose simple representations for the composition dependence of those integrals.

The Kirkwood-Buff formalism presents several advantages over other theories: (a) it gives the necessary relations between macroscopic (bulk) and microscopic (molecular) properties, (b) it is not restricted to particular forms of intermolecular potentials or to molecules of particules sizes and shapes, and (c) it applies to solutions containing any number of components.

This scheme presents two important advantages: (a) any new development starts from a hypothesis-free formalism (e.g., K-B formalism) , and (b) the interpretation of fluid phase equilibrium of real systems via K-B formalism allows setting well founded approximations, e.g., for the composition dependence of the K-B integrals, to obtain tractable mathematical expressions from the original exact formalism.

In this contribution we present a different approach to modeling activity coefficients of binary systems based on the K-B formalism. In Section 2 we show how a particular combination of K-B integrals can be used to interpret different kinds of phase behavior. Then in Section 3 we evaluate the K-B integrals at infinite dilution and relate them to composition derivatives of the activity coefficients at infinite dilution.

Using the insight gained in Sections 2 and 3, we study, in Section 4, how existing g^E-models for binary mixtures describe the K-B integrals. Such an analysis leads to an understanding of the success or failure of these correlators.

Finally a few simple and successful new correlators for binary activity coefficients are proposed and tested in Section 5, based on the ideas previously discussed.

2. PHASE BEHAVIOR IN TERMS OF $\Delta\,(x)_{TP}$ FOR BINARY MIXTURES

To relate the microscopic behavior of a mixture to macroscopic properties, K-B analyses the concentration fluctuations in an open system characterized by fixed temperature T, total volume V, chemical potential μ_i of each species (grand canonical ensemble). That is (Kirkwood and Buff,1951; Ben-Naim,1974)

$$G_{ij} = 4\pi\int_0^{\infty} \left[g_{ij}(r_{ij}) - 1\right] r_{ij}^2\, dr_{ij}$$

$$= kT\left[\kappa_T + \frac{\bar{v}_i \bar{v}_j}{V\mu_{ij}}\right] - v_i\,\delta_{ij} = V\left[\frac{\langle \delta N_i\, \delta N_j\rangle}{\langle \delta N_i\rangle\langle \delta N_j\rangle} - \frac{\delta_{ij}}{\langle N_i\rangle}\right] \tag{2.1}$$

where κ_T is the mixture isothermal compressibility, $\bar{v}_i$ and v_i the partial molar volume and the molar volume of species i respectively , and

$$\mu_{ij} = \left[\frac{\partial \mu_i}{\partial N_j}\right]_{TPN_{k\neq j}} \quad , \quad \mu_{ij} = \mu_{ji} \tag{2.2}$$

Angle brackets in eqn. (2.1) denote grand-canonical ensemble averages, δ_{ij} is the Kronecker delta and δN_i is the fluctuation of N_i ,i.e., $\delta N_i = N_i - \langle N_i \rangle$

This formalism provides the following exact relation between μ_i and the microscopic structure of the mixture (Ben-Naim,1974)

$$\beta\left(\frac{\partial \mu_i}{\partial x_i}\right)_{TP} = \left\{x_i\left[1 + x_i(1 - x_i)\,\rho\Delta\right]\right\}^{-1} \tag{2.3}$$

with β=1/kT, and

$$\Delta = G_{11} + G_{22} - 2G_{12} \tag{2.4}$$

From the definition of an excess property we have (Van Ness and Abbott,1982)

$$\left(\frac{\partial \mu_i^E}{\partial x_i}\right)_{TP} = \left(\frac{\partial \mu_i}{\partial x_i}\right)_{TP} - \left(\frac{\partial \mu_i^{is}}{\partial x_i}\right)_{TP} \tag{2.5}$$

and

$$\mu_i^E(x_i) = kT \ln \gamma_i(x_i) \tag{2.6}$$

where superscript "is" means ideal solution and γ_i is the activity coefficient of species i. For the case of an ideal solution given by the Raoult's law , eqns. (2.3), (2.5)-(2.6) yield (Chialvo,1988)

$$\ln \gamma_i(x_i) = -\int_1^{x_i} \left[\frac{\rho\Delta(1-x_i)}{1+x_i(1-x_i)\rho\Delta}\right] dx_i \tag{2.7}$$

Finally the excess Gibbs free energy equation becomes

$$g^E(x) = -kT \sum_{i=1}^{2} x_i \int_1^{x_i} \left[\frac{\rho\Delta(1-x_i)}{1+x_i(1-x_i)\rho\Delta}\right] dx_i \tag{2.8}$$

Eqns. (2.7) and (2.8) suggest that different types of phase equilibrium behavior may be interpreted in terms of $\Delta(x)_{TP}$ as done in what follows.

Thermodynamic Stability

At constant temperature, pressure and composition a phase of a binary mixture is stable if (Prigogine and Defay,1954)

$$\left(\frac{\partial \mu_1}{\partial x_1}\right)_{TP} > 0 \tag{2.9}$$

At the critical point of incipient immiscibility the system must satisfy (Prigogine and Defay,1954)

$$\left(\frac{\partial \mu_1}{\partial x_1}\right)_{TP} = \left(\frac{\partial^2 \mu_1}{\partial x_1^2}\right)_{TP} = 0 \tag{2.10}$$

and

$$\left(\frac{\partial^3 \mu_1}{\partial x_1^3}\right)_{TP} > 0 \tag{2.11}$$

or in terms of free energy of mixing g^m (TPx)

$$\left(\frac{\partial^2 g^m}{\partial x_1^2}\right)_{TP} = \left(\frac{\partial^3 g^m}{\partial x_1^3}\right)_{TP} = 0 \tag{2.12}$$

and

$$\left(\frac{\partial^4 g^m}{\partial x_1^4}\right)_{TP} > 0 \tag{2.13}$$

Then from eqns. (2.3) and (2.10) we have

$$\beta\left(\frac{\partial \mu_1}{\partial x_1}\right)_{TP} = \left[\frac{(\rho\Delta)^{-1}}{(\rho\Delta)^{-1} + x_1(1-x_1)}\right]\left(\frac{1}{x_1}\right) = 0 \tag{2.14}$$

After defining $\Omega \equiv (\rho\,\Delta)^{-1}$, from eqn. (2.10) follows

$$\beta\left(\frac{\partial^2 \mu_1}{\partial x_1^2}\right)_{TP} = -\left[\frac{\Omega}{\Omega + x_1(1-x_1)}\right]\left(\frac{1}{x_1^2}\right) + \left[\frac{\Omega'}{\Omega + x_1(1-x_1)} - \frac{\Omega(\Omega' + 1 - 2x_1)}{(\Omega + x_1(1-x_1))^2}\right]\left(\frac{1}{x_1}\right) = 0 \tag{2.15}$$

with $\Omega' = (\partial\,\Omega / \partial\, x_1)_{TP}$, then

$$(\rho\Delta)^{-1} = \left(\frac{\partial\left[(\rho\Delta)^{-1}\right]}{\partial x_1}\right)_{TP} = 0 \tag{2.16}$$

and finally eqn. (2.11) becomes

$$\left(\frac{\partial^2\left[(\rho\Delta)^{-1}\right]}{\partial x_1^2}\right)_{TP} > 0 \tag{2.17}$$

Thus, the stability condition is not satisfied when

$$\rho\Delta < -\left[x_1(1-x_1)\right]^{-1} < 0 \tag{2.18}$$

i.e., the system splits into two phases.

Although $\Delta < 0$, the integrand of eqn. (2.7) remains positive in the unstable region, and the corresponding activity coefficients will be greater than unity. Since the pair correlation functions cannot be defined for heterogeneous systems (Münster,1969), e.g., LLE, in eqn. (2.18) Δ has to be interpreted as that given by eqn. (2.4) for the hypothetical one-phase mixture at the state condition and global composition of the two-phase system.

Extrema in the Composition Dependence of Activity Coefficients

Activity coefficients show extrema with respect to composition when

$$\left(\frac{\partial \ln \gamma_i}{\partial x_i}\right)_{TP} = 0 \quad \text{at} \quad x_i = x_i^* \tag{2.19}$$

$$\left(\frac{\partial^2 \ln \gamma_i}{\partial x_i^2}\right)_{TP} \neq 0 \quad \text{at} \quad x_i = x_i^* \tag{2.20}$$

where x_i^* represents the composition where $\gamma_i(x_i)$ shows the extremum. In terms of $\rho\Delta\,(x)_{TP}$ and recalling eqn. (2.7), the conditions for the existence of extrema (eqns. (2.19) and (2.20)) become

$$\rho\Delta(x_i = x_i^*) = 0 \tag{2.21}$$

and

$$\left[\frac{\partial(\rho\Delta)}{\partial x_i}\right]_{TP} \neq 0 \quad \text{at} \quad x_i = x_i^* \tag{2.22}$$

Vapor-Liquid Equilibrium (VLE)

Four different types of VLE behavior can be expected for totally miscible mixtures, depending on the sign of $\Delta(TPx)$:

a) If $\Delta(TPx) = 0$ over the entire range of state conditions and compositions, then the mixture is an ideal solution, i.e., either $G_{ij} = 0$ or $2\,G_{12} = G_{11} + G_{22}$.

b) If $\Delta(TPx) = 0$ at a particular composition, when the state condition is fixed e.g., $[\Delta(x_i^*)]_{TP} = 0$, then the mixture has extrema in the actvity coefficients with respect to composition at $x_i = x_i^*$.

c) If $\Delta(TPx) > 0$ over the entire range of compositions when the state condition is fixed, e.g., $[\Delta(x_i)]_{TP} > 0$, then the mixture exhibits positive deviation from an ideal solution.

d) If $\Delta(TPx) < 0$ over the entire range of compositions when the state condition is fixed, e.g., $[\Delta(x_i)]_{TP} < 0$, then the mixture exhibits negative deviation from an ideal solution.

Note that a positive deviation in the whole range of composition means a positive integrand in eqn. (2.7) in the same range. This condition could be satisfied by

$$\rho\Delta < 0 \tag{2.23}$$

and

$$1 + \rho\Delta x_1(1 - x_2) < 0 \tag{2.24}$$

but the stability condition, eqn. (2.9), prevents this (see also eqn. (2.18)).

Likewise, for negative deviation, a similar analysis eliminates the following case

$$\rho\Delta > 0 \tag{2.25}$$

and

$$1 + \rho\Delta x_1(1 - x_2) < 0 \tag{2.26}$$

since eqns. (2.25) and (2.26) cannot be satisfied simultaneously.

3. INFINITE DILUTION BEHAVIOR IN BINARY MIXTURES

Here we relate the infinite dilution values of $\rho\Delta(TPx)$ to derivatives of activity coefficients with respect to composition at infinite dilution. At fixed temperature T, pressure P and composition x, the limiting conditions for G_{ij}'s can be obtained from eqn. (2.1) (Chialvo, 1988; Matteoli and Lepori,1984)

$$\lim_{x_1 \to 0} G_{11}\Big|_{TP} = \left(\frac{\kappa_T^{(2)}}{\beta}\right) - 2\bar{v}_1^{\infty} + v_2^0\left[1 - \left(\frac{\partial \ln\gamma_1}{\partial x_1}\right)^{\infty}_{TP}\right] \tag{3.1}$$

$$\lim_{x_1 \to 1} G_{11}\Big|_{TP} = \left(\frac{\kappa_T^{(2)}}{\beta}\right) - v_1^0 \tag{3.2}$$

$$\lim_{x_2 \to 0} G_{22}\Big|_{TP} = \left(\frac{\kappa_T^{(1)}}{\beta}\right) - 2\bar{v}_2^{\infty} + v_1^0\left[1 - \left(\frac{\partial \ln\gamma_2}{\partial x_2}\right)^{\infty}_{TP}\right] \tag{3.3}$$

$$\lim_{x_2 \to 1} G_{22}\Big|_{TP} = \left(\frac{\kappa_T^{(2)}}{\beta}\right) - v_2^0 \tag{3.4}$$

$$\lim_{x_1 \to 0} G_{12}\Big|_{TP} = \left(\frac{\kappa_T^{(2)}}{\beta}\right) - \bar{v}_1^{\infty} \tag{3.5}$$

$$\lim_{x_2 \to 0} G_{12}\Big|_{TP} = \left(\frac{\kappa_T^{(1)}}{\beta}\right) - \bar{v}_2^{\infty} \tag{3.6}$$

where $\bar{v}_i^{\infty}$ is the partial molar volume of component i at infinite dilution, v_i^0 is the molar volume of pure component i , and $\kappa_T^{(i)}$ is the isothermal compressibility of pure component i. Then from either eqn. (2.7) or eqns. (3.1)-(3.6) we obtain

$$\lim_{x_2 \to 0} \Delta(x_2)_{TP} = -v_1^0 \left(\frac{\partial \ln\gamma_2}{\partial x_2}\right)_{TP}^{\infty} = v_1^0 \left(\frac{\partial \ln\gamma_2}{\partial x_1}\right)_{TP}^{\infty}$$

$$= G_{11}^0 + G_{22}^{\infty} - 2G_{12}^{\infty} \tag{3.7}$$

$$\lim_{x_1 \to 0} \Delta(x_1)_{TP} = -v_2^0 \left(\frac{\partial \ln\gamma_1}{\partial x_1}\right)_{TP}^{\infty} = G_{11}^{\infty} + G_{22}^0 - 2G_{12}^{\infty} \tag{3.8}$$

Improper application of the limiting conditions to the definition of $\Delta(x)_{TP}$, eqn. (2.4), leads to erroneous results and misinterpretation of the physical picture of infinite dilution (Chialvo, 1988).

The derived limiting conditions for $\Delta(x)_{TP}$ have several applications which will be discussed in the following sections. However, we mention here two examples from the available literature in which eqns. (3.1)-(3.6) can be used to detect inconsistencies. Ben-Naim (1977) has found, using his inversion method that

$$\lim_{x_1 \to 0} \Delta(x_1) < 0 \tag{3.9}$$

for ethanol (1)-water (2) at 298 K. However, literature (Ochi and Kojima,1987) shows that for this mixture

$$\left(\frac{\partial \ln\gamma_1}{\partial x_1}\right)_{TP}^{\infty} < 0 \tag{3.10}$$

that is, by recalling eqn. (3.8)

$$\lim_{x_1 \to 0} \Delta(x_1) > 0 \tag{3.11}$$

so that Ben-Naim's result is not consistent (wrong sign) with the experimental evidence given by eqn. (3.11). The same system has been studied by Donkersloot (1979) and his results agree with eqns. (3.10)-(3.11). However, he has reported that

$$\lim_{x_1 \to 0} \Delta(x_1) < 0 \tag{3.12}$$

for methanol (1)-water (2) at 298 K, for which experimental data (Christian et al,1981) show that

$$\left(\frac{\partial \ln \gamma_1}{\partial x_1}\right)^{\infty}_{TP} < 0 \tag{3.13}$$

and eqn. (3.8) requires again

$$\lim_{x_1 \to 0} \Delta(x_1) > 0 \tag{3.14}$$

Usually each G_{ij} contribution is calculated separately (Matteoli and Lepori, 1984). But $\Delta(x)_{TP}$ is sensitive to errors in the G_{ij}'s to the extent that the computed $\Delta(x)_{TP}$ may have the wrong sign. A check for consistency of computed $\Delta(x)_{TP}$ can be made by observing its sign and the kind of deviation undergone by the mixture as shown in Table 1.

The limiting conditions given by eqns. (3.7)-(3.8) can shed light on the composition dependence of $\Delta(x)_{TP}$ for a mixture exhibiting either positive or negative deviation or extrema in γ's with respect to composition, as we will show next.

Positive deviation

If the activity coefficients show no extrema with respect to composition (see Figure 1) then, from eqns. (3.7)-(3.8)

$$\lim_{x_2 \to 0} \Delta(x_2) > 0 \qquad \text{since} \qquad \left(\frac{\partial \ln \gamma_2}{\partial x_1}\right)^{\infty}_{TP} > 0 \tag{3.15}$$

$$\lim_{x_1 \to 0} \Delta(x_1) > 0 \qquad \text{since} \qquad \left(\frac{\partial \ln \gamma_1}{\partial x_1}\right)^{\infty}_{TP} < 0 \tag{3.16}$$

In addition, if no phase split occurs (i.e., no LLE), then $\Delta(x)_{TP}$ is always positive and bounded. Otherwise, $\Delta(x)_{TP}$ may take negative values in the region where two phases coexist. At the incipient immiscibility condition $\Delta(x)_{TP}$ becomes unbounded, i.e., $\Delta(x)_{TP} \rightarrow \infty$ · The three cases are sketched in Figure 2.

Table 1. Types of behavior of $\Delta(x)_{TP}$ for binary mixtures in terms of deviations from the ideal solution.

#	Deviation from an IS	Extrema in $\gamma(x)$	Sign of $\Delta(x)_{TP}$	Discontinuity in $\Delta(x)_{TP}$
1	positive	no	positive	no
2	positive	yes	positive to negative (vice versa)	no
3	positive (LLE)	no	positive	yes
4	positive (LLE)	yes	positive to negative (vice versa)	yes
5	negative	no	negative	no
6	negative	yes	positive to negative (vice versa)	no
7	negative (LLE)	yes	positive to negative (vice versa)	yes
8	positive-negative	yes	positive to negative (vice versa)	no
9	positive-negative (LLE)	yes	positive to negative (vice versa)	yes

Now if the activity coefficients show extrema with respect to composition at $x_1 = x_1^*$, say $\gamma_1(x_1^*)$ is a maximum, then (see Figure 3)

$$\lim_{x_2 \to 0} \Delta(x_2) > 0 \qquad \text{since} \qquad \left(\frac{\partial \ln\gamma_2}{\partial x_1}\right)_{TP}^{\infty} > 0 \tag{3.17}$$

$$\lim_{x_1 \to 0} \Delta(x_1) < 0 \qquad \text{since} \qquad \left(\frac{\partial \ln\gamma_1}{\partial x_1}\right)_{TP}^{\infty} > 0 \tag{3.18}$$

and from eqn. (2.21) , $\Delta(x_1^*) = 0$. As a matter of fact, two extrema are present at $x_1 = x_1^*$, a maximum in $\gamma_1(x_1^*)$ and a minimum in $\gamma_2(x_1)$ (Chialvo, 1988).

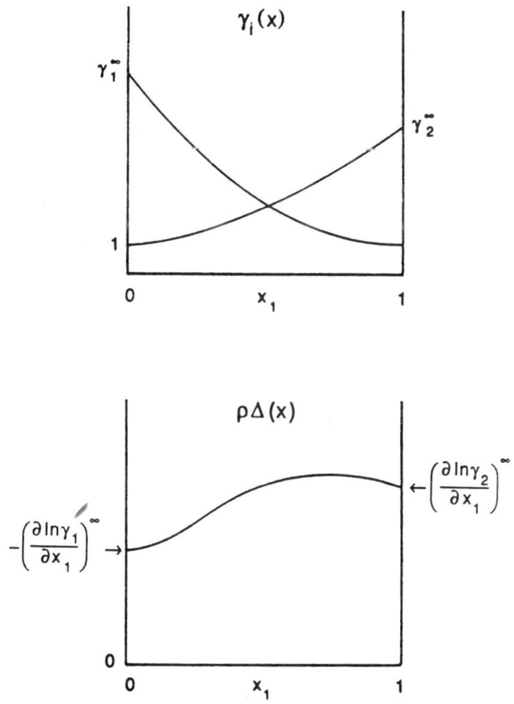

Figure 1. Composition dependence of the activity coefficients of a binary mixture showing positive deviation from the ideal solution, without extrema. The lower diagram shows the expected behavior of the corresponding dependence of $\rho\Delta$.

<u>Negative deviation</u>

If the activity coefficients present no extrema with respect to composition, then (see Figure 4.a)

$$\lim_{x_2 \to 0} \Delta(x_2) < 0 \qquad \text{since} \qquad \left(\frac{\partial \ln \gamma_2}{\partial x_1}\right)^{\infty}_{TP} < 0 \tag{3.19}$$

$$\lim_{x_1 \to 0} \Delta(x_1) < 0 \qquad \text{since} \qquad \left(\frac{\partial \ln \gamma_1}{\partial x_1}\right)^{\infty}_{TP} > 0 \tag{3.20}$$

In addition, if extrema are present, say $\gamma_1(x_1^*)$ is a maximum with respect to composition, then (see Figure 4.b)

$$\lim_{x_2 \to 0} \Delta(x_2) > 0 \qquad \text{since} \qquad \left(\frac{\partial \ln\gamma_2}{\partial x_1}\right)^{\infty}_{TP} > 0 \tag{3.21}$$

$$\lim_{x_1 \to 0} \Delta(x_1) < 0 \qquad \text{since} \qquad \left(\frac{\partial \ln\gamma_1}{\partial x_1}\right)^{\infty}_{TP} > 0 \tag{3.22}$$

and from eqn. (2.21) $\Delta\left(x_1^*\right) = 0$.

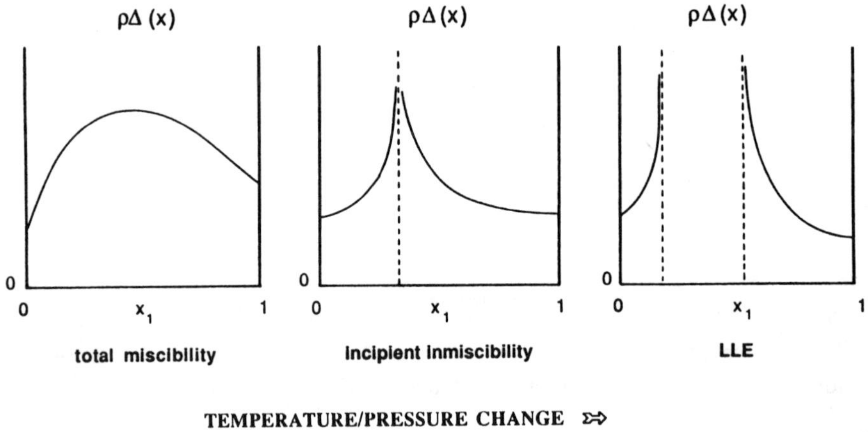

Figure 2. Effect of state conditions on the composition dependence of ρΔ of a binary mixture showing positive deviation from the ideal solution, without extrema. The initial system is a one-phase binary mixture (left). By changing temperature and/or pressure the system might reach the incipient instability condition (center), to finally split into two phases (right)

Positive-negative deviation

A less common case is found when $\gamma_1^{\infty} > 1$ and $\gamma_2^{\infty} < 1$ or vice versa (Rowlinson and Swinton,1982). This constitutes another case of extrema in γ's with respect to composition. For instance, if $\gamma_1^{\infty} > 1$ then (see Figure 5)

$$\lim_{x_2 \to 0} \Delta(x_2) < 0 \qquad \text{since} \qquad \left(\frac{\partial \ln\gamma_2}{\partial x_1}\right)^{\infty}_{TP} < 0 \tag{3.23}$$

$$\lim_{x_1 \to 0} \Delta(x_1) > 0 \qquad \text{since} \qquad \left(\frac{\partial \ln\gamma_1}{\partial x_1}\right)^{\infty}_{TP} < 0 \tag{3.24}$$

and from eqn. (2.21) , $\Delta(x_1^*) = 0$.

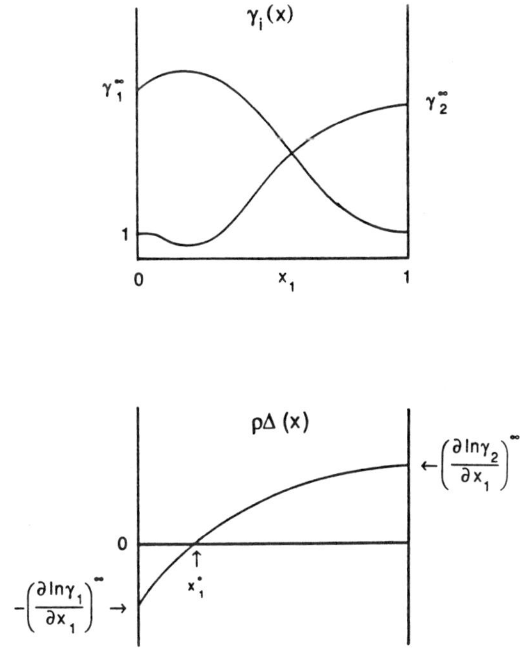

Figure 3. Composition dependence of the activity coefficients of a binary mixture showing positive deviation from the ideal solution, with extrema. The lower diagram shows the expected behavior of the corresponding dependence of $\rho\Delta$.

Discussion

It is generally accepted that LLE occurs when activity coefficients are greater than one and g^E is large and positive (Wisniak,1983; Modell and Reid, 1983; Prausnitz et al,1986). Consequently immiscibility results in systems with large positive deviations from ideal solution. Unfortunately there is no categorical definition for positive or negative deviation for a system. Indeed, if we say that positive (negative) deviation from ideal behavior means activity coefficients greater (smaller) than one, we also have to say that g^E must be greater (smaller) than zero. However, real systems do not always behave that way. In fact, we often find mixtures whose activity coefficients show opposite deviations (Rowlinson and Swinton,1982; Wallas,1985; Prausnitz et al,1986) either in a portion or in the whole range of composition (like those shown in Figures 6 and 7). In these cases the classical definition of positive (negative) deviation becomes meaningless and any statement about the link between LLE and a particular deviation is misleading.

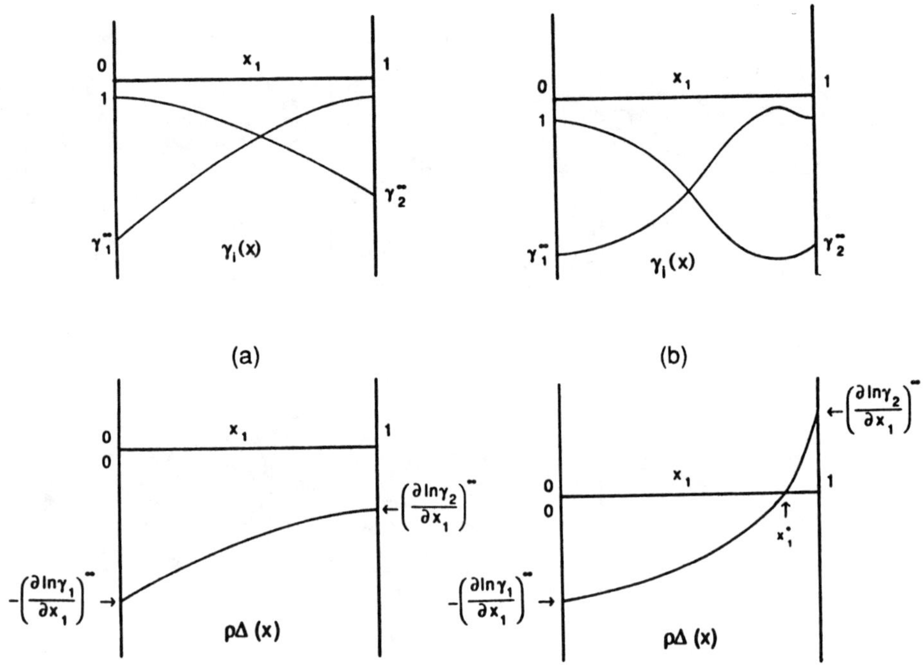

Figure 4. Composition dependence of the activity coefficients of a binary mixture showing negative deviation from the ideal solution: (a) without extrema, (b) with extrema. The lower diagrams show the expected behavior of the corresponding dependence of $\rho\Delta$.

The excess Gibbs free energy is directly related to γ's through eqns. (2.6)-(2.7), i.e.,

$$g^E = kT\left(x_1 \ln\gamma_1 + x_2 \ln\gamma_2\right) \tag{3.25}$$

so $g^E(x) > 0$ (< 0) over the entire range of composition for positive (negative) deviation, if and only if $\gamma(x)$'s present no extrema. Then, it follows

$$\Delta(x)_{TP} > 0 \quad \rightarrow \quad \gamma(x) > 1 \quad \rightarrow \quad g^E(x) > 0 \tag{3.26}$$

and

$$\Delta(x)_{TP} < 0 \quad \rightarrow \quad \gamma(x) < 1 \quad \rightarrow \quad g^E(x) < 0 \tag{3.27}$$

since the integrands of eqn. (2.8) are monotonic functions of composition and do not change signs. Now, if the γ's present extrema, then the integrands of eqn. (2.8) change signs at $x_1 = x_1^*$, i.e., $\Delta(x_1^*) = 0$, and consequently the sizes of γ's do not indicate the sign of either $g^E(x)$ or $\Delta(x)_{TP}$ (Chialvo,1988).

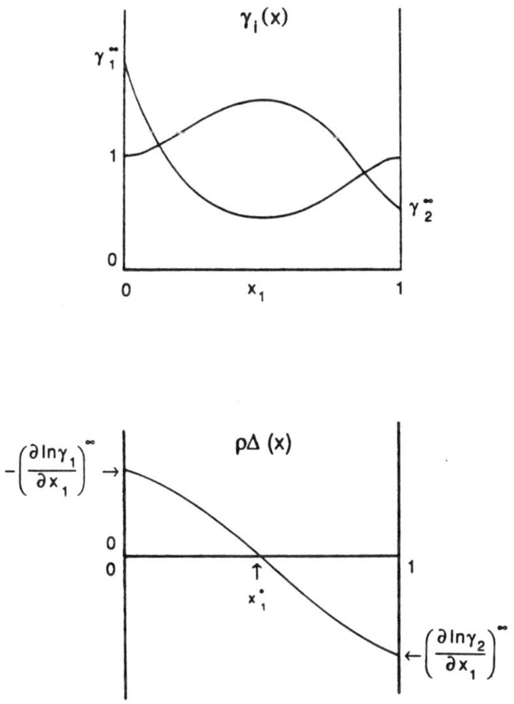

Figure 5. Composition dependence of the activity coefficients of a binary mixture showing positive-negative deviation from the ideal solution (i.e., extrema). The lower diagram shows the expected behavior of the corresponding dependence of $\rho\Delta$.

Mathematically, extrema in the γ's are necessary conditions to have LLE, in systems with either negative or positive-negative deviations (see Figures 4.b and 5). For these cases $g^E(x)$ could be negative over the whole range of composition (eventually over a portion). However $\Delta(x)_{TP}$ takes positive (unbounded) values at least about the composition where incipient immiscibility could occur (see Figures 6 and 7). $\Delta(x) > 0$ means stronger attraction among particles of the same species, i.e., $G_{12} < 0.5\,(G_{11} + G_{22})$, which promotes phase separation. Then LLE may occur in binary mixtures with $g^E(x)$ either positive or negative depending on: (a) whether both

γ's are larger than one over the entire range of composition (class 3 in Table 1), i.e., $g^E(x) > 0$, (b) whether the system shows extrema in γ's (classes 4,7,9 in Table 1), i.e., $g^E(x) \lessgtr 0$, and (c) whether $\Delta(x)_{TP} > 0$ becomes unbounded at a particular composition.

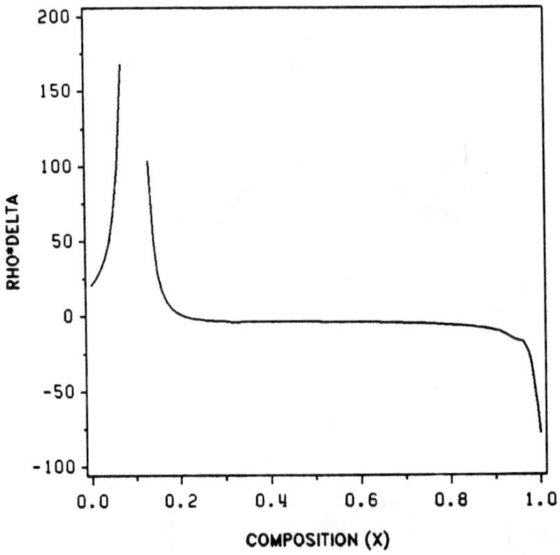

Figure 6. Composition dependence of $\rho\Delta$ predicted by Margules equation for a binary system showing negative deviation from the ideal solution, with extrema. A=-5.967 , B=-22.428. At x_1=0.1 and x_1=0.21 the system shows incipient instability and extrema in the activity coefficients, respectively. The corresponding $\gamma(x)$ diagram is given by Figure 4.b

4. EXCESS GIBBS FREE ENERGY MODELS IN TERMS OF K-B INTEGRALS

In this section we express several existing models of $g^E(x)$ in terms of $[\rho\Delta(x)]_{TP}$ to show how the previous analysis can be applied: (a) to bound adjustable parameters for each kind of phase behavior of Table 1, and (b) to interpret the success of those models in representing VLE and/or LLE.

In what follows, we will study: (a) the behavior of $[\rho\Delta(x)]_{TP}$ rather than $\Delta(x)_{TP}$, since the latter always appear multiplied by the total density ρ, and (b) the functional form of $\{[\rho\Delta(x)]_{TP}\}^{-1}$ to avoid singularities, e.g., in the case of LLE.

From eqn. (2.8) we can write for any binary g^E-model the following expression

$$\beta\left(\frac{\partial^2 g^E}{\partial x_1^2}\right)_{TP} = -\left[(\rho\Delta)^{-1} + x_1 x_2\right]^{-1} \equiv F(TPx) \tag{4.1}$$

so that

$$\left[(\rho\Delta)_{TP}\right]^{-1} = -\frac{(1 + F x_1 x_2)}{F} \tag{4.2}$$

For traditional g^E-models which involve only logarithmic and power functions of composition, F(PTx) reduces to rational functions, i.e.,

$$[F(x)]_{TP} = \frac{f(x)}{g(x)} \tag{4.3}$$

and then, from eqns. (4.1) and (4.2)

$$\left[(\rho\Delta)_{TP}\right]^{-1} = \frac{O(x)}{P(x)} \tag{4.4}$$

where O(x) and P(x) are polynomials of order **n** and **m** respectively with $\mathbf{m} \le \mathbf{n} \le \mathbf{m+2}$. It may happen that the residue of the ratio in eqn. (4.4) is zero, then $\{[\rho\Delta(x)]_{TP}\}^{-1}$ reduces to a simple polynomial in composition (see Table 2).

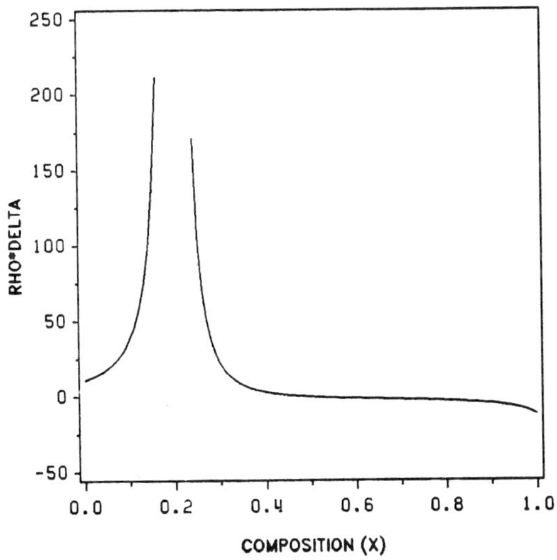

Figure 7. Composition dependence of $\rho\Delta$ predicted by Margules equation for a binary system showing positive-negative deviation from the ideal solution. A=1.5625 , B=-2.3437. At x_1=0.2 and x_1=0.466 the system shows incipient instability, and extrema in the activity coefficients, respectively. The corresponding $\gamma(x)$ diagram is given by Figure 5.

Table 2. $[\rho\Delta(x)_{TP}]^{-1}$ representation for existing g^E-models (Chialvo, 1988)

g^E model	$[\rho\Delta(x)_{TP}]^{-1}$ representation: Rational function [order O(x)/P(x)]	polynomial (order)	# of Coeffs. for $[\rho\Delta(x)_{TP}]^{-1}$ representation	# of model parameters
Two-suffix Margules	---	2nd.	1	1
Three-suffix Margules	3rd./1st.	----	3	2
Van Laar	----	3rd.	4	2
Redlich-Kister (2 parameters)	3rd./1st.	----	3	2
Wilson	5th./3rd.	----	10	2
NRTL	6th./4th.	----	14	3

Now we apply eqn. (4.2) to three representative g^E-models, namely: (a) Margules, (b) Wilson and (c) NRTL. By using the resulting $\{[\rho\Delta\,(x)]_{TP}\}^{-1}$ equations we analyse their limiting conditions, incipient instability , and the existence of extrema in γ's.

Margules : $g^E/RT = (Ax_2 + Bx_1)\,x_1x_2$

$$(\rho\Delta)^{-1} = \frac{1 - 2(2A - B)x_1 + (10A - 8B)x_1^2 + 6(B - A)\,x_1^3}{2(2A - B) + 6(B - A)\,x_1} \tag{4.5}$$

and its limiting conditions are

$$\lim_{x_2 \to 0} \rho\Delta = 2(2B - A) \tag{4.6}$$

$$\lim_{x_1 \to 0} \rho\Delta = 2(2A - B) \tag{4.7}$$

Setting eqns. (4.6) and (4.7) to zero, defines four different regions in the A-B plane corresponding to the classes 1-9 in Table 1 as shown in Figure 8.

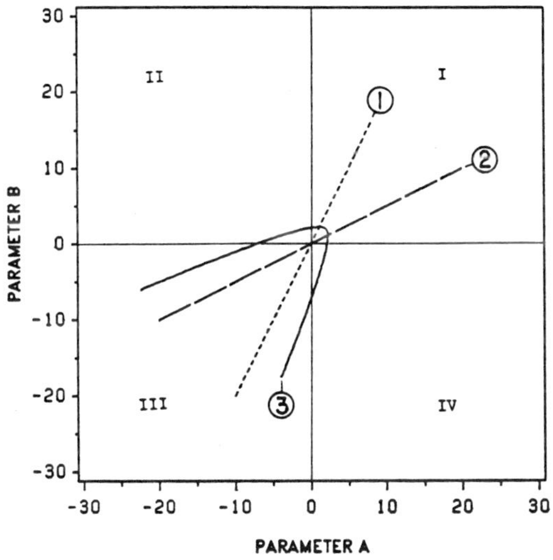

Figure 8. Plane of variability of parameters of Margules equation. The plane is divided in four quadrants according to the relative sizes of γ_1^∞ and γ_2^∞, i.e. , **I**: $\gamma_1^\infty>1$ and $\gamma_2^\infty>1$; **II**: $\gamma_1^\infty<1$ and $\gamma_2^\infty>1$; **III**: $\gamma_1^\infty<1$ and $\gamma_2^\infty<1$; and **IV**: $\gamma_1^\infty>1$ and $\gamma_2^\infty<1$. To the left of line 1 and to the right of line 2 in **I** lie systems showing positive deviation with extrema. To the right of line 1 and to the left of line 2 in **III** lie systems showing negative deviation with extrema. In the region enclosed by lines 1 and 2 lie systems showing: positive deviation in **I** and negative deviation in **III**, without extrema, respectively. Systems lying to the right of curve 3 show LLE.

To locate the curve in the A-B plane where incipient immiscibility occurs, we apply eqns. (2.16) and (2.17); that is

$$1-2(2A-B)x_1+(10A-8B)x_1^2-6(A-B)x_1^3 = 0 \tag{4.8}$$

$$(B-2A)+(10A-8B)x_1+9(B-A)x_1^2 = 0 \tag{4.9}$$

with

$$2(2B-A)+6(B-A)x_1 \neq 0 \tag{4.10}$$

whose solution is (Chialvo, 1988)

$$A=\frac{-9x_1^2+8x_1-1}{6x_1^2x_2^2} \tag{4.11}$$

$$B = \frac{-9x_1^2+10x_1-2}{6x_1^2x_2^2} \tag{4.12}$$

Simultaneous solution of eqns. (4.11) and (4.12) results in A = f(B). This curve divides the A-B plane into two regions corresponding to one- and two-phase systems (see curve 3 in Figure 8).

From eqns. (2.21)-(2.22) and (4.5) extrema in γ's are expected at

$$x_1^* = \frac{2A-B}{3(A-B)} \tag{4.13}$$

<u>Wilson:</u> $g^E/RT = -x_1 \ln(x_1 + \Lambda_{12} x_2) - x_2 \ln(x_2 + \Lambda_{21} x_1)$

Eqn. (4.2) becomes

$$(\rho\Delta)^{-1} = -\frac{\dfrac{x_2\Lambda_{12}^2}{(x_1+x_2\Lambda_{12})^2} + \dfrac{x_1\Lambda_{21}^2}{(x_2+x_1\Lambda_{21})^2}}{\dfrac{\Lambda_{12}^2}{x_1(x_1+x_2\Lambda_{12})^2} + \dfrac{\Lambda_{21}^2}{x_2(x_2+x_1\Lambda_{21})^2} - \dfrac{1}{x_1x_2}} \tag{4.14}$$

and its limiting conditions

$$\lim_{x_2\to 0} \rho\Delta = \frac{2-\Lambda_{21}\left(1+\Lambda_{12}^2\right)}{\Lambda_{21}} \tag{4.15}$$

$$\lim_{x_1\to 0} \rho\Delta = \frac{2-\Lambda_{12}\left(1+\Lambda_{21}^2\right)}{\Lambda_{12}} \tag{4.16}$$

Eqns. (4.15) and (4.16) set to zero define four regions in the Λ_{12}–Λ_{21} plane, two of them correspond to systems with positive and negative deviation respectively and the other two regions correspond to systems with extrema in γ's. Moreover, each region where extrema in $\gamma(x)$ occurs can be divided into three subregions (separated by the curves $\gamma_1^\infty = 1$ and $\gamma_2^\infty = 1$, regions 3,4 and 5 in Figure 9) corresponding to the classes 2 (positive deviation), 6 (negative deviation), and 8 (positive-negative deviation) of Table 1.

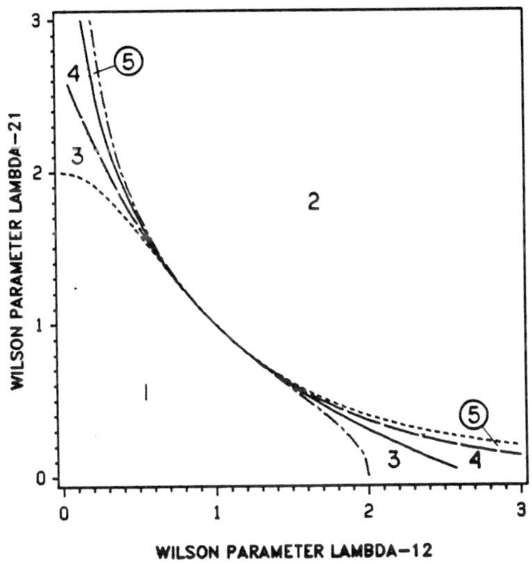

Figure 9. Plane of variability of parameters of Wilson equation. Five regions are defined in the positive quadrant according to the relative sizes of γ_1^∞ and γ_2^∞, i.e. , **1**: positive deviation ; **2**: negative deviation; **3**: positive deviation with extrema; **4**: positive-negative deviation (i.e., extrema) and **5**: negative deviation with extrema.

The simultaneous solution of eqns. (4.15) and (4.16) for positive-negative deviation, i.e., when they have different signs (class 8 in Table 1) results in

$$(\Lambda_{12} - \Lambda_{21})(1 - \Lambda_{12}\Lambda_{21}) > 0 \tag{4.17}$$

with

$$\lim_{x_2 \to 0} \rho\Delta > 0 \tag{4.18}$$

$$\lim_{x_1 \to 0} \rho\Delta < 0 \tag{4.19}$$

so that

$$\Lambda_{12} > \Lambda_{21} \tag{4.20}$$

$$\Lambda_{12}\Lambda_{21} > 1 \tag{4.21}$$

or

$$\Lambda_{12} < \Lambda_{21} \tag{4.22}$$

$$\Lambda_{12}\Lambda_{21} < 1 \tag{4.23}$$

Then from eqns. (4.18)-(4.20) we have

$$2 - \Lambda_{21}\left(1 + \Lambda_{12}^2\right) > 2 - \Lambda_{21}\left(1 + \Lambda_{21}^2\right) > 0 \tag{4.24}$$

$$\Lambda_{21} < 1 \tag{4.25}$$

and

$$2 - \Lambda_{12}\left(1 + \Lambda_{12}^2\right) > 2 - \Lambda_{12}\left(1 + \Lambda_{21}^2\right) > 0 \tag{4.26}$$

$$\Lambda_{12} > 1 \tag{4.27}$$

but the product of eqns. (4.25) and (4.27) has to satisfy eqn. (4.21). Λ_{12} and Λ_{21} given by eqns. (4.25) and (4.27) correspond to the region on the right of Λ_{12}–Λ_{21} plane (Figure 9). By means of a similar analysis, from eqns. (4.22) and (4.23) we find

$$\Lambda_{12} < 1 \tag{4.28}$$

$$\Lambda_{21} > 1 \tag{4.29}$$

which correspond to the region on the left of the Λ_{12}–Λ_{21} plane.

Contrary to the accepted idea in literature (Orye and Prausnitz, 1965; Prausnitz et al, 1967; Prausnitz et al, 1986) the existence of extrema in $\gamma(x)$ is not restricted to systems close to ideal solutions, i.e., $\gamma_i^\infty = 1$. For instance, for a system showing a positive-negative deviation (class 8 in Table 1) we can have $\Lambda_{12} = 0.01$ and $\Lambda_{21} = 3.0$, so that $\gamma_1^\infty = 13.53$ and $\gamma_2^\infty = 0.897$ with extrema at $x_1 = 0.24$ as shown in Figure 10 ; or the case of a positive deviation with $\Lambda_{12} = 0.05$ and $\Lambda_{21} = 2.5$, so that $\gamma_1^\infty = 4.46$ and $\gamma_2^\infty = 1.034$ with extrema at $x_1 = 0.43$ as shown in Figure 11.

The composition at which the extrema occurs can be calculated from eqns. (2.21) and (4.14), that is

$$F = \frac{\Lambda_{12}^2}{x_1(x_1 + x_2\Lambda_{12})^2} + \frac{\Lambda_{21}^2}{x_2(x_2 + x_1\Lambda_{21})^2} - \frac{1}{x_1x_2} = 0 \tag{4.30}$$

or (i.e., excluding the trivial roots x_2=1 and x_2=0)

$$A^2B^2x_2^3 + \left\{A\left[2B\left(1 + \Lambda_{21}(A-1)\right) + \Lambda_{21}^2 A\right]\right\}x_2^2 +$$

$$\left\{B\left[B + 2\Lambda_{21}\left(2A - \Lambda_{12}^2\right)\right] + 2A\Lambda_{21}^2\right\}x_2 +$$

$$\Lambda_{21}\left[1 - B\left(1 + \Lambda_{12}^2\right)\right] = 0 \quad , \quad A = \Lambda_{12} - 1 \quad , \quad B = 1 - \Lambda_{21} \tag{4.31}$$

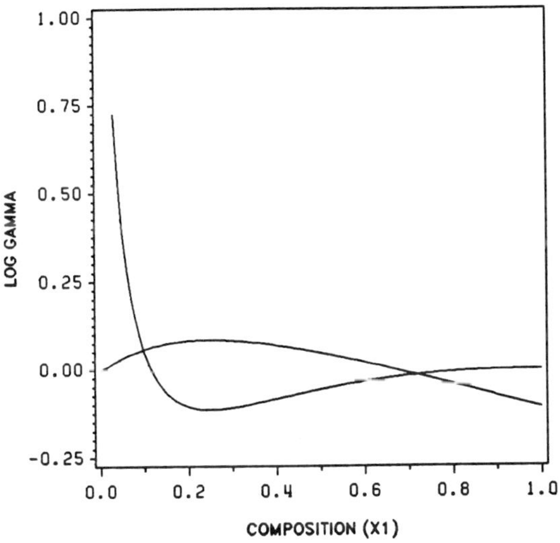

Figure 10. Composition dependence of the activity coefficients given by Wilson equation for a binary system showing positive-negative deviation (extrema) from the ideal solution, i.e., Λ_{12}=0.010 and Λ_{21}=3.0, corresponding to $\gamma_1^\infty = 13.53$ and $\gamma_2^\infty = 0.897$, respectively.

Now from equation (2.16) and (4.2) a necessary condition for incipient immiscibility is

$$Fx_1x_2 + 1 = \frac{x_2\Lambda_{12}^2}{(x_1 + x_2\Lambda_{12})^2} + \frac{x_1\Lambda_{21}^2}{(x_2 + x_1\Lambda_{21})^2} = 0 \qquad (4.32)$$

Obviously eqn. (4.32) can never be satisfied by nonzero values of Λ_{12} and Λ_{21} (finite values of γ_i^∞), since the lhs of eqn. (4.32) is positive definite. Thus, Wilson model cannot handle unbounded values of $\Delta(x)_{TP}$, i.e., LLE.

Non-random Two liquid (NRTL): $g^E/RT = x_1x_2\left[\frac{\Gamma_{21}\vartheta_{21}}{x_1 + \vartheta_{21}x_2} + \frac{\Gamma_{12}\vartheta_{12}}{x_2 + \vartheta_{12}x_1}\right]$

$$\vartheta_{ij} = \exp(-\alpha\Gamma_{ij}) \qquad (4.33)$$

so that

$$(\rho\Delta)^{-1} = \frac{1 - 2x_1x_2\left[\dfrac{\Gamma_{21}\vartheta_{21}^2}{(x_1 + x_2\vartheta_{21})^3} + \dfrac{\Gamma_{12}\vartheta_{12}^2}{(x_2 + x_1\vartheta_{12})^3}\right]}{2\left[\dfrac{\Gamma_{21}\vartheta_{21}^2}{(x_1 + x_2\vartheta_{21})^3} + \dfrac{\Gamma_{12}\vartheta_{12}^2}{(x_2 + x_1\vartheta_{12})^3}\right]} \tag{4.34}$$

Then the limiting conditions for $[\rho\Delta]_{PT}$ become

$$\lim_{x_2 \to 0} \rho\Delta = 2\left[\frac{\Gamma_{12}}{\vartheta_{12}} + \Gamma_{21}\vartheta_{21}^2\right] \tag{4.35}$$

$$\lim_{x_1 \to 0} \rho\Delta = 2\left[\frac{\Gamma_{21}}{\vartheta_{21}} + \Gamma_{12}\vartheta_{12}^2\right] \tag{4.36}$$

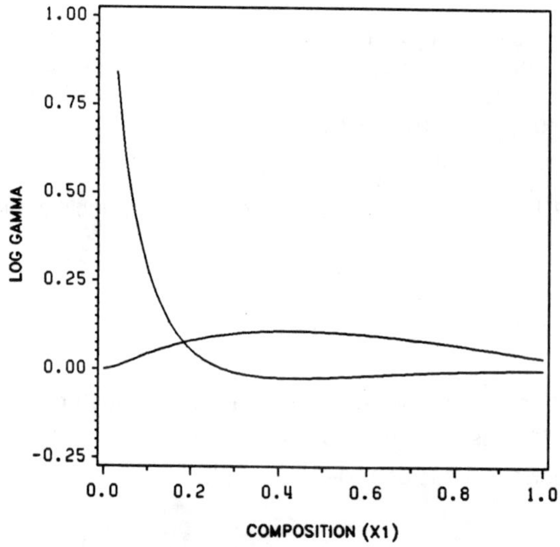

Figure 11. Composition dependence of the activity coefficients given by Wilson equation for a binary system showing positive deviation with extrema from the ideal solution, i.e., $\Lambda_{12} = 0.05$ and $\Lambda_{21} = 2.5$, corresponding to $\gamma_1^\infty = 4.46$ and $\gamma_2^\infty = 1.034$, respectively.

By equating eqns. (4.35) and (4.36) to zero we define four regions in the $\Gamma_{12} - \Gamma_{21}$ plane, parametric in α (see Figure 12). Two of those regions correspond to systems with positive and negative deviations and the others to systems with extrema in $\gamma(x)$. Each region where extrema in $\gamma(x)$ occurs can be divided in three additional subregions by means of the curves $\gamma_1^\infty = 1$ and $\gamma_2^\infty = 1$ (regions 3,4 and 5 in Figure 12) corresponding to classes 2 (positive deviation) , 6 (negative deviation) and 8 (positive-negative deviation) of Table 1.

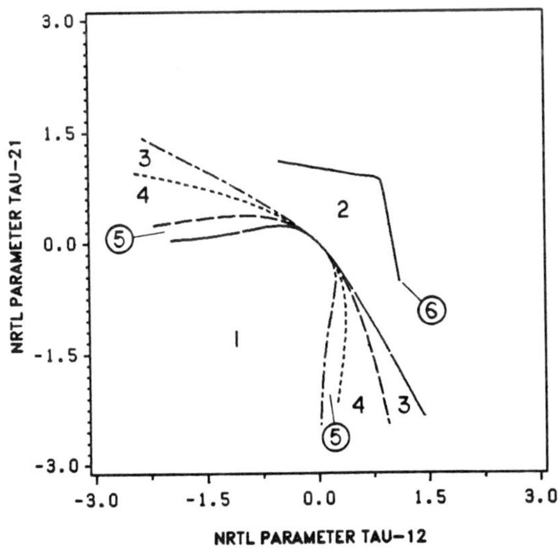

Figure 12. Plane of variability of parameters of NRTL equation with α=-1. The five labeled regions are defined according to the relative sizes of γ_1^∞ and γ_2^∞, i.e. , **1**: negative deviation ; **2**: positive deviation; **3**: positive deviation with extrema; **4**: positive-negative deviation (i.e., extrema) and **5**: negative deviation with extrema. Systems lying to the right of curve 6 show LLE.

The simultaneous solution of eqns. (4.35) and (4.36) for positive-negative deviation (class 8 in Table 1) yields

$$\vartheta_{12}\vartheta_{21} = \exp[-\alpha(\Gamma_{12} + \Gamma_{21})] < 1 \tag{4.37}$$

with

$$\lim_{x_2 \to 0} \rho\Delta > 0 \tag{4.38}$$

$$\lim_{x_1 \to 0} \rho\Delta < 0 \tag{4.39}$$

so that there are two possible sets of $\vartheta_{12} - \vartheta_{21}$ that satisfy eqn. (4.37)(recall eqn. (4.33))

$$\vartheta_{12} > 1 \quad , \quad \Gamma_{12} < 0 \tag{4.40}$$

$$\vartheta_{21} < 1 \quad , \quad \Gamma_{21} > 0 \tag{4.41}$$

corresponging to the region on the left of the $\Gamma_{12} - \Gamma_{21}$ plane (Figure 12) and

$$\vartheta_{12} < 1 \quad , \quad \Gamma_{12} > 0 \tag{4.42}$$

$$\vartheta_{21} > 1 \quad , \quad \Gamma_{21} < 0 \tag{4.43}$$

corresponging to the region on the right of the $\Gamma_{12} - \Gamma_{21}$ plane as shown in Figure 12 with α = -1 (Marina and Tassios, 1973; Wallas, 1985)

The composition at which extrema occur can be evaluated from eqns. (2.21), that is

$$F = -2\left[\frac{\Gamma_{21}\vartheta_{21}^2}{(x_1 + x_2\vartheta_{21})^3} + \frac{\Gamma_{12}\vartheta_{12}^2}{(x_2 + x_1\vartheta_{12})^3}\right] = 0 \tag{4.44}$$

or (Chialvo, 1988)

$$x_1^{*} = \frac{A - \vartheta_{21}}{(1-\vartheta_{21}) + A(1-\vartheta_{12})} \quad , A = \sqrt[3]{\frac{\Gamma_{21}\vartheta_{21}^2}{\Gamma_{12}\vartheta_{12}^2}} \tag{4.45}$$

Finally the condition of incipient immiscibility, eqn. (2.16) becomes

$$1 + Fx_1x_2 = (1 - 2x_1)F^2 - \left(\frac{\partial F}{\partial x_1}\right)_{TP} = 0 \tag{4.46}$$

The numerical solution of eqn. (4.46) is given in Figure 12 (curve 6).

Discussion:

The analysis performed in this section shows that simple g^E-models have rather complex $\rho\Delta$-representation (rational functions) with only a few independent coefficients (equal to the number of g^E-adjustable parameters ; seeTable 2). For instance, the Margules' model has only two adjustable parameters, A and B, but its $\rho\Delta$-representation has three: c_1 = -2 (2A - B), c_2 = (10A - 8B) and c_3 = -6 (A - B) with $c_1+c_2+c_3$ = 0 (see eqn. (4.5)). This raises two important questions: (a) whether a complicated $\rho\Delta$-functional form is needed to build a reasonably flexible g^E-model, and (b) whether a simple g^E-model satisfies the limiting conditions of infinite dilution, i.e., γ_i^∞ and $(\partial \ln\gamma_i/\partial x_i)^\infty$, which are directly related to the coefficients of the $\rho\Delta$-representation.

Usually the $\rho\Delta$-representation (derivative formulation) is a consequence of an existing correlation because correlations for excess functions typically start with the formulation of either a new or a modified g^E-expression (integral formulation). Since g^E and $\rho\Delta$ are related by a second derivative (see eqn. (4.1)-(4.2)) the g^E-expression must be flexible enough to map the g^E-experimental data onto the $\rho\Delta$-experimental data, avoiding unexpected points of inflexion and extrema (Kato et al, 1982). For instance, Gaw and Swinton (1968) have described the system hexafluorobenzene (1) - benzene (2) by a third order Redlich-Kister model

$$g^E = x_1 x_2\left[A + B(x_2 - x_1) + C(x_2 - x_1)^2\right] \tag{4.47}$$

with

$$A = -176 + 6.033\,(T - 313.15) - 0.0134\,(T - 313.15)^2$$
$$B = -486 + 2.913\,(T - 313.15) - 0.0196\,(T - 313.15)^2$$
$$C = -161 + 1.540\,(T - 313.15)$$

where g^E is in Joule/gmol and T in Kelvin. This system shows a positive-negative deviation (class 8 in Table 1), i.e., γ's present extrema, but no LLE is found over the range of temperature studied. Although the system is not strictly isobaric, the maximum variation of total pressure is very small, i.e., ~ 0.002 MPa. Thus $\rho\Delta$ must be zero at some composition, but it has to be always bounded. However, the corresponding $\rho\Delta$-representation of eqn. (4.47) at T=313.15 K shows not only a zero value for $\Delta(x_1)$, i.e., $\Delta(0.6018) = 0$, but also discontinuities at $\Delta(0.6023) \sim \infty$ and $\Delta(0.9997) \sim \infty$ (see Figure 13).

One reason for the unwanted behavior found for $\rho\Delta(x)$ can be because the common practice in data reduction is to mimic experimental data by using a statistical method, e.g., maximum likelihood, to obtain the best fit without giving attention to the limiting conditions of infinite dilution. In that way, for a set of experimental data, infinite dilution quantities become model-dependent. For instance, Zou and Prausnitz (1987) have studied the system 2-butanone (1) - water (2) using several g^E-models and compared the set of γ_i^∞ correlated against the corresponding experimental values, as shown in Table 3.

From the chemical engineering point of view it does not matter whether or not a particular correlation agrees with the experimental values of γ_i^∞ and / or $(\partial \ln\gamma_i/\partial x_i)^\infty$ as long as the model used fits the data within an acceptable statistical interval of confidence. From the thermodynamic standpoint, the lack of agreement in the infinite dilution quantities constitutes an inconsistency. The activity coefficients and their slopes at infinite dilution are thermodynamic constraints as $\gamma_i = 1$ and $(\partial \ln\gamma_i / \partial x_i) = 0$ are for the limit behavior of pure component i when the standard state is given by the Raoult's law. Therefore, the infinite dilution quantities have to be used, when available, to constraint the data reduction and avoid inconsistencies.

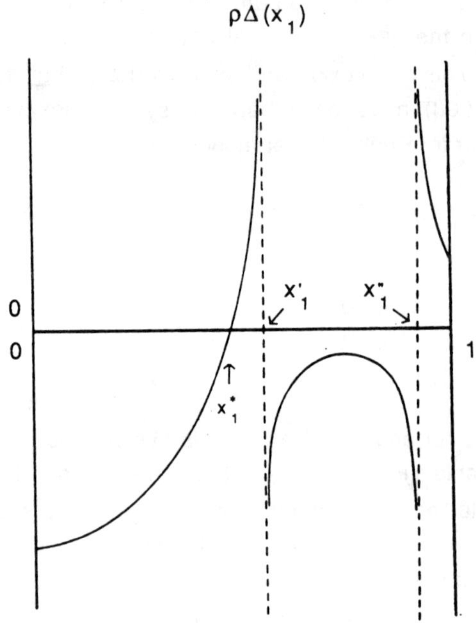

Figure 13. Composition dependence (no scale drawing) of $\rho\Delta$ at T=313K for hexafluorobenzene(1)-benzene(2) (Gaw and Swinton, 1968). At x_1=0.6018 the system shows extrema in $\gamma(x)$, i.e., $\rho\Delta(0.6018)=0$. At x_1=0.6023 and x_1=0.9997, $\rho\Delta(x_1)$ becomes unbounded, i.e., LLE.

In recent years emphasis has been placed on collecting reliable data not only for γ_i^∞ (Valentin and Guiochon, 1975; Kikic and Renon, 1976; Loblen and Prausnitz, 1982; Sager and Lau, 1986; Shen <u>et al.</u> 1986; Ochi and Kojima, 1987 ; Wobst and Hradetzky, 1988) but also for $(\partial \ln\gamma_i / \partial x_i)^\infty$ (Valentin and Guiochon, 1975; Kikic and Renon, 1976; Shen <u>et al.</u> 1986; Wobst and Hradetzky, 1988). However, the goal of that work has been comparison of experimental with correlated data. Unfortunately data for $\gamma_i(x)$ are usually not published together with the corresponding γ_i^∞ (and vice versa) so that any attempt to determine $(\partial \ln\gamma_i / \partial x_i)^\infty$ from them is impossible.

Table 3. Activity coefficients at infinite dilution of 2-butanone (1)-water (2) correlated by several models.

Model	γ_1^∞	γ_2^∞	Ref.
Redl ich-Kister (4 parameters)	31.7	7.26	(a)
Van Laar	30.0	6.04	(a)
NRTL	32.4	7.18	(a)
UNIQUAC	31.6	6.52	(a)
Feix et al.	32.1	6.50	(b)
Tsuboka and Katayama	33.3	7.01	(c)
Kreglewski and Hall	31.5	7.46	(d)
Wang-Chao (I)	37.0	7.04	(e)
Wang-Chao (II)	33.7	6.54	(e)
Van Laar modified	30.02	6.04	(e)
Experimental	29.5± 2.5	6.4±0.2	(e)

a) Wallas, (1985) b) Feix et al. (1980) c) Tsuboka and Katayama, (1975)
d) Kreglewski et al. (1985) e) Zou and Prausnitz, (1986)

5. SIMPLE $\rho\Delta(x)$ REPRESENTATIONS FOR BINARY MIXTURES

An exact description of $g^E(x)$ will necessarily provide an exact mapping onto $\rho\Delta(x)$; that is, a successful description of $\rho\Delta(x)$ depends strongly on the accuracy of $g^E(x)$ because $\rho\Delta(x)$ and $g^E(x)$ are related through a second derivative with respect to composition (eqns. (4.1)-(4.2)). However the inverse is not necessarily true. Indeed a simple and crude $\rho\Delta$-functional form may result in a useful g^E-model, provided the infinite dilution quantities are used to constrain the data reduction. There are two reasons to expect success: (a) because the mathematical relation between $\rho\Delta(x)$ and $g^E(x)$ is exact, and (b) because the double integration involved should smooth differences between the actual $\rho\Delta(x)$ (unknown) and the crude approximation used (model). Paradoxically a simple $\rho\Delta(x)$ functional form becomes a rather complicated $g^E(x)$ expression.

Rational Function Approach

We adopt the general rational function approximation to $[\rho\Delta(x)]_{TP}$ as

$$[\rho\Delta(x)]_{TP} = \frac{\sum_{i=0}^{n} a_i x_1^i}{\sum_{j=0}^{m} b_j x_1^j} \tag{5.1}$$

Three simple forms from eqn. (5.1) are the following

$$[\rho\Delta(x)]_{TP} = \frac{a_0}{b_0} \tag{5.2}$$

$$[\rho\Delta(x)]_{TP} = \frac{a_0}{b_0 + b_1 x_1} \tag{5.3}$$

$$[\rho\Delta(x)]_{TP} = \frac{a_0}{b_0 + b_1 x_1 + b_2 x_1^2} \tag{5.4}$$

which result in $\left(\partial^2 g^E/\partial x_1^2\right)_{TP} = \left(\eta + \varphi x_1 + \sigma x_1^2\right)^{-1}$ with η, φ and $\sigma \neq 0$.

The first $\rho\Delta$-representation is very restrictive; it results in the special behavior $\gamma_1^\infty = \gamma_2^\infty$ with the symmetry $\gamma_1(x_1) = \gamma_2(x_2)$ analogous to that found in the two suffix Margules (Wallas, 1985). However, the equations are rather complex, that is,

$$\ln\gamma_i = -\left(\frac{0.5}{A}\right)\ln\left(1 + Ax_j - Ax_j^2\right) - \left(\frac{1}{\sqrt{\delta}}\right)\left[\tan^{-1}\left(\frac{A - 2Ax_j}{\sqrt{\delta}}\right) - \tan^{-1}\left(\frac{A}{\sqrt{\delta}}\right)\right] \tag{5.5}$$

for negative deviation, and

$$\ln\gamma_i = -\left(\frac{0.5}{A}\right)\ln\left(1 + Ax_j - Ax_j^2\right) - \left(\frac{1}{\sqrt{\delta}}\right)\ln\left[\frac{\left(\sqrt{-\delta} - A + 2Ax_j\right)\left(\sqrt{-\delta} + A\right)}{\left(\sqrt{-\delta} + A - 2Ax_j\right)\left(\sqrt{-\delta} - A\right)}\right] \tag{5.6}$$

for positive deviation, with $\delta = -A(4+A)$ and $A = a_0 / b_0$.

The second $\rho\Delta$-representation, eqn. (5.3), can be recast as

$$[\rho\Delta(x)]_{TP} = \frac{A}{(1 + Bx_1)} \tag{5.7}$$

with A= a_o / b_o and B= b_1 / b_o . This expression accounts for both positive and negative deviations, though it cannot describe extrema in $\gamma(x)$ or LLE. The corresponding expressions for γ's are

$$\ln\gamma_1 = -\,0.5\,\ln\left[\frac{1+(A+B)x_1 - Ax_1^2}{(1+B)}\right]$$

$$+\left(\frac{A-B}{2\sqrt{-\delta}}\right)\ln\left[\frac{(\sqrt{-\delta}+A+B-2Ax_1)(\sqrt{-\delta}+A-B)}{(\sqrt{-\delta}-A-B+2Ax_1)(\sqrt{-\delta}-A+B)}\right] \tag{5.8}$$

$$\ln\gamma_2 = -\,0.5\,\ln\left[1+B+(A-B)x_2 - Ax_2^2\right]$$

$$+\left(\frac{A-B}{2\sqrt{-\delta}}\right)\ln\left[\frac{(\sqrt{-\delta}+A-B-2Ax_2)(\sqrt{-\delta}+A+B)}{(\sqrt{-\delta}-A+B+2Ax_2)(\sqrt{-\delta}-A-B)}\right] \tag{5.9}$$

for positive deviations, and

$$\ln\gamma_1 = -\,0.5\,\ln\left[\frac{1+(A+B)x_1 - Ax_1^2}{(1+B)}\right]$$

$$+\left(\frac{A-B}{\sqrt{\delta}}\right)\left[\tan^{-1}\left(\frac{2Ax_1 - A - B}{\sqrt{\delta}}\right) - \tan^{-1}\left(\frac{A-B}{\sqrt{\delta}}\right)\right] \tag{5.10}$$

$$\ln\gamma_2 = -\,0.5\,\ln\left[1+B+(A-B)x_2 - Ax_2^2\right]$$

$$+\left(\frac{A+B}{\sqrt{\delta}}\right)\left[\tan^{-1}\left(\frac{2Ax_2 - A + B}{\sqrt{\delta}}\right) - \tan^{-1}\left(\frac{A+B}{\sqrt{\delta}}\right)\right] \tag{5.11}$$

for negative deviations, with $\delta = -4A - (A+B)^2$. The bound between positive and negative deviations is given by the condition $\delta = 0$ (Chialvo, 1988).

The third $\rho\Delta$-representation, eqn. (5.4), can be recast as

$$[\rho\Delta(x)]_{TP} = \frac{A}{\left(1+Bx_1+Cx_1^2\right)} \tag{5.12}$$

with $C = c_1 / b_o$. This functional form describes positive and negative deviations as well as LLE. The corresponding equations for the activitiy coefficients are

$$\ln \gamma_1 = \left(\frac{0.5A}{C-A}\right) \ln \left[\frac{1+(A+B)x_1+(C-A)x_1^2}{1+B+C}\right] - \left(\frac{A[(A+B)+2(C-A)]}{2(C-A)\sqrt{-\delta}}\right) \times$$

$$\ln \left\{\frac{[\sqrt{-\delta}+(A+B)+2(C-A)x_1][\sqrt{-\delta}-(A+B)-2(C-A)]}{[\sqrt{-\delta}-(A+B)-2(C-A)x_1][\sqrt{-\delta}+(A+B)+2(C-A)]}\right\} \quad (5.13)$$

$$\ln \gamma_2 = \left(\frac{0.5A}{C-A}\right) \ln \left[(1+B+C)-[(A+B)-2(C-A)]x_2+(C-A)x_2^2\right]$$

$$-\left(\frac{A(A+B)}{2(C-A)\sqrt{-\delta}}\right) \times$$

$$\ln \left\{\frac{[\sqrt{-\delta}-(A+B)-2(C-A)+2(C-A)x_2][\sqrt{-\delta}+(A+B)]}{[\sqrt{-\delta}+(A+B)+2(C-A)+2(C-A)x_2][\sqrt{-\delta}-(A+B)]}\right\} \quad (5.14)$$

for positive deviations, and

$$\ln \gamma_1 = \left(\frac{0.5A}{C-A}\right) \ln \left\{\frac{1+(A+B)x_1+(C-A)x_1^2}{1+B+C}\right\} - \left(\frac{A[(A+B)+(C-A)]}{2(C-A)\sqrt{\delta}}\right) \times$$

$$\left\{\tan^{-1}\left[\frac{-(A+B)-2(C-A)x_1}{\sqrt{\delta}}\right] - \tan^{-1}[-(A+B)-2(C-A)]\right\} \quad (5.15)$$

$$\ln \gamma_2 = \left(\frac{0.5A}{C-A}\right) \ln \left\{(1+B+C)-[(A+B)-2(C-A)x_2+(C-A)x_2^2]\right\}$$

$$-\left(\frac{A+B}{(C-A)\sqrt{\delta}}\right) \times$$

$$\left\{\tan^{-1}\left[\frac{(A+B)+2(C-A)-2(C-A)x_2^2}{\sqrt{\delta}}\right] - \tan^{-1}\left[\frac{A+B}{\sqrt{\delta}}\right]\right\} \quad (5.16)$$

for negative deviations, with $\delta = 4(C\text{-}A) - (A\text{+}B)^2$ (Chialvo, 1988).

These new correlations have been tested for several binary systems whose behavior range from negative to high positive deviations from ideal solution. For models related to the linear representation of $[\rho\Delta(x)]_{TP}$, eqn. (5.7), their parameters (A and B) were obtained by solving the pair of nonlinear coupled equations for the activity coefficients at infinite dilution. Then there is no statistical adjustment involved at all.

In the case of the quadratic representation of $[\rho\Delta(x)]_{TP}$, eqn. (5.12), the third parameter was set using one of the following additional constraints

$$A = -\left(\frac{\partial \ln\gamma_1}{\partial x_1}\right)_{TP}^{\infty} \tag{5.17}$$

$$\frac{A}{1+B+C} = \left(\frac{\partial \ln\gamma_2}{\partial x_1}\right)_{TP}^{\infty} \tag{5.18}$$

In practice due to the lack of reliable experimental data, the right-hand side of either eqn. (5.17) or (5.18) was tunned, starting with the corresponding theoretical values given by Margules equation, i.e.,

$$\left(\frac{\partial \ln\gamma_1}{\partial x_1}\right)_{TP}^{\infty} = -2(2A - B) \tag{5.19}$$

$$\left(\frac{\partial \ln\gamma_2}{\partial x_1}\right)_{TP}^{\infty} = 2(2B - A) \tag{5.20}$$

with

$$A = \ln\gamma_1^{\infty} \tag{5.21}$$

$$B = \ln\gamma_2^{\infty} \tag{5.22}$$

The VLE results given by the new correlations were compared with experimental data and with the two-parameter Wilson equation for the same set of γ_i^{∞} as shown in Figures 14-17 and Table 4. Wilson parameters were determined using γ_i^{∞}'s along with the equations (constraints)

$$\ln\gamma_1^{\infty} = -\ln\Lambda_{12} + 1 - \Lambda_{21} \tag{5.23}$$

$$\ln\gamma_2^{\infty} = -\ln\Lambda_{21} + 1 - \Lambda_{12} \tag{5.24}$$

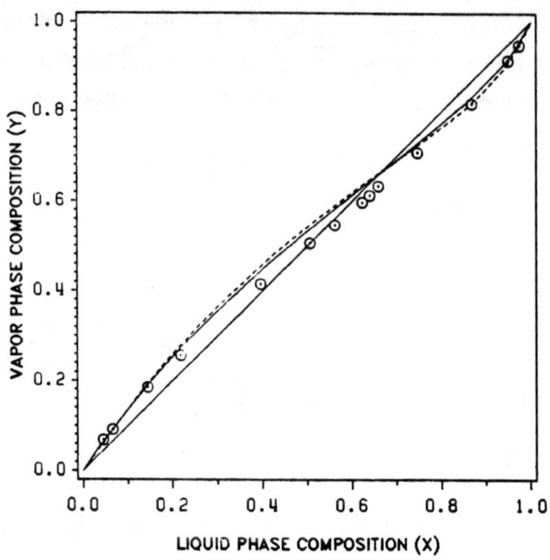

Figure 14. Y-X diagram for the system acetone(1)-methanol(2) at T=373K described by Wilson equation (---) and our linear model (—) with $\gamma_1^\infty = 1.405$ and $\gamma_2^\infty = 2.485$. Open circles are experimental data (Hirata et al, 1976)

Table 4. Parameters used in the correlation of VLE data with Wilson and proposed equations

System	T(K)	γ_1^{∞} (a)	γ_2^{∞}	Λ_{12}	Λ_{21}	A	B	C(b)	Ref
Acetone (1)-methanol (2)	373	1.405	2.485	1.52479	0.23810	-0.9634	0.4397	0.0	(c)
Acetone (1)-ethanol (2)	328	1.869	1.525	0.48574	1.09667	2.4487	2.4790	0.0	(c)
Acetone (1)-chloroform (2)	318	0.439	0.489	1.25295	1.59775	0.4958	-1.4942	0.0	(c)
Propanol (1)-water (2)	363	17.10	3.24	0.07287	0.7800	22.412	55.370	0.0	(c)
Propanol (1)-water (2)	363	17.10	3.24	0.07287	0.7800	15.203	-9.422	22.375	(c)

(a) γ_i^∞ are extrapolations of experimental data. (b) C=0 represents the linear model.
(c) Hirata et al (1976).

Results in Figures 14-17 show that our linear model is as flexible as Wilson's in correlating VLE of binary mixtures. Since neither model can predict LLE, their predictions become poor as mixtures approach immiscibility. Neither model can describe an unbounded peak in $[\rho\Delta(x)]_{TP}$ (see Figure 17). However, the quadratic representation (eqn. (5.12)) describes very well the X-Y behavior of a binary mixture close to immiscibility as shown in Figure 17.

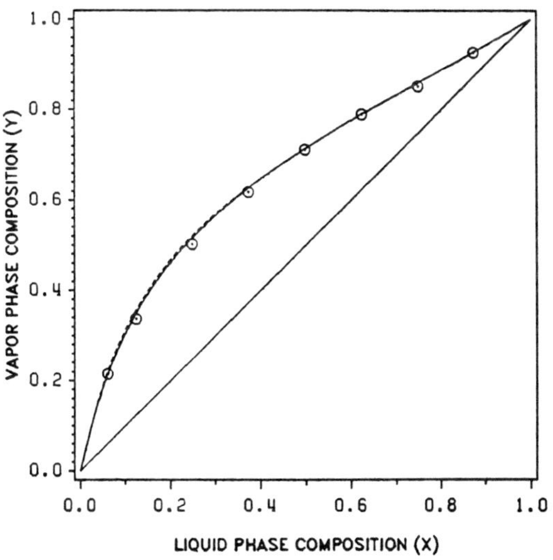

Figure 15. Y-X diagram for the system acetone(1)-ethanol(2) at T=328K described by Wilson equation (---) and our linear model (—) with $\gamma_1^\infty = 1.869$ and $\gamma_2^\infty = 1.525$. Open circles are experimental data (Hirata et al, 1976)

For systems showing LLE we have correlated the model parameters using either solubility data (Figure 18) or γ_i^∞ (Figure 19) as shown in Table 5. By using solubilities to correlate parameters we predict γ_i^∞ and Y-X curve, while the use of γ_i^∞ allows prediction of solubility as well as the Y-X curve. Results in Table 5 show that our quadratic representation can describe reasonably well liquid-liquid equilibria and vapor-liquid equilibria simultaneously.

Discussion:

We have illustrated with three new sets of equations how a rigorous formalism can be used to propose successful models for activity coefficients of liquid mixtures. One oustanding feature of these new equations is that their composition dependence depends on whether the models deal with positive or negative deviations. This results from the microscopic behavior of $[\rho\Delta(x)]_{TP}$: (a) it is always negative and bounded for negative deviation with no extrema in $\gamma(x)$, but (b) it might be unbounded positively for positive deviation (LLE). For simple rational functional forms (like eqns. (5.7) and (5.12)) (a) and (b) cannot be described by the same integral (eqn. (2.7)). This becomes an advantage because we automatically constrain the correlation to a particular deviation, avoiding unwanted behavior. For instance the system tetrachloride (1)-heptane (2) at 50 °C (Smyth and Engel, 1929) shows a negative deviation with no extrema in $\gamma(x)$; however, the best fit of the Wilson equation (system 647 in Hirata et al, 1976) predicts a positive-negative deviation.

Table 5. Comparison of predicted LLE results from quadratic model with experimental data. When experimental γ_i^∞ are used to correlate the model parameters, the solubilities are predicted and vice versa.

System	State condition	Experimental(a) γ_1^∞	γ_2^∞	x'_1	x''_1	Predicted γ_1^∞	γ_2^∞	x'_1	x''_1	Ref.
Butanol(1)-water (2)	760 mm Hg.	57.45	32.0	0.019	0.37	-	-	0.019	0.46	(b)
Acetonitrile (1)-water (2)	318 K	25.1	32.4	0.101	0.937	22.82	35.02			(c)

(a) γ_i^∞ are extrapolations of experimental data.

(b) Ellis and Garbett (1960)

(c) Palmer and Smith (1972)

Only an exact representation of $[\rho\Delta(x)]_{TP}$ will satisfy all the constraints given by the infinite dilution quantities, so that the more constraints we set in the data reduction, the more severe will be the correlation. However, any model should at least be constrained by the infinite dilution activity coefficients. Note that the eqns. (3.7) and (3.8) written in the following forms

$$\lim_{x_2 \to 0} (\rho\Delta)_{TP} = \left(\frac{\partial \ln \gamma_2}{\partial x_1}\right)_{TP}^{\infty} \qquad (5.25)$$

$$\lim_{x_1 \to 0} (\rho\Delta)_{TP} = -\left(\frac{\partial \ln \gamma_1}{\partial x_1}\right)^{\infty}_{TP} \qquad (5.26)$$

constitute a stringent pair of constraints. Indeed, by applying the Gibbs-Duhem equation, eqns. (5.25) and (5.26) can also be written as

$$\lim_{x_2 \to 0} (\rho\Delta)_{TP} = \lim_{x_1 \to 1} \left(\frac{\partial^2 \ln \gamma_1}{\partial x_1^2}\right)_{TP} \qquad (5.27)$$

$$\lim_{x_1 \to 0} (\rho\Delta)_{TP} = \lim_{x_2 \to 1} \left(\frac{\partial^2 \ln \gamma_2}{\partial x_1^2}\right)_{TP} \qquad (5.28)$$

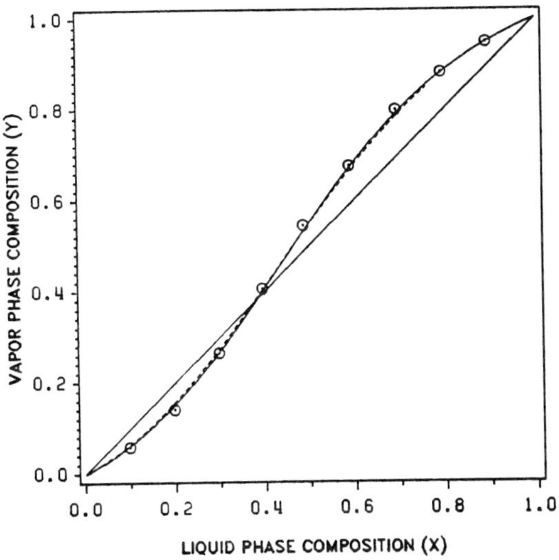

Figure 16. Y-X diagram for the system acetone(1)-chloroform(2) at T=318K described by Wilson equation (---) and our linear model (—) with γ_1^∞ = 0.439 and γ_2^∞ = 0.489. Open circles are experimental data (Hirata et al, 1976)

Then each limiting condition, either eqn. (3.7) or (3.8) constitutes a pair of constraints for $\gamma_i(x)$, i.e., one for the slope of $\gamma_i(x)$ at infinite dilution and the other for the second derivative of $\gamma_i(x)$ with respect to composition at the corresponding limit of pure component (recall we are using the Raoult's law for the ideal solution). From the geometrical standpoint the application of the limiting conditions, eqns. (5.25)-(5.28), means that we are constraining the curvatures of $\gamma_i(x)$ in both extrema, i.e., at x=0 and x=1. Since we are also constraining the γ_i^{∞}, it is easy to realize how crucial the choice of a simple functional form $[\rho\Delta(x)]_{TP}$ (few adjustable parameters) is to succeed in the correlation of activity coefficients.

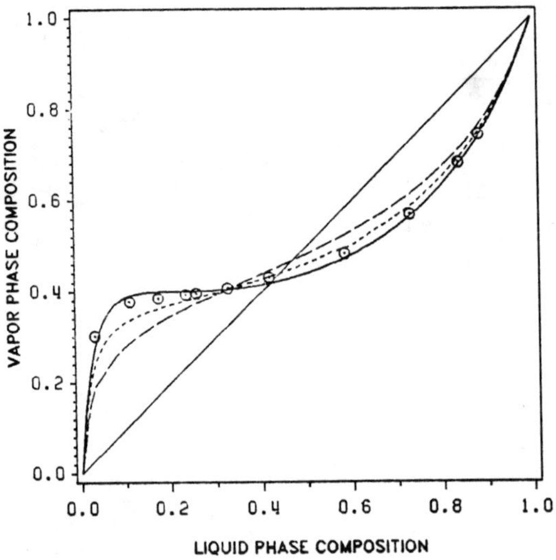

Figure 17. Y-X diagram for the system propanol(1)-water(2) at T=363K described by Wilson equation (---), our linear model (····) and our quadratic model (—) with γ_1^{∞} = 17.10 and γ_2^{∞} =3.24. The quadratic model is constrained by $(\partial \ln\gamma_2/\partial x_1)^{\infty}$ = 1.09. Open circles are experimental data (Hirata et al, 1976)

6. CONCLUSIONS

In this contribution we have illustrated how the K-B formalism can be used to obtain rigorous thermodynamic constraints for the activity coefficients and their slopes at infinite dilution. These constraints were related to a particular linear combination of K-B integrals, i.e., $[\rho\Delta(x)]_{TP}$. Application of the derived constraints allowed interpreting particular equilibrium behavior predicted by existing models in terms of $[\rho\Delta(x)]_{TP}$ and determining the corresponding regions in the space of their adjustable parameters. As a result we have shown, e.g., that the Wilson equation can describe binary mixtures whose $\gamma(x)$ present extrema far from the ideal solution condition (Raoult's law).

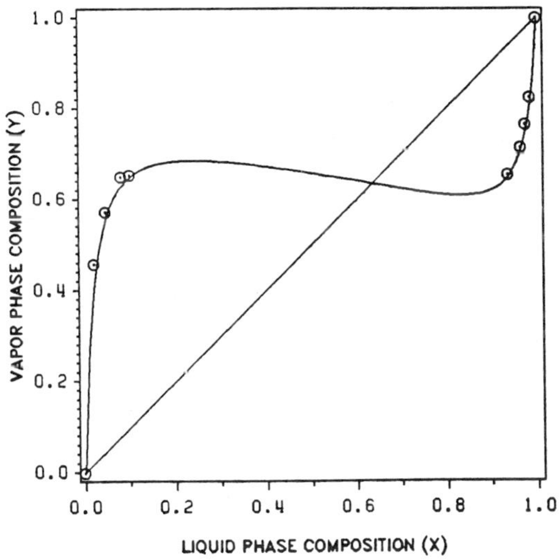

Figure 18. Y-X diagram for the system acetonitrile(1)-heptane(2) at T=318K described by our quadratic model (—). Parameters were correlated using the experimental solubility $x'_1 = 0.1016$, $x''_1 = 0.9372$ and $(\partial \ln\gamma_2/\partial x_1)^\infty = 18.16$. Open circles are experimental data (Palmer and Smith, 1972)

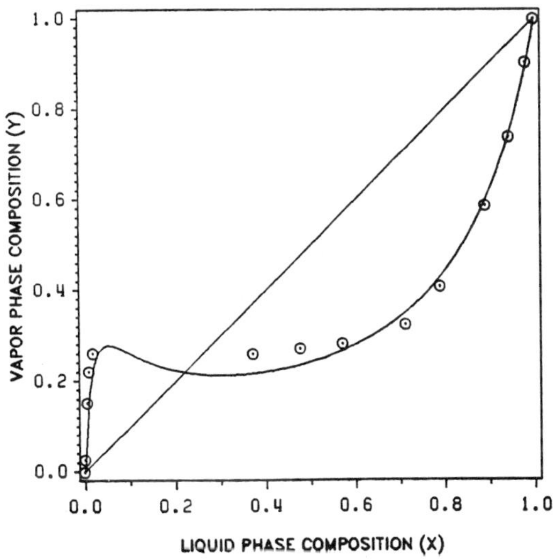

Figure 19. Y-X diagram for the system butanol(1)-water(2) at P=760 mm Hg described by our quadratic model (—) with $\gamma_1^\infty = 57.45$ and $\gamma_2^\infty = 32.0$ and $(\partial \ln\gamma_2/\partial x_1)^\infty = 0.850$. Open circles are experimental data (Ellis and Garbett, 1960)

At this point we can predict, by only knowing the infinite dilution quantities γ_i^∞ and $(\partial \ln\gamma_i/\partial x_i)^\infty$, (a) whether a liquid binary system shows positive, negative or positive-negative deviations, (b) whether the liquid system presents extrema in $\gamma(x)$, and (c) whether the liquid system is one- or two-phases.

The proposed analysis is also particularly suited for deciding which root is appropriate when multiple set of parameters are mathematically possible, e.g., in NRTL, Wilson or UNIQUAC models. Literature (Ladurelli et al, 1975; Silverman and Tassios, 1977; Tassios, 1979) shows many recipies for choosing the correct root; however none is rigorously founded. It might happen that a pair of roots are equally probable from the statistical standpoint. This raises the question as to which roots should be chosen. Again it does not matter if only quality of fit is the goal, but it does matter if we look for a physical interpretation of the phenomenon. An unambiguous choice could be based not only on the values of γ_i^∞ (not enough) but also on, at least, one slope $(\partial \ln\gamma_i/\partial x_i)^\infty$. In this way we might know the phase behavior associated with those infinite dilution quantities (say negative deviation with extrema, etc.). Finally by using the diagrams already developed (say Figure 9) for the space of parameters, the choice is straightforward. For example, for the system pentane (1) - squalane (2) the literature reports (Ladurelli et al, 1975)

$$\gamma_1^\infty = 0.37 \qquad\qquad \gamma_2^\infty = 0.50$$

so that three sets of Wilson's parameters are possible

$$\Lambda_{12} = 0.0403 \qquad\qquad \Lambda_{21} = 5.2$$
$$\Lambda_{12} = 1.0 \qquad\qquad \Lambda_{21} = 2.0$$
$$\Lambda_{12} = 7.36 \qquad\qquad \Lambda_{21} = 0.0034$$

From Figure 9 we learn that the first and the third sets describe negative deviations with extrema, while the second set describes negative deviation only. The data reduction shows that the first two sets of parameters are equally probable but the experimental evidence that $(\partial \ln\gamma_i/\partial x_i)^\infty > 0$ implies that $\gamma_2(x)$ has a minimum; thus the choice must be the first set.

We have related a variety of phase equilibrium behavior to the qualitative functional form of $[\rho\Delta(x)]_{TP}$. Consequently, we proposed three crude although successful $[\rho\Delta(x)]_{TP}$ expressions and the corresponding sets of $\gamma(x)$ correlations. Although the most complete of these correlations can describe successfully VLE and LLE, none of them can handle extrema in $\gamma(x)$. Although in principle the rational function ρΔ-representation can handle the case of extrema in $\gamma(x)$, in practice the analytical solution of eqn. (2.7) for this case becomes very cumbersome, if not impossible, so that another ρΔ-representation would be necessary. For instance, the extension of the present approach to ternary or even n-component mixtures constitutes an entire new step, yet to be taken. The motivation has been to demonstrate how the K-B formalism can be succesfully applied in developing new models, but we have by no means exhausted the study in this matter.

ACKNOWLEDGMENTS

I thank Professors J.M. Haile and G.A. Mansoori for fruitful discussions and Professor Abel C. Chialvo for illuminating comments regarding infinite dilution quantities.

REFERENCES

Ben-Naim, A. 1977. Inversion of the Kirkwood-Buff theory of solutions: Application to the water-ethanol system. J. Chem. Phys.,67:4884-90

Ben-Naim, A. 1974. Water and Aqueous Solutions, Plenum Press, New York

Chialvo, A.A. 1988. Computer simulation of diatomic fluids and their mixtures. Doctoral Dissertation. Clemson University.

Christian, S., E. Lane and E. Tucker. 1981. Liquid-vapor equilibrium of dilute aqueous solutions of methanol. J. Solution Chem., 10: 181-88

Donkersloot, M.C. 1979. The structure of binary liquids: The Kirkwood-Buff theory of liquid mixtures, illustrated on the basis of the systems water/methanol, water/ethanol, and cyclohexene/2,3-dimethylbutane, as link between thermodynamic data and x-ray and neutron scattering results. J. Solution Chem., 8: 293-307

Ellis, S. and R. Garbett.,1960. A new equilibrium still for the study of partially miscible systems. I&EC, 52: 385-88

Feix, G., H. Bittrich and D. Lempe. 1980. A two-parameter equation for the description of phase equilibriums in partially miscible systems. Collect. Czech. Chem. Commun., 45: 2035-41

Gaw , W. and F.L. Swinton. 1986. Thermodynamic properties of binary systems containing hexafluorobenzene. Trans. Faraday Soc., 64: 2023-34

Hirata, M., S. Ohe and K. Nagahama. 1976. Computer-Aided Data Book of Vapor-Liquid Equilibria, Elsevier, Amsterdam

Jonah, D. 1986. On the method of numerical differentiation and its application to the evaluation of limiting activity coefficients and their first derivatives. Chem. Eng. Sci., 41: 2261-72

Kato, T., T. Fujiyama and H. Nomura. 1982. Estimation of parameters G_{11}, G_{22}, and G_{12} in the Kirkwood-Buff solution theory on the basis of the concentration fluctuation data obtained from Rayleigh scattering. Bull. Chem. Soc. Japan, 55: 3368-72

Kikic, I. and H. Renon. 1976. Extension of chromatographic method of determination of thermodynamic properties. Sep. Sci., 11: 45-63

Kirkwood, J.G. and F. Buff. 1951. The statistical mechanics theory of solutions. I. J. Chem. Phys.,19: 774-77

Kreglewski, A., K. Marsh and K. Hall. 1985. A simple relation for the excess functions of non-random mixtures. Fluid Phase Equil., 21, 25-37

Ladurelli, A., C. Eon and G. Guiochon. 1975. Fallibilities inherent in the Wilson equation applied to systems having a negative excess Gibbs energy. I&EC, 14: 191-95

Lee, L.L., T. Chung and K. Starling. 1983. A molecular theory for the thermodynamic behavior of polar mixtures. I. Statistical-mechanical local composition model. Fluid Phase Equil., 12:105-24

Loblen, G. and J.M. Prausnitz. 1982. Infinite-dilution activity coefficients from differential ebulliometry. I&EC Fund., 21: 109-13

Marina, J. and D. Tassios. 1973. Effective local compositions in phase equilibrium correlations. I&EC Process Des. Dev., 12: 67-71

Matteoli , E. and L. Lepori. 1984. Solute-solute interaction in water. II. An analysis through the Kirkwood-Buff integrals for 14 organic solutes. J. Chem. Phys., 80: 2856-63

Modell, M. and R. Reid. 1983. Thermodynamics and its Applications, 2nd. Ed., Englewood Cliffs, New Jersey:Prentice-Hall.

Münster, A. 1969. Statistical Thermodynamics, Vol 1, 370. New York: Academic Press.

Ochi, K. and K. Kojima. 1987. A measurement of vapor-liquid equilibria at extreme dilution. J. Chem. Eng. Japan, 20: 6-10

Orye, R. and J.M. Prausnitz. 1965. Multicomponent equilibria with the Wilson equation. I&EC, 57: 18-26

Palmer, D. and B. Smith. 1972. Thermodynamic excess property measurement for acetonitrile-benzene-n-heptane system at 45º C. J. Chem. Eng. Data, 17: 70-76

Prausnitz, J.M., R. Lichtenthaler and E. Gomez de Azevedo. 1986. Molecular Thermodynamics of Fluid-Phase Equilibria. 2nd. Ed., Englewood Cliffs, New Jersey: Prentice-Hall.

Prausnitz, J.M., C. Eckert, R. Orye and J.P. O'Connell. 1967. Computer Calculations for Multicomponent Vapor-Liquid Equilibria, Englewood Cliffs, New Jersey: Prentice-Hall.

Prigogine, I. and R. Defay. 1954. Chemical Thermodynamics, D.H. Everett (Transl). London: Longmans.

Rowlinson, J.S. and F.L. Swinton. 1982. Liquid and Liquid Mixtures, 3rd. Ed. London: Butterworth.

Sager , N. and D. Lau. 1986. Limiting activity coefficients for butyl alcohol in water, n-octane, carbon tetrachloride. J. Chem. Eng. Data, 31: 475-78

Shen, S., S. Wang, J. Zhang and J. Shi. 1986. Study of parameters in models of excess Gibbs free energy by the chromatographic slope method. J. Chem. Ind. & Eng. (China), 2: 140-50

Shukla, K.P., A.A. Chialvo and J.M. Haile. 1988. Thermodynamic excess properties in binary mixtures. I&EC Research, 27: 664-71

Silverman, N. and D. Tassios. 1977. The number of roots in the Wilson equation and its effect on vapor-liquid equilibrium calculation. I&EC Process Des. Dev, 16: 13-20

Smyth, C. and E. Engel. 1929. Molecular orientation and the partial vapor pressures of binary mixtures. I. Systems composed of normal liquids. J. Am. Chem. Soc., 51: 2646-60

Tassios, D. 1979. The number of roots in the NRTL and LEMF equations and the effect on their performance. I&EC Process Des. Dev, 18: 182-86

Tsuboka, T. and T. Katayama. 1975. Modified Wilson equation for vapor-liquid and liquid-liquid equilibria. J. Chem. Eng. Japan, 8: 181-87

Valentin, P. and G. Guiochon. 1976. Determination of gas-liquid and gas-solid equilibrium isotherms by chromatography. II. Apparatus, specifications, and results. J. Chromatographic Sci., 14: 132-39

Van Ness, H. and M. Abbott. 1982. Classical Thermodynamics of Non-Electrolyte Solutions, New York: McGraw Hill

Wallas, S. 1985. Phase Equilibria in Chemical Engineering. Boston: Butterworth.

Wilson, G. 1964.Vapor-liquid equilibrium. XI. A new expression for the excess free energy of mixing. J. Am. Chem. Soc., 86: 127-30

Wisniak, J. 1983. Liquid-liquid phase splitting. I. Analytical models for critical mixing and azeotropy. Chem. Eng. Sci., 38: 969-78

Wobst , M. and G. Hradetzky. 1988. Modelling of binary vapor-liquid equilibriums using data of the range of high dilution. Chem. Techn., 40: 244-48

Zou, M. and J.M. Prausnitz. 1987.Vapor-liquid and liquid-liquid equilibria in binary aqueous systems. J. Chem. Eng. Data, 32: 34-37

Relations between and Estimations of Fluctuation Integrals and Direct Correlation Function Integrals

L. Lepori and E. Matteoli
Istituto di Chimica Quantistica ed Energetica Molecolare del CNR, Via Risorgimento 35 56126 Pisa, Italy

E. Z. Hamad and G. A. Mansoori
Department of Chemical Engineering University of Illinois, Chicago, IL 60680

ABSTRACT

New relations among the mixture direct correlation function integrals (or fluctuation integrals) in terms of concentration variables are developed. These relations indicate that, for example, for a binary mixture only one of the three direct correlation function integrals (or one of the three fluctuation integrals) is independent. Different closure expressions for mixture cross direct correlation function integrals are suggested and they are joined with the exact relations to calculate all the direct correlation function integrals and fluctuation integrals in a mixture.

The results indicate the possibility of introduction of simple closure expressions relating unlike- and like-interaction direct correlation integrals. It is demonstrated that the relation between the direct correlation integrals of hard-sphere mixtures can be satisfactorily represented by a simple geometric mean closure. For Lennard-Jones mixtures with the same intermolecular energy parameters ($\varepsilon_{12}/\varepsilon_{22}=1$) simple geometric mean closure is sufficient for the relation between the direct correlation function integrals so far as values of $\sigma_{12}/\sigma_{22}<1$, while the weighted arithmetic mean closure is necessary for $\sigma_{12}/\sigma_{22}>1$. Generally, a weighted arithmetic mean closure is sufficient in order to represent the relation between the direct correlation function integrals of real (simple or complex) mixtures.

INTRODUCTION

The knowledge about fluctuation integrals and their relations in mixtures is important in the development of mixture theories of complex molecules (Hamad, et al. 1989; Lepori and Matteoli 1988; Matteoli and Lepori 1984). Studies on concentration fluctuations in mixtures have resulted in numerous advances in the fundamental understanding of behavior of mixtures at, and away from, equilibrium (Gibbs, 1873; Landau and Lifshitz 1980). Definition of properties of mixtures in terms of concentration fluctuations has always been a viable empirical and theoretical route for analysis, correlation and prediction of mixture properties (Hamad, et al. 1987; Kirkwood and Buff 1951; Mansoori and Ely 1985). The engineering concept of "local compositions" is recently linked to the statistical mechanical theory of concentration fluctuations (Mansoori and Ely 1985). Combination of the conformal solution theory and concentration fluctuation theory have resulted in new equation of state mixing rules which have found special utility for analysis and prediction of equilibrium behavior of asymmetric mixtures of interest in such processes as supercritical fluid extraction, retrograde condensation, and miscible flood enhanced oil recovery techniques.

Among the concentration fluctuation theories of statistical mechanics the Kirkwood-Buff solution theory is the one which is developed specifically for mixtures at equilibrium (Kirkwood and Buff 1951). Obtaining thermodynamic properties from this theory requires knowledge of the fluctuation integrals, G_{ij}, or the direct correlation function integrals, C_{ij}. Evaluating these integrals from microscopic information such as the intermolecular potential energy has not been easy. This is because our knowledge about the intermolecular potential energy functions of real fluids is incomplete and the present techniques for calculating these integrals from potential energy functions are not accurate enough (McQuarrie 1975). Also the relations presented in the first chapter of the present monograph, although exact, can not be used, as they are, to obtain mixture properties. This is because there are too many such integrals, G_{ij}'s or C_{ij}'s, for a mixture to deal with. In this paper new and exact relations between direct correlation function integrals are reported. Also, since the direct correlation function integrals have simple shapes and short ranges we report here the use of simple combining rules (closures), which are relations between unlike and like interaction direct correlation integrals, for calculation of such integrals and reduction of their number.

THEORY OF CALCULATING C_{ij}'S

In what follows we present the technique for calculating C_{ij}'s and G_{ij}'s for binary mixtures. Extension of this technique to multicomponent mixtures is straightforward. To derive the relations among the C_{ij} integrals we use the mathematical fact that the mixed second derivatives of a function of two variables are equal at all points where the derivatives are continuous. Let us consider the mole fractions x_i, total pressure P, and the absolute temperature T as the independent variables in a mixture in one phase. Then, for the chemical potential, μ_i, of component i in a binary mixture one has (Hamad, et al. 1989):

$$(\partial^2\mu_i/\partial x_1\partial P)_T = (\partial^2\mu_i/\partial P\partial x_1)_T \qquad i=1,2 \qquad (1)$$

The derivative on the right-hand side of the above equation is equal to $(\partial\overline{v}_i/\partial x_1)_{T,P}$

where $\overline{v}_i$ is the partial molar volume of component i. As a result equation (1) now becomes

$$(\partial^2\mu_i/\partial x_1\partial P)_T = (\partial\overline{v}_i/\partial x_1)_{T,P} \qquad i=1,2 \tag{2}$$

Substituting the expressions for the partial molar volume and the chemical potential with respect to the direct correlation function integrals C_{ij} (see Chapter 1),

$$\rho\overline{v}_i = \left(1-\rho\sum_{j=1}^{2}x_jC_{ij}\right) \Big/ \left(1-\rho\sum_{j=1}^{2}\sum_{k=1}^{2}x_jx_kC_{jk}\right) \tag{3}$$

$$x_1[\partial\mu_1/\partial x_1]_{P,T} = kT\ \frac{1-x_1\rho C_{11}-x_2\rho C_{22}+x_1x_2\rho^2(C_{11}C_{22}-C_{12}^2)}{1-\Sigma\Sigma x_i x_j\rho C_{ij}} \tag{4}$$

in Eq. (2) gives the following two equations:

$$\begin{aligned} &kT\partial\{[1-x_1\rho C_{11}-x_2\rho C_{22}+x_1x_2\rho^2(C_{11}C_{22}-C_{12}^2)]/[1-\Sigma\Sigma x_i x_j\rho C_{ij}]\}/\partial p \\ &\quad = x_1\,\partial\{[1-x_1\rho C_{11}-x_2\rho C_{12}]/[\rho(1-\Sigma\Sigma x_i x_j\rho\, C_{ij})]\}/\partial x_1 \end{aligned} \tag{5}$$

and

$$\begin{aligned} &kT\partial\{[1-x_1\rho C_{11}-x_2\rho C_{22}+x_1x_2\rho^2(C_{11}C_{22}-C_{12}^2)]/[1-\Sigma\Sigma x_i x_j\rho\, C_{ij}]\}/\partial p \\ &\quad = x_2\,\partial\{[1-x_2\rho C_{22}-x_1\rho C_{12}]/[\rho(1-\Sigma\Sigma x_i x_j\rho\, C_{ij})]\}/\partial x_2 \end{aligned} \tag{6}$$

These two equations constitute two independent expressions relating C_{11}, C_{22} and C_{12} in a binary mixture. In principle Eq.s (5) and (6), together with another expression relating C_{12} to C_{11} and C_{22}, can be used to solve for the three quantities C_{11}, C_{22} and C_{12}.

An Alternative Derivation: Consider a system which is described by the following set of independent variables $\{T, \rho_1, \rho_2, \ldots, \rho_n\}$, where T is the absolute temperature, $\rho_i = N_i/V$, N_i is the number of molecules i in the system, V is the system volume, and n is the number of components in the mixture. In the grand canonical ensemble the following expression for $\partial\mu_i/\partial\rho_j$ is derived (O'Connell 1971; Landau and Lifshitz 1980)

$$[\partial\mu_i/\partial\rho_j]_{T,\rho_j'} = kT\,\{\,\delta_{ij}/\rho_i - C_{ij}\,\} \tag{7}$$

where ρ_j' stands for the set $\rho_1, \rho_2, \ldots, \rho_n$ variables excluding ρ_j,

$$C_{ij} = \int c_{ij}(r)dr \tag{8}$$

and $c_{ij}(r)$ is the direct correlation function of species i and j. Using Eq. (7) exact relations can be derived among the direct correlation function integrals, C_{ij}, by differentiating this equation with respect to ρ_i and equating the mixed second derivatives of the chemical potentials (Hamad, et al. 1989). For example, for a binary mixture with T, ρ_1, ρ_2 as the independent variables one can write:

$$\partial C_{12}/\partial \rho_1 = \partial C_{11}/\partial \rho_2 \quad (9)$$

and

$$\partial C_{12}/\partial \rho_2 = \partial C_{22}/\partial \rho_1 \quad (10)$$

The above two equations are exact relating C_{11}, C_{22}, and C_{12}. In order to solve these equations for C_{11}, C_{22}, and C_{12} one needs a closure relation among C_{11}, C_{22}, and C_{12} to solve for the three direct correlation function integrals. In the next section a number of closure expressions for C_{12} are considered and the direct correlation integrals are solved for the hard-sphere fluid mixture, the Lennard-Jones fluid mixture, and real mixtures consisting of simple and complex molecules.

CHOICE OF THE C_{12} CLOSURE

Simultaneous solution of Eq.s (5) and (6) for C_{11}, C_{22} and C_{12} is not possible without another expression relating C_{12} to C_{11} and C_{22} so that

$$C_{12} = C_{12}\,(C_{11}, C_{22}; T, p, x_1, x_2, \ldots x_{n-1}) \quad (11\text{-}1)$$

Similarly, simultaneous solution of Eq.s (9) and (10) for C_{11}, C_{22} and C_{12} is not possible without another expression relating C_{12} to C_{11} and C_{22} so that

$$C_{12} = C_{12}\,(C_{11}, C_{22}; T, \rho_1, \rho_2, \ldots, \rho_n) \quad (11\text{-}2)$$

We choose to call Eq (11-1), or (11-2), the C_{12} closure. The exact C_{12} closure expression is not presently available. There exist variety of ways of assuming the closure for the direct correlation function integrals. With an appropriate choice for the functional form of the closure expression, Eq.s (5, 6, 11-1), or Eq.s (9, 10, 11-2), can be solved simultaneously for C_{11}, C_{22} and C_{12}. Then the fluctuation integrals G_{11}, G_{22} and G_{12} can be derived from C_{11}, C_{22} and C_{12} using the following equations (Pearson and Rushbrooke 1957; O'Connell 1971, 1981).

$$\rho G_{11} = \frac{\rho C_{11}-(1-x_1)\rho^2(C_{11}C_{22}-C_{12}^2)}{1-x_1\rho C_{11}-x_2\rho C_{22}+x_1x_2\rho^2(C_{11}C_{22}-C_{12}^2)} \quad (12)$$

$$\rho G_{22} = \frac{\rho C_{22}-(1-x_2)\rho^2(C_{11}C_{22}-C_{12}^2)}{1-x_1\rho C_{11}-x_2\rho C_{22}+x_1x_2\rho^2(C_{11}C_{22}-C_{12}^2)} \tag{13}$$

$$\rho G_{12} = \frac{\rho C_{12}}{1-x_1\rho C_{11}-x_2\rho C_{22}+x_1x_2\rho^2(C_{11}C_{22}-C_{12}^2)} . \tag{14}$$

Before making any assumption about the mathematical expression of the C_{ij} closure it is helpful to study the behavior of certain known direct correlation function integrals.

The direct correlation functions for a mixture of hard-spheres have been obtained from the Percus-Yevick approximation (Lebowitz 1964). The direct correlation function integrals were then calculated from Lebowitz's work (Hamad 1988). It was found that for a binary mixture of hard-spheres C_{12} always lies in between C_{11} and C_{22}. This suggests that C_{12} can be approximated by some kind of an average of C_{11} and C_{22}.

Another source of information about the mathematical form of the C_{ij} closure is based on the fact that the virial expansion of the direct correlation function integral can be written in the following form (Hamad 1988)

$$C_{ij} = -\left(2B^{(2)}{}_{ij} + 3\sum_{k=1}^{n} B^{(3)}{}_{ijk}\rho_k + 4\sum_{k=1}^{n}\sum_{\ell=1}^{n} B^{(4)}{}_{ijk\ell}\rho_k\rho_\ell + \ldots\right) \tag{15}$$

where $B^{(2)}{}_{ij}$, $B^{(3)}{}_{ijk}$ and $B^{(4)}{}_{ijk\ell}$ are the second, third and fourth virial coefficients, respectively. From this equation we can conclude,

$$\text{Limit } C_{ij} = -2\,B^{(2)}{}_{ij}(T) \text{ as } \rho \text{ ---> } 0, \tag{16}$$

where $B^{(2)}{}_{ij}(T)$ is the second virial coefficient of the pair i and j. At low densities the sign of C_{ij} is opposite of that of the second virial coefficient, $B^{(2)}{}_{ij}$. It is well known that, while the second virial coefficient of hard-sphere fluids is always positive, the second virial coefficient of real fluids can be positive or negative depending on the temperature. This implies that C_{11}, C_{22} and C_{12} of real mixtures can have positive or negative signs.

Considering the above facts we can assume C_{12} is an average of C_{11}, C_{22}. Then, some of the possible expressions which may be chosen for Eq. (11) are as the following:

1) Arithmetic mean closure: By postulating that C_{12} is an arithmetic mean of C_{11} and C_{22} one may write

$$C_{12} = (C_{11} + C_{22})/2 \tag{17}$$

Eq. (17) implies that the mixture under consideration is an ideal solution (Hamad 1988). Instead of Eq. (17) one may write the following weighted arithmetic mean C_{12} closure for non-ideal solutions (Hamad, et al. 1989)

$$C_{12} = \alpha_{21}C_{11} + \alpha_{12}C_{22} \tag{18}$$

Parameters α_{21} and α_{12} are adjustable interaction parameters which have to be determined for the interacting species under consideration. By joining Eq.s (16) and (18) and choosing $\alpha_{21}\alpha_{12}=1/4$ (for the sake of simplicity) we can derive the following limiting expression for α_{21} with respect to the second virial coefficients.

$$\alpha_{21} = B_{12}/B_{11} \pm \left[(B_{12}/B_{11})^2 - B_{22}/B_{11}\right]^{1/2} \tag{16-1}$$

As it is demonstrated by Hamad and Mansoori (1989) the choice of $\alpha_{21}\alpha_{12}=1/4$ allows us to derive analytic expressions for chemical potentials of components in the mixture.

2) Geometric mean closure: This closure is suggested based on a relation among the contact values of the direct correlation functions in a mixture of one-dimensional hard-rods. It is shown that the contact c_{ij} values are related by (Lebowitz 1964)

$$c_{12}^2(\sigma_{12}) = c_{11}(\sigma_{11})\ c_{22}(\sigma_{22}) \tag{19}$$

where σ_{ij} is the distance of closest approach between centers of particles i and j, and $\sigma_{ij}=(\sigma_{ii}+\sigma_{jj})/2$. For the direct correlation integrals of a mixture of hard-spheres a relation similar to Eq. (19) may be assumed (Hamad, et al. 1989)

$$C_{12}^{\ 2} = C_{11}C_{22} \tag{20}$$

In this equation C_{12} is assumed to be the geometric mean of C_{11} and C_{22}. For molecules other than hard-spheres Eq. (20) may be replaced with the following expression:

$$C_{12}^{\ 2} = \beta\, C_{11}C_{22}, \tag{21}$$

where β is an adjustable interaction parameter.

3) A combined arithmetic and geometric mean closure: In the above mentioned closures parameters α_{21}, α_{12}, and β are adjustable parameters. In order to make the closure more versatile (with more adjustable parameters) one may combine the

arithmetic and geometric means in order to produce the following closure

$$C_{12} = \alpha_{21}C_{11} + \alpha_{12}C_{22} \pm (\beta C_{11}C_{22})^{1/2}, \tag{22}$$

However, as it will be demonstrated later for all the binary mixtures studied here, either of the geometric or arithmetic mean closures are sufficient enough to represent C_{12} with respect to C_{11} and C_{22}. It is also possible to come up with variety of other C_{12} closure expressions. However, the above mentioned C_{12} closures are the simplest possible relations that one may suggest. The simplicity of these closures makes it easy to use them in calculating the G_{ij} and C_{ij} integrals.

TEST OF C_{12} CLOSURES

1) <u>hard-Sphere Mixture</u>: The validity of the proposed closures can be tested by comparing the fluctuation integrals or direct correlation function integrals calculated by the present technique with their exact values. For this purpose we first consider a model fluid mixture (hard-sphere mixture) for which accurate direct correlation function integrals are already available. The Percus-Yevick theory (Percus and Yevick 1958) gives quite accurate expressions for the hard-sphere mixture direct correlation function integrals (Lebowitz 1964). In Table 1 the C_{12}/C_{11} values calculated by the Percus-Yevick theory are compared with the result of geometric mean closure, Eq. (20) and the arithmetic mean closure, Eq. (17) for a binary equimolar mixture of hard-spheres with σ_{11}/σ_{22}=1.5.

<u>Table 1.</u>Values of C_{12}/C_{11} for an equimolar hard-sphere mixture (σ_{11}/σ_{22}=1.5) calculated according to simple geometric mean (SGM), Eq. (20), and weighted arithmetic mean (WAM), Eq. (18), closures and compared with the Percus-Yevick (P-Y) theory (Hamad and Mansoori 1989).

$(\pi/6)\rho\sigma_{22}^3$	P-Y	SGM	% Dev.	WAM	% Dev.
0.05	0.5265	0.5047	4.1	0.5243	0.4
0.10	0.4806	0.4677	2.7	0.4772	0.7
0.20	0.4066	0.4034	0.8	0.4039	0.7
0.30	0.3530	0.3526	0.1	0.3537	-0.2
0.40	0.3141	0.3142	0.02	0.3202	-1.9
0.50	0.2844	0.2844	0.0	0.2969	-4.4

For the weighted arithmetic mean closure, $\alpha_{21}\alpha_{12}$=1/4 is chosen. At σ_{11}/σ_{22}=1.5, the negative sign in Eq. (16-1) gives α_{21}=0.1911. See Hamad and Mansoori (1989).

This table shows clearly that both geometric and arithmetic mean closures are accurate enough for hard-sphere mixture. However, the geometric mean closure is more accurate at high densities while the weighted arithmetic mean closure is more accurate at low densities.

2) Real Mixtures: Data on fluctuation integrals are available for variety of binary mixtures (Matteoli and Lepori, 1984; Lepori and Matteoli, 1988). Such data can be introduced into inverted Eqs. (12)-(14) to calculate C_{ij} and then used for the test of different closure expressions. The validity of C_{12} closures can be tested for real mixtures by plotting C_{12}/C_{11} data versus C_{22}/C_{11} data and studying the slope of the resulting curve. Figures 1 and 2 show the plot of C_{12}/C_{11} versus C_{22}/C_{11} for four different binary mixtures. The linearity of these plots suggests that the weighted arithmetic mean closure, Eq. (18), is quite sufficient for real mixtures. As a further

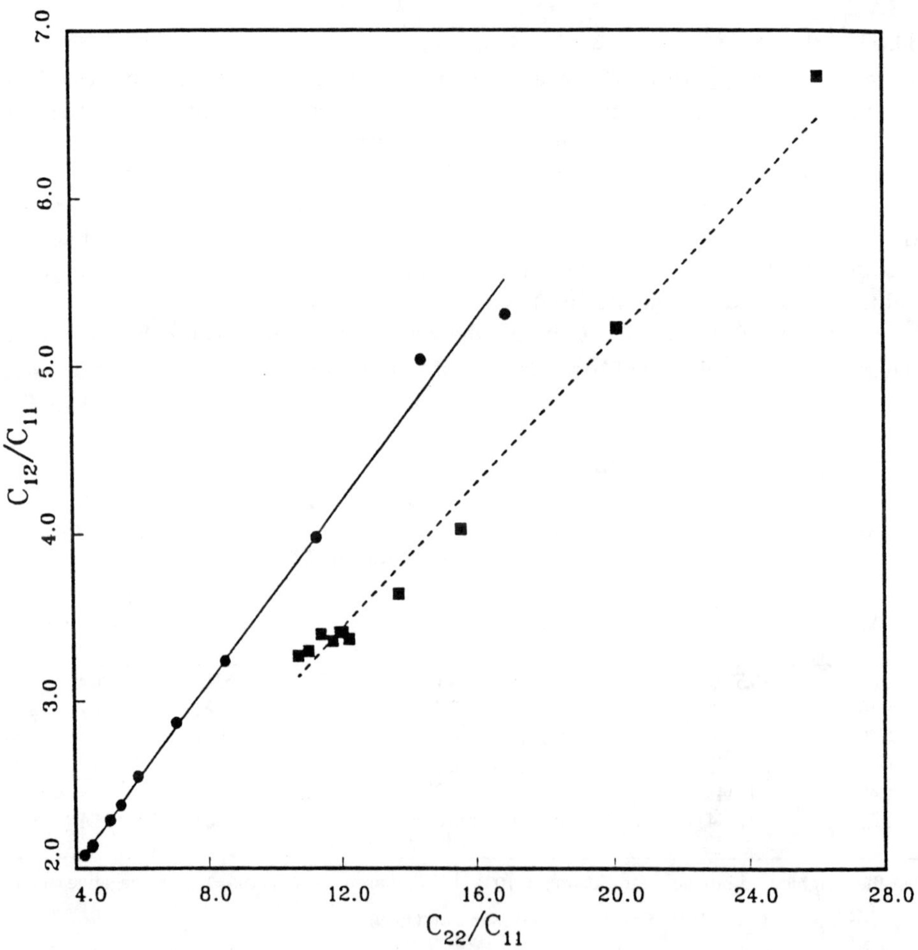

Figure 1. The test of the linearity of C_{12}/C_{11} versus C_{22}/C_{11} for water(1) + methanol(2) (circles) and water (1) + aminoethanol (2) (squares) binary mixtures. The C_{ij} data has been calculated from G_{ij} values taken from Matteoli and Lepori (1984). The curves represent the weighted arithmetic mean closure, Eq. (18).

test of the validity of the weighted arithmetic mean closure for other real mixtures the deviations of C_{12} data from the C_{12} value calculated using Eq.(18) are reported in Table 2.

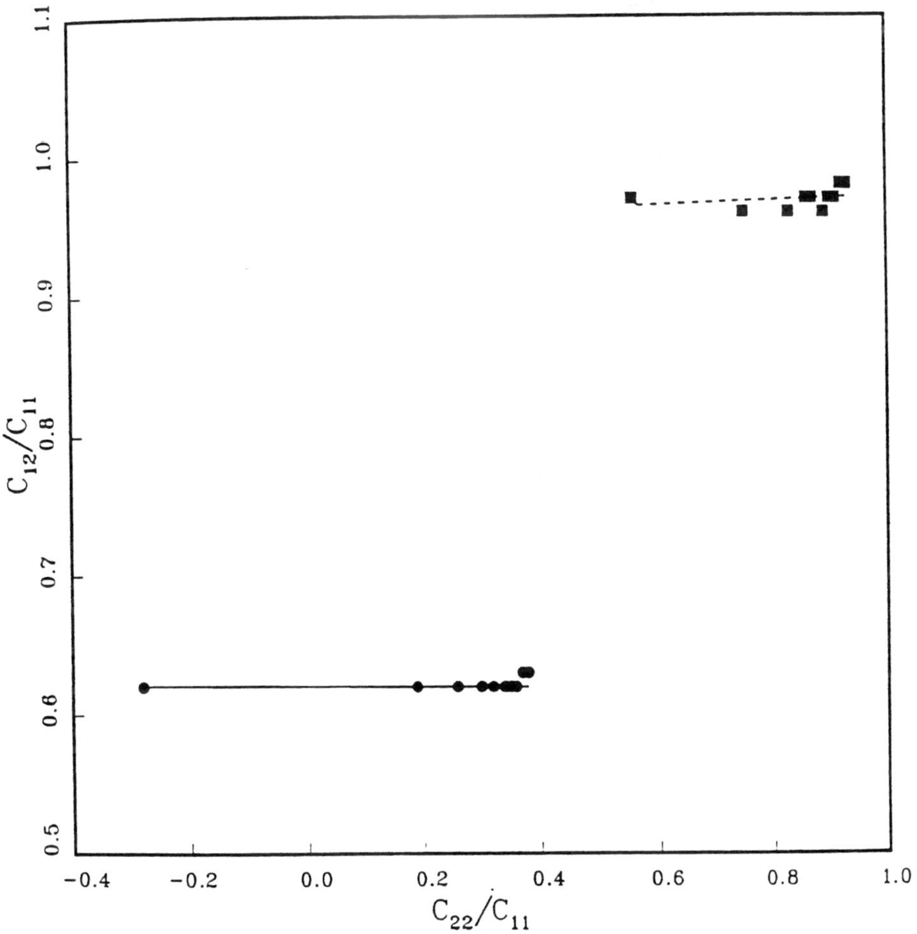

Figure 2. The test of the linearity of C_{12}/C_{11} versus C_{22}/C_{11} for tetrachloromethane (1) + ethanol (2) (circles) and tetrachloromethane (1) + n-butanol (2) (squares) binary mixtures. The C_{ij} data has been calculated from G_{ij} values taken from Lepori and Matteoli (1988). The curves represent the weighted arithmetic mean closure, Eq. (18).

Table 2. The deviations of real fluid direct correlation integrals from the relation $C_{12}=\alpha_{21}C_{11}+\alpha_{12}C_{22}$

System	Temperature [°C]	S X100	Max Dev. [%]	x_1 at Max Dev.
water+methanol	0.0	2.1	7.7	0.0
" "	25.	1.9	7.5	0.0
" "	60.	2.2	-2.6	0.4
water+ethanol	25.	14.5	22.	0.0
" "	50.	3.2	4.1	0.1
" "	90.	4.1	6.7	0.1
water+propanol	25.	2.7	7.4	0.0
water+tert-butanol	25.	1.5	-4.6	1.0
water+acetonitrile	30.	2.3	-11.	1.0
water+acetone	25.	1.9	-4.4	1.0
water+dimethylsulfoxide	25.	5.9	-8.7	0.0
water+tetrahydrofuran	25.	4.3	-8.8	1.0
water+1,4-dioxane	25.	4.0	-5.9	1.0
water+2-aminoethanol	25.	6.9	-6.2	1.0
CCl4+methanol	25.	0.32	-2.4	0.0
CCl4+ethanol	25.	0.28	-2.1	0.0
CCl4+propanol	25.	0.12	-1.3	0.0
CCl4+n-butanol	25.	0.12	-1.2	0.0
CCl4+tetrahydrofuran	25.	0.0004	.005	0.9
CCl4+1,4-dioxane	25.	0.00003	0.003	0.0
CY-hexane+diethylether	25.	0.014	-0.4	1.0
CY-hexane+dipropylether	25	0.030	-0.6	0.0
CY-hexane+dibuthylether	25.	0.002	0.2	0.0
CY-hexane+ethylbutylether	25.	0.020	-0.5	0.0
CY-hexane+dimethoxymethane	25.	0.24	-2.1	1.0
CY-hexane+diethoxymethane	25.	0.0005	-0.02	1.0
CY-hexane+diethoxyethane	25.	0.19	-1.8	0.0
CY-hexane+diglyme	25.	0.16	1.0	1.0

In this table, $S=\sum\{(C_{12,pred}-C_{12,exp})/C_{12,exp}\}^2$ and $\%Dev.=\{(C_{12,pred}-C_{12,exp})/C_{12,exp}\}X100$.
Experimental C_{ij} values were evaluated from experimental G_{ij}'s using inverted Eq's (12)-(14). G_{ij}'s were taken from Matteoli and Lepori (1984), Lepori and Matteoli (1988),or calculated using activity coefficient data (Lepori, et al. 1988). Density data were taken from Berti, et al., (1988) and isothermal compressibility data were taken from Brostow and Maynadier (1979) and Handbook of Chemistry and Physics (1989).

The maximum deviation for all studied systems is about 22%. This large error happens where the C_{ij} goes to zero and crosses the horizontal axis. Since some of the data has uncertainties up to 30-40% (Lepori and Matteoli, 1988), the present results demonstrate that the weighted arithmetic mean closure is a good closure approximation for real mixtures.

CALCULATION OF MIXTURE C_{IJ}'S FROM PURE FLUID DATA

To avoid having to solve numerically the nonlinear partial differential equations, Eq.s (5 & 6) or (9 & 10), one can make use of the fact that for liquids at low to moderate pressures the direct correlation function integrals are week functions of pressure. Under this assumption, for example Eq.s (5) and (6) will give:

$$\left(1-x_1\rho C_{11}-x_2\rho C_{12}\right) / \left(\rho-\rho^2\sum\sum x_j x_k C_{jk}\right) = v_1 \tag{23}$$

$$\left(1-x_1\rho C_{12}-x_2 v C_{22}\right) / \left(\rho-\rho^2\sum\sum x_j x_k C_{jk}\right) = v_2 \tag{24}$$

where v_1 and v_2 are the molar volumes of pure components 1 and 2, respectively. Comparisons of Eq.s (23) and (24) with Eq. (3) imply that the partial molar volumes are set equal to the pure component volumes at all concentrations. This is a result of assuming that C_{ij}'s are pressure-independent. This approximation, which is also made in lattice models of liquid mixtures, does not have a pronounced effect on the C_{ij}'s as will be seen later. Joining Eq.s (23) and (24) with the equation for the isothermal compressibility,

$$(\rho kT\kappa_T)^{-1} = 1-\rho\sum_{i=1}^{2}\sum_{j=1}^{2} x_i x_j C_{ij} \tag{25}$$

gives:

$$x_1\rho C_{11}+x_2\rho C_{12} = 1 + W[(v_1/v_2)x_1+x_2] \tag{26}$$

$$x_1\rho C_{12}+x_2\rho C_{22} = 1 + W[x_1+(v_2/v_1)x_2] \tag{27}$$

where

$$W = -\left[kT(x_1\kappa_{T_1}/v_2 + x_2\kappa_{T_2}/v_1)\right]^{-1} \tag{28}$$

and κ_{T_1} and κ_{T_2} are the isothermal compressibilities of the pure components 1 and 2, respectively. The direct correlation integrals, C_{ij}'s, in a binary mixture can now be solved for by combining Eq.s (18), (26) and (27) with the following result:

$$\rho C_{11} = \frac{W(1-\alpha_{12}v_2/v_1)x_2^2 + Wx_1(\alpha_{12}x_1+x_2)v_1/v_2 + (1-\alpha_{12})x_2 + \alpha_{12}x_1}{\alpha_{21}x_2^2 + x_1x_2 + \alpha_{12}x_1^2} \tag{29}$$

$$\rho C_{22} = \frac{W(1-\alpha_{21}v_1/v_2)x_1^2 + Wx_2(\alpha_{21}x_2+x_1)v_2/v_1 + (1-\alpha_{21})x_1 + \alpha_{21}x_2}{\alpha_{21}x_2^2 + x_1x_2 + \alpha_{12}x_1^2} \tag{30}$$

$$\rho C_{12} = \frac{W(\alpha_{12}x_1^2 + \alpha_{21}x_1x_2v_1/v_2 + \alpha_{12}x_1x_2v_2/v_1 + \alpha_{21}x_2^2) + \alpha_{12}x_1 + \alpha_{21}x_2}{\alpha_{21}x_2^2 + x_1x_2 + \alpha_{12}x_1^2} \tag{31}$$

With the availability of pure fluid thermodynamic data and interaction parameters α_{21} and α_{12} Eqs (29-31) can be used for calculation of mixture direct correlation function integrals.

RESULTS AND DISCUSSION

The C_{12} closures discussed in the previous sections can be used to predict the properties of hard-sphere mixtures, Lennard-Jones mixtures and real mixtures using pure fluid properties (Hamad, et al. 1989; Hamad and Mansoori 1989).

For hard-spheres the Carnahan-Starling (1969) equation of state is used to describe the behavior of the pure hard-sphere fluid. In Table 3 the chemical potential predictions using the two simple C_{12} closures, Eq.s (17) and (20), are compared to the Mansoori et al. (1971) hard-sphere mixture equation, which is known to reproduce the computer simulation data accurately.

Table 3 The variation of the chemical potential of component 1 in a mixture of hard-spheres (σ_{11}/σ_{22}=1.5, $\rho\sigma_{22}^3$=0.25) with composition based on different closures and compared with the MCSL (Mansoori, et al. 1971) equation of state (Hamad and Mansoori 1989).

x_1	μ_{1r}/kT				
	MCSL	$C_{12}=(C_{11}C_{22})^{1/2}$	%Dev.	$C_{12}=\alpha_{21}C_{11}+\alpha_{12}C_{22}$	%Dev.
0.177	3.60	3.49	2.9	3.56	1.1
0.340	4.49	4.40	2.0	4.44	1.1
0.491	5.53	5.46	1.3	5.47	1.1
0.631	6.73	6.67	0.8	6.67	0.9
0.761	8.12	8.09	0.4	8.07	0.6
0.883	9.75	9.74	0.2	9.72	0.3

For the weighted arithmetic mean closure $\alpha_{21}\alpha_{12}=1/4$. is chosen. At σ_{11}/σ_{22}=1.5 and the negative sign in Eq. (16-1) gives α_{21}=0.1911. See Hamad and Mansoori (1989).

The chemical potential calculations reported in this table are for component 1 of a binary mixture with σ_{11}/σ_{22}=1.5 and $\rho\sigma_{22}^3$=0.25. The dimensionless density $\rho\sigma_{22}^3$=0.25 corresponds to 95% of the solidification density of component 1. According to Table 3, for prediction of the hard-sphere mixture chemical potentials either of the two simple C_{12} closures, Eq.s (17) or (20), are quite sufficient. Similar results are also obtained for the chemical potential of component 2, and at other conditions which will not be reported here.

The hard-sphere intermolecular potential energy function is purely repulsive. To test the proposed closure models for mixtures of molecules with intermolecular potentials which possess, both, attractive and repulsive parts, the Lennard-Jones potential model is considered here. The chemical potential of a mixture of Lennard-Jones fluids at infinite dilution is predicted here and it is compared with the available computer simulation data. The necessary pure component Lennard-Jones fluid data is calculated using the Nicolas et al. (1979) Lennard-Jones fluid equation of state. The chemical potential predictions are compared to the infinite dilution simulation data of Shing (1982) in Table 4.

Table 4 .The chemical potential at infinite dilution of Lennard-Jones mixture. (kT/ε_{22} =1.2, $\rho\sigma_{22}^3$=0.7). (Hamad and Mansoori, 1989).

$(\sigma_{12}/\sigma_{22})^3$	$\varepsilon_{12}/\varepsilon_{22}$	μ^{∞}_{1r}/kT		
		Simulation Data	$C_{12}=(C_{11}C_{22})^{1/2}$	$C_{12}=\alpha_{21}C_{11}+\alpha_{12}C_{22}$
0.3	1.0	-1.30	-1.75	+4.14
0.5	1.0	-1.63	-1.80	-1.29
0.75	1.0	-1.83	-1.86	-1.92
1.0	1.0	-1.93	-1.99	-1.99
1.5	1.0	-1.87	-2.26	-1.65
2.0	1.0	-1.55	-2.58	-1.52

In this table, the simulation data are taken from Shing (1982), and Shing et al, (1988).

According to this table for molecules with the same intermolecular energy parameters, $\varepsilon_{12}/\varepsilon_{22}=1$, the weighted arithmetic mean closure shows better agreement with the simulation data at $\sigma_{12}/\sigma_{22}>1$. Geometric mean closure shows better agreement with the simulation data at $\sigma_{12}/\sigma_{22}<1$.

A critical test of the present technique for calculating C_{ij} and G_{ij} integrals is its application to real mixtures. We already have demonstrated that the weighted arithmetic mean closure is sufficient to correlate the direct correlation function integrals of real mixtures (see Table 2). Eqs (5), (6), and (18) are joined together to fit the experimental C_{ij} data by optimizing parameters α_{21} and α_{12}. The experimental C_{ij} integrals for real mixtures are calculated from experimental G_{ij} integrals (Matteoli and Lepori, 1984; Lepori and Matteoli, 1988; Lepori, et al., 1988; Berti, et al.,

1989; and Hamad, et al., 1989) inverting Eqs (12) - (14).

In Figures 3 - 13 the C_{11}, C_{22} and C_{12} integrals resulting from this fit are reported for twenty two different binary mixtures and they are compared with the experimental C_{11}, C_{22} and C_{12} data.

According to Figures 3 - 13, the weighted arithmetic mean closure, Eq. (18), is quite sufficient for application to real mixtures. Table 2 lists the results of fitting Equation (18) to twenty-eight binary mixtures that include the twenty-two systems reported in Figures 3-13 and comprise a wide variety of non-polar, polar, and associating compounds. In Table 2 the sum of squares of the deviations, the maximum percentage deviation and the mole fraction of component 1 at the maximum deviation are also reported. According to this table, the maximum deviation for all the systems studied is 22% (water + ethanol mixture at 25 C), and usually it occurs at the infinite dilution region of one of the two components of the binary mixture (far right column in Table 2). Keeping in mind that some of the C_{ij} data have uncertainty of up to 30 - 40% (Lepori and Matteoli 1988) we conclude that Eq. (18) is an excellent closure for the relation between C_{12}, C_{11}, and C_{22}. All the results reported in this table are at constant temperature and at low pressures. It should be pointed out that parameters α_{21} and α_{12} are expected to be dependent on temperature and, to a lower degree, to pressure.

Figures 14-19 represent comparisons of G_{ij} values calculated using the present technique (Eqs. 28-31 and 12-14) and the experimental values for water + organic and tetrachloromethane (TCM) + organic mixtures. The experimental G_{ij} values were taken from Matteoli and Lepori (1984) for the aqueous systems and from Lepori and Matteoli (1988) for mixtures containing TCM. The agreement between the experimental and calculated data ranges from excellent as in the case of TCM+dioxane, to poor as in the case of water+dioxane. The largest deviations are seen at the extrema (maxima or minima) of every curve where the experimental uncertainties are highest at these locations. Overall, according to these figures the calculated G_{ij} integrals are in good agreement with the experimental values.

Table 5 shows the values of the molar volumes and the isothermal compressibilities of the pure components used in Eq.s (29)-(31).

Parameters α_{21} and α_{12} for different binary mixtures are calculated and they are reported in Table 6. These parameters are obtained by minimizing the following objective function

$$\text{O.F.} = \sum \left| G_{11}/G_{11,ex} - 1 \right| + \left| G_{12}/G_{12,ex} - 1 \right| + \left| G_{22}/G_{22,ex} - 1 \right| \tag{32}$$

where $G_{ij,ex}$ is the experimental G_{ij} and the summation runs over all the experimental points.

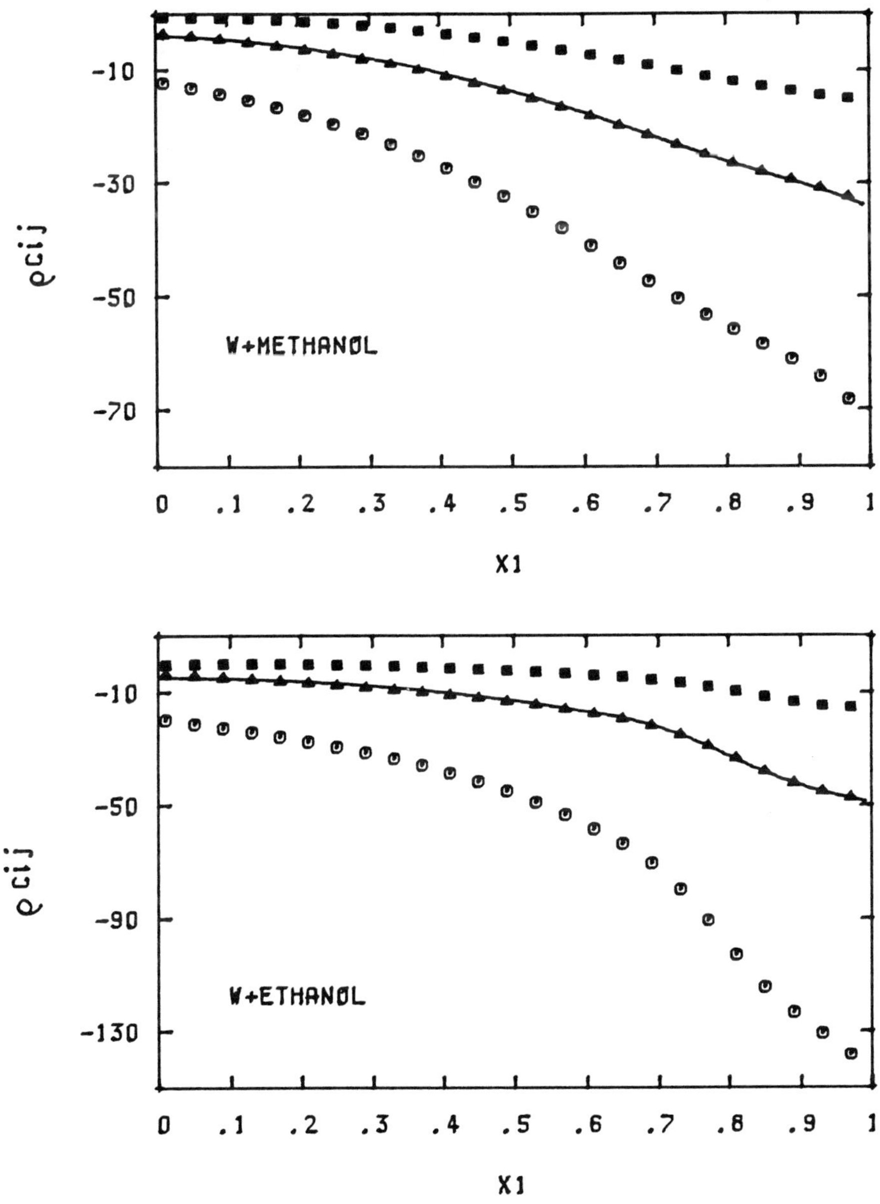

Figure 3. The Variation of ρC_{ij} with composition for water (1) + methanol (2) and for water (1) + ethanol (2) at 25 °C. The points represent experimental ρC_{12} (triangle), ρC_{11} (square), and ρC_{22} (circle) as calculated from data reported by Matteoli and Lepori (1984), Lepori and Matteoli (1988), Lepori, et al. (1988), Berti, et al. (1989) and Hamad, et al. (1989). The solid curve (———) represents the ρC_{12} prediction by the present theory using weighted arithmetic-mean C_{12} closure.

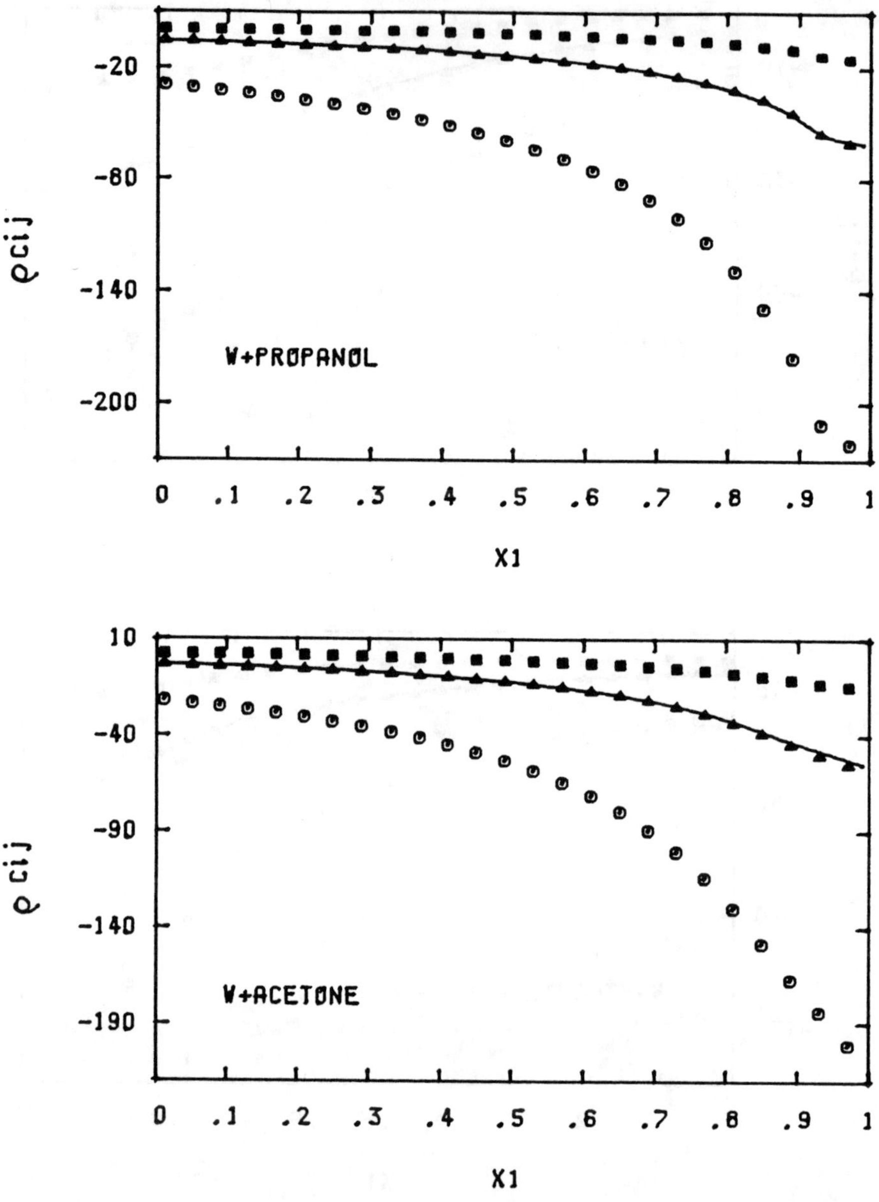

Figure 4. The variation of ρC_{ij} with composition for water (1) + propanol (2) and for water (1) + acetone (2) at 25 °C. For explanation see caption to Figure 3.

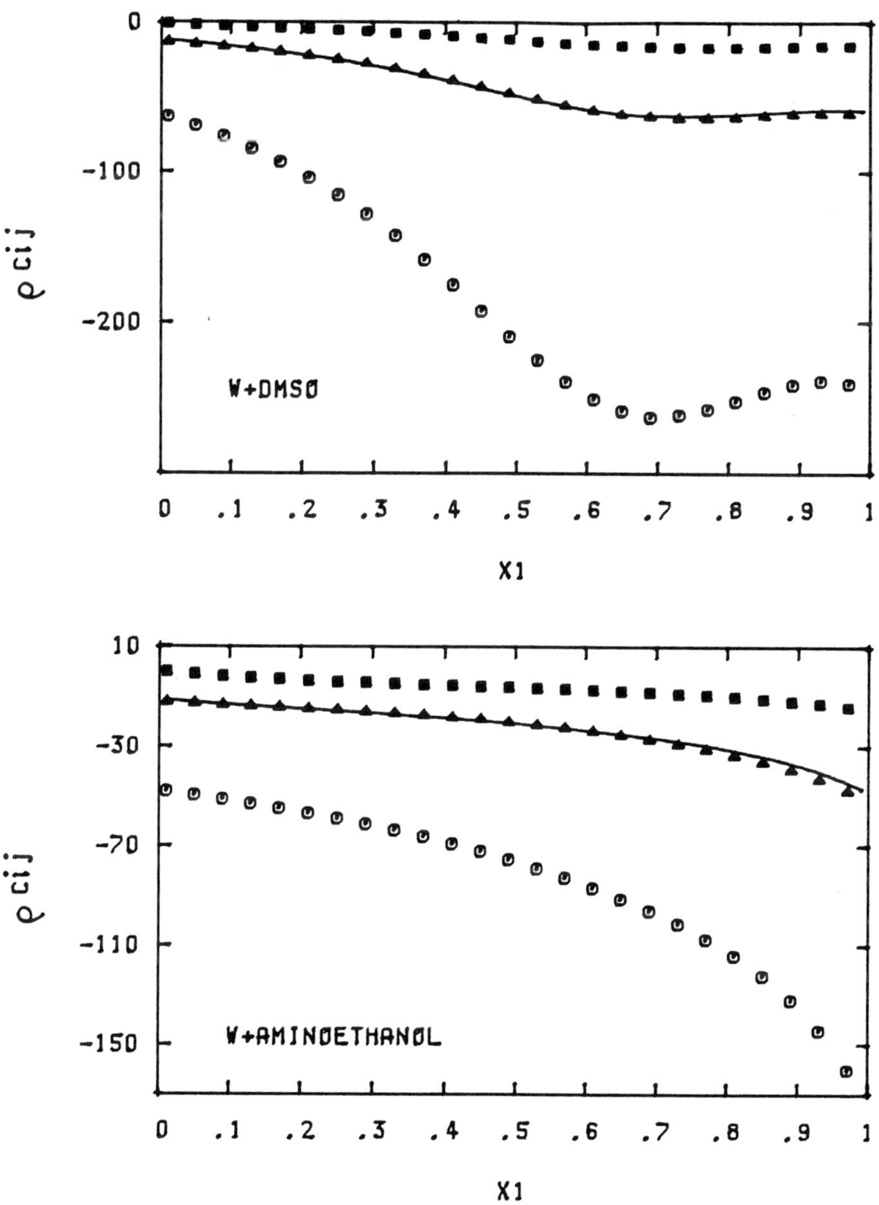

Figure 5. The Variation of ρC_{ij} with composition for water (1) + dimethylsulphoxide (DMSO) (2) and for water (1) + aminethanol (2) at 25 °C. For explanation see caption to Figure 3.

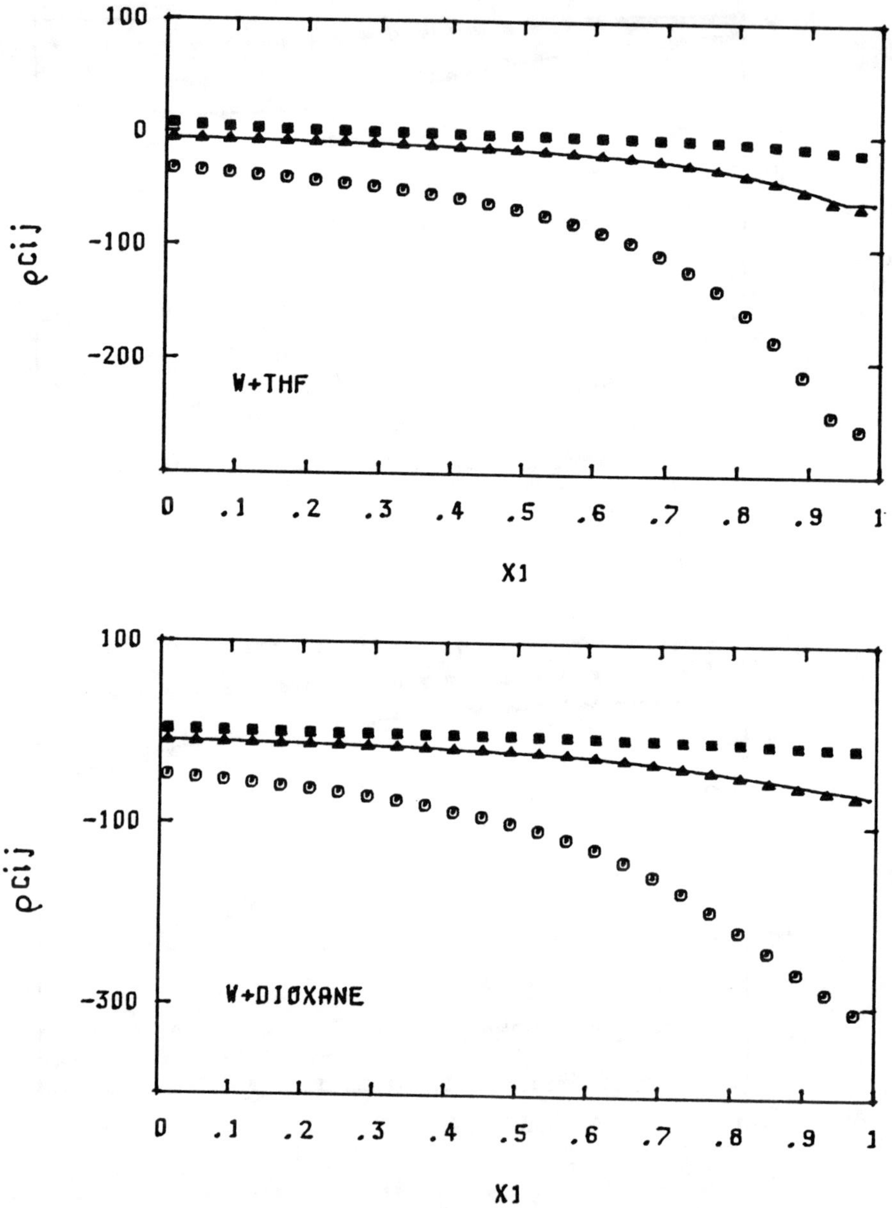

Figure 6. The Variation of ρC_{ij} with composition for water (1) + tetrahydrofuran (THF) (2) and for water (1) + dioxane (2) at 25 °C. For explanation see caption to Figure 3.

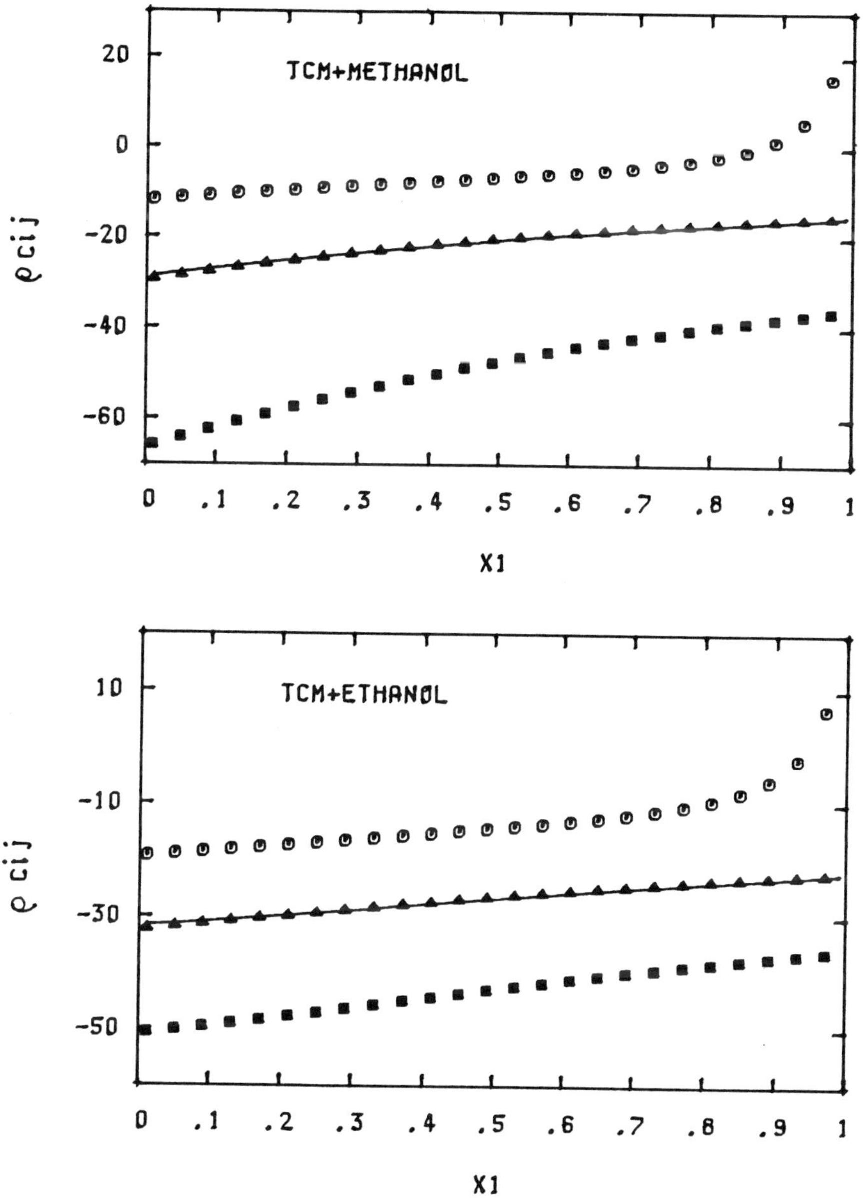

Figure 7. The Variation of ρC_{ij} with composition for tetrachloromethane (TCM) (1) + methanol (2) and for TCM (1) + ethanol (2) at 25 °C. For explanation see caption to Figure 3.

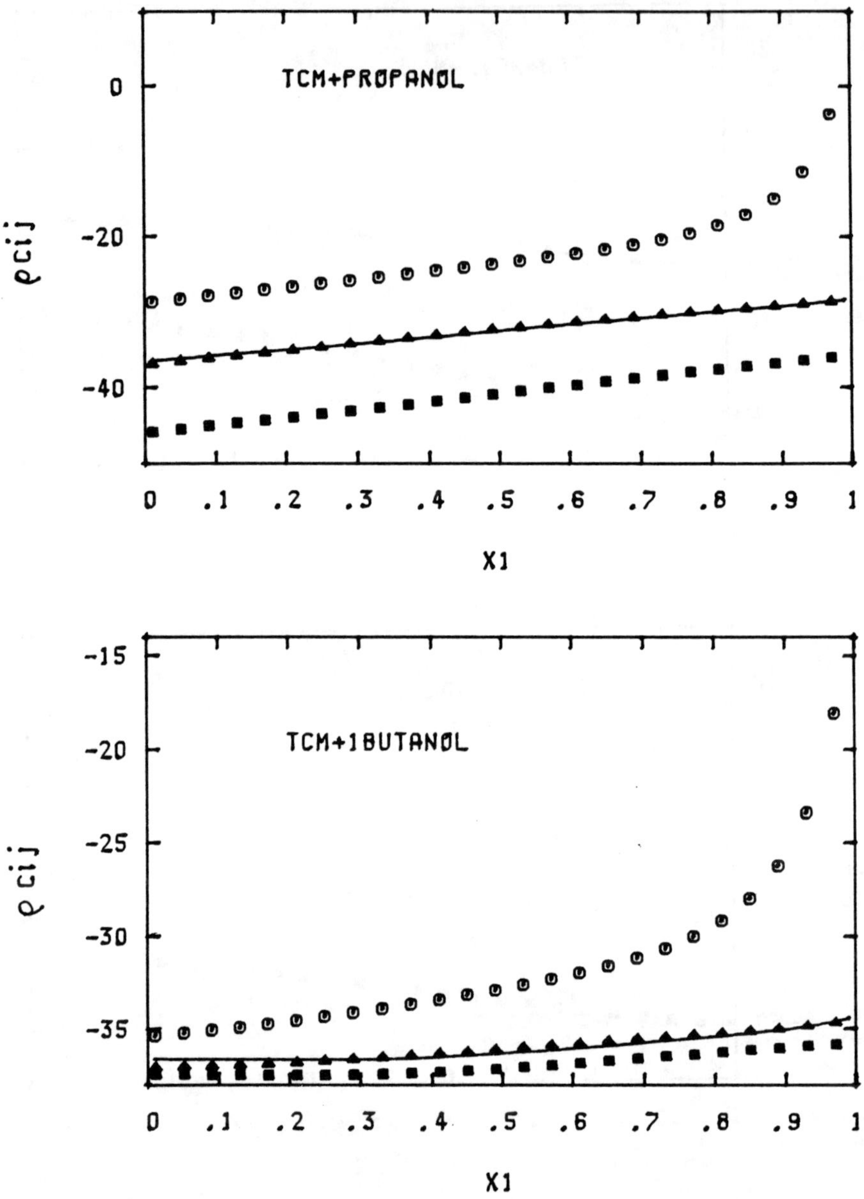

Figure 8. The Variation of ρC_{ij} with composition for TCM (1) + propanol (2) and for TCM (1) + 1-butanol (2) at 25 °C. For explanation see caption to Figure 3.

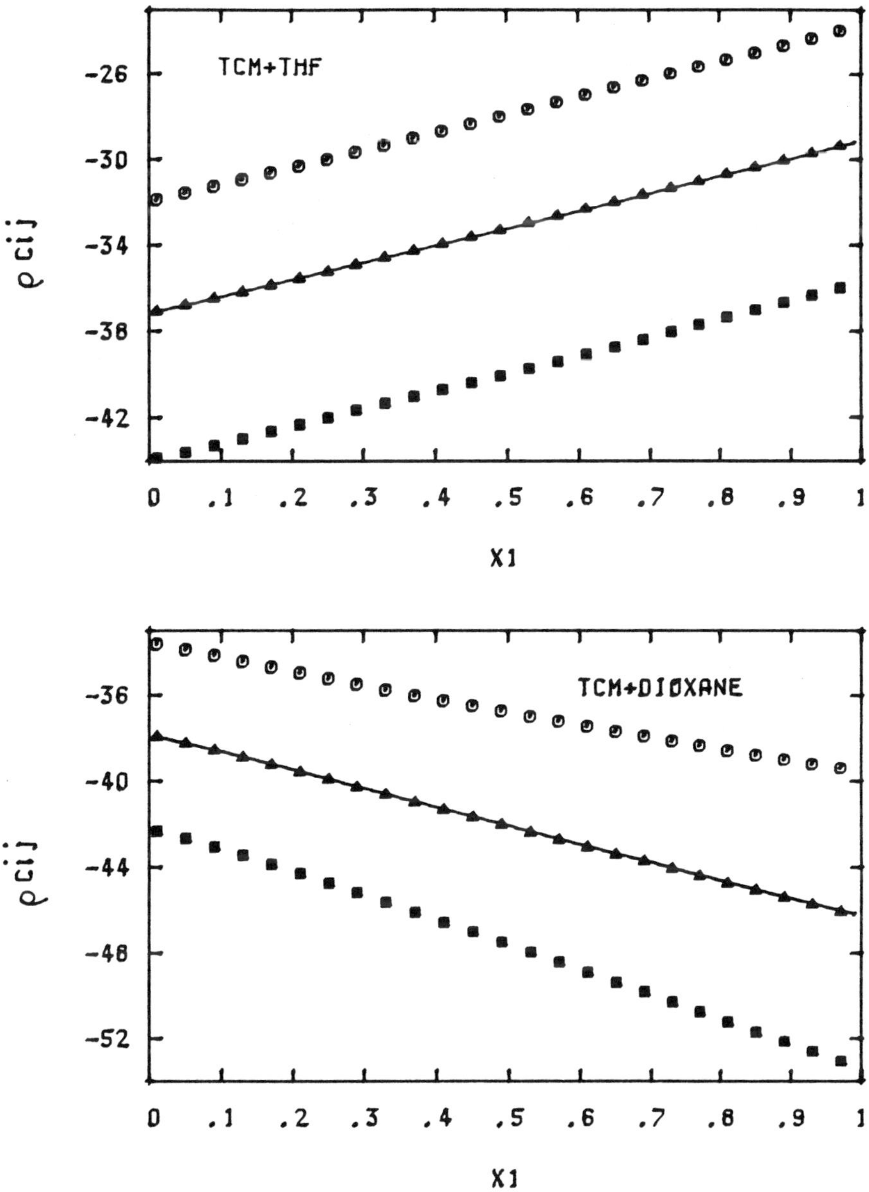

Figure 9. The Variation of ρC_{ij} with composition for TCM (1) + THF (2) and for TCM (1) + dioxane (2) at 25 °C. For explanation see caption to Figure 3.

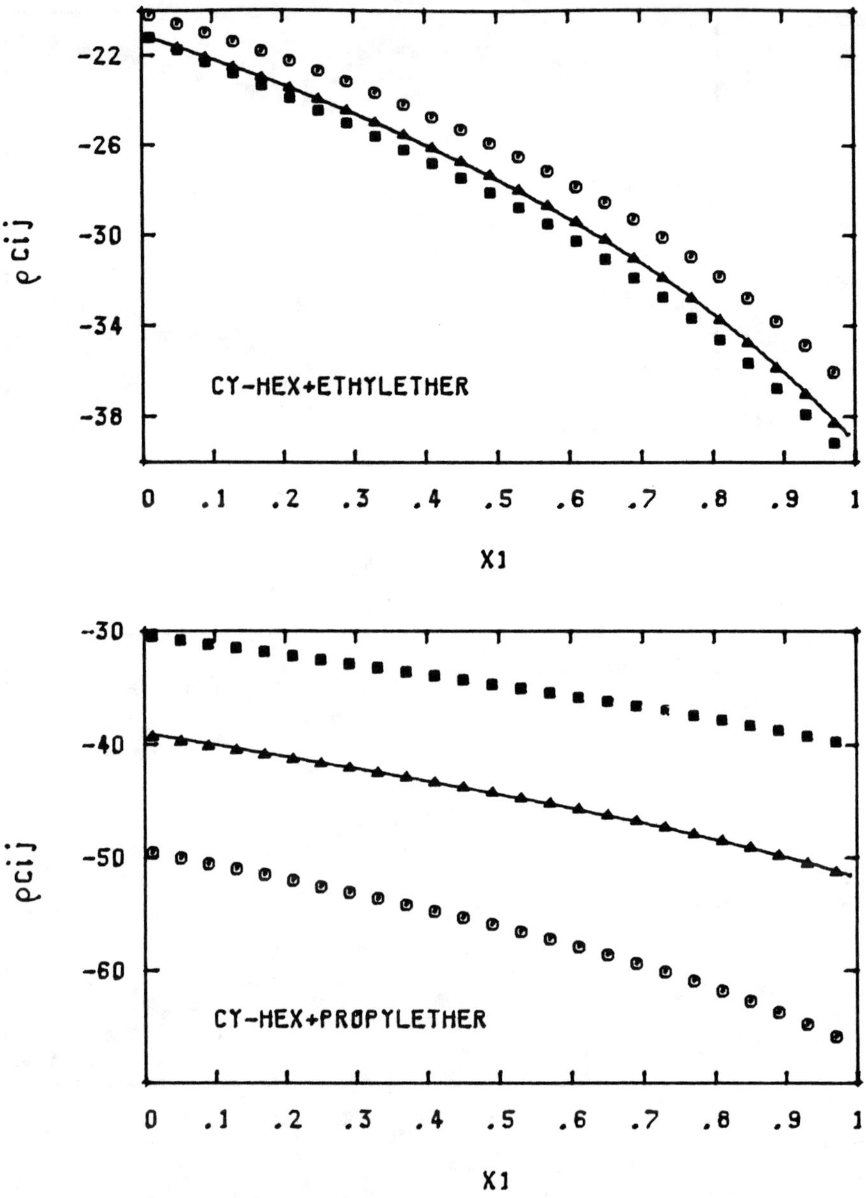

Figure 10. The Variation of ρC_{ij} with composition for cyclohexane (C-HEX) (1) + ethylether (2) and for cyclohexane (1) + propylether (2) at 25 °C. For explanation see caption to Figure 3.

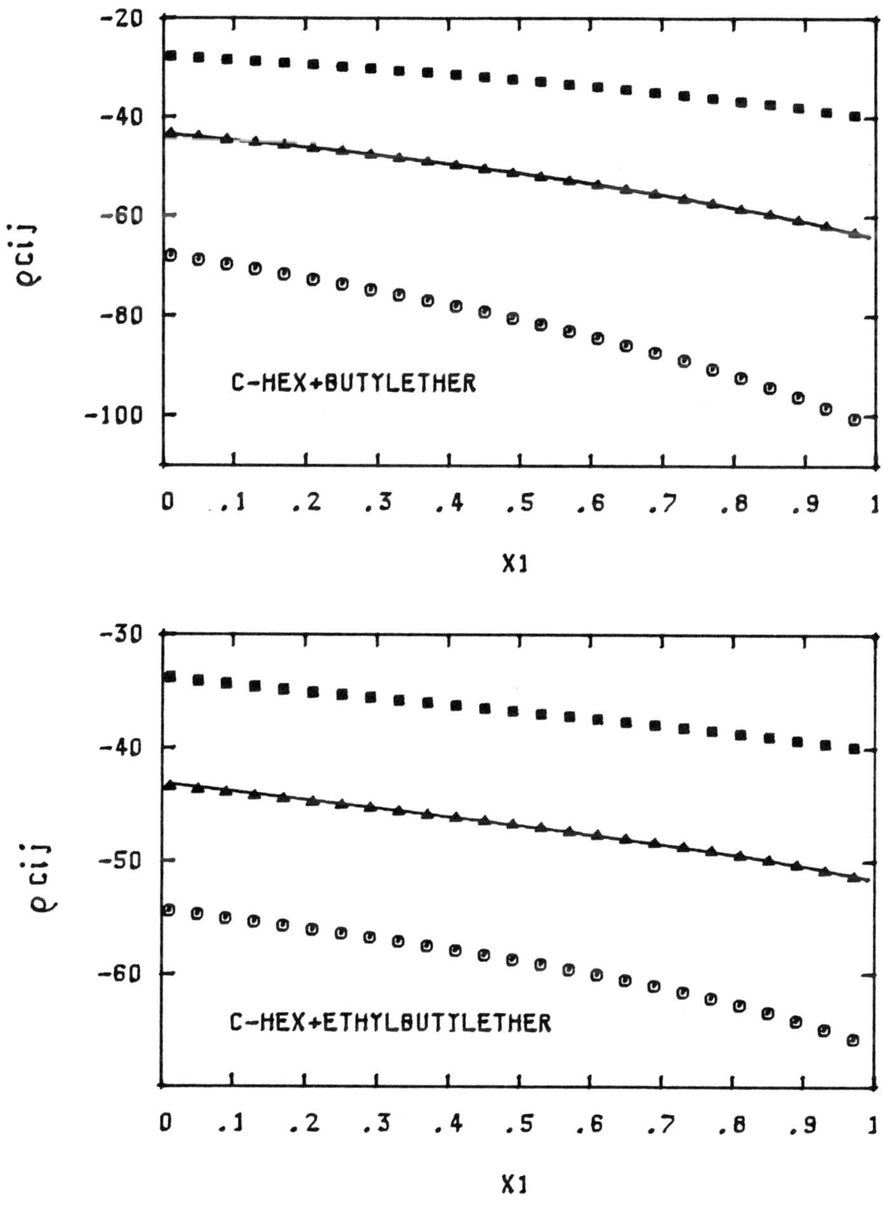

Figure 11. The Variation of ρC_{ij} with composition for cyclohexane (C-HEX) (1) + butylether (2) and for cyclohexane (1) + ethylbutylether (2) at 25 °C. For explanation see caption to Figure 3.

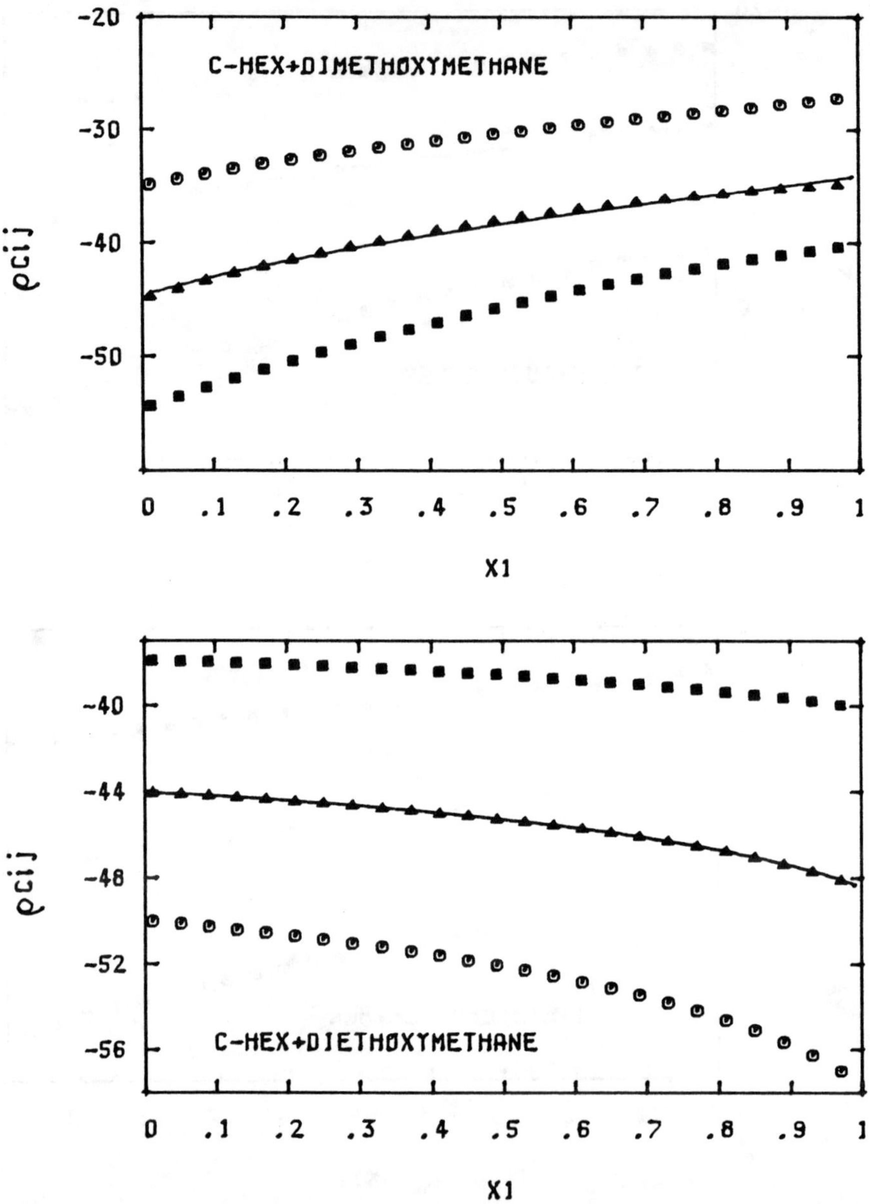

Figure 12. The Variation of ρC_{ij} with composition for cyclohexane (C-HEX) (1) + dimethoxymethane (2) and for cyclohexane (1) + diethoxymethane (2) at 25 °C. For explanation see caption to Figure 3.

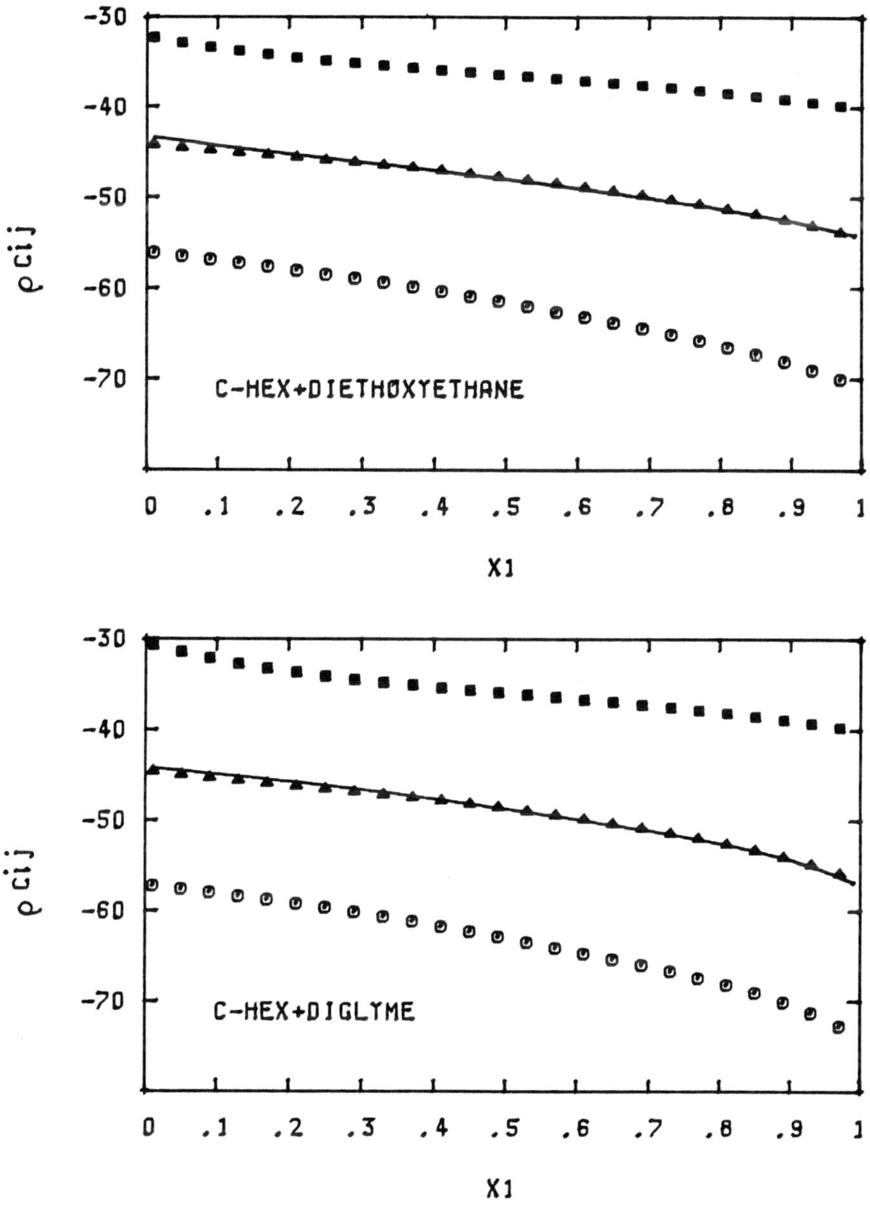

Figure 13. The Variation of ρC_{ij} with composition for cyclohexane (C-HEX) (1) + diethoxyethane (2) and for cyclohexane (1) + diethylenglycoldimethylether (diglyme) (2) at 25 °C. For explanation see caption to Figure 3.

Table 5. Molar volumes and isothermal compressibilities of the pure liquids used in this report.

liquid	v_i [cm^3 mol^{-1}]	T [°C]	κ_T [10^{-4}MPa^{-1}]
water	18.07	0.0	5.01
		25	4.43
		50	4.40
		60	4.43
		90	4.68
tetrachloromethane	96.50	25	10.67
methanol	40.73	0.0	10.62
		25	12.52
		60	15.70
ethanol	58.68	25	12.7
		50	13.7
		90	18.0
propanol	75.14	25	10.2
n-butanol	91.53	25	10.2
tert-butanol	94.88	25	12.0
THF	91.53	25	10.2
DMSO	70.94	25	4.60
1,4-dioxane	85.25	25	7.20
2-aminoethanol	60.01	25	5.0
acetone	74.05	25	12.39
acetonitrile	52.25	30	11.1

Data obtained as reported in Matteoli and Lepori (1984) o taken from Brostow and Maynadier (1979) or Weast et al, (1989).

For the systems water+alcohol, both α_{21} and α_{12} decrease with increasing the alcohol chain length, at the same temperature. The opposite trend is seen for the α_{21} and no trend for α_{12} of the TCM+alcohol systems. The parameter α_{21} and α_{12} show monotonic variation with temperature for water+methanol, but not for water + ethanol.

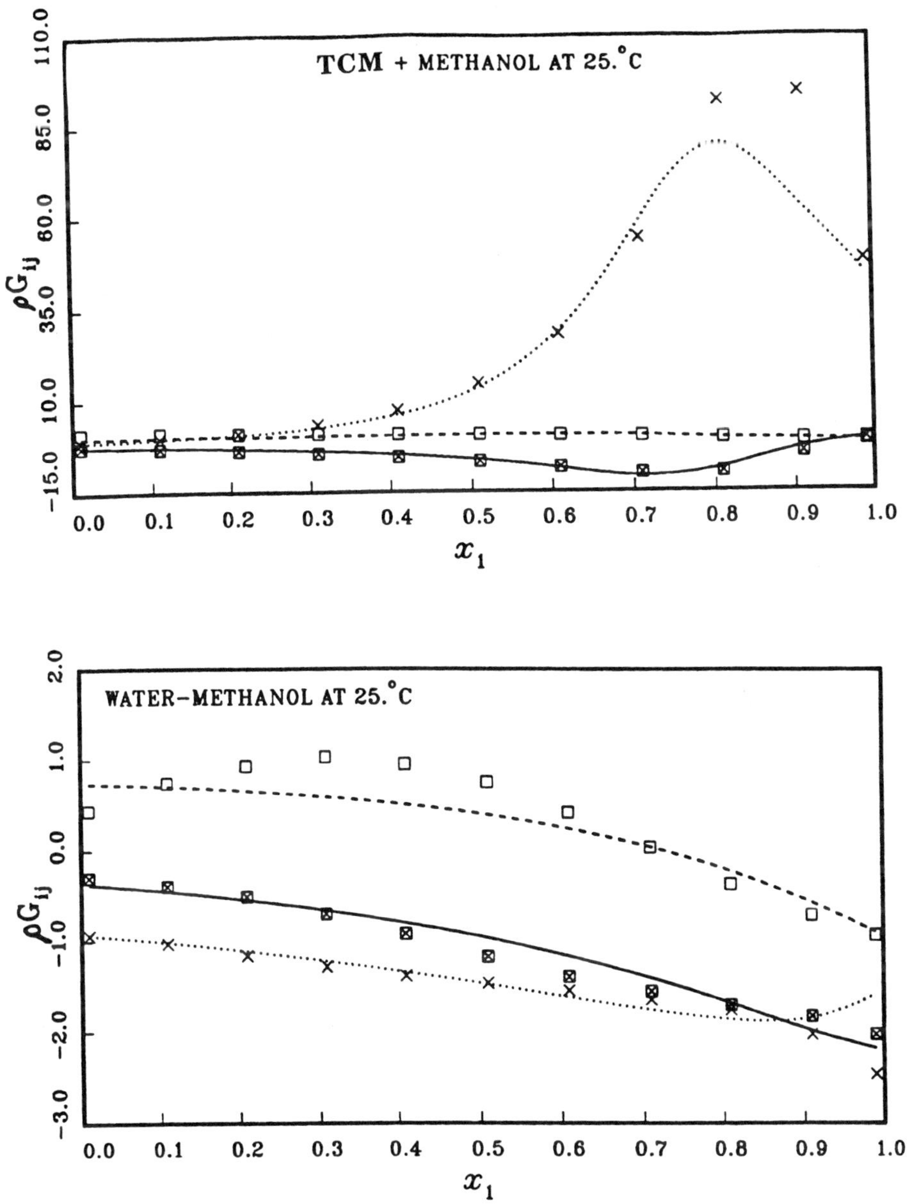

Figure 14. The Variation of ρG_{ij} with composition for tetrachloromethane (TCM) (1) + methanol (2) and for water (1) + methanol (2) at 25 °C. The points represent the experimental data (Matteoli and Lepori 1984; Lepori and Matteoli 1988; Hamad, et al. 1989): □, ρG_{11}, ⊠, ρG_{12}, and X, ρG_{22}. The curves (- - - - ρG_{11}, ——— ρG_{12}, and .... ρG_{22}) represent the predictions by the present theory using weighted arithmetic-mean C_{12} closure.

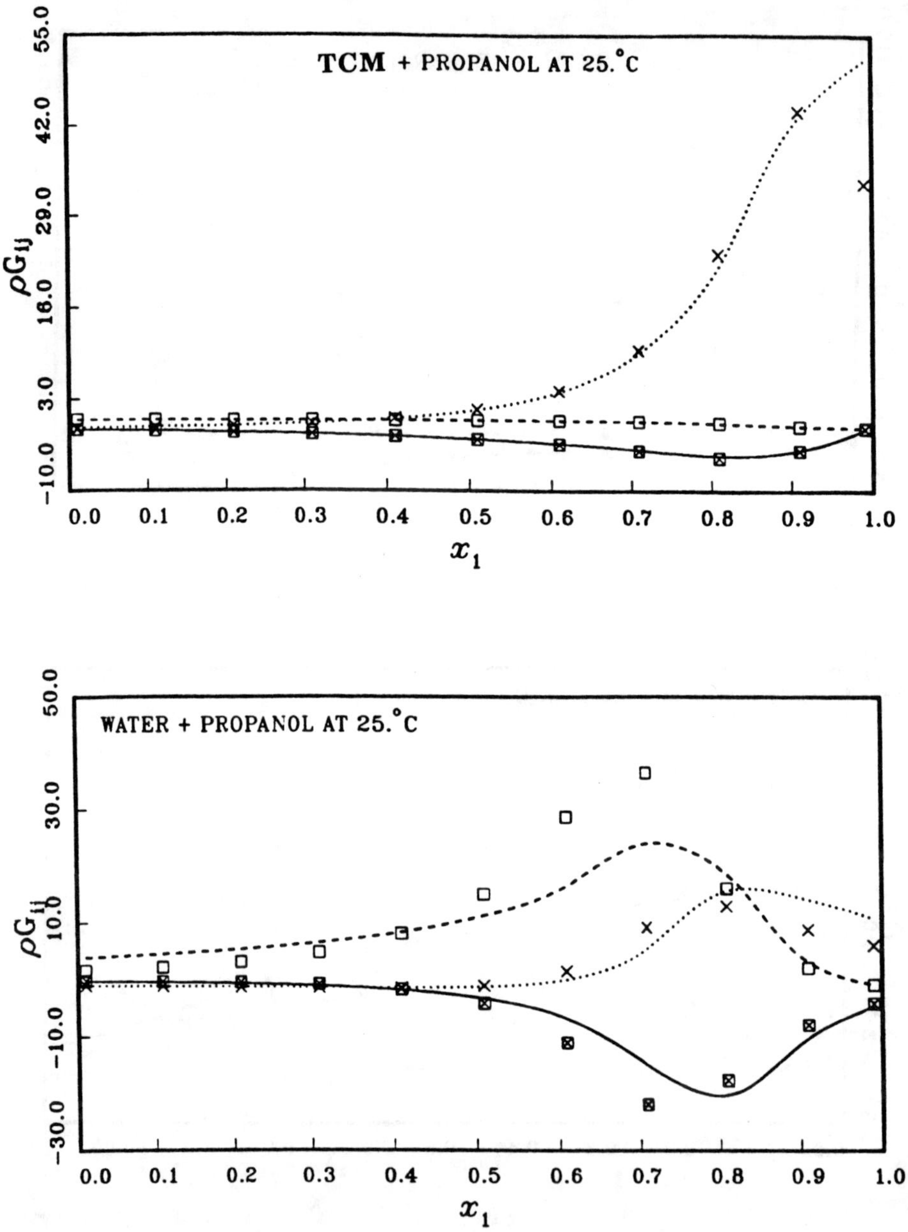

Figure 15. The Variation of ρG_{ij} with composition for tetrachloromethane (TCM) (1) + propanol (2) and for water (1) + propanol (2) at 25 °C. For explanation see caption to Figure 14.

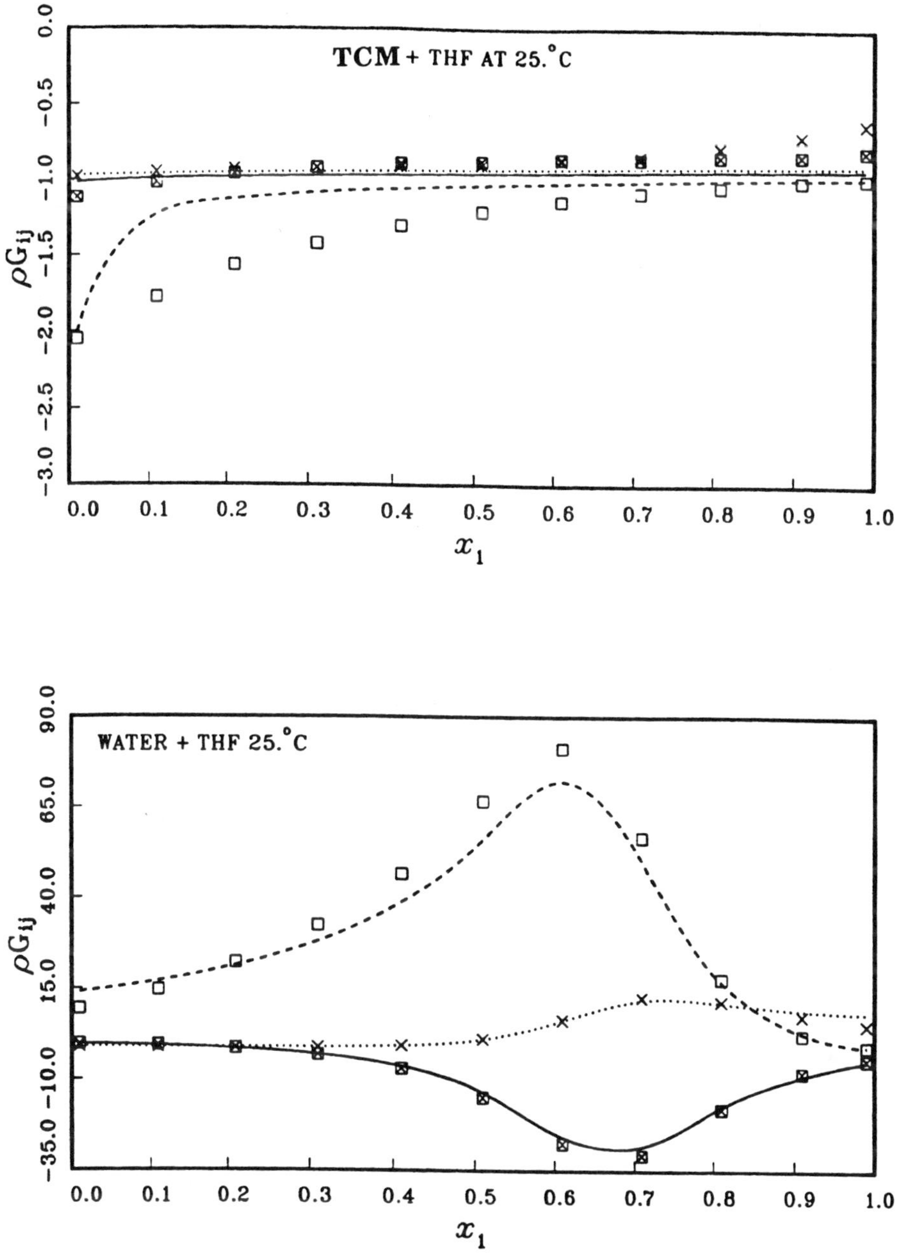

Figure 16. The Variation of ρG_{ij} with composition for tetrachloromethane (TCM) (1) + tetrahydrofuran (THF) (2) and for water (1) + THF (2) at 25 °C. For explanation see caption to Figure 14.

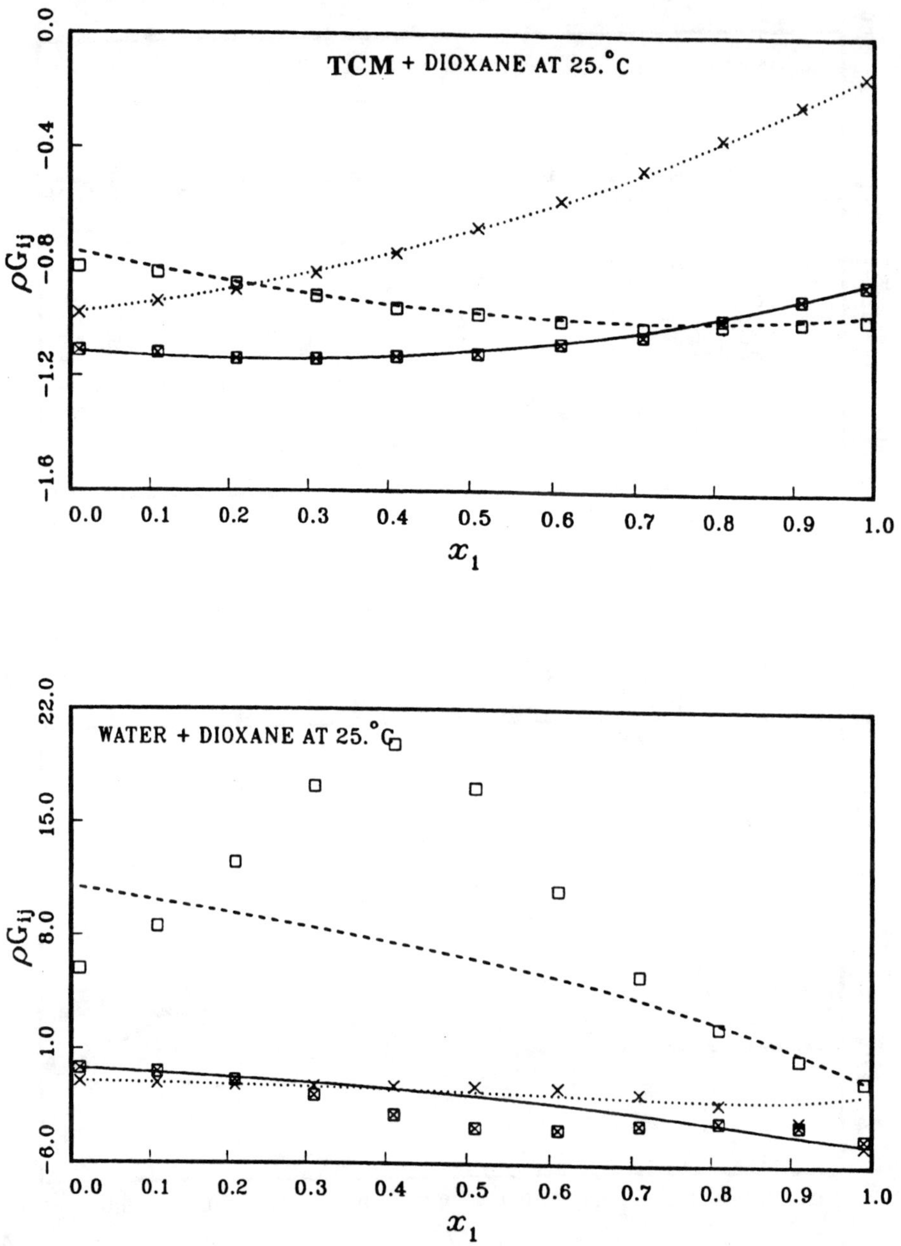

Figure 17. The Variation of ρG_{ij} with composition for tetrachloromethane (TCM) (1) + Dioxane (2) and for water (1) + Dioxane (2) at 25 °C. For explanation see caption to Figure 14.

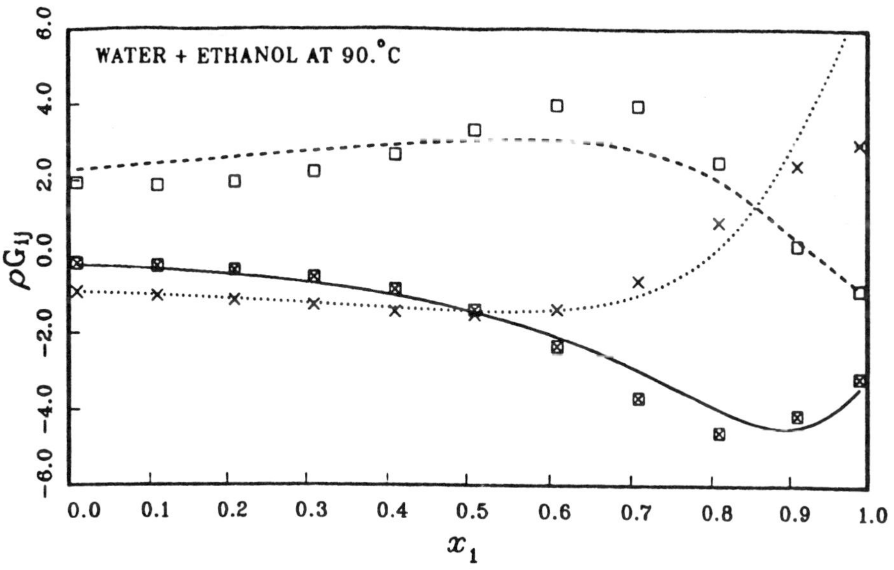

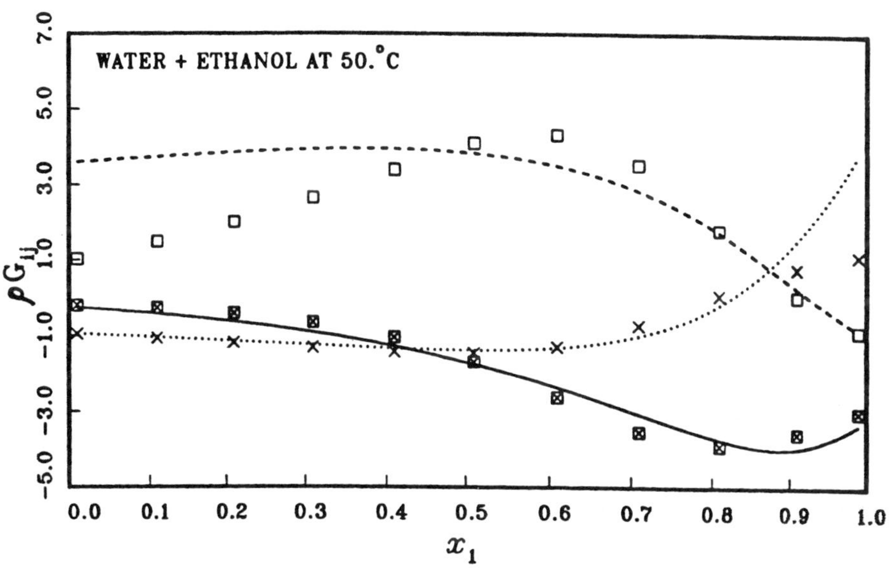

Figure 18. The variation of ρG_{ij} with composition for water (1) + ethanol (2) at two temperatures. For other details see caption to Figure 14.

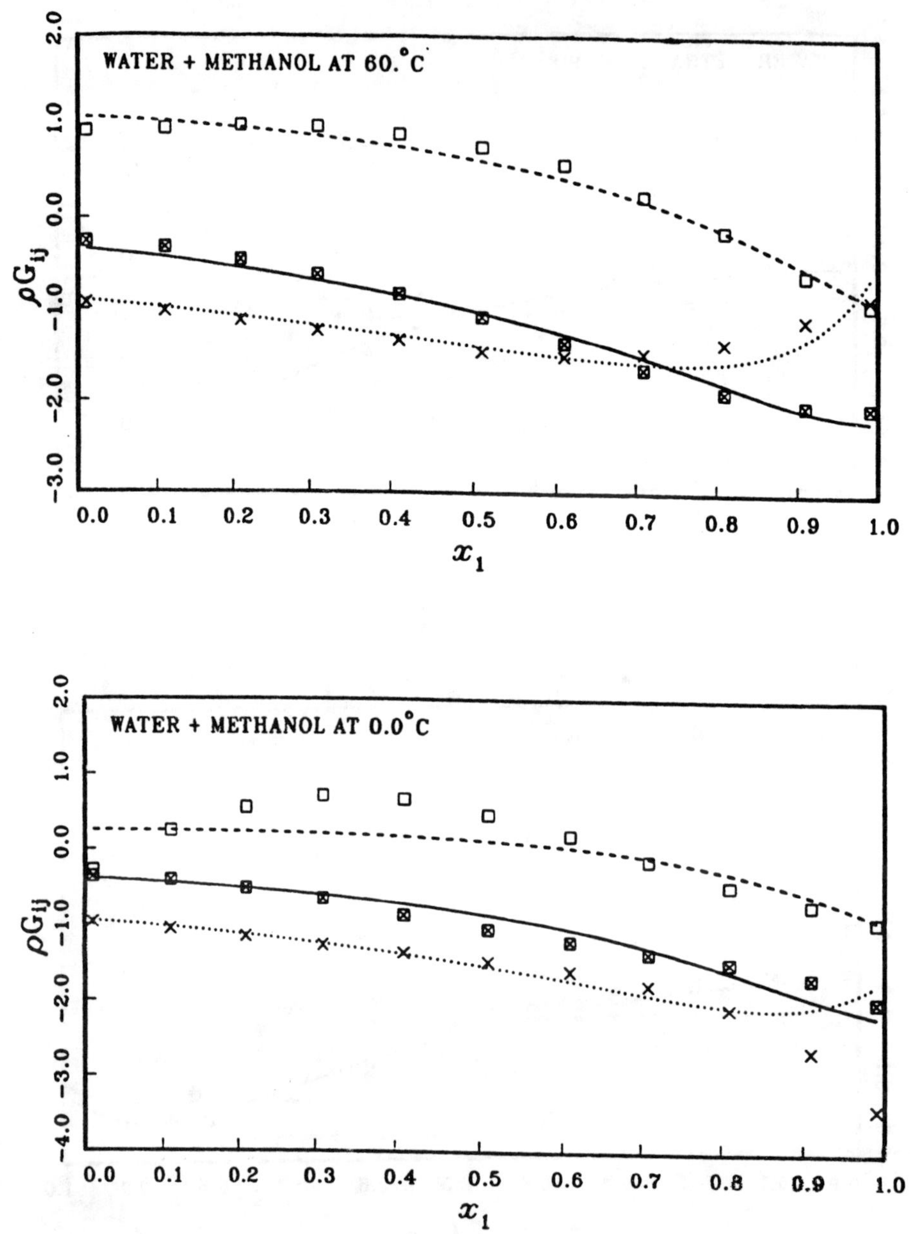

Figure 19. The variation of ρG_{ij} with composition for water (1) + methanol (2) at two temperatures. For other details see caption to Figure 14.

Table 6. Values of the interaction-parameters α_{21} and α_{12}.

organic compound	T[°C]	water + organic		TCM + organic	
		α_{21}	α_{12}	α_{21}	α_{12}
methanol	0	0.7633	0.3008	-	-
"	25	0.4691	0.3577	0.4322	0.0253
"	60	0.3890	0.3790	-	-
ethanol	25	0.4672	0.2727	0.6161	0.0349
"	50	0.2268	0.3015	-	-
"	90	0.3494	0.3007	-	-
propanol	25	0.2657	0.2382	0.7892	0.0196
n-butanol	25	-	-	0.9317	0.0521
tert-butanol	25	0.2436	0.1891	-	-
THF	25	0.0749	0.2282	0.0182	1.0348
DMSO	25	2.4922	0.0932	-	-
1,4-dioxane	25	0.0774	0.2105	0.6348	0.3241
2-aminoethanol	25	1.1289	0.1912	-	-
acetone	25	0.1203	0.2427	-	-
acetonitrile	30	0.1885	0.3537	-	-

CONCLUSION

The relations between the direct correlation function (or fluctuation) integrals reported here, Eq.s (5) & (6) when T, p, and x_1 are the independent variables and Eq.s (9) & (10) when T, ρ_1, and ρ_2 are the independent variables, are mathematically exact. The choice of a closure expression relating the cross direct correlation integral to the other two integrals in a binary mixture allows us to calculate these integrals simply and analytically using only the pure fluid thermodynamic data. It is demonstrated here that simple geometric and arithmetic mean closures are satisfactory for simple model fluids. It is also demonstrated that a weighted arithmetic mean closure expression is sufficient to represent the relation between the cross direct correlation integral to the other correlation integrals of binary real mixtures. Success of this theory in its application to variety of mixtures comprising of components with polar and associating intermolecular potential energies is indicative of its promise as a strong technique for prediction of properties of complex mixtures of practical interest. In the report by Hamad and Mansoori in the present monograph, applicability of this theory for prediction of phase behavior of complex mixtures is demonstrated.

ACKNOWLEDGMENT: This research is supported by the NATO Advanced Study Program.

REFERENCES

Berti, P., L. Lepori and E. Matteoli. 1989. Excess volumes of cyclohexane + a linear ether. Fluid Phase Equilib. 44: 285-294.

Brostow, W. and P. Maynadier. 1979. Temperature dependence of the isothermal compressibility of liquids. High Temp. Sci. 11: 7-21.

Carnahan, N.F. and K.E. Starling. 1969. Rigid sphere equation of state. J. Chem. Phys. 51: 635-638.

Gibbs, J.W. 1948. Elementary principles in statistical mechanics. In The Collected Works of J. Willard Gibbs, Vol. II, New Haven: Yale University Press.

Hamad, E.Z., G.A. Mansoori, and J.F. Ely. 1987. Conformality of the Kirkwood-Buff solution theory of statistical mechanics. J. Chem. Phys. 86: 1478- 1484.

Hamad, E.Z. 1988. Statistical Mechanical Modeling of Complex Mixtures: Distribution Function Approach. PhD Thesis, University of Illinois, Chicago, IL.

Hamad, E.Z., G.A. Mansoori, E. Matteoli and L. Lepori. 1989. Relations among concentration fluctuation integrals in mixtures (theory and experiments). Z. Phys. Chem. Neue Folge. 162: 27-45.

Hamad, E.Z. and G.A. Mansoori. (to appear) 1989. Model calculations of thermodynamic mixture properties from direct correlation integrals. Z. Phys. Chem. Neue Folge.

Kirkwood, J.G. and F.P. Buff. 1951. The statistical mechanical theory of solutions:I. J. Chem. Phys. 19: 774-777.

Landau, L. and E. M. Lifshitz. 1980. Course of Theoretical Physics. Vol. 5, Part 1, Oxford: Pergamon Press.

Lebowitz, J.L. 1964. Exact solution of generalized Percus-Yevick equation for a mixture of hard spheres. Phys. Rev. A 133: 895-899.

Lepori, L. and E. Matteoli. 1988. Solute-solute-solvent interactions studied through the Kirkwood-Buff theory. 3. Effect of the pressure on G_{ij} for some aqueous and non aqueous mixtures. J. Phys. Chem. 92: 6997-7001.

Lepori, L., E. Matteoli and B. Marongiu. 1988. Isothermal liquid-vapor equilibria of mixtures containing organic compounds. 3. Excess Gibbs free energies of Cyclohexane + a linear ether. Fluid Phase Equilib. 42: 229-240.

Mansoori, G.A., N.F. Carnahan, K.E. Starling and T.W. Leland. 1971. Equilibrium thermodynamic properties of mixtures of hard spheres. J. Chem. Phys. 54: 1523-1525.

Mansoori, G.A. and J.F. Ely. 1985. Statistical mechanical theory of local compositions. Fluid Phase Equilib. 22: 253 275.

Matteoli, E. and L. Lepori. 1984. Solute-solute interactions in water. II. An analysis through the Kirkwood-Buff integrals for 14 organic solutes. J. Chem. Phys. 80: 2856-2863.

McQuarrie, D.A. 1975. Statistical Mechanics. New York: Harper and Row.

Nicolas, J.J., K.E. Gubbins, W.B. Street, and D.J. Tildesley. 1979. Equation of state for Lennard-Jones fluid. Mol. Phys. 37: 1429-1454.

O'Connell, J.P. 1971. Thermodynamic properties of solutions based on correlation functions. Mol. Phys. 20: 27-33; 1981. Thermodynamic properties of solutions and the theory of fluctuations. Fluid Phase Equilib. 6: 21-38.

Pearson F.J. and G. S. Rushbrooke. 1957. On the theory od binary fluid mixtures. Proc. Roy. Soc. Edinburgh, Sec A, A64, 305-317.

Percus, J.K. and G.J. Yevick. 1958. Analysis of classical statistical mechanics by means of collective coordinates. Phys. Rev. 110: 1-13.

Shing, K.S. 1982. Chemical Potential of Dense Fluid Mixtures from Computer Simulation. Ph. D. Thesis, Cornell University.

Shing, K.S., K.E. Gubbins and K. Lucas. 1988. Henry constants in non-ideal fluid mixtures: computer simulation and theory. Mol. Phys. 65: 1235-1252.

Weast R.C., M.J. Astle and Boca Raton 1989. Handbook of Chemistry and Physics, eds. , Florida: CRC Press, Inc.

Inversion of the Kirkwood-Buff Theory of Solutions and Its Applications

A. Ben-Naim
Department of Physical Chemistry
The Hebrew Univ. of Jerusalem, Jerusalem, Israel

Abstract

The Kirkwood-Buff theory of solutions was originally formulated to obtain thermodynamic quantities from molecular distribution functions. This formulation is useful whenever molecular distribution functions are available either by analytical calculations or from computer simulations. The inversion procedure of the same theory reverses the role of the thermodynamic and molecular quantities – i.e., it enables the evaluation of integrals over the pair correlation functions from thermodynamic quantities. These integrals, G_{ij} referred to as the Kirkwood-Buff integrals, were found useful to characterize mixtures on a molecular level. They are also used in the theory of preferential solvation to characterize the local environment of a molecule in a mixture of a two component solvent.

1. Introduction

The Kirkwood-Buff (KB) theory was published in 1951 (Kirkwood and Buff, 1951) about six years after the publication of the McMillan-Mayer theory of solutions (McMillan and Mayer, 1945). Although both theories are founded on the firm basis of statistical mechanics, the KB theory was quite dormant for a long time, and did not enjoy the same popularity as the McMillan-Mayer theory did.

The KB theory, as originally formulated, gave a recipe for computing some thermodynamic quantities, from integrals over the pair correlation functions, in multicomponent systems. The relevant thermodynamic quantities are: the partial molar volumes, V_i, the isothermal compressibility κ_T, and the derivatives of the chemical potentials with respect to concentrations, say $\partial\mu_i/\partial\rho_j$. The molecular quantities G_{ij} are defined by

$$G_{ij} = \int_0^\infty [g_{ij}(R)-1]4\pi R^2 dR \tag{1.1}$$

where g_{ij} (R) is the pair correlation function for the pair of species i and j and the integral extends over the entire range of distances $0 \leq R \leq \infty$. These integrals are sometimes referred to as the KB integrals.

The interpretation of these quantities is best obtained by considering the meaning of $\rho_i G_{ij}$, where ρ_i is the number density of the i species (Ben-Naim, 1974). The quantity $\rho_i g_{ij}(R) 4\pi R^2 dR$ measures the average number of i molecules in a spherical shell of width dR at a distance R from the center of a j molecule. On the other hand, the quantity $\rho_i 4\pi R^2 dR$ measures the average number of i molecules in the same spherical shell, the origin of which has been chosen at random. Thus the quantity $\rho_i [g_{ij}(R)-1] 4\pi R^2 dR$ is a measure of the excess (or deficiency) in the average number of i molecules in the spherical shell at distance R from a j molecule relative to the number obtained by eliminating the condition that a j molecule be placed at the center of the sphere.

Clearly, $\rho_i G_{ij}$ reflects the total average excess (or deficiency) of i molecules in the entire surrounding of a j molecule. Finally, by dividing this quantity by ρ_i we get a measure, per unit density, which is now symmetrical in the two indices, namely, $G_{ij} = G_{ji}$. These quantities convey information on the average "affinity" of i molecules towards j molecules and vice versa.

The main result of the KB theory can be symbolically written as:

$$\{G_{ij}\} \longrightarrow \{V_i, \kappa_T, \partial\mu_i/\partial\rho_j\} \tag{1.2}$$

Having information on G_{ij}'s one can compute the thermodynamic quantities. However, since the quantities G_{ij} were not available from experiments, the theory could have been used only in rare cases where G_{ij} were obtained from theoretical work. In principle, having an approximate theory for computing the various pair correlation functions $g_{ij}(R)$, it is possible to evaluate the integrals G_{ij}, and then compute the thermodynamic quantities through the KB theory. Comparison between the thermodynamic quantities, thus obtained and the corresponding experimental data could serve as a test of the theory that provides the pair correlation functions.

In 1977 an inversion procedure was suggested that reversed the role of the thermodynamic quantities and the molecular quantities (Ben-Naim, 1977). This may be written symbolically as:

$$\{V_i, \kappa_T, \partial\mu_i/\partial\rho_j\} \longrightarrow \{G_{ij}\} \tag{1.3}$$

In this form the thermodynamic quantites are used as input to compute the molecular quantities G_{ij}. Since it is relatively easier to measure the required thermodynamic quantities the inversion procedure provides a new tool to investigate the characteristics of the local environments of each species in a multicomponent system.

A brief outline of the KB results as well as the inversion procedure is given below. We consider a two component system, say of water (W) and ethanol (E) (this system was the first to be analyzed by the inversion of the KB theory (Ben-Naim, 1977)).

Let ρ_w and ρ_E be the number densities of W and E respectively. We now introduce the two auxiliary quantities:

$$\eta = \rho_W + \rho_E + \rho_W \rho_E (G_{WW} + G_{EE} - 2G_{WE}) \tag{1.4}$$

$$\xi = 1 + \rho_W G_{WW} + \rho_E G_{EE} + \rho_W \rho_E (G_{WW} G_{EE} - G_{WE}^2) \tag{1.5}$$

The isothermal compressibility κ_T, the partial molar volumes V_E, V_W, and the derivatives of the chemical potentials may be expressed in terms of the $G_{\alpha\beta}$'s by the following relations (Kirkwood and Buff, 1951, Ben-Naim, 1974):

$$\kappa_T = \xi/kT\eta \tag{1.6}$$

$$V_E = [1 + \rho_W(G_{WW} - G_{WE})]/\eta \tag{1.7}$$

$$V_W = [1 + \rho_E(G_{EE} - G_{WE})]/\eta \tag{1.8}$$

$$\mu_{EE} = \rho_W kT/\rho_E\eta V,\ \mu_{WW} = \rho_E kT/\rho_W\eta V,\ \mu_{EW} = \mu_{WE} = -kT/\eta V \tag{1.9}$$

where $\mu_{\alpha\beta} = (\partial\mu_\alpha/\partial N_\beta)_{T,P,N}$. Here we have written six equations, although only three of these are independent, since we have the following thermodynamic identities:

$$\rho_W\mu_{WW} + \rho_E\mu_{WE} = 0 \tag{1.10}$$

$$\rho_W\mu_{WE} + \rho_E\mu_{EE} = 0 \tag{1.11}$$

$$\rho_E V_E + \rho_W V_W = 1 \tag{1.12}$$

Thus in essence we have three independent equations relating the thermodynamic quantities to the three quantities G_{WW}, G_{EE}, and $G_{EW} = G_{WE}$ (for any given temperature and composition of the system). An inversion procedure is possible in which the three $G_{\alpha\beta}$'s may be computed from the above-mentioned thermodynamic quantities.

First we assume that the vapor above the mixture, at room temperature, may be considered to be an ideal gas mixture. Thus, for the chemical potential of the water we write:

$$\mu_W^{\,l} = \mu_W^{\,g} = \mu_W^{\,og} + kT\ln P_W \tag{1.13}$$

Using the equation for μ_{WW} in (1.9) and the thermodynamic identity

$$\begin{aligned}\left(\frac{\partial\mu_W}{\partial N_W}\right)_{T,P,N_E} &= \left(\frac{\partial\mu_W}{\partial x_W}\right)_{T,P}\left(\frac{\partial x_W}{\partial N_W}\right)_{T,P,N_E} \\ &= \left(\frac{\partial\mu_W}{\partial x_W}\right)_{T,P}\frac{N_E}{(N_E+N_W)^2}\end{aligned} \tag{1.14}$$

we obtain

$$\left(\frac{\partial\mu_W}{\partial x_W}\right)_{T,P} = \frac{kT\varrho^2}{\varrho_W\eta} \tag{1.15}$$

where $\rho = (N_W + N_E)/V$ is the total number density of molecules in the system. Introducing the assumption of ideality for the gaseous phase we obtain, from (1.13) and (1.15):

$$x_W\frac{\partial\ln P_W}{\partial x_W} = \frac{\varrho}{\eta} \tag{1.16}$$

from which we can compute η if we have the partial vapor pressure of one component as a function of the composition.

Note that we can use information on the vapor pressure of either component, since the corresponding derivatives are related by the identity:

$$x_W \frac{\partial \ln P_W}{\partial x_W} = x_E \frac{\partial \ln P_E}{\partial x_E} \tag{1.17}$$

Having computed η from (1.16), we use relation (1.6) to compute ξ from the experimental values of the isothermal compressibilities.

Next we use the following identity, which can be easily obtained from (1.7) and (1.8):

$$\bar{V}_W \bar{V}_E = \frac{\xi - \eta G_{WE}}{\eta^2} \tag{1.18}$$

and which permits the evaluation of G_{WE} in terms of η, ξ, and the product of the two partial molar volumes of E and W. Finally from (1.7) and (1.8) we may get G_{WW} and G_{EE}. This concludes the inversion procedure.

We note that the inversion procedure can, in principle be applied to any multicomponent system. To the best of the author's knowledge, this has been applied only to two and three component systems (Patil, 1981, Donkersloot, 1979, Matteoli and Lepori, 1984, Kato, 1984, Kato et al., 1982, Rubio, et al., 1987, Nishikawa et al., 1987, Ben-Naim, 1988, Ben-Naim, 1989).

2. Application to the theory of preferential solvation of a solute in a two component solvent (Ben-Naim, 1988)

The problem of preferential solvation (PS) arises almost in any physical chemical study of solutes in mixed solvents. The study could be thermodynamic, spectroscopic, or kinetic (Engberts, 1979). The behaviour of the solutes (e.g., chemical shift, diffusion, reactivity, and so on) depends on the composition of the solvent. However, in order to understand how the solvent composition affects the solute behavior, one needs to know the composition that the solute "sees", i.e., the local composition in the immediate vicinity of the solute.

The simplest approach to the problem of PS is to follow some property of a solute in a mixed solvent. For example, if δ_A is the NM*R* chemical shift (or other spectroscopic quantity) characteristic of the solute *S* in pure solvent *A*, and δ_B the corresponding quantity for pure solvent *B*, then one might relate the observed chemical shift of *S* in a mixture of *A* and *B*, $\delta_{A,B}$, to δ_A and δ_B by the equation

$$\delta_{A,B} = x_A(\text{local})\, \delta_A + [1 - x_A(\text{local})]\delta_B \tag{2.1}$$

where x_A(local) defined in Eq. (2.1) is a measure of the local composition of the solution near the solute. This may or may not be different from the bulk composition x_A of the solvent mixture, x_A being the mole fraction of the component A in the mixture.

Although Eq. (2.1) can serve as an operational definition of x_A(local), it does not really tell us what is the local composition in the vicinity of the solute S. We should not be surprised to find that different properties of S, used in Eq. (2.1), will result in different values of x_A (local). The reason is that there is no theoretical support to the assumption that $\delta_{A,B}$ is an average of δ_A and δ_B, as implied in Eq. [2.1]. Therefore, the approximation involved in using Eq. [2.1] will, in general, be different for different properties of S in mixtures of A and B.

What we need is an unambiguous definition of, and a method of measuring, the local composition of the solvent, that is independent of a specific measurable property of S.

Perhaps the first thermodynamic treatment of the problem of PS was presented by Grunwald et al. (Grunwald et al., 1960), and further developed by Covington and Newman (Covington and Newman, 1976). We present here a different approach based on the KB theory of solutions (Ben-Naim, 1988).

Consider a mixture of two components, N_A molecules of A and N_B molecules of B, at some temperature T and pressure P. In such a system the composition measured by the mole fraction $x_A = N_A/(N_A + N_B)$ will be the same at any point $\mathbf{R}_o$ within the system.

We shall refer to x_A as the *bulk* composition of the system. Next consider a very dilute solution of a solute S in our two-component solvent. (In principle, we could have discussed also high concentrations of S in the system. However, this will require consideration of three components, S, A, and B in the surroundings of S. For simplicity, we treat only the case of very dilute solutions, so that S "sees" practically only As and Bs).

Qualitatively, the question we pose now is the following. What is the composition of the liquid in the immediate vicinity around the solute S? Clearly, since the affinity of S toward A might be different from its affinity toward B, we should expect that the composition near the solute S will differ from the bulk composition x_A.

The main question that concerns us is how to define the *local* region in the vicinity of S in which the composition is expected to be affected by the presence of S.

Consider first a simple spherical solute, say argon, in a two-component solvent, say water and ethanol. Let $dR' = dx'dy'dz'$ be an element of volume at a distance R' from the center of S. The average number densities of A and B in this element of volume will be

$$\rho_A(R') = \rho_A(\text{bulk})g_{AS}(R') \tag{2.2}$$

$$\rho_B(R') = \rho_B(\text{bulk})g_{BS}(R') \tag{2.3}$$

where ρ_A(bulk) and ρ_B(bulk) are the bulk densities of A and B, respectively, and $g_{AS}(R')$ and $g_{BS}(R')$ are the radial distribution functions for the pair of species A,S and B,S, respectively.

Clearly, if we had the full information on these two radial distribution functions, we could have defined the local composition at any distance R' from the center of S by

$$x_A(R') = \rho_A(R')/[\rho_A(R') + \rho_B(R')] \qquad (2.4)$$

Furthermore, this local composition will be different at different distances, say R' and R'' (see Fig. 1). If the solute is not spherical, e.g., S is a protein molecule, then the local composition would be dependent on both distance and relative orientation of the point at which we choose our element of volume.

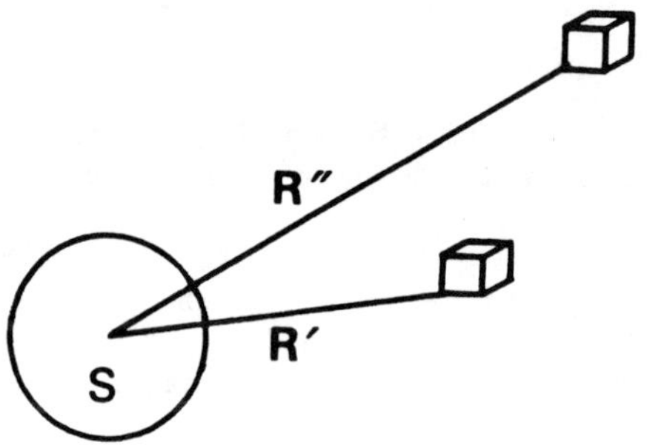

Fig. 1: The average local density of the solvent around a spherical solute S depends only on the distances |**R'**| and |**R''**| from the center of S.

We know from the general theory of liquids that the radial distribution functions normally will tend to unity at distances of the order of magnitude of a few molecular diameters. Thus, at these distances from the center of S, all local densities will be identical to the bulk densities.

At short distances, however, a large deviation from the bulk densities are expected. A typical form of the radial distribution function for a one-component simple liquid is depicted in Fig. 2. We observe that at very short distances the local density will oscillate above and below the bulk density. Clearly, similar behavior is expected in a multicomponent system.

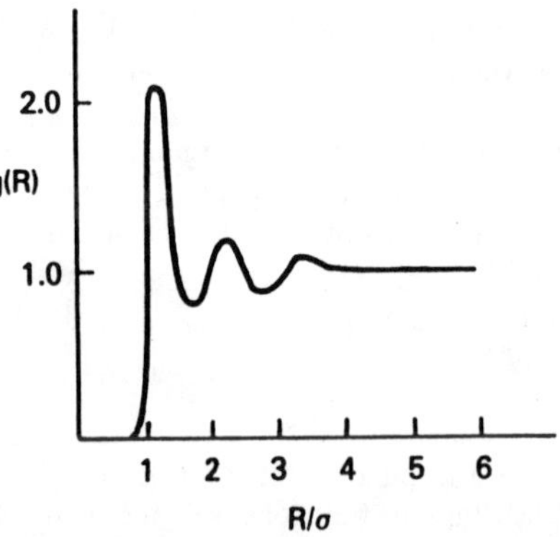

Fig. 2: Schematic form of the pair correlation function $g(R)$ for pure liquid of simple spherical particles. Note that $g(R)$ is practically unity at distances of a few molecular diameters σ.

Unfortunately, there is no experimental data on the separate radial distribution functions in two or more component systems. Even if we had such information, it would have been too detailed to be useful for practical purposes. Instead, we are interested in the overall composition in the local neighborhood of the solute, which roughly coincides with the region in which $g(R)$ is significantly different from unity. Fortunately, this information may be obtained from thermodynamic quantities through the inversion of the KB theory. To do that we first exploit the fact that $g_{AB}(R)$ decays to unity beyond some distance $R \geq R_{AB}$ where R_{AB} may be referred to as the correlation distance for the species A and B. Let R_c be the largest correlation distance for any pair of species in the system. We define the correlation volume as

$$V_{cor} = 4\pi R_c^3/3 \tag{2.5}$$

Since all the pair correlation functions are practically equal to unity at $R \geq R_c$ we may write the average number of A particles in the correlation volume around S as

$$\begin{aligned} \overline{N}_{A,S}(R_c) &= \rho_A \int_0^{R_c} g_{AS}(R) 4\pi R^2 dR \\ &= \rho_A \int_0^{\infty} [g_{AS}(R) - 1] 4\pi R^2 dR + \rho_A \int_0^{R_c} 4\pi R^2 dR \\ &= \rho_A G_{AS} + \rho_A V_{cor} \end{aligned} \tag{2.6}$$

Equation (2.6) simply means that the average number of As in the correlation volume is the sum of the aveage number of As in the same region, before placing S at its center, plus the change in the number of As in the same region caused by placing S in the center of this region.

Using a similar definition for $N_{B,S}(R_c)$, we obtain

$$N_{B,S}(R_c) = \rho_B G_{BS} + \rho_B V_{cor} \tag{2.7}$$

we can define the local composition in the correlation region around S as

$$\begin{aligned} x_{A,S}(\text{local}) &= N_{A,S}(R_c)/[N_{A,S}(R_c) + N_{B,S}(R_c)] \\ &= (x_A G_{AS} + x_A V_{cor})/(x_A G_{AS} + x_B G_{BS} + V_{cor}) \end{aligned} \tag{2.8}$$

where x_A is the bulk composition in the system.

The local composition $x_{A,S}(\text{local})$ can now be compared with x_A to determine the preferential solvation of S. If $x_{A,S}(\text{local}) > x_A$, we may say that S is preferentially solvated by A, and if $x_{A,S}(\text{local}) < x_A$, S is preferentially solvated by B.

Thus, we define the preferential solvation of S with respect to A simply by

$$\begin{aligned} \delta_{A,S} &= x_{A,S}(\text{local}) - x_A \\ &= x_A x_B (G_{AS} - G_{BS})/(x_A G_{AS} + x_B G_{BS} + V_{cor}) \end{aligned} \tag{2.9}$$

Clearly, the sign and extent of preferential solvation might depend on the composition of the solvent. Figure 3 depicts a few possible cases where there are positive, negative, or mixed signs of preferential solvation according to whether $x_{A,S}$(local) is above or below the diagonal line.

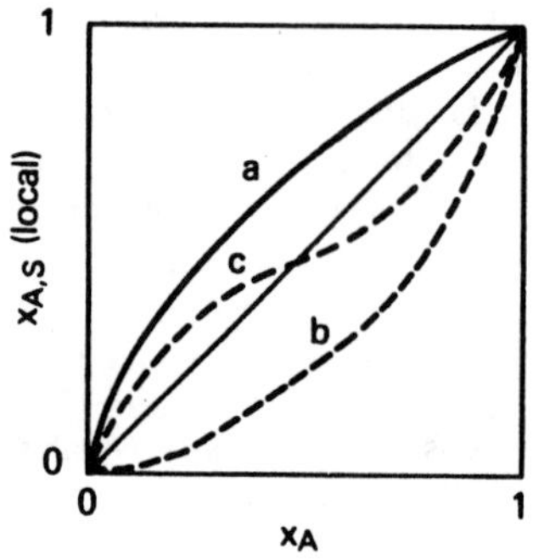

Fig. 3: Schematic dependence of the local composition $x_{A,S}$(local) as a function of the bulk composition x_A. The diagonal line corresponds to the case when there is no preferential solvation of *S*. Curves *a* and *b* correspond to positive and negative preferential solvation. In curve *c*, the preferential solvation changes sign as x_A changes.

Note that in all cases

$$\delta_{A,S} \longrightarrow 0 \quad \begin{cases} \text{if } x_A \longrightarrow 0 \\ \text{or } x_B \longrightarrow 0 \\ \text{or } G_{AS} - G_{BS} \longrightarrow 0 \end{cases} \tag{2.10}$$

The only quantity that is left ambiguous in Eq. (2.9) is V_{cor}. Clearly, if we take very large correlation volume, we obtain

$$\delta_{A,S} \longrightarrow 0 \qquad \text{for } V_{cor} \longrightarrow \infty \tag{2.11}$$

On the other hand, for too small V_{cor}, the approximate equality of $g_{AS}(R) \simeq 1$ presumed in Eq. (2.6) (for $R \geq R_c$) will not hold. In practice, we can choose, for each specific system, a reasonable R_c (and hence V_{cor}) according to the behavior of the functions g_{ij} at large distances. Theoretically, however, we can get rid of V_{cor} by taking the first order term in the expansion of $\delta_{A,S}$ in Eq. (2.9) in power series about $\epsilon \equiv V^{-1}_{cor}$, thus

$$\delta_{A,S} = 0 + \epsilon x_A x_B (G_{AS} - G_{BS}) + \ldots \tag{2.12}$$

We define the limiting linear preferential solvation as

$$\delta^0_{A,S} = \left. \frac{\partial \delta_{A,S}}{\partial \epsilon} \right|_{\epsilon = 0} = x_A x_B (G_{AS} - G_{BS}) \tag{2.13}$$

Since $x_A x_B > 0$, the sign of $\delta^0_{A,S}$ is determined by the sign of $G_{AS} - G_{BS}$, and this is, of course, independent of the correlation volume. Thus, we have defined in Eq. (2.13) a quantity that unambiguously measures the preferential solvation of *S* with respect to a two-component solvent.

3. Expressing G_{AS} and G_{BS} in terms of experimental quantities

Let S be a solute highly diluted in a solvent mixture of A and B at composition x_A. Let l denote the liquid mixture of A and B. In this system the chemical potential of the solute S is given by (Ben-Naim, 1974, Ben-Naim, 1987).

$$\mu_s^l(T,P,\rho_s,x_A) = \mu_s^{*l}(T,P,x_A) + kT \ln q_s \Lambda_s^3 q_s^{-1} \tag{3.1}$$

The quantity μ_s^{*l} is referred to as the pseudochemical potential (but note that this is not the conventional standard chemical potential (Ben-Naim, 1987). This quantity depends on the composition of the solvent mixture x_A but not on the solute density $\rho_s = N_s/V$. This follows from the assumption that the system is very dilute with respect to S. Λ_s^3 is the momentum partition function of S and q_s is the internal partition function (including rotational, vibrational, etc.) of S. In the following we shall assume that neither Λ_s^3 nor q_s depends on the composition of the solvent mixture.

The derivative of μ_s^l with respect to, say, N_A is

$$(\partial\mu_S^l/\partial N_A)_{T,P,N_B,N_S} = (\partial\mu_S^{*l}/\partial N_A)_{T,P,N_B,N_S} - (kT/V)\bar{V}_A \tag{3.2}$$

where V_A is the partial molar volume of A in the mixture. The derivative in Eq. (3.2) may be expressed in terms of Kirkwood-Buff integrals. However, the application of the KB theory to systems of more than two components is quite cumbersome. A considerable simplification has been achieved in the expressions for the derivatives of the chemical potential (Ben-Naim, 1975). The result is:

$$\left(\frac{\partial\mu_S^l}{\partial N_A}\right)_{T,P,N_B,N_S} = \frac{kT}{\rho_S\rho_A V}\frac{|E(S,A)|}{|D|} \tag{3.3}$$

where $|E(S,A)|$ and $|D|$ are two determinants, the definition of which for any mixture is available (Ben-Naim, 1975). Here we shall be interested only in the case of very dilute solutions of S in the mixture. For this case we have

$$|E(S,A)| = \begin{vmatrix} 1 & 0 & 1 \\ G_{SA} & 1 & G_{AB} \\ G_{SB} & 1 & G_{BB}+\rho_B^{-1} \end{vmatrix} = \frac{1}{\rho_B}(1 + \rho_B G_{BB} - \rho_B G_{AB}) + G_{SA} - G_{SB}$$
$$= \eta\bar{V}_A\rho_B^{-1} + G_{SA} - G_{SB} \tag{3.4}$$

where we have used the Kirkwood-Buff result for the partial molar volume in the pure mixture of A and B, i.e.,

$$V_A = (1 + \rho_B G_{BB} - \rho_B G_{AB})/\eta \tag{3.5}$$

and where η is defined for any mixture of A and B as

$$\eta = \rho_A + \rho_B + \rho_A\rho_B(G_{AA} + G_{BB} - 2G_{AB}) \tag{3.6}$$

The determinant in the denominator of (3.3) is defined by

$$|D| = \begin{vmatrix} 0 & 1 & 1 & 1 \\ 1 & G_{SS}+\rho_S^{-1} & G_{SA} & G_{SB} \\ 1 & G_{SA} & G_{AA}+\rho_A^{-1} & G_{AB} \\ 1 & G_{SB} & G_{AB} & G_{BB}+\rho_B^{-1} \end{vmatrix} \tag{3.7}$$

which has the limiting form

$$|D| \xrightarrow[\rho_S \to 0]{} -\eta/\rho_S\rho_A\rho_B \tag{3.8}$$

Substituting (3.4) and (3.8) in (3.3), we obtain

$$\left(\frac{\partial \mu_S^l}{\partial N_A}\right)_{T,P,N_B,N_S} = \frac{-kT}{V\eta}[\eta\bar{V}_A + \rho_B(G_{SA} - G_{SB})] \tag{3.9}$$

From (3.2) and (3.9) we obtain

$$\left(\frac{\partial \mu_S^{*l}}{\partial N_A}\right)_{T,P,N_B,N_S} = \frac{-kT}{V}\frac{\eta\bar{V}_A + \rho_B(G_{SA} - G_{SB})}{\eta} + \frac{kT\bar{V}_A}{V} \tag{3.10}$$

However, for practical purposes, it is more convenient to transform into a derivative with respect to the mole fraction x_A:

$$\frac{\partial \mu_S^{*l}}{\partial N_A} = \frac{\partial \mu_S^{*l}}{\partial x_A}\frac{\partial x_A}{\partial N_A} = \frac{\partial \mu_S^{*l}}{\partial x_A}\frac{N_B}{(N_A + N_B)^2} \tag{3.11}$$

and hence we have the final result

$$\lim_{\rho_S \to 0}\left(\frac{\partial \mu_S^{*l}}{\partial x_A}\right)_{T,P} = \frac{kT(\rho_A + \rho_B)^2}{\eta}(G_{BS} - G_{AS}) \tag{3.12}$$

Let μ_s^{*g} be the pseudochemical potential of S in an ideal gas phase. Clearly, μ_s^{*g} is independent of x_A. Therefore, we may rewrite Eq. (3.12) as

$$\lim_{\rho_S \to 0}\left(\frac{\partial \Delta G_S^{*l}}{\partial x_A}\right)_{P,T} = \frac{kT(\rho_A + \rho_B)^2}{\eta}(G_{BS} - G_{AS}) \tag{3.13}$$

where $\Delta G_s^* = \mu_s^{*l} - \mu_s^{*g}$ is the solvation free energy of S in our system (Ben-Naim, 1987).

Thus, by measuring the slope of the solvation free energy as a function of x_A, we can extract the difference $G_{AS} - G_{BS}$. As we have seen, this is a measure of the limiting preferential solvation of S with respect to A.

Note that η is a calculable quantity through the inversion of the Kirkwood-Buff theory (Ben-Naim, 1974). However, since $\eta > 0$, the entire quantity $kT(\rho_A + \rho_B)^2/\eta$ is always positive (Ben-Naim, 1974). Therefore, the sign of the derivative is the same as the sign of $G_{BS} - G_{AS}$.

In the following we shall derive another equation relating G_{BS} and G_{AS} with experimental quantities. The latter, together with relation (3.13) may be used to compute G_{BS} and G_{AS} individually.

The second equation that we shall now derive is for the partial molar volume of S at infinite dilution in the solvent l, of composition x_A.

$$\mu_{\alpha\beta} = \left(\frac{\partial \mu_\alpha}{\partial N_\beta}\right)_{P,T,N'} = \frac{kTB^{\alpha\beta}}{V|B|} - \frac{\bar{V}_\alpha\bar{V}_\beta}{\kappa_T V} \tag{3.14}$$

where N' includes all N_i's except N_β, the determinant $|B|$, for our case, is defined by

$$|B| = \begin{vmatrix} \rho_S + \rho_S^2 G_{SS} & \rho_S\rho_A G_{AS} & \rho_S\rho_B G_{BS} \\ \rho_S\rho_A G_{AS} & \rho_A + \rho_A^2 G_{AA} & \rho_A\rho_B G_{AB} \\ \rho_S\rho_B G_{BS} & \rho_A\rho_B G_{AB} & \rho_B + \rho_B^2 G_{BB} \end{vmatrix} \tag{3.15}$$

which has the following limiting behavior at $\rho_s \longrightarrow 0$

$$|B| \longrightarrow \rho_s \rho_A \rho_B \xi \tag{3.16}$$

with ξ defined by

$$\xi = 1 + \rho_A G_{AA} + \rho_A \rho_B (G_{AA} G_{BB} - G^2_{AB}) \tag{3.17}$$

The quantities $B^{\alpha\beta}$ in (3.14) are the cofactors of the determinant B, which for the limit $\rho_s \longrightarrow 0$ are

$$B^{SA} \longrightarrow - \rho_S \rho_A \rho_B [G_{SA}(1 + \rho_B G_{BB}) - G_{SB} \rho_B G_{AB}] \tag{3.18}$$

$$B^{SB} \longrightarrow - \rho_S \rho_A \rho_B [G_{SB}(1 + \rho_A G_{AA}) - G_{SA} \rho_A G_{AB}] \tag{3.19}$$

$$B^{SS} \longrightarrow \rho_A \rho_B \xi \tag{3.20}$$

Using the Gibbs-Duhem equation

$$\rho_A \mu_{SA} + \rho_B \mu_{SB} + \rho_S \mu_{SS} = 0 \tag{3.21}$$

and the Kirkwood-Buff results for the isothermal compressibility and the partial molar volumes of A and B

$$\kappa_T = \xi / kT\eta \tag{3.22}$$

$$V_A = [1 + \rho_B (G_{BB} - G_{AB})/\eta \tag{3.23}$$

$$V_B = [1 + \rho_A (G_{AA} - G_{AB})/\eta \tag{3.24}$$

we obtain the following equality

$$\begin{aligned} &\rho_A \left[\frac{kTB^{SA}}{V|B|} - \frac{V_S V_A}{\kappa_T V} \right] + \rho_B \left[\frac{kTB^{SB}}{V|B|} - \frac{V_S V_A}{\kappa_T V} \right] \\ &\qquad + \rho_S \left[\frac{kTB^{SS}}{V|B|} - \frac{V_S^2}{\kappa_T V} \right] = 0 \end{aligned} \tag{3.25}$$

which may be reduced to

$$\xi - \rho_A G_{SA} V_A \eta - \rho_B G_{SB} V_B \eta - \rho_A V_S V_A \eta - \rho_B V_S V_B \eta = 0 \tag{3.26}$$

From (3.26) we may eliminate V_S (for $\rho_S \longrightarrow 0$) to obtain the final result:

$$V_S^0 = \lim_{\rho_s \to 0} V_S = kT\kappa_T - \rho_A V_A G_{SA} - \rho_B V_B G_{SB} \tag{3.27}$$

In Eqs. (3.13) and (3.27) all the quantities κ_T, η, ρ_A, ρ_B, V_A, V_B, V_S^0, and $\partial \Delta G_S^* / \partial X_A$ are experimentally determinable. Hence, these two equations may be used to eliminate the required quantities G_{SB} and G_{SA}. (Note that $(\rho_A + \rho_B)^2/\eta$ in Eq. (3.13) may be computed directly from the vapor pressure as a function of composition in the mixture of A and B. Thus, denoting

$$a = \frac{\partial \Delta G_S^*}{\partial X_A} \quad b = \frac{kT(\rho_A + \rho_B)^2}{\eta} \quad \text{and } c = kT\kappa_T \tag{3.28}$$

we may solve for G_{AS} and G_{BS}. The results are

$$G_{BS} = c - V_S^0 + \frac{a}{b}\, \rho_A V_A \tag{3.29}$$

$$G_{AS} = c - V_S^0 + \frac{a}{b}\, \rho_B V_B \tag{3.30}$$

which are the required quantities.

4. Preferential solvation in a two component system

In the previous section we discussed the theory of preferential solvation of a solute S in a two component system. In the realm of the traditional concept of solvation thermodynamics only very dilute solutions could be treated. Therefore the question of PS could have been asked only for a solute S in a very dilute solution in any solvent. Therefore the minimum number of components required from such a study is three; a solute and two solvents.

However, the question of PS can also be asked in a two component system, say of A and B. At any composition x_A, we may focus on one A molecule and ask what is the PS of A with respect to the two components, A and B in its immediate vicinity. Likewise, we may focus on one B-molecule and ask the same, but independent, question of the PS of B with respect to the two components A and B.

Since our mixture may have an arbitrary composition, the last two questions concerning the PS of A and B could not be dealt with by the traditional concept of solvation. Fortunately the new concept of solvation (Ben-Naim, 1987) can be easily applied to these systems, a feature that opens the way for investigating two-component systems in the entire range of compositions. In a two component system we define the average number of A and B molecules around an A molecule by (note eq. 2.6)

$$N_{A,A} = \rho_A \int g_{A,A}(R)\, 4\pi R^2 dR \tag{4.1}$$

$$N_{B,A} = \rho_B \int g_{B,A}(R)\, 4\pi R^2 dR \tag{4.2}$$

where ρ_A and ρ_B are the number densities of A and B respectively, and $g_{\alpha\beta}$ is the, angular averaged pair correlation function for the pair of species α and β. In the following treatment we focus on a single A molecule to which we refer to as an A-solvaton (Ben-Naim 1987). A similar treatment applies to a B-solvaton.

For any radius R_c, we define the local mole fraction of A-molecules around an A-solvaton by

$$x_{A,A}(R_c) = N_{A,A}(R_c)/[N_{A,A}(R_c) + N_{B,A}(R_c)] \tag{4.3}$$

Next we define the PS of an A solvaton with respect to A- molecules, simply by the deviation of the local from the bulk composition, i.e.,

$$\delta_{A,A} = x_{A,A,}(R_c) - x_A \tag{4.4}$$

As in section 3, we rewrite $N_{A,A}$ as

$$\begin{aligned} N_{A,A} &= \rho_A \int g_{AA}(R) 4\pi R^2 dR = \rho_A \int [g_{AA}(R)-1] 4\pi R^2 dR + \rho_A V_{cor} \\ &\simeq \rho_A \int [g_{AA}(R)-1] 4\pi^2 dR + \rho_A V_c = \rho_A G_{AA} + \rho_A V_{cor} \end{aligned} \tag{4.5}$$

This approximation is rendered possible by the choice of R_c sufficiently large, but still undefined, such that for $R > R_c$ the integrand in (4.5) is negligibly small. (In the actual calculation of $G_{\alpha\beta}$ through an inversion of the Kirkwood-Buff theory, we require that all $g_{\alpha\beta}$ be defined in an open system where $g \longrightarrow 1$ at $R \longrightarrow \infty$). We now introduce (4.5) into our definition of the PS of A, equation (4.4) to obtain

$$\delta_{A,A} = \frac{N_{AA}}{N_{A,A} + N_{B,A}} - x_A = \frac{x_A x_B (G_{AA} - G_{BA})}{x_A G_{AA} + x_B G_{AB} + V_{cor}} \tag{4.6}$$

similarly for the PS of B in the same system we have

$$\delta_{A,B} = \frac{N_{A,B}}{N_{A,B} + N_{BB}} - x_A = \frac{x_A x_B (G_{AB} - G_{BB})}{x_A G_{AB} + x_B G_{BB} + V_{cor}} \tag{4.7}$$

Note that $G_{AB} = G_{BA}$, and that $\delta_{A,A}$ and $\delta_{A,B}$ are, in general independent quantities but $\delta_{A,A} = -\delta_{B,A}$ and $\delta_{B,B} = -\delta_{A,B}$ which follows from the definitions (4.6) and (4.7).

Since all the $G_{\alpha\beta}$ are computable from thermodynamic quantities, using the inversion of the Kirkwood-Buff theory of solution, one can compute $\delta_{A,A}$ and $\delta_{A,B}$ for any choice of V_{cor}.

As in section 3 we define the linear coefficient of the PS as

$$\delta^o_{A,A} = x_A x_B (G_{AA} - G_{AB}) \tag{4.8}$$

$$\delta^o_{A,B} = x_A x_B (G_{AB} - G_{BB}) \tag{4.9}$$

Thus, besides the product $x_A x_B$ the difference $G_{AA} - G_{AB}$ characterizes the linear coefficients of the PS of A, and likewise $G_{AB} - G_{BB}$ characterizes the linear coefficient of the PS of B. By using these quantities as our measures of the PS we have eliminated the need to make a choice of R_c. In fact these quantities include the effect of the solvaton-molecule on the entire range for which the correlation between the various molecules extends.

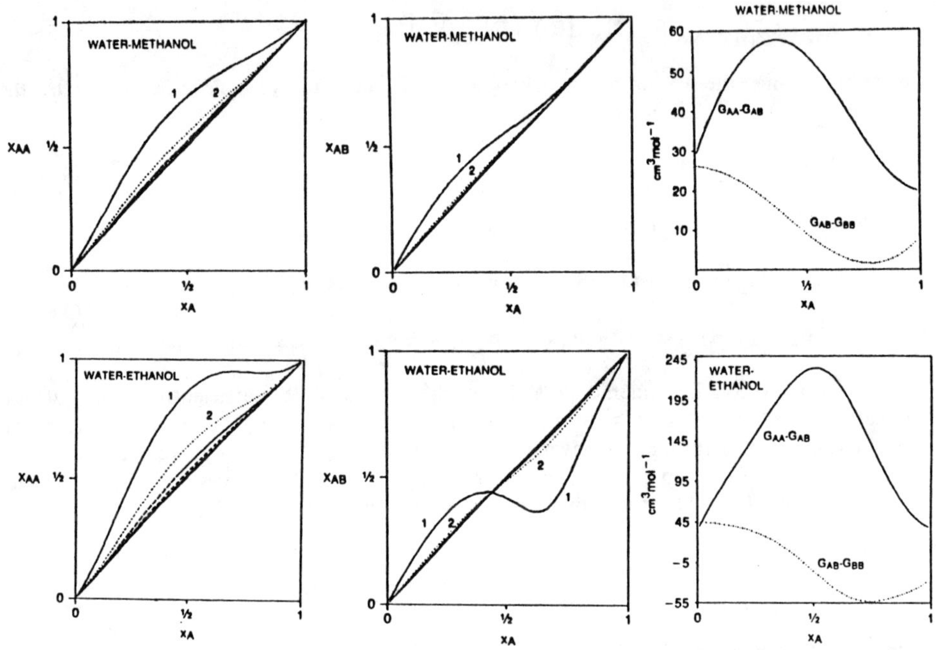

Fig. 4: Preferential solvation of water-methanol and water-ethanol as a function of the mole fraction of the water (component A).

In figure 4 we present some results for the system of water methanol and water ethanol. A series of R_c values were chosen (Ben-Naim, 1989), roughly corresponding to successive coordination spheres about the solvation. Thus, n = 1 corresponds to the first coordination sphere. We used values n = 2,3,4,5 to compute $x_{A,A}$ (R_c) as defined in (4.3). These were plotted in figure 4 as a function of the bulk mole fraction of water, x_A in the various mixtures.

Although we have assigned no particular significance to the correlation distances R_c it is quite clear from figure 4 that as R_c becomes of the order of five or six, σ_{AB}, all the curves of $x_{A,A}$ (R_c) and $x_{A,B}$ (R_c) approach the diagonal line. In all cases it is found that for n = 6 the relative deviation $[x_{A,A}]$ (R_c) - x_A)/x_A becomes less than 0.01, which we consider to be an effective limit of the correlation distance. In other words, for $R_c \geq 5$ σ_{AB} the local composition approaches the bulk composition.

For these systems, we also reported in figure 4 the linear coefficient of the PS of both A and B.

5. Conclusion

One of the most outstanding features of the KB theory of solution is its very generality. It can be applied to any mixture of simple or complex molecules and does not rely on the commonly assumed pairwise additivity of the interaction potential. These features make it particularly appealing for the study of aqueous solutions.

We have outlined the inversion procedure of the KB theory and demostrated some applications. We believe that this powerful theoretical tool will find many more applications in the field of solution chemistry.

In particular we expect that extension of this theory to electrolyte solutions will be developed. In these solutions, if we are interested in the local environment of each individual ion, then we have more G_{ij}'s than thermodynamic quantities. Therefore the inversion procedure needs some modification and perhaps some further assumptions regarding the way we split the partial molar quantities of a solute into the contributions of the individual ions. However, recent development in the theory of ionic solutions (Kusalik and Patey, 1987, Newman, 1988), indicates that such an extension of the KB theory might be achieved in the near future.

REFERENCES

Ben-Naim, A. 1974. Water and aqueous solutions. Plenum Press. New York.

Ben-Naim, A. 1975. Solute and solvent effects on chemical equilibria. J Chem. Phys. 63: 2064-2073.

Ben-Naim, A. 1977. Inversion of the Kirkwood-Buff theory of solutions: Application to the water-ethanol system. J. Chem. Phys. 67: 4884-4890.

Ben-Naim, A. 1987. Solvation thermodynamics. Plenum Press, New York.

Ben-Naim, A. 1988. Theory of preferential solvation of non electrolytes. Cell Biophys. 12: 3694-3699.

Ben-Naim, A. 1989. J. Chem. Phys. 93: 3809-3813

Covington, A.K. and K.E. Newman. 1976. Thermodynamics of preferential solvation of electrolytes in binary solvent mixtures. Adv. Chem. Ser. 155: 153-196

Donkersloot, M.C.A. 1979. The structure of binary liquids. The Kirkwood-Buff theory of liquid mixtures. J. Solution Chem. 8: 193-307

Engberts J.B.F.N. 1979. Mixed aqueous solvent effects on kinetics and mechanism of organic reaction in: Water, a comprehensive Treatise, F. Franks, ed., Plenum Press, New York, Vol. 6, Chapt. 4, 139-171.

Grunwald, E., G. Baughman and G. Kohnstam. 1960. The solvation of electrolytes in dioxane-water mixtures, as deduced from the effect of solvent change on the standard partial molar free energy. J. Am. Chem. Soc. 82: 5801-5811.

Kato, T., T. Fujiyama, H. Nomura. 1982. Estimation of parameters, G_{11}, G_{22} and G_{12} in the Kirkwood-Buff solution theory on the basis of the concentration fluctuation data obtained from Rayleigh scattering. Bull. Chem. Soc. Japan, 55: 3368-3372.

Kato, T. 1984. Kirkwood-Buff parameters and correlation length in aqueous solutions of n-alkoxyethanols. J. Phys. Chem. 88: 1248-1252.

Kirkwood, J. G. and F. P. Buff. 1951. Statistical mechanical theory of solutions. J. Chem. Phys. 19: 774-778.

Kusalik, P. G. and G. N. Patey. 1987. The thermodynamic properties of electrolyte solutions: Some formal results. J. Chem. Phys. 86: 5110-5116.

Matteoli E. and L. Lepori. 1984. Solute-solute interactions in water. II. An analysis through the Kirkwood-Buff integrals for 14 organic solutes. J. Chem. Phys. 80: 2856-2863.

McMillan, W. G. and J. E. Mayer. 1945. The statistical thermodynamics of multicomponent system. J. Chem. Phys. 13: 276-305.

Newman, K.E. 1988. Interpretation of free energies of transfer and solute partial molar volumes in mixed solvents using Kirkwoo-Buff theory. J. Chem. Soc. Faraday Trans. I. 84: 1387-1391.

Nishikawa, K., Y. Kodera and T. Iijima. 1987. Fluctuations in the particle number and concentration and the Kirkwood-Buff parameters of the tert-butyl alcohol and water mixtures studied by small-angle X-ray scattering. J. Phys. Chem. 91: 3694-3699.

Patil, K.J. 1981. Application of Kirkwood-Buff theory of liquid mixtures to water-butanol system. J. Solution Chem. 10: 315-320.

Rubio, R.G., M.G. Prolongo, M. Diaz-Pena and J.A.R. Renuncio. 1987. Local compositions in real mixtures of simple molecules. J. Phys. Chem. 91: 1177-1184.

Light Scattering Study of Concentration Fluctuations in Liquid Mixtures

Tadashi Kato
Dept. of Chemistry, Tokyo Metropolitan Univ.
Setagaya-ku, Tokyo 158, Japan

ABSTRACT
Mean-square concentration fluctuations were measured on associated solutions of organic solvents and aqueous solutions of alcohols by means of Rayleigh-Brillouin scattering. It is shown that the concentration fluctuation is a useful quantity for studying molecular association. Relations with thermodynamic quantities and molecular distribution functions (Kirkwood-Buff integrals) are also discussed. For one of the above systems, which has a lower critical consolute point, correlation lengths of concentration fluctuations and self-diffusion coefficient of each molecule were measured also by using dynamic light scattering and pulsed-gradient NMR, respectively, at different temperatures in order to elucidate effects of critical phenomena on molecular association.

INTRODUCTION

It is well known that the solution near the critical consolute point scatters light very strongly and looks opalescent. This arises from the fact that molecules of the same species gather together and hence concentration fluctuations become extremely large. Such opalescence is not observed for solutions far from the critical point. However, concentration fluctuations exist even in an ideal solution. If molecules of the same species tend to gather together, the concentration fluctuation takes larger values than those of an ideal solution as can be expected from the above example, i. e., the critical phenomena. We can consider another extreme case where molecules of different species form a 1:1 complex. In an equimolar mixture, the concentration fluctuation is absent because the local mole fraction is always 0.5. Thus the concentration fluctuation is a useful quantity for understanding mixing states intuitively.

This review concerns itself with our light scattering studies on the concentration fluctuation for binary solutions. First, the theoretical background of our methods is briefly described. Then, we show the results for binary mixture of organic solvents and aqueous solutions of alcohols. Finally, comments on the so-called Kirkwood-Buff integrals are added.

METHODS

Mean-Square Concentration Fluctuation and Rayleigh-Brillouin Scattering Intensity

Light scattering for a dense system occurs due to the fluctuation of the dielectric constant which is induced by the fluctuations of temperature, pressure, concentration, and so on (Fabelinskii 1968). As the time scales of these fluctuations are different, information on each fluctuation can be obtained by observing light scattering spectra.

Rayleigh-Brillouin scattering is one of such light scattering spectra. For a binary solution composed of A and B, the Rayleigh ratio for a central peak, R_{cen}, at a scattering angle of 90° is expressed as (Iwasaki et al. 1976a)

$$R_{cen} = R_S + R_C$$

$$R_S = \frac{4\pi^2 n^2}{\lambda^4}\left(\frac{\partial n}{\partial T}\right)^2_{P,x_\alpha}\frac{kT^2}{dC_{pg}} \qquad (1)$$

$$R_C = \frac{4\pi^2 n^2}{\lambda^4}\left(\frac{\partial n}{\partial x_\alpha}\right)^2_{T,P} V\langle(\Delta x)^2\rangle$$

where n is the refractive index, λ, the wave length of the incident light, T, the absolute temperature, P, the pressure, k, the Boltzmann constant, d, the density, C_{pg}, the specific heat, x_α the mole fraction of component α (α = A or B), and $\langle(\Delta x)^2\rangle$ the mean-square fluctuation of the mole fraction in the volume V. If data on the heat capacity, density, and temperature and concentration derivatives of the refractive index are available, $V\langle(\Delta x)^2\rangle$ values can be obtained from the observed Rayleigh ratio, R_{cen}.

Equation 1 is the same as that derived by Miller (Miller 1967) if $\langle(\Delta x)^2\rangle$ is replaced by

$$\langle(\Delta x)^2\rangle = \frac{(1 - x_\alpha)kT}{N(\partial\mu_\alpha/\partial x_\alpha)_{T,P}} \qquad (2)$$

where μ_α is the chemical potential of component α and N is the average number of total molecules in the volume V. However, Miller used Equation 1 for obtaining activity coefficients of binary solutions. This method needs an integration of $(\partial\mu_\alpha/\partial x_\alpha)_{T,P}$ which leads to large errors.

In order to compare the mean-square concentration fluctuations for different systems, it is more useful to use the dimensionless quantity

$$N\langle(\Delta x)^2\rangle = \rho V\langle(\Delta x)^2\rangle \qquad (3)$$

where ρ is the average number density. If the fluctuation of the mole fraction under a constant volume is considered, $\langle(\Delta x)^2\rangle$ can be expressed in terms of the number fluctuation of each component in the volume V, ΔN_α, as

$$<(\Delta x)^2> = \left(\frac{x_B}{N}\right)^2 <(\Delta N_A)^2> + \left(\frac{x_A}{N}\right)^2 <(\Delta N_B)^2> - \frac{2x_A x_B}{N^2} <\Delta N_A \Delta N_B> \quad (4)$$

The number fluctuations in Equation 4 are related to the radial distribution function $g_{\alpha\beta}(r)$ as (Kirkwood and Buff 1951)

$$<\Delta N_\alpha \Delta N_\beta> = (N_\alpha N_\beta/V)G_{\alpha\beta} + \delta_{\alpha\beta}N_\beta$$

$$G_{\alpha\beta} \equiv \int_0^\infty [g_{\alpha\beta}(r)-1]4\pi r^2 dr \quad (5)$$

where N_α is the average number of molecules α. Substituting Equation 5 into Equation 4, we obtain (Kato et al. 1980b)

$$N<(\Delta x)^2> = x_A x_B[1 + \rho x_A x_B(G_{AA} + G_{BB} - 2G_{AB})] \quad (6)$$

For an ideal solution ($G_{AA} = G_{BB} = G_{AB}$), $N<(\Delta x)^2>$ is equal to $x_A x_B$ as can be seen from Equation 6 (needless to say, the same result can be derived also by substituting $\mu_\alpha \propto kT\ln x_\alpha$ into Equation 2). If the same species have a tendency to gather together ($G_{AA} + G_{BB} > 2G_{AB}$), $N<(\Delta x)^2>$ takes larger values than $x_A x_B$. If molecules of different species have a tendency to gather together ($G_{AA} + G_{BB} < 2G_{AB}$), on the other hand, $N<(\Delta x)^2>$ takes smaller values than $x_A x_B$. So it is sometimes useful to define the quantity

$$F \equiv N<(\Delta x)^2>/x_A x_B$$
$$= 1 + \rho x_A x_B(G_{AA} + G_{BB} - 2G_{AB}) \quad (7)$$

In the case where solute molecules are self-associated, F is equal to the mean association number in the infinite dilution (see later). The quantity F can be expressed in terms of thermodynamic quantities as

$$\frac{1}{F} = \frac{\partial \ln a_\alpha}{\partial \ln x_\alpha} = 1 + \frac{\partial \ln \gamma_\alpha}{\partial \ln x_\alpha} = 1 + \frac{x_A x_B}{RT}\frac{\partial^2 G^E}{\partial x_\alpha^2} \quad (8)$$

where a_α and γ_α are the activity and activity coefficient of the species α, respectively, and G^E is the excess free energy (see also Figures 2a and 2b)

For binary alloys, Bhatia and Thornton defined three structure factors, $S_{NN}(q)$, $S_{NC}(q)$, and $S_{CC}(q)$, which are derived from the Fourier transforms of the correlation function of local number density and concentration (Bhatia and Thornton 1970). These structure factors can be used directly for liquid mixtures. The quantity $N<(\Delta x)^2>$ is equal to $S_{CC}(0)$.

Correlation Length of Concentration Fluctuation and Dynamic Light Scattering

In a dynamic light scattering experiment, one can measure the autocorrelation function of the scattered light (Berne and Pecora 1976)

$$\Phi(q,\tau) = <I^*(q,t)I(q,t+\tau)> \quad (9)$$

where I(q,t) is the scattered intensity at time t and q is the absolute value of the scattering vector which is related to the scattering angle θ as

$$q = (4\pi n/\lambda)\sin(\theta/2) \tag{10}$$

A rigorous formula for the power spectrum or the time correlation function of the scattered electric field for a binary solution was given by Mountain and Deutch (Mountain and Deutch 1969). As the correlation time of the concentration fluctuation is much longer than those of other fluctuations which induce the fluctuation of the refractive index, $\Phi(q,\tau)$ for homodyne optical mixing can be approximately written as

$$\Phi(q,\tau) = A\exp(-2Dq^2\tau) + B \tag{11}$$

where D is the mutual diffusion coefficient and A and B are the constants. The mutual diffusion coefficient may be expressed as (Kawasaki 1970, Ferrel 1970)

$$D = kT/(6\pi\eta\xi') \tag{12}$$

where η is the viscosity of the solution and ξ' is the apparent correlation length. Near the critical point, ξ' is equal to the correlation length defined by the equation

$$\xi \equiv \frac{1}{6}\int_0^\infty r^2 \langle\Delta x(0)\Delta x(r)\rangle 4\pi r^2\,dr \Big/ \int_0^\infty \langle\Delta x(0)\Delta x(r)\rangle 4\pi r^2\,dr \tag{13}$$

where $\langle\Delta x(0)\Delta x(r)\rangle$ is the space correlation function of the concentration fluctuation.

RESULTS

Associated Solutions Composed of Organic Solvents

Figure 1 shows the concentration dependences of $N\langle(\Delta x)^2\rangle$ observed for several binary mixtures of organic solvents (Kato and Fujiyama 1981). The dashed line in the figure represents $N\langle(\Delta x)^2\rangle$ for an ideal solution, $x_A x_B$. It is seen from the figure that $N\langle(\Delta x)^2\rangle$ values for the ethanol-carbon tetrachloride system (Fujiyama and Tanaka 1978) are larger than $x_A x_B$. This arises from the fact that ethanol molecules are self-associated by hydrogen bonding ($G_{AA} + G_{BB} > 2G_{AB}$). On the other hand, the $N\langle(\Delta x)^2\rangle$ values for the diethyl ether-chloroform system (Iwasaki et al. 1976b) are smaller than $x_A x_B$ because diethyl ether and chloroform molecules form a hydrogen bonded complex ($G_{AA} + G_{BB} < 2G_{AB}$). In the ethanol-chloroform system (Kato and Fujiyama 1976), hydrogen bonds are formed between ethanol and chloroform molecules as well as between ethanol molecules. So the $N\langle(\Delta x)^2\rangle$ takes values between those for the other two systems.

In classical analyses of thermodynamic quantities of solutions in which strong intermolecular interactions such as hydrogen bonding are expected, an ideal associated solution model has often been used (Prigogine and Defay 1954). In this model, it is assumed that molecules which interact strongly with each other form an associated complex and that the solution is regarded to be an ideal solution whose components are free molecules and the associated complexes. If these associated complexes are

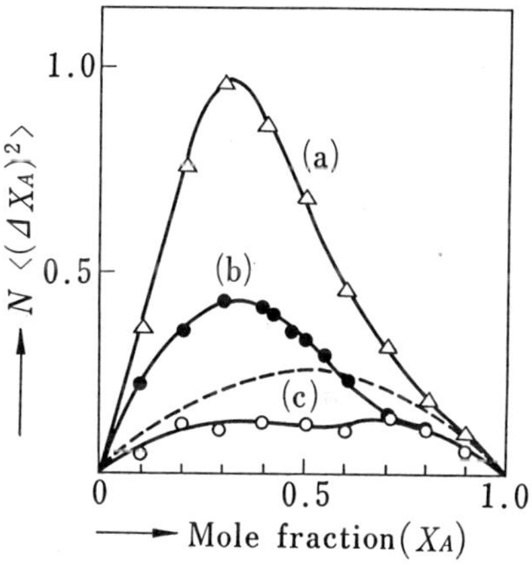

Figure 1. $N<(\Delta x)^2>$ for binary mixtures of (a):ethanol(A)-carbon tetrachloride(B), (b):ethanol(A)-chloroform(B), and (c):diethyl ether(A)-chloroform(B) at room temperature. The abscissa is mole fraction of component A and the dashed line corresponds to an ideal solution. (Kato and Fujiyama 1981).

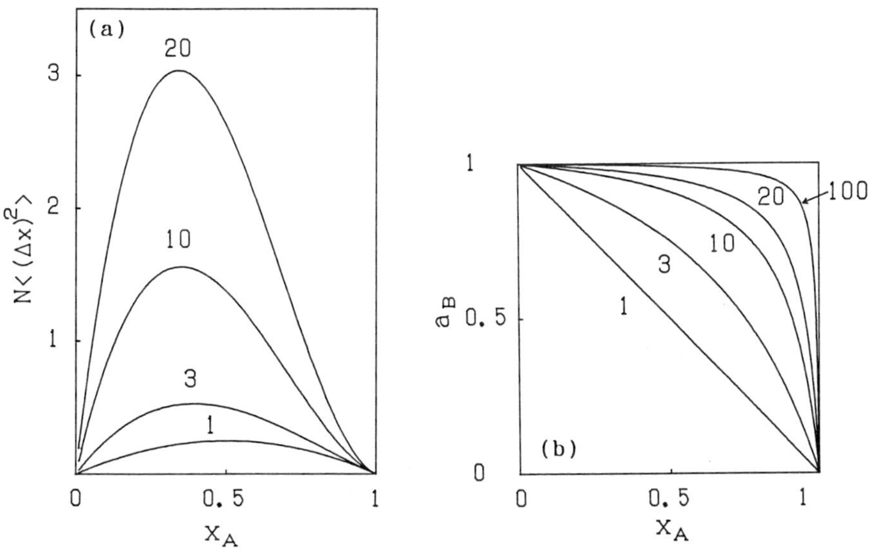

Figure 2. Theoretical calculations on the basis of an ideal associated solution model. (a) $N<(\Delta x)^2>$ calculated from Equation 15. The numbers in the figure indicate the weight-averaged association number of A, $<i>$. (b) activity of species B calculated from Equation 17. The numbers in the figure indicate the number-averaged association number of A, $<i>_n$.

represented by A_i, B_i, and A_iB_j (i and j are integers), we have (Kato et al. 1978, Kato 1980)

$$F = \frac{x_B}{N_A}\left(\sum_i i^2 N_{A_i} + \sum_i\sum_j i^2 N_{A_iB_j}\right) + \frac{x_A}{N_B}\left(\sum_i i^2 N_{B_i} + \sum_i\sum_j j^2 N_{A_iB_j}\right)$$

$$- \frac{2}{N_A + N_B}\sum_i\sum_j ij N_{A_iB_j} \qquad (14)$$

where $N_{A_iB_j}$ is the number of the complex A_iB_j. Prior to our work, Bhatia et al. have derived similar equations for binary alloys (Bhatia et al. 1974). In the case where only A_i-type molecules exist, Equation 14 is reduced to

$$F = x_A + (1 - x_A)\langle i\rangle$$

$$\langle i\rangle \equiv \sum_i i^2 N_{A_i} / \sum_i i N_{A_i} \qquad (15)$$

where $\langle i\rangle$ is the weight-averaged association number of A. Figure 2a shows the concentration dependences of $N\langle(\Delta x)^2\rangle$ calculated for different $\langle i\rangle$ values. In the infinite dilute solution ($x_A \to 0$), we have

$$F = \langle i\rangle \qquad (16)$$

Thus F becomes the mean association number in this case.

By using the same model, the activity of B can be expressed as (Prigogine and Defay 1954)

$$a_B = \frac{1}{1 + \dfrac{x_A}{\langle i\rangle_n x_B}} \qquad (17)$$

$$\langle i\rangle_n \equiv \sum_i i N_{A_i} / \sum_i N_{A_i}$$

where $\langle i\rangle_n$ is the number-averaged association number. In Figure 2b, are shown the concentration dependences of a_B calculated for different $\langle i\rangle_n$ values. Figure 2b demonstrates that the activity of B depends on the association number only slightly for large association numbers. On the other hand, the mean-square concentration fluctuation $N\langle(\Delta x)^2\rangle$ is nearly proportional to the association number as can be seen from Figure 2a.

We have analyzed concentration and temperature dependences of $N\langle(\Delta x)^2\rangle$ for several systems using the ideal associated solution model. Figure 3a shows the concentration dependences of $N\langle(\Delta x)^2\rangle$ for an acetic acid-carbon tetrachloride system at different temperatures (Kato et al. 1980a). As can be seen from the figure, $N\langle(\Delta x)^2\rangle$ takes larger values than $x_A x_B$. This comes from the fact that acetic acid molecules are associated each other in carbon tetrachloride solutions. Figure 3a demonstrates also that $N\langle(\Delta x)^2\rangle$ slightly increases with the elevation of temperature. Such temperature dependences are different from those for ordinary self-associated solutions. In the ethanol-carbon tetrachloride system, in fact, $N\langle(\Delta x)^2\rangle$ decreases with increasing temperature (Fujiyama and Tanaka 1978). We have shown that the temperature

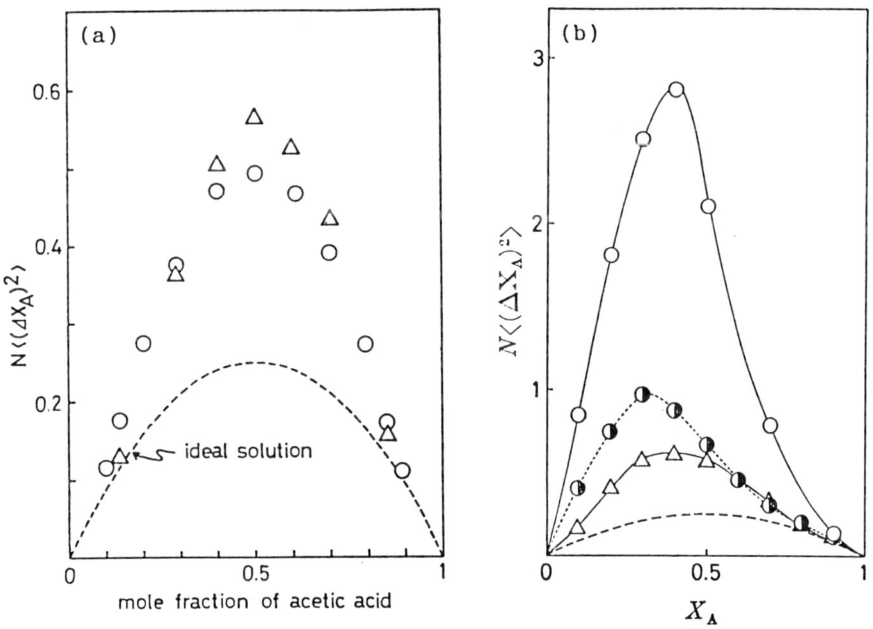

Figure 3. $N\langle(\Delta x)^2\rangle$ for associated solutions composed of organic solvents. (a) acetic acid-carbon tetrachloride system at 22.5°C (○) and 51°C(△). (b) methanol-carbon tetrachloride(○), ethanol-carbon tetrachloride(◑), and 1-propanol-carbon tetrachloride(△) systems at room temperature. The abscissa is mole fraction of alcohols.(Kato et al. 1980a and b).

dependence of $N\langle(\Delta x)^2\rangle$ for the acetic acid system can be explained by association equilibria involving both cyclic dimers and linear polymers.

Figure 3b presents the $N\langle(\Delta x)^2\rangle$ values for methanol, ethanol, and 1-propanol-carbon tetrachloride systems (Kato et al. 1980b). It can be seen from the figure that as the length of the hydrocarbon chain of alcohol increases, $N\langle(\Delta x)^2\rangle$ decreases. This arises from the fact that the association tendency is in the order methanol > ethanol > 1-propanol. We have analyzed these results assuming that the hydrogen-bonded alcohol molecules can be treated as a linear and flexible polymer chain.

Aqueous Solutions of Alcohols

Figure 4 shows concentration dependences of $F = N\langle(\Delta x)^2\rangle/(x_A x_B)$ for aqueous solutions of ethanol, t-butyl alcohol(Iwasaki and Fujiyama 1977, 1979), 2-ethoxyethanol (Kato 1984), and 2-butoxyethanol (Ito et al. 1983). The F values for the ethanol system was calculated from the data of Donkersloot (Donkersloot

1979). It is seen from the figure that F is nearly equal to unity in the low concentration range and begins to increase at a certain concentration. This figure demonstrates also that the F values dramatically increase as the hydrophobicity of alcohols increases. In Figure 5a, are shown the concentration dependences of $N\langle(\Delta x)^2\rangle$ for aqueous solutions of 2-butoxyethanol at different temperatures (Ito et al. 1983). As illustrated in the figure, $N\langle(\Delta x)^2\rangle$ values are close to those of an ideal solution in the concentration range $0 < x_{BE} < 0.02$ (x_{BE}: the mole fraction of 2-butoxyethanol). In the range $0.02 < x_{BE} < 0.2$, on the other hand, $N\langle(\Delta x)^2\rangle$ takes values much larger than those of an ideal solution and increases rapidly with temperature. These results come from the fact that phase separation occurs at higher temperatures in the range $0 < x_{BE} < 0.02$ (see Figure 5b). Similar results have been obtained for aqueous solutions of 2,6-dimethyl pyridine (Ito and Kato 1984).

Figure 6a illustrates concentration dependences of the apparent correlation length for the 2-butoxyethanol system obtained by dynamic light scattering (Kato 1985). It can be seen from the figure that the apparent correlation length is much longer than the molecular radii.

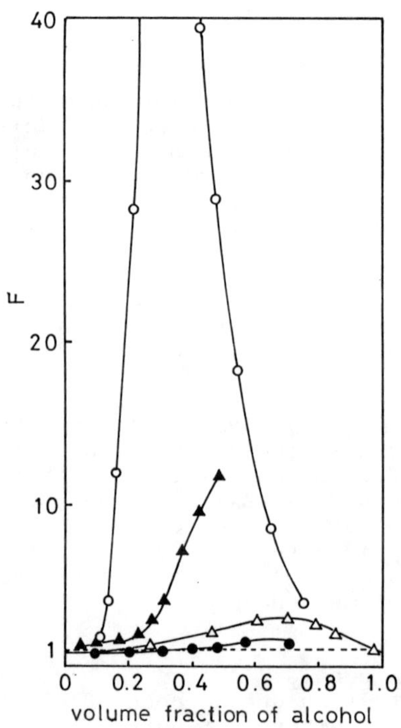

Figure 4. $F = N\langle(\Delta x)^2\rangle/(x_A x_B)$ for aqueous solutions of ethanol (△), t-butyl aclohol(▲), 2-ethoxyethanol(●), and 2-butoxyethanol(○) at room temperature. The dashed line corresponds to an ideal solution (F = 1).

The results in Figures 5b and 6a suggest that even at the temperatures far from the critical consolute point, there exist some sort of clusters whose sizes are much larger than molecular sizes and that these clusters grow in size with increasing temperature. In order to examine the relation between the formation of these clusters and translational motion of each molecule, we have measured the self-diffusion coefficient by using pulsed-gradient FT-NMR (Kato 1985). Figure 6b shows concentration dependences of hydrodynamic radii of H_2O (R_W) and 2-butoxyethanol (R_{BE}) obtained from the following equation

$$D_\alpha = kT/4\pi\eta R_\alpha \qquad (18)$$

where D_α is the self-diffusion coefficient of component α. The figure demonstrates that R_α depends on concentration and temperature only slightly. Moreover, these R_α values are nearly equal to or smaller than the molecular radius of α (It is well known that the water molecule can be approximately regarded as a sphere of radius 0.14 nm and the radius of the 2-butoxyethanol molecule, r, is calculated to be about 0.37 nm, if the relation $(4\pi/3)r^3\rho = 1$ is used). These results indicate that each molecule diffuses almost independently even in the large clusters.

2-Butoxyethanol may be regarded as a homologue of alkylpolyoxyethylene surfactants, $C_nH_{2n+1}(OC_2H_4)_mOH$ (n=4 and m=1

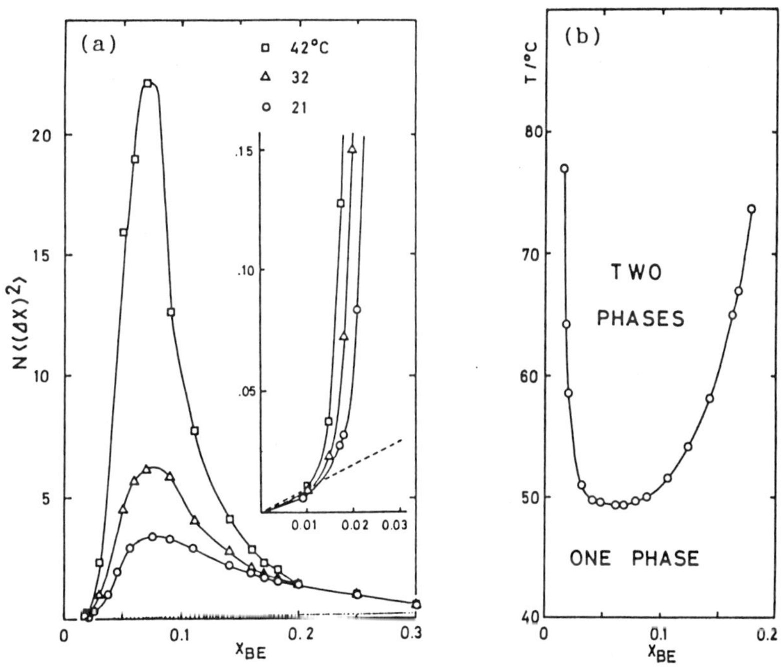

Figure 5. $N\langle(\Delta x)^2\rangle$ (a) and mutual solubility curve (b) for 2-butoxyethanol(BE)-water system. The dashed line corresponds to an ideal solution. (Ito et al. 1983).

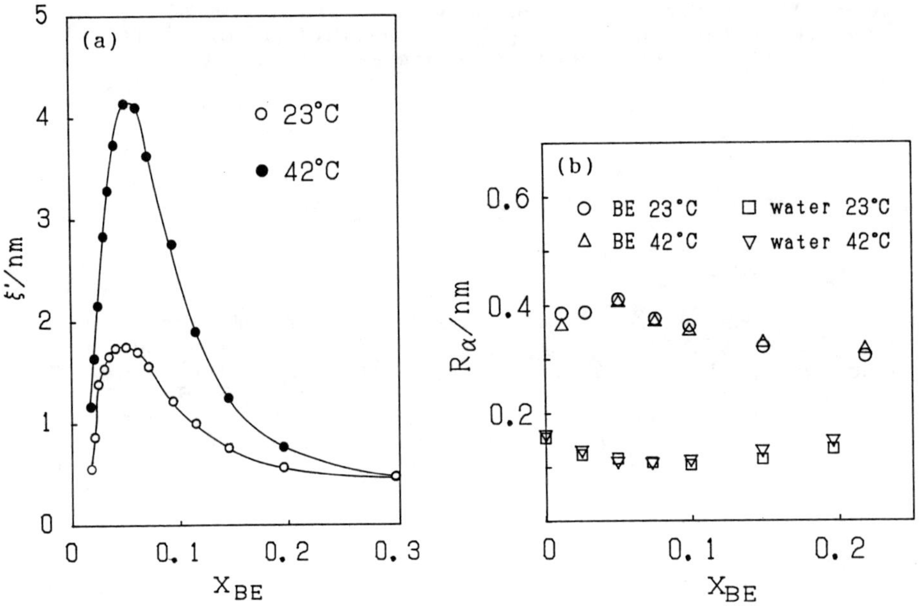

Figure 6. Apparent correlation lengths obtained from the mutual diffusion coefficients using Equation 12 (a) and hydrodynamic radii obtained from the self-diffusion coefficients using Equation 18 (b) for 2-butoxyethanol(BE)-water system.

for 2-butoxyethanol). For surfactants with high n values, the hydrodynamic radii are much longer than the molecular radii. Above a certain temperature, moreover, the hydrodynamic radii increase with increasing temperature (Kato and Seimiya 1986, Kato et al. 1989). These results arise from the formation of long-lived clusters, i. e., micelles.

Kirkwood-Buff Integrals

For a binary solution, three kinds of Kirkwood-Buff integrals can be obtained from the relations (Donkersloot 1979, Kato 1984)

$$G_{AA} = kT\kappa_T + \frac{x_B}{x_A}\rho v_B{}^2 F - \frac{1}{\rho x_A} \tag{19a}$$

$$G_{BB} = kT\kappa_T + \frac{x_A}{x_B}\rho v_A{}^2 F - \frac{1}{\rho x_B} \tag{19b}$$

$$G_{AB} = kT\kappa_T - \rho v_A v_B F \tag{19c}$$

where κ_T is the isothermal compressibility and v_α is the partial molecular volume of component α. Usually, the first term is negligible compared to other terms. For example, the ratio of the first term to the third term in Equation 19a or Equation 19b,

$\rho x_\alpha kT\kappa T$, is $(10^{-2} \sim 10^{-1})x_\alpha$ in ordinary solutions at room temperature (note that $x_\alpha < 1$). As the change of v_α with concentration is much smaller than v_α itself, concentration dependence of $G_{\alpha\beta}$ is determined by F values alone.

After the work of Ben-Naim (Ben-Naim 1978), several studies concerning determination of $G_{\alpha\beta}$ have been reported. In such studies, F values are calculated usually from the data on activity coefficient using Equation 8. However, it is difficult to determine the concentration derivative of the activity coefficient with high accuracy. As the Rayleigh-Brillouin scattering gives F values directly, this technique is useful also for determining $G_{\alpha\beta}$. We have calculated $G_{\alpha\beta}$ values for alcohols-carbon tetrachloride systems (Kato et al. 1982) and aqueous solutions of alkoxyethanols (Kato 1984) by using our data on Rayleigh-Brillouin scattering.

Although $G_{\alpha\beta}$ is useful for comparison between thermodynamic data and radial distribution functions calculated theoretically, $G_{\alpha\beta}$ itself does not give much more information than $N\langle(\Delta x)^2\rangle$. The reason is as follows. Neglecting the first terms in Equation 19, we obtain

$$\frac{\phi_A G_{AA} + v_A}{\phi_B} = \frac{\phi_B G_{BB} + v_B}{\phi_A} = - G_{AB} \qquad (20)$$

where ϕ_α is the volume fraction of α. The same equation has been already reported by Bale et al. (Bale et al. 1968). Equation 20 demonstrates that the three kinds of integrals, G_{AA}, G_{BB}, and G_{AB} are not independent and the relation among them depends on volume fraction alone (remember that the change of v_α is much smaller than v_α itself). Moreover, it is difficult to define a reference system such as an ideal solution. Even if we define a reference system by putting $F = 1$, $v_\alpha = v_\alpha^0$ (v_α^0: molecular volume in a pure liquid), the $G_{\alpha\beta}$ values of the reference system depend on the molecular size. So comparison of $G_{\alpha\beta}$ among different systems are not so easy as $N\langle(\Delta x)^2\rangle$ or F (when F is high, $|G_{\alpha\beta}|$ values becomes much larger than those of reference system. Then, the choice of reference system does not affect the results very much. However, comparisons among $G_{\alpha\beta}$ of different systems are still not easy because $G_{\alpha\beta}$ values themselves depend on the molecular volume of each species explicitly (see Equation 19)).

ACKNOWLEDGEMENT

I am grateful to (the late) Professor Tsunetake Fujiyama who introduced me to the light scattering study of concentration fluctuations. I am also indebted to Professor Hiroyasu Nomura of Nagoya University for helpful discussion. Thanks are given to Professor Yasuo Udagawa of Institute for Molecular Science for his continuous support.

REFERENCES

Bale, H. D., R. E. Shepler, and D. K. Sorgen. 1968. Small angle X-ray scattering from tertiary-butyl alcohol-water mixtures. Phys. Chem. Liq. 1: 181-190.

Ben-Naim, A. 1977. Inversion of the Kirkwood-Buff theory of

solutions: Application to the water-ethanol system. J. Phys. Chem. 37: 4884-4890.

Berne, B. J. and R. Pecora. 1976. Dynamic Light Scattering, Wiley-Interscience, New York.

Bhatia, A. B. and D. E. Thornton. 1970. Structural aspects of the electrical resistivity of binary alloys. Physical Review B 2: 3004-3012

Bhatia, A. B., W. H. Hargrove, and D. E. Thornton. 1974. Concentration fluctuations and partial structure factors of compound-forming binary molten alloys. Phys. Rev. B 9: 435-444.

Donkersloot, M. C. A. 1979. The structure of binary liquids. The Kirkwood-Buff theory of liquid mixtures, illustrated on the basis of the systems water/methanol, water/ethanol, and cyclohexane/2,3-dimethylbutane, as a link between thermodynamic data and x-ray and neutron scattering results. J. Solution Chem. 8: 293-307.

Fabelinskii, I. L. 1968. Molecular Scattering of Light, Plenum Press, New York.

Ferrel, R. A. Decoupled-mode dynamical scaling theory of the binary-liquid phase transition. Phys. Rev. Lett. 24: 1169-1172.

Fujiyama, T. and M. Tanaka. 1978. Light scattering study of local structures in solutions. Structure-forming effect of non-associative solvent on ethanol aggregate. Bull. Chem. Soc. Jpn. 51: 1655-1658.

Ito, N., T. Fujiyama, and Y. Udagawa. 1983. Bull. Chem. Soc. Jpn. 56: 379-385.

Ito, N. and T. Kato. 1984. Orientational pair correlation of pyridine and 2,4-dimethylpyridine in water by depolarized Rayleigh scattering and nuclear magnetic resonance spectra. J. Phys. Chem. 88: 801.

Iwasaki, K., M. Tanaka, and T. Fujiyama. 1976a. Concentration fluctuation and light scattering spectrum. Bull. Chem. Soc. Jpn. 49: 2719-2723.

Iwasaki, K., Y. Katayanagi, and T. Fujiyama. 1976b. Associated complex formation in a binary solution and the concentration fluctuation: The chloroform-diethyl ether system. Bull. Chem. Soc. Jpn. 49: 2988-2994.

Iwasaki, K. and T. Fujiyama. 1977. Light scattering study of clathrate hydrate formation in binary mixtures of tert-butyl alcohol and water. J. Phys. Chem. 81: 1908-1912.

Iwasaki, K. and T. Fujiyama. 1979. Light scattering study of clathrate hydrate formation in binary mixtures of tert-butyl alcohol and water. 2. Temperature effect. J. Phys. Chem. 83: 463-468.

Kato, T. and T. Fujiyama. 1976. Light scattering study of local structures in solutions: Chloroform-ethanol system. J. Phys. Chem. 80, 2771-2777.

Kato, T., S. Hyodo, and T. Fujiyama. 1978. Study of local structures formed in ethanol solutions. Determination of equilibrium constants by light scattering and infrared spectra. J. Phys. Chem. 82: 1010-1015.

Kato, T. 1980. Thesis, Tokyo Metropolitan University, Tokyo, 1980.

Kato, T., K. Yano, and T. Fujiyama. 1980a. Light scattering of local structures in solutions. Importance of temperature dependency measurements. Bull. Chem. Soc. Jpn. 53, 53-57.

Kato, T., T. Nakanishi, and T. Fujiyama. 1980b. Light scattering study of local structures in solutions. Mean association numbers and concentration fluctuations for alcohol-carbon tetrachloride systems. Bull. Chem. Soc. Jpn. 53: 2173-2180.

Kato, T. and T. Fujiyama. 1981. Light Scattering and Molecular Association. Yukagaku (in Japanese), 30, 742-748.

Kato, T., T. Fujiyama, and H. Nomura. 1982. Estimation of parameters, G11, G22, and G12 in the Kirkwood-Buff solution theory on the basis of the concentration fluctuation data obtained from Rayleigh scattering. Bull. Chem. Soc. Jpn. 55: 3363-3372.

Kato, T. 1984. Kirkwood-Buff parameters and correlation length in aqueous solutions of n-alkoxyethanols. J. Phys. Chem. 88: 1248-1252.

Kato, T. 1985. Determination of velocity correlation coefficients in aqueous solutions of 2-butoxyethanol by quasi-elastic light scattering, pulsed-gradient FT-NMR, and Rayleigh-Brillouin scattering. J. Phys. Chem. 89: 5750-5755.

Kato, T. and T. Seimiya. 1986. Study on intermicellar interactions and micelle size distribution in aqueous solutions of nonionic surfactants by measurements of mutual diffusion and self-diffusion coefficients. J. Phys. Chem. 90: 3159-3167.

Kato, T., S. Anzai, S. Takano, and T. Seimiya. 1989. Intermicellar interactions and micelle size distribution in aqueous solutions of polyoxyethylene surfactants. J. Chem. Soc. Faraday Trans. I. 85: 2499-2506.

Kawasaki, K. 1970. Kinetic equations and time correlation functions of critical fluctuations. Ann. Phys. 61: 1-56.

Kirkwood, J. G. and F. P. Buff. 1951. The statistical mechanical theory of solutions. I. J. Chem. Phys. 19: 774-777.

Miller, G. A. 1967. Equations for the calculation of activity coefficients of solutions from the intensity of Brillouin scattering. J. Phys. Chem. 71: 2305-2308.

Mountain, R. D. and J. M. Deutch. 1969. Light scattering from binary solutions. J. Chem. Phys. 50: 1103-1109.

Prigogine, I. and R. Defay. 1954. Chemical Thermodynamics, London: Longmans Green and Co.

Local Composition and Concentration Fluctuations in Binary Mixtures of Non-Electrolytes

Ramón G. Rubio and Mateo Diaz Peña
Dept. Química Física, Fac. Químicas
Univ. Complutense, 28040 Madrid, Spain

Donald Patterson
Dept. of Chemistry, McGill University
Montreal, PQ H3A 2K6, Canada

1 Abstract

An expression for the local mole fraction has been obtained from the Kirkwood-Buff integrals. When applied to simple equal-sized molecules, the results show that the quasichemical hypothesis leads to qualitatively wrong results. The local composition of mixtures of fluids with size mismatch are discussed in terms of the concentration- concentration correlation function $S_{CC}(0)$. The results show that, at least for some region of the phase diagram, packing effects are not the only responsible of the nonrandomness in the solution. Finally, the W-shape C_p^E curves found in some binary mixtures has been related to $S_{CC}(0)$, showing that $S_{CC}(0)$ can be used as a good predictor, and thus discarding a previous hypothesis that the W-shape was due to solvent induced conformational changes.

2 Introduction

It is well known that, far from critical points, only short-range order is present in the liquid state (Hansen and McDonald, 1986) , i.e., molecules are not correlated at large distances. Moreover, under those conditions, the structure of fluids is mostly due to packing effects. Since the thermodynamic properties depend primarily upon such structure, one can understand the big success of Perturbation theories in predicting them (Gray and Gubbins, 1984).

In multicomponent fluids, besides a molecular model for the pure components, one has to specify the combination rules which relate the interaction between unlike molecules to those of the like ones (Bohn et al., 1986). Unfortunately, rigorous statistical mechanical theories can only be applied to relatively simple molecular liquids (Gray and Gubbins, 1984), and more or less drastic physical or mathematical approximations ought to be introduced in order to develop models that can be applied to complex molecular fluids (Prausnitz, 1985). In these models one has to specify the so-called mixing-rules that relate the structure of the mixture to that of a pure component (Prausnitz et al., 1986). The most frequently used mixing rules have been the van der Waals'ones, which assume a random distribution of the molecules in the multicomponent fluid. This assumption is not compatible with the information arising from the radial distribution function g_{ij} obtained either from theoretical calculations (Chandler, 1982), or computer simulations (Allen and Tildesley, 1987). The g_{ij}'s show that the composition in the immediate neighborhood of a molecule is, in general, different to the overall composition. The difference between the local and the overall composition arise from size and shape differences (packing effects), and from differences in the interaction between molecules of different species. In general, such differences will depend upon composition, temperature and density (Lee et al., 1985; Gierycz and Nakanishi, 1984). A knowledge of such dependences seems necessary for establishing better mixing rules in mixtures.

Experimental information also points toward the existence of local composition different from the overall one: X-ray and neutron scattering in alloys (Ruppersberg, 1987), liquid mixtures (Nishikawa et al., 1987), and preferential sorption in polymer-solvent 1-solvent 2 systems (Masegosa et al., 1984).

Once the g_{ij}'s are known, the number of j molecules around an i molecule, within a spherical volume of radius L_{ij} is given by

$$N_{ij} = \frac{4\pi N_j}{V} \int_0^{L_{ij}} g_{ij}(r) r^2 \, dr \tag{2.1}$$

So far there is no way of obtaining $g_{ij}(r)$ for complex fluids from statistical mechanical principles, which makes the calculation of the local composition a very difficult task for any theoretician. Even though there is a wealth of simulations of simple fluids, data for complex fluids are scarce, specially if the molecules have internal degrees of freedom, which is the usual case for mixtures of practical use; therefore, no N_{ij} results are available for complex mixtures. As a consequence, most frequently, the local composition has been calculated using models that were either phenomenological (Renon,1985), or based on the quasichemical hypothesis (Panayiotou and Vera, 1980; Pouchly, 1989). The test of such models against computer simulations for simple fluids (Lee et al.,1985; Gierycz and Nakanishi, 1984) and Perturbation theory (Hoheisel and Kohler,1984) shows that these models are not qualitatively correct. Some new models that have recently been proposed are based on correlations of simulation results for square-well fluids (Lee and Sandler, 1987), however it is difficult to know whether they can be applied to more complex fluids. Hence information on local compositions in complex mixtures from experimental data could be helpful in clarifying the extent of the local compositions in mixtures, and its influence on the thermodynamic properties. Fluctuation theory of mixtures (Ben Naim, 1974) allows us to obtain such information.

This work is organized as follows. In section 3 an expression for the local composition is worked out. Section 4 discusses some results for mixtures of almost equal-sized molecules. In section 5 mixtures of molecules with size mismatch are studied in terms of the concentration-concentration correlation function. Finally, in section 6 the peculiar behaviour of C_p^E for some complex mixtures is related to nonrandomness.

3 Local mole fractions and Kirkwood-Buff integrals

The relationship between both microscopic and macroscopic magnitudes can be established using the Kirkwood-Buff theory (Kirkwood and Buff, 1951),which relates the thermodynamic properties to the so-called Kirkwood-Buff integrals G_{ij}, defined such as the product

$$\rho_j G_{ij} = \rho_j \int_0^\infty 4\pi r^2 [g_{ij}(r) - 1]\, dr \tag{3.1}$$

gives the average of j molecules around a central i molecule over the bulk average. The above integral can be split

$$G_{ij} = -\int_0^{\gamma_{ij}} 4\pi r^2\, dr + \int_{\gamma_{ij}}^{L_{ij}} 4\pi r^2 [g_{ij} - 1]\, dr + \int_{L_{ij}}^\infty 4\pi r^2 [g_{ij} - 1]\, dr \tag{3.2}$$

where γ_{ij} is an effective distance of closest approach between molecules i and j such that $g_{ij}(r) = 0$, $r < \gamma_{ij}$, and L_{ij} is the smallest distance beyond which

$$\int_{L_{ij}}^\infty 4\pi r^2 [g_{ij} - 1]\, dr \approx 0 \tag{3.3}$$

The first integral in Eq. (3.2) is a volume which should be almost independent of the state of the liquid (ρ, T), except perhaps for water or other highly hydrogen-bonded liquids (Ben Naim, 1977). To identify L_{ij} is a major problem, and much discussion has been devoted to it (Kojima et al., 1984; Mansoori and Ely, 1984;Rubio et al., 1986; Pfund et al., 1988; Zaitsev et al., 1989). In addition, Hoheisel and Kohler (Hoheisel and Kohler, 1984) have shown that it is adventageous to use a unique value L_{ij} for all the $i - j$ pairs, thus we have assumed that taking

$$L_{ij} = L = 5 \times max(\sigma_{ii}, \sigma_{jj}) \tag{3.4}$$

Eq. (3.3) is satisfied (Rubio et al., 1987).

Combining Eqs. (2.1) and (3.1) one finds

$$\rho_j G_{ij} = N_{ij} - N_{ij}^0 \tag{3.5}$$

where

$$N_{ij}^0 = x_j \rho \int_0^{L_{ij}} 4\pi r^2\, dr = \langle n_j(L_{ij}) \rangle \tag{3.6}$$

is the bulk average of j molecules in a spherical volume of radius L_{ij} and may be written as $N_{ij}^0 = 4\pi\rho x_j L_{ij}^3/3$.

The local mole fraction around an i molecule can be written as

$$x_{ij} = \frac{x_j \rho G_{ij} + N_{ij}^0}{N_{ij}^0 + N_{ii}^0 + \rho(x_j G_{ij} + x_i G_{ii})} \tag{3.7}$$

Even though the calculation of the G_{ij}'s from theory or computer simulations has the problems discussed in the previous section, inversion of the Kirkwood-Buff theory allows their calculation from experimental data (Ben Naim, 1977; Matteoli and Lepori, 1984). In effect

$$\begin{aligned} G_{ij}(i \neq j) &= G_{ji} = RT\kappa_T - \frac{\overline{V_i}\,\overline{V_j}}{DV} && (3.8) \\ G_{ii} &= G_{ij} + [(V_j/D_\mu) - V]/x_i && (3.9) \\ D_\mu &= \frac{x_i}{RT}\left(\frac{\partial \mu_i}{\partial x_i}\right)_{p,T} && (3.10) \end{aligned}$$

where κ_T is the isothermal compressibility, $\overline{V_i}$ the partial molar volumes, and μ_i the chemical potential of component i. The problems involved in the estimation of $\overline{V_i}$ and D_μ from excess volume and G^E or light scattering data have been discussed in detail elsewhere (Rubio et al., 1986; Matteoli and Lepori, 1984; Kato et al.,1982).

Eq. (3.7) has been recently extended to calculate the local composition in the vapor-liquid interface of liquid mixtures (Tronel-Peyroz et al., 1989) In this case, in addition to choose L_{ij}, one has to estimate the area occupied by a molecule in the interface.

4 Binary mixtures of almost equal-sized molecules

When the size difference between the molecules of the two components is small, the local composition around an i molecule is not qualitatively changed by the cut-off distance L_{ij} of the G_{ij} integrals (Hoheisel and Kohler, 1984) therefore a different choice in L_{ij} would modify the absolute values of x_{ij}, the larger L_{ij} the smaller the difference between x_{ij} and x_j. However, for a semiquantitative discussion of x_{ij}, Eq. (3.4) will be valid. Figure 1 shows

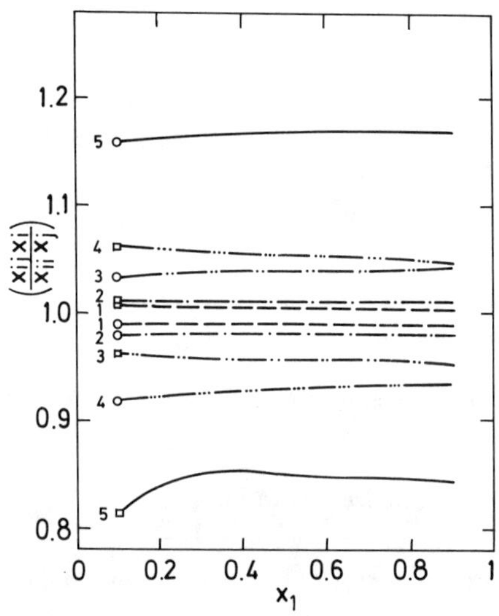

Figure 1: Nonrandom factors for systems of almost equal-sized molecules. Curves labeled □ correspond to Q_{21}, and those labeled ∘ to Q_{12}. (1) $Ar + O_2$; (2) $CO + N_2$; (3) $Kr + CH_4$; (4) $C_2H_4 + Xe$; (5) $C_2H_6 + C_2H_4$. x is the mole fraction of the first component

$Q_{ij} = (x_{ij}x_i)/(x_{ii}x_j)$ for some mixtures of simple fluids for which the ratio of their diameters (as given by the Lennard-Jones σ_{ii}'s) is smaller than 1.05 (Rubio et al., 1987). The first point to be noted is that the random distribution hypothesis $Q_{ij} = 1$ is quite wrong even for such simple mixtures . Also the Q_{ij}'s show a weak composition dependence, which is in agreement with the results of Lee et al. (1986) for square-well spheres .

As already said, most local composition models are based on the quasichemical hypothesis (Guggenheim, 1952).

$$x_{ij}/x_{ii} = (x_j/x_i)\exp(-\epsilon_{ij}/RT) \tag{4.1}$$

where ϵ_{ij} is an energetic parameter which depends on the difference between $i-j$ and $i-i$ interactions . According to Eq. (4.1) Q_{ij} should be composition independent, which is close to the results shown in Figure 1. Panayiotou and Vera (1980) have given analytical expressions for the local composition as a solution of the quasichemical equations. In terms of the mole fractions x_{ij} reads

$$x_{ij} = \frac{2x_j}{1+(1-4x_ix_jK)^{1/2}} \quad ; \quad x_{ii} = 1 - x_{ij} \tag{4.2}$$

$$K = 1 - \exp(\Delta\epsilon/RT) \tag{4.3}$$

where $\Delta\epsilon$ characterizes the difference of of the $i-i$ and $i-j$ interactions. According to Eq. (4.2) the Q_{ij}'s must be composition dependent. Finally, the strictly regular solution model leads to

$$x_{ij}/x_{ii} = (x_j/x_i)[1-(\Delta\epsilon/RT)x_j] \tag{4.4}$$

which for Lennard-Jones spheres has been shown to give better results than Eq. (4.1). Figure 2 shows the predictions of Eqs. 4.2 and 4.4 for some of the systems discussed in Figure 1. $\Delta\epsilon/RT$ has been calculated in such a way that $Q_{12}(x=0.5)$ is well reproduced. The models predict $Q_{12}(x=0.5) = Q_{21}(x=0.5)$ which is in disagreement with the values calculated from the G_{ij}'s. Also if $\Delta\epsilon/RT \geq 0$ both Q_{12} and Q_{21} are ≥ 1 (the opposite holds if $\Delta\epsilon < 0$), which is reasonable within a lattice theory. This is also in disagreement with the Q_{ij}'s calculated from G_{12} and G_{21}. The latter can be understood considering that, in terms of the distribution-function formalism, the coordination number of the two components are not equal. Furthermore, computer simulation results indicate that the local compositions are density dependent (Gierycz and Nakanishi, 1984; Lee et al., 1985), a fact that cannot be included in Eqs. (4.1) to (4.4). Unfortunatly, the lack of precise data does not allow to obtain the G_{ij}'s as a function of density for the systems of Figure 1.

Hu et al. (1983) have proposed a local composition model for mixtures of quasi-spherical fluids. The model specifies quite clearly the cut-off distance beyond which there is a random distribution of molecules L_{ij}^m, and provides an expression for the local mole fraction $(x_{ij})_m$. If the model is correct

$$(x_{ij})_m = \frac{x_j\rho G_{ij} + (N_{ij}^0)_m}{(N_{ij}^0)_m + (N_{ii}^0)_m + \rho(x_jG_{ij} + x_iG_{ii})} \tag{4.5}$$

should be satisfied. In Eq. (4.5) the subscript m has been written to stress that the corresponding magnitudes have been calculated according to the assumptions of the local composition model. We have calculated both terms of Eq. (4.5) for some of the systems studied by Hu et al. (1983) at $x=0.5$. Table 1 shows the results as well as the differences between both terms δ_{ij}. For the first two systems the agreement is quite good, the δ_{ij}'s being of the same magnitude and sign than those found between the calculations of Hu's model and the

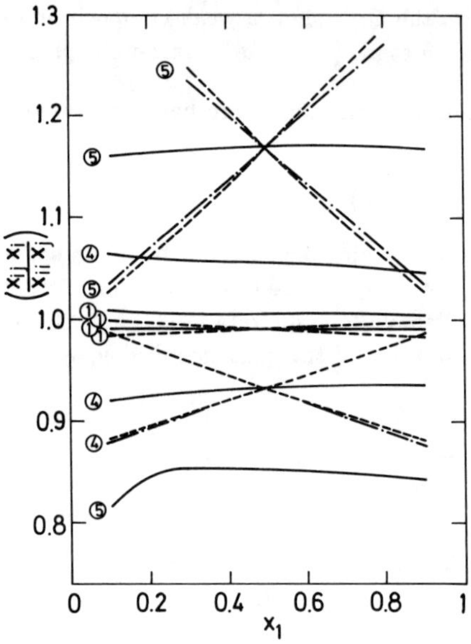

Figure 2: Nonrandom factors for systems of almost equal-sized molecules:(—) from the G_{ij} integrals; (- - -) from the Panayiotou and Vera's model; (-·-), from the strictly regular model. Labels as in Figure 1

computer simulations of Sandler (1985). For $Ar + N_2$ one obtains $x_{12} > x_2$ according to Hu's model and $x_{12} < x_2$ from the G_{ij}'s. Clearly such discrepancy cannot disappear by changing the L_{ij}'s since increasing them would lead to a value x_{12} closer to but not larger than $x_2 = 0.5$.

Hoheisel and Kohler (1984) have found that the Q_{ij}'s show strong oscillations around 1 as the cut-off distance is increased when the two components differ in size, therefore Eq. (3.7) should not be used unless being sure that one is beyond the region of oscillations. This is unfortunate, since the G_{ij}'s present sharp changes with composition for some systems with molecules of different sizes and/or shapes (Matteoli and Lepori, 1984; Rubio et al., 1986; 1987; Ben Naim, 1989).

Figure 3 shows the results for $Ar + C_2H_6$ and $Xe + CF_4$, and indicates that in these systems each molecule tends to segregate those of different species, i.e., the local composition should differ substantially from the overall one. In the next section we will consider how to deal with this type of systems.

Table 1: Local composition according to Hu's model (Hu et al., 1983) and to Eq. (3.5), and residuals δ_{ij} (see text)

System	T/K	Hu et al. (1983)		Eq. (4.5)		$10^3\delta_{21}$	$10^3\delta_{12}$
		x_{21}	x_{12}	x_{21}	x_{12}		
$CO(1) + N_2(2)$	84	0.4999	0.4951	0.4995	0.4990	0.4	-3.9
$Ar(1) + O_2(2)$	84	0.4983	0.4946	0.4981	0.4995	0.2	-4.9
$Ar(1) + N_2(2)$	84	0.4780	0.5240	0.4936	0.4801	-15.6	—

5 Mixtures of molecules with different size and shape. Concentration fluctuations

Hoheisel and Kohler (1984) have shown that at liquid-like densities Perturbation theory accounts for the local composition in mixtures of Lennard-Jones spheres. In order to study the influence of packing effect, and eventually that of the attractive interactions in mixtures of real fluids, it would be useful to define a reference system to which a random distribution could be assessed. This is a difficult problem whithin the framework of the Kirkwood-Buff integrals since the G_{ij}'s depend upon the size of each molecule, while for mixtures of hard spheres packing effects arise from the differences in size of the two components. Nevertheless, it is possible to define a random system using the concentration-concentration correlation function $S_{CC}(0)$ introduced by Bathia and Thornton (1970). $S_{CC}(0)$ is related to the long wavelength limit of the partial structure factors of the mixture, and may be written as

$$S_{CC}(0) = \frac{x_i x_j}{D_\mu} \tag{5.1}$$

with D_μ given by Eq. (3.10).For a binary mixture of equal-sized hard spheres $S_{CC}(0) = x_1 x_2$, thus being a suitable random distribution reference. Since D_μ is closely related to the condition of thermodynamic stability, $S_{CC}(0)$ diverges to $+\infty$ at any critical point. Values of $S_{CC}(0) > S_{CC}^{id}(0)$ indicate some tendency to homocoordination in the mixture, while $S_{CC}(0) < S_{CC}^{id}(0)$ indicates heterocoordination. $S_{CC}(0)$ allows us to compare the degree of nonrandomness in two systems regardless of differences in size, shape or intermolecular potentials, and can be calculated from experimental data of G^E or light scattering as discussed above (Rubio el al., 1987; Matteoli and Lepori, 1984; Kato el al., 1982).

Figure 4 shows $S_{CC}(0)$ for the systems of Figure 3, confirming what was said in the previous section. It also shows the values calculated for mixtures of hard-spheres with different diameter ratios at packing fraction $\eta = 0.49$. The equation of state of Mansoori et al. (1971) was used for the calculations. Increasing the size mismatch always prevents concentration fluctuations increasing the tendency to homocoordination, therefore how to explain the behaviour found for the systems $Ar + C_2H_6$ or $Xe + CF_4$? Using a mean-field treatment for the attractive forces, Joarder and Silbert (1985) found $S_{CC}(0) > S_{CC}^{id}(0)$ for spheres with permanent electrical moments, thus perhaps the attractive forces could play a non-negligeable role in the nonrandomness of mixtures, at least in conditions different from those contemplated by Hoheisel and Kohler (1984). Figure 5 shows $S_{CC}(0)$ for mixtures of Lennard-Jones spheres

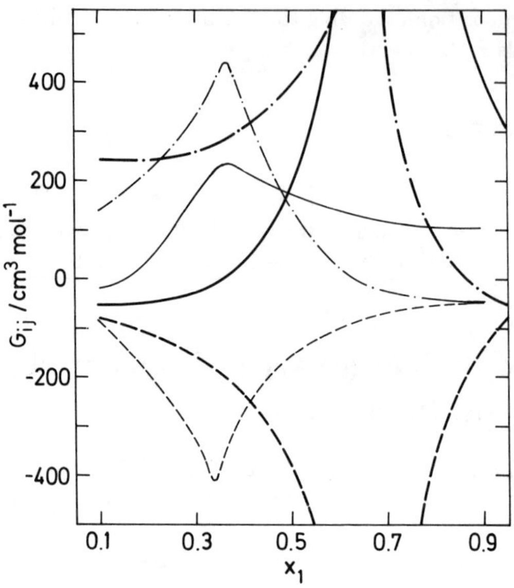

Figure 3: G_{ij} integrals for two systems with size mismatch. (—) G_{22}; (- - -) G_{12}; (-·-) G_{11}. Thin lines $C_2H_6(1) + Ar(2)$ at 91K. Thick lines $Xe(1) + CF_4(2)$ at 159K

as calculated from the Percus-Yevick approximation for different values of R, S and Q:

$$R = \sigma_{11}/\sigma_{22} \tag{5.2}$$

$$S = \frac{(\epsilon_{11}\epsilon_{22})^{1/2}}{\epsilon_{12}} \tag{5.3}$$

$$Q = \epsilon_{22}/\epsilon_{11} \tag{5.4}$$

As it can be observed increasing the difference in interaction energies between the pure components increases $S_{CC}(0)$ above $S_{CC}^{id}(0)$ for equal- size spheres and assuming the Lorentz-Berthelot (LB) combination rule (curves 1 and 2). Curve 4 corresponds to equal-sized spheres for which the reduced temperature of the pure components are intermediate cases between curves 1 and 2, but there is a noticeable departure from the LB rule. Curve 3 corresponds to a mixture in which both the differences between the reduced temperatures and departures from the LB rule are larger than in curve 4, however the introduction of size mismatch strongly prevents the tendency to homocoordination. Figs. 4 and 5 show that, at least for some regions of the phase diagram, not only packing effects but also the attractive forces are important in determining the distribution of molecules in the solution. This is specially interesting for the liquid-liquid equilibrium, since it is strongly dependent on the values of the activity coefficients, and they are affected by the mixing rules and thus by the local compositions.

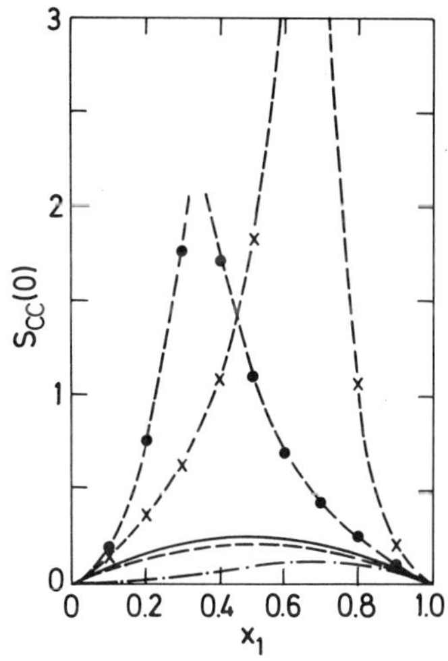

Figure 4: Concentration-concentration correlation function : (•) $Ar(1)+C_2H_6(2)$ at 91K ; (×) $Xe(1)+CF_4(2)$ at 159K ; (-) ideal random system ; (- -) hard spheres with a diameter ratio of 0.9 ; (-·-) hard spheres with a diameter ratio of 0.3.

6 Concentration fluctuations and W-shape C_p^E curves

Recent experimental work has revealed the existence of many binary mixtures for which the concentration dependence of the excess heat capacity C_p^E has a "W-shape", i.e., two regions of positive curvature separated by a region of negative curvature (Lainez et al., 1985) For some years this behavior was supposed to be due to conformational changes in one component induced by other one (Wilhem, 1985). Recently Saint Victor and Patterson (1986), making use of a simple lattice model and the quasichemical hypothesis, have suggested that nonrandomness could also explain such a behavior.

We have calculated $S_{CC}(0)$ from light scattering experiments for several mixtures that show W-shape C_p^E curves (Rubio et al., 1989; Andreolli-Ball et al., 1989). Figures 6 and 7 show the $S_{CC}(0)$ and C_p^E curves, respectively Figure 6 shows the $S_{CC}(0)$ for several systems that show W-shape C_p^E curves: (-+-) ideal random system ; (-×-) 1,4-dichlorobutane (14ClBu) + n-heptane (nC_7) ; (- · -) cyclohexane (cC_6) + 1,4-dioxane ; (- - -) nitrobenzene (NBz) + nC_7 ; (- ·· -) acetone (Ac) + n-dodecane (nC_{12}) ; (—) 2-butanone + nC_{12}. Figure 7 shows the C_p^E curves: (- ·· -) 14ClBu + nC_7 ; (- ···· -) 14ClBu + n-tetradecane (nC_{14}) ; (- -) NBz + nC_7 ; (— —) Ac + nC_{12}; (- · -) cC_6 + 1,4-dioxane ; (- • -) 1,4-dioxane + dicyclohexyl ;

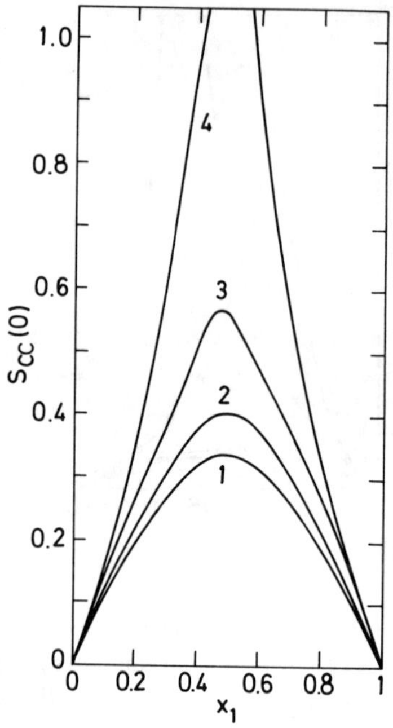

Figure 5: Concentration-concentration correlation function for mixtures of Lennard-Jones spheres : (1) $R = 1$, $S = 1$, $Q = 2$; (2) $R = 1$, $S = 1$, $Q = 2.5$; (3) $R = 0.8$, $S = 1.3$, $Q = 1.7$; (4) $R = 1$, $S = 1.2$, $Q = 1.5$

(—) 14ClBu + nC_{10}. As it can be observed $S_{CC}(0) > S_{CC}^{id}(0)$ for all the systems, i.e., there is a clear tendency to homocoordination. Also there exists a qualitative correlation between the degree of nonrandomness and the intensity of the W-shape effect. Moreover, there also is a rough correspondence between the concentration of the maxima of the $S_{CC}(0)$ curves, and that at which the negative curvature region appears in the C_p^E curves.

The systems with 1,4-dioxane and with 14ClBu point out a rather frequent characteristic of the W-shape effect: in general this effect is more intense when the difference in size between the components is increased, and the concentration at which the W-effect appears shifts to larger concentration of the smaller component. Using the simple lattice model of Flory-Huggins (Flory, 1971) this behavior can be related to that of $S_{CC}(0)$.

The Flory-Huggins expression for the excess Gibbs energy G^E is

$$\frac{G^E}{RT} = x_1 \ln \frac{\phi_1}{x_1} + x_2 \ln \frac{\phi_2}{x_2} + \chi_v (x_1 V_1 + x_2 V_2) \phi_1 \phi_2 \tag{6.1}$$

where ϕ_i is the volume fraction, V_i the molar volume of component i and χ_v the intensive quantity

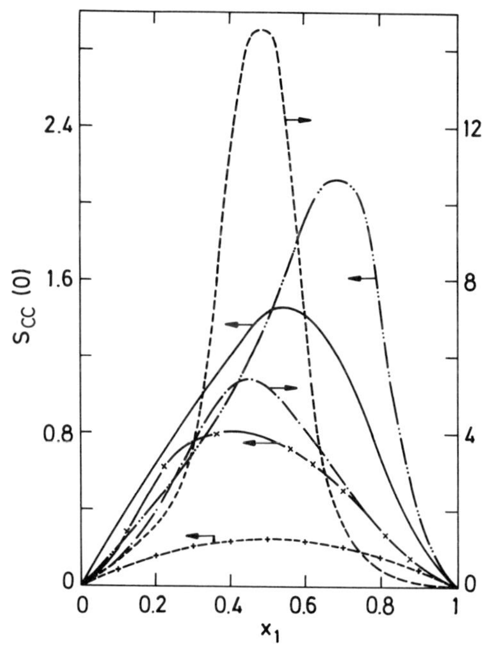

Figure 6: Concentration-concentration correlation function for some systems that show "W-shape" C_p^E curves. For symbols see text.

$$\chi_v = \frac{z\Delta\omega}{V_s RT} \tag{6.2}$$

with z being the coordination number, $\Delta\omega$ the interchange free energy per mole of segments, and V_s the segmental molar volume. Within this model, $S_{CC}(0)$ is given by

$$\frac{S_{CC}(0)}{x_1x_2} = \left[1 + \frac{x_1x_2(r-1)^2}{(x_1+x_2r)^2} - \frac{2\chi_v V_1 x_1 x_2 r^2}{(x_1+x_2r)^3}\right]^{-1} \quad ; \quad r = \frac{V_2}{V_1} \tag{6.3}$$

From Eqs. (6.1) and (6.2), $S_{CC}(0)$ increases with decreasing T. Figure 8 shows that for hexafluorobenzene (HFB) + cC_6, and + dicyclohexyl the W-shape C_p^E follows the same trend (Andreolli-Ball et al.,1989). The dependence of $S_{CC}(0)$ with V_2, i.e., r is more complex since it depends upon the magnitude of χ_v. For small χ_v's the first term of Eq. (6.3) dominates and $S_{CC}(0)$ decreases as V_2 increases. If χ_v is large, the second term of Eq. (6.3) dominates and $S_{CC}(0)$increases with r. Besides, the maximum in $S_{CC}(0)$occurs at high concentrations of the smaller component. This behavior is similar to that discussed after Figs. 6 and 7. If r is increased $S_{CC}(0)$ grows up until for sufficiently large a value $S_{CC}(0) \to \infty$, the critical point being reached at

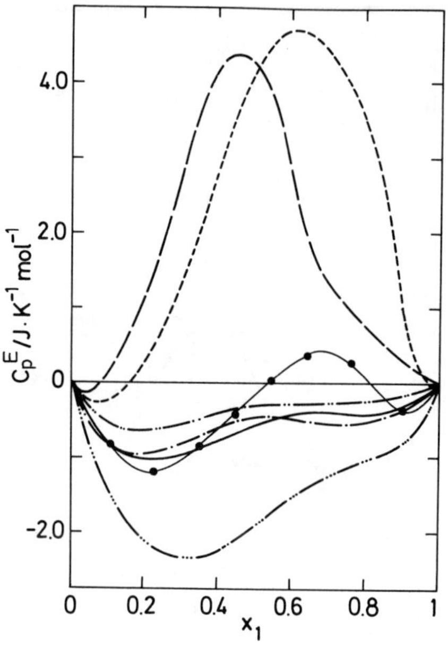

Figure 7: Some "W-shape" C_p^E curves . For symbols see text.

$$\chi_v V_1 = \frac{1}{2}\left(1 + \frac{1}{\sqrt{r}}\right)^2 \quad ; \quad \frac{x_1}{x_2} = r^{2/3} \tag{6.4}$$

Figure 9 shows the critical locus of the Flory-Huggins model, as well as the points corresponding to the systems: 14ClBu + nC_7, (○); Ac + nC_{12}, (◇); C_6H_6 + nC_{12}, (▽); NBz + nC_7, (□); methylethylketone (MEK) + nC_{10}, (△); 1-chloronaphtalene (ClN) + 2,2,4,4,6,8,8-heptamethylnonane (brC_{16}), (□); ClN + nC_{16}, (•); HFB + 2,2,4-trimethylpentane (iC_8), (△) and HFB + cC_6, (◇). The χ_v values are those that fit the $S_{CC}(0)$ curves with Eq. (6.3). As it can be observed, the systems that show W-shape lie relatively close to the critical curve,hence small changes in r, i.e., V_2 make the system more critical. Since at the critical point $S_{CC}(0) \to \infty$ and $C_p^E \to \infty$ one can easily understand the increase of the W-shape with r. A glance to Figure 9 allows one to see that for small χ_v's the effect of V_2 is different, since the critical locus is not approached with increasing r. Even though it explains the qualitative behavior of $S_{CC}(0)$ with T and V_2, and hence that of the W-shape effect, the Flory-Huggins model is too simple for quantitative predictions, which can explain that some points of Figure 9 lie slightly to the right of the critical locus.

We have used the Guggenheim model (Guggenheim, 1958), of which that of Flory-Huggins is

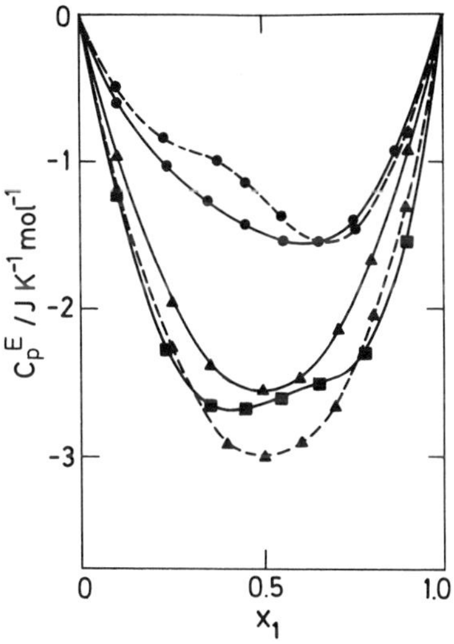

Figure 8: C_p^E curves at different temperatures for some systems containing hexafluorobenzene as component 1. Continuous lines results at 298.15K . Dashed lines at 283.15K . (•) + cyclohexane ; (■) + dicyclohexyl ; (▲) + 2,2,4-trimethylpentane

a particular case for $z = \infty$, z being the coordination number of the lattice. In what follows we will refer to the Guggenheim model with $z = 4$. In order to make quantitative predictions we have defined ΔC_p^W as the maximum height of the C_p^E above a straight line drawn tangent to the minima of the C_p^E curve. Table 2 shows ΔC_p^W and the concentration at which the W-shape effect is maximum for four pairs of systems for which C_p^E is available.$\chi_v V_1$ has been calculated from the $S_{CC}(0)$of the first system of each pair, and then the $S_{CC}(0)$ curve for the second system of each pair was calculated with Guggenheim's model ($z = 4$). In each case the maxima of the $S_{CC}(0)$ curves, and the composition at which they appear are shown. It can be observed that there exist correlations between the maxima of $S_{CC}(0)$ and ΔC_p^W, and between the concentrations at which they appear. Assuming the correlation shown in Table 2, ΔC_p^W should reach zero when $max.S_{CC}(0) \sim 0.7$, therefore this value could be considered as the threshold above which nonrandomness would give rise to W-shape C_p^E curves. As a test of such a hypothesis we have calculated $S_{CC}(0)$ for the system HFB + cC_6 at 298.15K using the G^E data of Gaw and Swinton (1968), the results being shown in Figure 10.

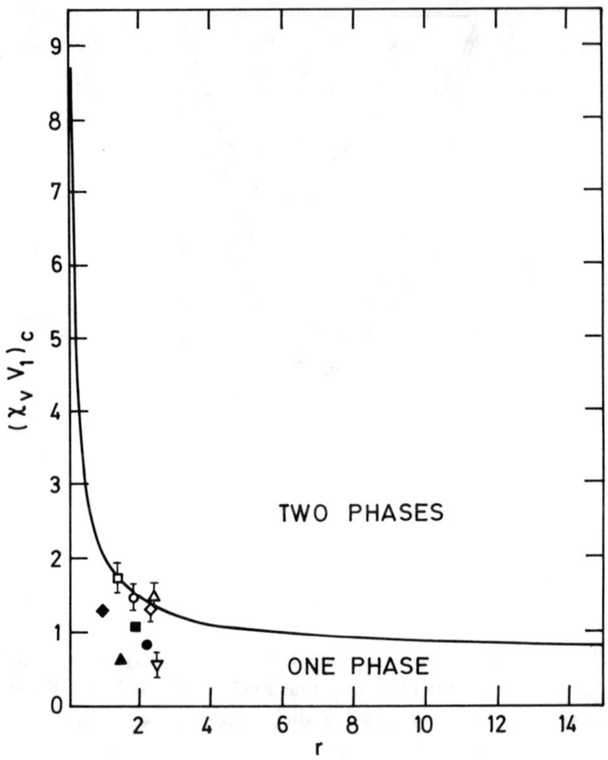

Figure 9: Critical locus according to Flory-Huggins theory. For symbols see text. T = 298.15K

Even though this system does not present W-shape effect at 298.15K, the proximity of the $max.S_{CC}(0)$ (~ 0.65) to the threshold value, and the temperature dependence deduced from the above mentioned models, suggested that the W-shape should appear in this system at lower T, and also for another system with larger V_2 at 298.15K. In Figure 8 the C_p^E curves of $HFB + cC_6$ at 283.15K and $HFB + dicyclohexyl$ at 298.15K are shown, both curves show the W-shape effect. Figure 10 shows $S_{CC}(0)$ for HFB+dicyclohexyl calculated from the $\chi_v V_1$ of $HFB + cC_6$. As it can be observed $S_{CC}(0)$ is a reasonable good predictor of W-shape effect, pointing out its close relationship to nonrandomness. In addition the fact that the W-shape effect is found (Andreolli-Ball et al., 1989) in systems containing a rigid plate-like molecule such as HFB or ClN and a molecule like cC_6, for which both isomers are equilibrated at room temperature, or highly branched alkanes like brC_{12}, whose steric hindrance is very high, allows us to discard solvent induced conformational changes as the reason for the W-shape effect on C_p^E.

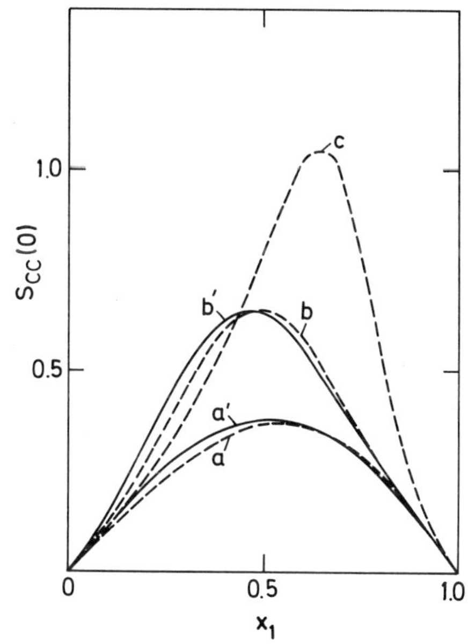

Figure 10: $S_{CC}(0)$ at 298.15 for: HFB + (a) iC_8; (b) cC_6; (c) dicyclohexyl. Continuous lines: experimental; dashed lines: calculated from the Flory-Huggins model.

Table 2: $\Delta C_p^W / J.mol^{-1}.K^{-1}$ and calculated $S_{CC}(0)$ according to Guggenheim's model ($z = 4$), and compositions at which their maxima appear. $T = 298.15K$.

System	x_1	ΔC_p^W	x_1	$S_{CC}(0)$
$1,4\ ClBu + nC_7$	0.50	0.15	0.54	0.80
$1,4\ ClBu + nC_{14}$	0.65	0.7	0.65	0.94
$dioxane + cC_6$	0.50	0.3	0.53	1.00
$dioxane + dicyclo$	0.70	1.1	0.65	1.45
$MEK + nC_7$	0.50	1.8	0.60	1.28
$MEK + nC_{10}$	0.60	1.0	0.65	1.45
$Ac + nC_6$	0.50	3.5	0.60	1.28
$Ac + nC_{12}$	0.60	4.7	0.72	2.10

7 Acknowledgments

This work was supported in part by CICYT (Spain) under grant PB86-0159 , and by Univ. Complutense under a grant for *Grupos Precompetitivos*.

8 References

Allen, M.P. and D.J. Tildesley, 1987. Computer Simulation of Liquids, Clarendon Pres, Oxford.

Andreolli-Ball, L. , M. Costas, D. Patterson, R.G. Rubio, R.M. Masegosa, and M. Cáceres , 1989. Solution nonrandomness and the W-shape concentration dependence of C_p^E for hexafluorobenzene and chloronaphtalene in hydrocarbons. Ber. Bunsenges. Phys. Chem.; 93,882-887.

Bathia, A.B. and D.E. Thornton, , 1970. Structural aspects of the electrical resistivity of binary alloys. Phys. Rev. B; 2, 3004-3012.

Ben Naim, A., 1974. Water and Aqueous Solutions. Plenum, N.Y.

Ben Naim, A., 1977. Inversion of the Kirkwood-Buff theory of solutions. Applications to the water-ethanol system. J.Chem.Phys.; 67,4884-4890.

Ben Naim, A., 1989. Preferential solvatation in two-component systems. J. Phys. Chem.; 93,3809-3813.

Bohn, J., J. Fischer, and F. Kohler, F., 1986. Prediction of excess properties for liquid mixtures: results from perturbation theory for mixtures with linear molecules. Fluid Phase Equilibria; 31,233-252.

Chandler, D., 1982. The Liquid State Fluids, Simple and Complex . Montrol,E.W. and Lebowitz, J.L.,Eds., North Holland, Amsterdam.

Flory, P., 1971. Principles of Polymer Chemistry. Cornell Univ. Press, Ithaca, N.Y.

Gaw, W.J. and F.L. Swinton, 1968. Thermodynamic properties of binary systems containing hexafluorbenzene. Trans. Faraday Soc.; 64,2023-2034.

Gierycz, P. and N. Nakanishi, 1984. Local composition in binary mixtures of Lennard-Jones fluids with different sizes of components.Fluid Phase Equilibria; 16,255-273.

Gray, C.G. and K.E. Gubbins, 1984. Theory or Molecular Fluids. Vol 1, Clarendon Press, Oxford.

Guggenheim, E.A., 1952. Mixtures. Clarendon Press, Oxford.

Hansen, J.P.E. and I.R. McDonald, 1986. Theory of Simple Liquids . 2nd ed. , Academic Press, London.

Hoheisel, C. and F. Kohler, 1984. Local composition in liquid mixtures. Fluid Phase Equilibria; 16,13-24

Hu, Y., E.G. Azevedo, and J.M. Prausnitz, 1983. The molecular basis for local composition in liquid mixtures models. Fluid Phase Equilibria; 13,351-360.

Joarder, R.N. and M. Silbert, 1985. Concentration fluctuations in binary mixtures of model polar fluids. Chem.Phys.; 95,357-363.

Kato, T., T. Fujiyama and H. Nomura, 1982. Estimation of parameters G_{11}, G_{22} and G_{12} in the Kirkwood-Buff solution theory on basis of concentration fluctuation data obtained from Rayleigh scattering. Bull. Chem. Soc. Japan; 55,3368-3372.

Kirkwood, J.C. and F.J. Buff, 1951. The statistical mechanical theory of solutions. J.Chem.Phys.; 19,774-777

Kojima, K., T. Kato, and H. Nomura, 1984. Estimation of the parameters in the Kirkwood-Buff theory of solutions using Percus-Yevick fluid mixtures. J.Solution Chem.; 13,151-165.

Lainez, A., G. Roux-Desgranges, J-P.E. Grolier, and E. Wilhelm, 1985. Mixtures of alkanes with polar molecules showing internal rotation: An unusual composition dependence of C_p^E of 1,2-dichloroethane + an alkane.

Fluid Phase Equilibria; 20,47-56.
Lee, K.H., M. Lombardo, and S.I. Sandler, 1986. The generalized van der Waals partition function. II. Application to the square-well fluid. Fluid Phase Equilibria; 21,177-196.
Lee, K.H., S.I. Sandler, and P.A. Monson, 1986. Local composition and the square-well fluid. Int. J. Thermophys.; 7,367-379.
Lee,K.H. and S.I. Sandler, 1987. The generalized van der Waals partition function. IV. Local composition models for mixtures of unequal-sized molecules. Fluid Phase Equilibria; 34,113-147.
Mansoori, G.A., N.F. Carnahan, K.E. Starling, and T.W. Leland , 1971. Equilibrium thermodynamic properties of the mixture of hard spheres. J.Chem.Phys.; 54,1523-1525.
Mansoori,G.A. and J.F. Ely, 1984. Statistical mechanical theory of local compositions. Fluid Phase Equilibria; 22,253-275.
Masegosa, R., M.G. Prolongo, I. Hernandez Fuentes, and A. Horta, 1984. Preferential and total sorption of polymethyl methacrylate in the cosolvent formed by acetonitrile with penthyl acetate and with alcohols (1-butanol, 1-propanol, and methanol.) Macromolecules; 17,1181-1187.
Matteoli, E. and L. Lepori, 1984. Solute-solute interactions in water. II. An analysis through the Kirkwood-Buff integrals for 14 organic solutes. J.Chem.Phys.; 80,2856-2863.
Nishikawa,K., Y. Kodera, and T. Iijima, 1987. Fluctuations in the particle number and concentration and the Kirkwood-Buff parameters of tert-butyl alcohol and water mixtures studied by small-angle X-ray scattering. J. Phys. Chem.; 91,3694-3699.
Panayiotou, C. and J.H. Vera, 1980. The quasi-chemical approach for non-randomnes in liquid mixtures. Expressions for the local surfaces and local compositions with an application to polymer solutions. Fluid Phase Equilibria; 5,55-80.
Pfund, D.M., L.Ll. Lee and H.D. Cochran, 1988. Application of the Kirkwood-Buff theory of solutions to dilute supercritical mixtures. II. The excluded volume and local composition models. Fluid Phase Equilibria; 39,161-192.
Pouchly, J., 1989. Sorption equilibria in ternary systems : polymer/mixed solvent. Pure Appl. Chem.; 61,1085-1095.
Prausnitz, J.M., 1985. Equations of state from van der Waals theory : The legacy of Otto Redlich. Fluid Phase Equilibria; 24,63-76.
Prausnitz, J.M., R.N. Lichtenthaler, and E.G. Azevedo, 1986. Molecular Thermodynamics of Fluid Phase Equilibria. 2nd ed., Prentice Hall, Englewood Cliffs, N.J.
Renon, H., 1985. NRTL : An empirical equation or an inspiring model for fluid mixtures properties?. Fluid Phase Equilibria; 24,87-114.
Rubio, R.G., M.G. Prolongo, U. Cabrerizo, , M. Diaz Peña and J.A.R. Renuncio, 1986. Kirkwood-Buff integrals in non-electrolyte solutions. An evaluation of the local composition from experimental data. Fluid Phase Equilibria; 26,1-13.
Rubio, R.G., M.G. Prolongo, M. Diaz Peña, and J.A.R. Renuncio, 1987. Local composition in real mixtures of simple molecules. J. Phys. Chem.; 91,1177-1184.
Rubio, R.G., M. Cáceres, R.M. Masegosa, L. Andreolli-Ball, M. Costas, and D. Patterson, 1989. Mixtures with "W-shape" C_p^E curves. A light scattering study. Ber. Bunsenges. Phys. Chem.; 93,48-56.

Ruppersberg, H., 1987. Neutron diffraction investigation of liquid Li/Ba alloys. Phys. Chem. Liq.; 17,73-89

Saint-Victor, M.E. and D. Patterson, 1986. The W-shape concentration dependence of C_p^E and solution non-randomness. Fluid Phase Equilibria; 35,237-252.

Sandler, S.I., 1985. A basis for equation of state mixing rules. Proc. 2nd CODATA Symp. Critical Evaluation of Phase Equilibria in Multicomponent Systems, Paris (Lecture 3M-3, pg. 342-351).

Tronel-Peyroz, E., J.M. Douillard, R. Bennes and M. Privat, 1989. Application of the Kirkwood-Buff theory of solutions to the surface phase between the water-ethanol binary system. Langmuir; 5,54-59.

Wilhelm, E., 1985. Molecular thermodynamics of flexible-molecule fluids. Thermochimica Acta; 94,47-66.

Zaitsev, A.L., V.E. Petrenko, and Yu.M. Kesler, 1989. Solution structure and Kirkwood-Buff Theory: Informativity and sensitivity to specific interactions. J. Solution Chem.; 18,115-130.

Interactions in Mixtures Studied through the Kirkwood-Buff Theory

Enrico Matteoli and Luciano Lepori
Istituto di Chimica Quantistica ed Energetica Molecolare del CNR
Via Risorgimento 35, 56126 Pisa, Italy

ABSTRACT
The equations relating the Kirkwood-Buff (K-B) integrals, G_{ij}, to measurable thermodynamic quantities are reviewed and operative equations for calculating these quantities for binary and ternary mixtures are reported. The role and usefulness of the K-B integrals for a better understanding of the microscopic structure of solutions and the type of interactions involved are discussed with the aid of examples of the G_{ij} behaviour for some published as well as new systems.

INTRODUCTION
The availability of powerful computers has allowed the applications of the methods based on quantum mechanics, molecular dynamics, and Monte Carlo simulations to very complex problems encountered in the study of solutions and mixtures, yet the prediction of the radial distribution functions, g_{ij}, or of thermodynamic properties such as activity coefficients, γ_i, molal volumes, V_i, isothermal compressibility coefficients, κ_T, of real mixtures is still far away.

The knowledge of g_{ij} would provide the fundamental tool for the application of the Kirkwood-Buff (K-B) theory (Kirkwood and Buff, 1951), perhaps the most rigorous and general among theories of mixtures based on statistical thermodynamics. This theory in principle can be used to calculate the above thermodynamic properties even for complex mixtures such as aqueous solutions, thanks to a set of equations which explicitly correlate γ_i, V_i and κ_T to the integrals, G_{ij}, of the radial distribution functions; unfortunately, these latter fundamental quantities are not known even for very simple real mixtures, thus the theory is practically not yet applicable in this direct manner.

Some years ago, Ben-Naim (1977) was able to show that the relationships of the K-B theory between G_{ij} and thermodynamic properties could be inverted to get equations which allowed to calculate G_{ij} from experimental quantities, in this way opening a new road for the application of this theory. He emphasized the importance of this additional tool for the investigation and understanding of the structure of solutions, which comes from the fact that G_{ij} quantities, defined as

$$G_{ij} = \int_0^\infty \left(g_{ij} - 1\right) 4\pi r^2 dr \qquad (1)$$

convey information on the local environment of molecules, or, in other words, on the tendency of molecules j to be more or less concentrated than the average concentration in the entire sorroundings of species i taken as center. This can be realized by considering that the product $c_i\ (g_{ij}-1)4\pi r^2 dr$, where c_i is the molar concentration of species i, gives the excess (or deficiency) number of moles of i in the spherical crown at distance r and thickness dr from the central molecule with respect to the case when the center is taken at random; the product $c_i G_{ij}$ gives the average excess of i in the entire sorroundings of j. G_{ij} can be considered a sort of "normalized" measure of tendency of species i to be attracted by species j.

This type of approach for the study of interactions in mixtures presents the advantages of being applicable in the whole range of complete miscibility, of giving independent information on i-i, i-j and j-j interactions, and of providing the possibility to calculate, under certain approximations, the values of the local concentration of a given species around a given particle. As it will be evident, the approach described in this chapter relies on the examination and comparison of the G_{ij} behaviour of different systems and its limitations are substantially due to its qualitative nature and to the uncertainties which usually affect G_{ij} results (Matteoli and Lepori, 1984; Zaitsev et al, 1989).

Besides the above application, G_{ij} quantities also provide a link between thermodynamic data and X-ray, neutron or light scattering experiment results, and may turn out helpful for testing calculations of g_{ij} by theoretical methods. Recently, they have been used to test mixture models based on the local composition concept, to calculate the direct correlation function integrals and to develop a theory of preferential solvation. All these aspects are reviewed in other chapters of this book.

In the past years we have carried out a systematic determination of γ_i and V_i of a variety of systems (Lepori and Matteoli, 1986; Matteoli and Lepori, 1986; Lepori et al, 1988; Berti et al, 1989) from which we have calculated G_{ij} values for a number of mixtures selected in order to allow proper correlations between G_{ij} and molecular structure of components (Matteoli and Lepori, 1982; Matteoli and Lepori, 1984; Lepori and Matteoli, 1988). In this paper, we discuss the procedure for obtaining G_{ij} from experimental quantities, and report on the most interesting results of our work on G_{ij} of aqueous and non-aqueous mixtures, giving particular emphasis to the information on the microscopic structure of solutions which can be gained by examining G_{ij} data.

DERIVATION OF WORKING EQUATIONS AND TREATMENT OF EXPERIMENTAL DATA

According to Kirkwood and Buff (1951), the fundamental equation relating G_{ij} to thermodynamic quantities is the following (in the molar scale):

$$G_{ij} = \frac{1}{c_i c_j} \frac{A^{ij}}{|A|} - \frac{\delta_{ij}}{c_j} \tag{2}$$

where $c_i = N_i/V$, $V = \Sigma N_i V_i$;V is the volume of the sistem, N_i the number of moles of the ith species, δ_{ij} the Kronecker delta; the matrix element A_{ij} (whose cofactor in Equation (2) is indicated as A^{ij}) is given by:

$$A_{ij} = \frac{V\mu_{ij}}{RT} + \frac{V_i V_j}{RT\,\kappa_T} \tag{3}$$

In Equation (3), V_i is the partial molar volume, κ_T the isothermal compressibility coefficient, and μ_{ij} is defined as follows:

$$\mu_{ij} = \left(\frac{\partial \mu_i}{\partial N_j}\right)_{T,P,N_i}, \tag{4}$$

μ_i being the chemical potential in the rational scale.

In principle, by using the generalized Gibbs-Duhem equation, $\mu \cdot \mathbf{c} = \mathbf{0}$, (where μ is the matrix with element μ_{ij} and $\mathbf{c}$ is the concentration vector), to relate off-diagonal elements to the diagonal ones, and provided analytical expressions for $\mu_i = \mu_i(\mathbf{c})$, $V = V(\mathbf{c})$ and $\kappa_T = \kappa_T(\mathbf{c})$ are available (and reminding that matrices $\mathbf{G}, \mathbf{A}$ and μ are symmetrical), all matrix elements in Equations (2)-(4) can be calculated.

For a binary system, analytical expressions for all G_{ij}'s can be easily obtained. In fact, the Gibbs-Duhem equation implies that:

$$\mu_{22} = \mu_{11}\frac{c_2^2}{c_2^2} \quad \text{and} \quad \mu_{12} = \mu_{21} = -\mu_{11}\frac{c_1}{c_2} \tag{5}$$

If the function $\mu_i = \mu_i^\circ + RT\ln\gamma_i x_i$ is used for the chemical potential (x is the molar fraction) μ_{11} is given by Equation (6):

$$\mu_{11} = \left(\frac{\partial \mu_1}{\partial N_2}\right)_{T,P,N_1} = \left(\frac{\partial \mu_1}{\partial x_1}\right)_{T,P} \frac{N_2}{(N_1+N_2)^2} = \frac{x_2}{x_1}\frac{RT}{N_1+N_2}\left(1 + x_1\frac{\partial \ln\gamma_1}{\partial x_1}\right) \tag{6}$$

By substitution of the two previous equations into Equation (3) and indicating vith V* the volume per mole of mixture ($V^* = x_1V_1 + x_2V_2$, $c_i = x_i/V^*$) we obtain the following expression for the determinant of **A**:

$$|\mathbf{A}| = \frac{V\mu_{11}}{\kappa_T (RT)^2 c_2^2} = \frac{D_\mu V^{*3}}{RT\,\kappa_T x_1 x_2} \tag{7}$$

and finally, substituting Equations (7) and (3) into Equation (2):

$$G_{ij} = RT\kappa_T - V_i V_j / D_\mu V^* \quad (i \neq j) \tag{8}$$

$$G_{ii} = G_{ij} + (V_j/D_\mu - V^*) / x_i \tag{9}$$

$$D_\mu = 1 + x_i \left(\frac{\partial \ln \gamma_i}{\partial x_i} \right)_{T,P} \tag{10}$$

The limiting value of G_{ii} as $x_i \rightarrow 0$, G_{ii}°, is related to the pairwise virial coefficient of the McMillan-Mayer expansion of the osmotic pressure, B_2^*, (Matteoli and Lepori, 1982) according to the following equation (L is Avogadro's number and V_j^* the molar volume of j):

$$G_{ii}^o = -\,2\,LB_{2,i}^* = RT\kappa_{T,j} + V_j^* - 2\,V_i^o - V_j^* \left(\partial \ln \gamma_i / \partial x_i \right)_{x_i \rightarrow 0} \tag{11}$$

The final Equations (8) and (9) (and Equation (11) for G_{ii}°) are very convenient for the calculation of the G_{ij}'s of a binary system. The experimental quantities required are: 1) precise activity data so that by proper fitting it may be possible to obtain good analytical expression of $\ln\gamma = g(x)$, from which the quantity D_μ is obtained by differentiation; 2) good density data which allow to obtain an analytical expression of $V^* = f(x)$, from which by standard thermodynamics, V_1 and V_2 are obtained; 3) compressibility data; these latter quantities, if not available, can safely be estimated from the compressibility of the pure components or from correlation methods, because the contribution of the term $RT\kappa_T$in Equation (8) is usually negligible.

For ternary systems, the derivation of explicit equations similar to those for binary mixtures would be quite cumbersome. The recommended procedure would be to calculate all matrix elements A_{ij} and μ_{ij} and to directly apply Equations (2)-(4). In the appendix we report useful expressions for these calculations.

It should be pointed out that in treating activity data caution must be exercised in order to use proper fitting equations; in particular, these must be thermodynamically consistent and the number of adjustable parameters should not be too high with respect to the precision of the experimental data. All this would cause systematic errors on the derivative D_μ and consequently on G_{ij}. Even excluding this type of errors, when the value of D_μ approaches zero, as it happens in mixtures with strong deviation from ideality, uncertainties as high as 30-40% may propagate on G_{ij} values. For a thorough discussion of all sources of errors see Matteoli and Lepori (1984).

USE OF K-B INTEGRALS FOR INVESTIGATING THE STRUCTURE OF MIXTURES

In our first paper on this subject, (Matteoli and Lepori, 1982) we determined B_2* values of a variety of solutes in different solvents and then extended this study on the infinite dilution behavior of G_{ii} to the calculation of the G_{ij} values of a group of 14 binary mixtures of water with monofunctional as well as polyfunctional solutes (Matteoli and Lepori, 1984). Afterwords, G_{ij} data were computed for selected non-aqueous systems, and also the pressure coefficient of G_{ij} quantities for a number of cases was determined (Lepori and Matteoli, 1988).

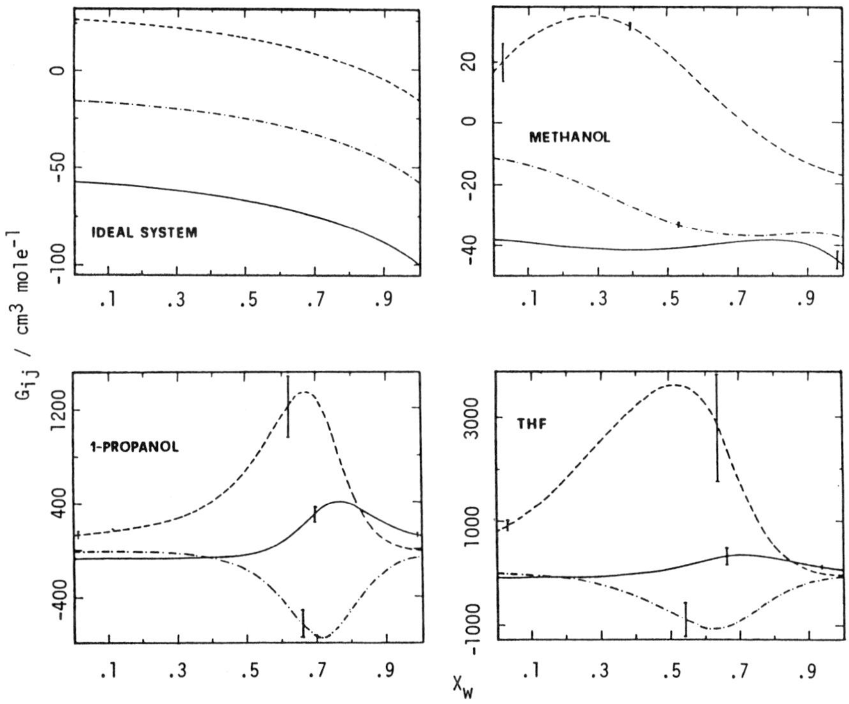

Figure 1. Plot of G_{ij} against x_w, the molar fraction of water, for three aqueous binary mixtures and a hypothetical ideal system. (- - -), G_{11}; (-·-·-), G_{12}; (——) G_{22}. Index 1 indicates water. T=298.15 K.

G_{ij} and molecular structure

As a general behaviour, it has been found that in mixtures of water (component 1) with polyfunctional or with small-size monofunctional compounds the trend of all three G_{ij} functions with respect to x approaches the monotonic behaviour characteristic of a symmetrical ideal system; for mixtures containing compounds with a prevailing hydrocarbon-like moiety, all three G_{ij}'s show extrema: G_{11} and G_{22} present more or less high maxima at about x_1=0.6-0.8, and G_{12} shows a minimum in the same range of concentration.

In Figure 1 are represented some examples illustrating the above described behaviour; by looking at the shapes as well as at the values of all G_{ij}'s, it is recognised that methanol+water system is quite different from mixtures of water with propanol or tetrahydrofuran (THF); moreover, the extrema of the G_{ij}'s of water+THF are larger in absolute value than those for water+propanol.

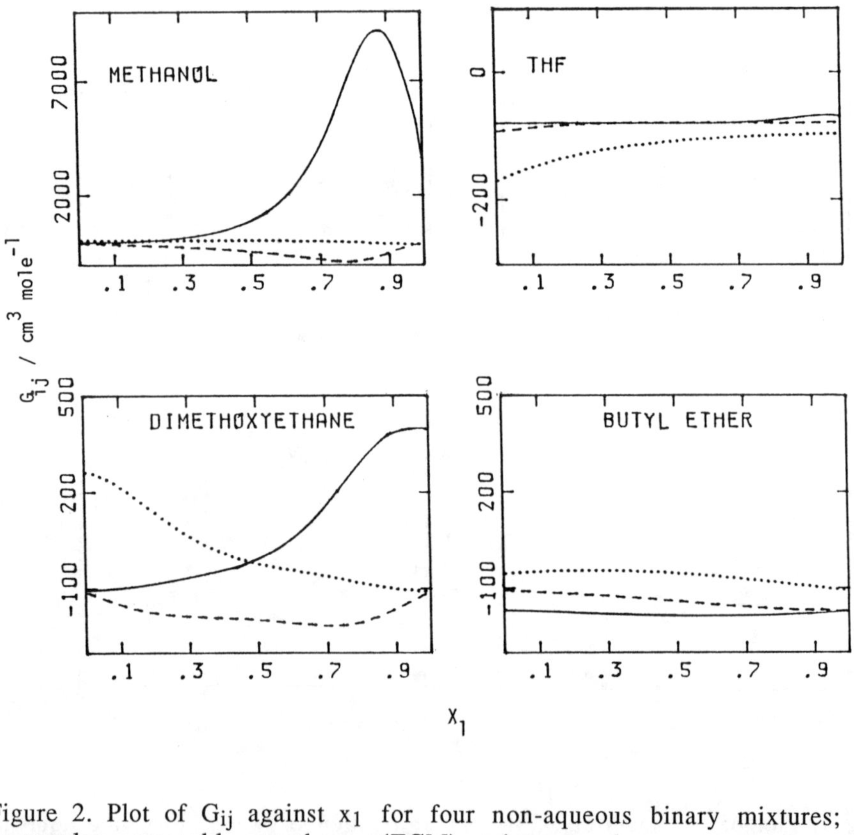

Figure 2. Plot of G_{ij} against x_1 for four non-aqueous binary mixtures; upper rectangles, tetrachloromethane (TCM) mixtures; lower rectangles, cyclohexane mixtures. (· · ·), G_{11}; (- - -), G_{12}; (——) G_{22}. Index 1 refers to TCM or cyclohexane. T=298.15 K.

The above behaviour, confirmed by other systems, suggests that in mixtures of water and hydrophobic solutes, these molecules are sorrounded in preference by particles of the same species, and at the same time, also water, more than the solute, shows this tendency to homocoordination.

In Figure 2 are shown some examples of the G_{ij} behaviour in non-aqueous systems. In the case of tetrachlorometane (TCM, component 1) containing mixtures, a behaviour opposite to the corresponding aqueous systems can be recognised: G_{ij} values of TCM + methanol show large extrema, whereas those of TCM + THF show monotonic trends. The two cyclohexane (component 1) containing mixtures (Matteoli and Lepori, 1989a) show monotonic trends, and the higher values for G_{22} of dimethoxyethane in the range $0<x_1<0.3$ are evidence of tendency to stronger ether-ether interactions.

The effect on G_{ij} behaviour of increasing the chain length of the organic co-solvent is shown in Figure 3, where striking differences between aqueous

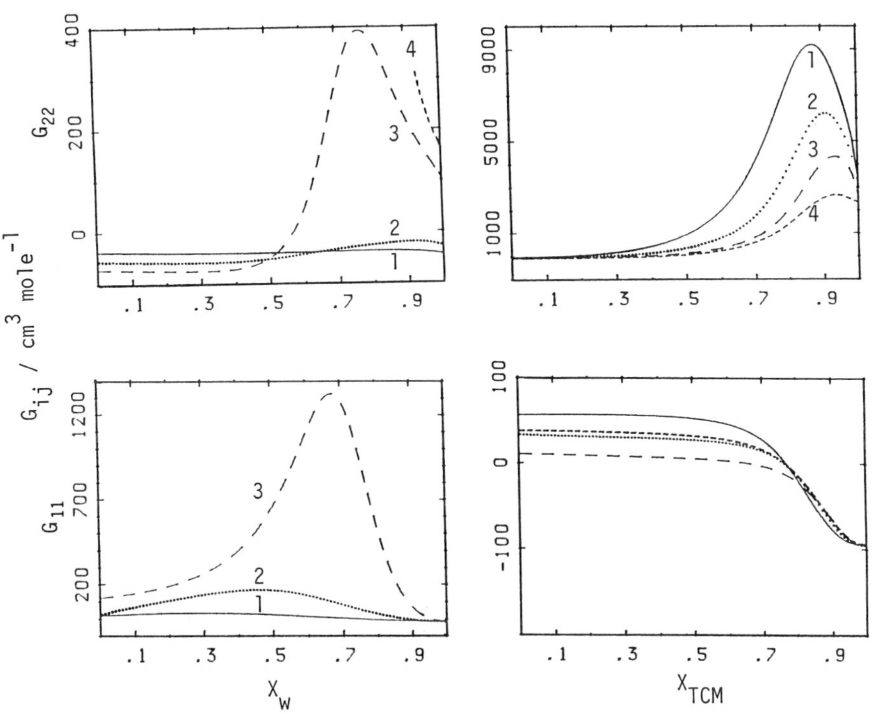

Figure 3. Comparison among G_{22} (upper rectangles) and G_{11} (lower rectangles) for different systems. Lhs., aqueous mixtures; rhs., TCM systems; x is the molar fraction. Index 1 refers to water or TCM; the inside numbers indicate: (1), methanol, (2), ethanol, (3), 1-propanol, (4), 1-butanol. T=298.15 K.

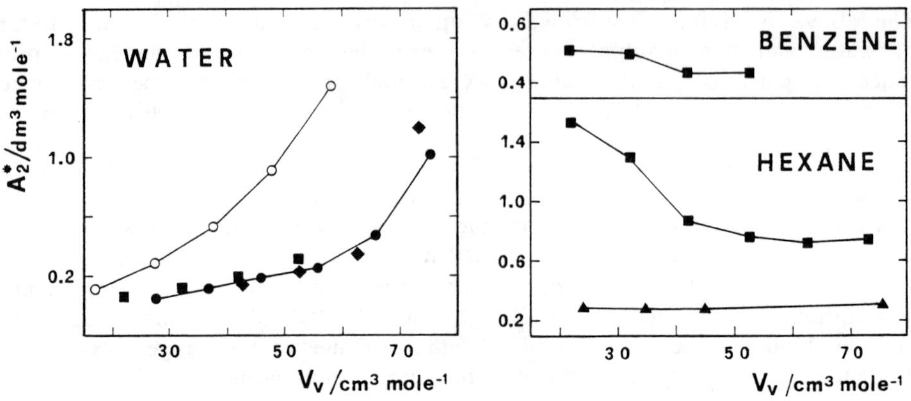

Figure 4. Attractive component $A_2{}^*$ of the MacMillan Mayer virial coefficient of homologous series of solutes in different solvents (indicated) vs. the solute van der Waals volume, V_V. (○), hydrocarbons in the gas phase; (■), primary alcohols, C_1-C_4; (◆), secondary alcohols, C_4-C_6; (●), cyclic imines, C_2-C_7; (▲), aliphatic linear amines, C_1-C_3, C_6. T=298.15 K.

and non-aqueous systems are evident. In aqueous mixtures, G_{11} and G_{22} increase with increasing the hydrocarbon-like portion of the molecules, both for systems containing alcohols and ethers; in TCM mixtures, on the contrary, G_{22} strongly decreases and G_{11} practically remains unchanged in going from lower to higher alcohols, and for ether containing mixtures (not shown) both G_{11} and G_{22} are practically constant. These findings on the whole suggest that the observed tendency to homocoordination of large-size monofunctional alcohols and ethers in water is due to interaction between hydrocarbon-like moieties and to a concomittant change of the water structure as evidenced by G_{11} maxima; in TCM containing mixtures, the tendency to association of alcohols is to be ascribed to OH-OH interactions; in this case the other component, TCM, does not show any tendency to homocoordination.

The increasing values of G_{ii} for aqueous solutions of monofunctional compounds in water as their hydrophobic chain increases is also observed at infinite dilution. In Figure 4 are reported the values of the attractive component, $A_2{}^*$, of $B_2{}^*$ for different series of homologous solutes in water and in organic solvents (Matteoli and Lepori, 1982) obtained from Equation (12) (V_v is the van der Waals volume):

$$A_2{}^* = 4\,V_v - LB_2{}^* \qquad (12)$$

The figure shows that, independently of the functional group, $A_2{}^*$ values of solutes in water are larger the larger the size of the solute, but they are practically constant in non-aqueous solvents.

Calculation of local compositions

By reasonably assuming that the contribution to the integral in Equation (1) becomes negligible at distances from the central molecule larger than about 4 or 5 molecular diameters, it has been possible to calculate the values of the local concentrations (Matteoli and Lepori, 1988); of course, these values are meaningful when used to make comparisons among different systems, less so as values in se, but certainly they help give a quantitative picture of the effect of molecular interactions on the local environment.

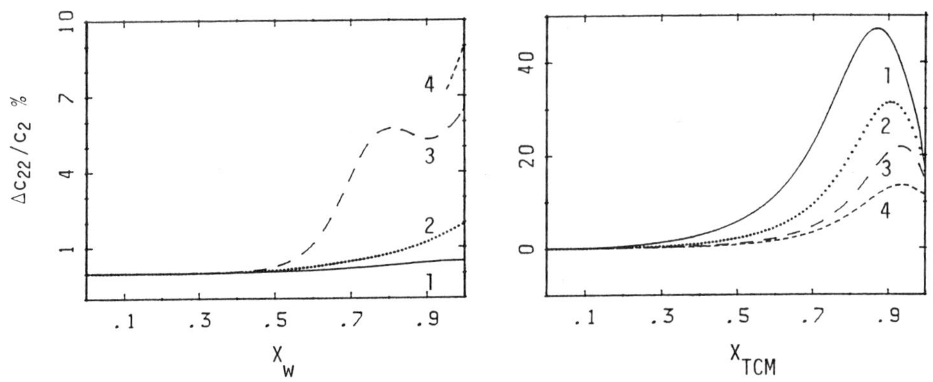

Figure 5. Plot vs. molar fraction of $\Delta c_{22}/c_2$ %, the relative change of molar concentration of species 2 (alcohol) around a given alcohol molecule, with respect to its bulk concentration c_2. Other explanations as in Figure 3.

In Figure 5 are reported the values of $\Delta c_{ij}/c_i$, the relative difference between the local concentration and the average concentration, for some of the previously considered mixtures. The trends are obviously similar to the G_{ij} from which they are calculated, except for aqueous alcohols, which show higher values in the dilute region. The figure shows that the strong tendency of alcohols to associate in TCM solution at x_{TCM} = 0.85 brings about a concentration of alcohol molecules inside a sphere of radius about 4 molecular diameters centered at a given methanol particle much larger than the average concentration, f.i., for methanol by about 45%.

Dependence of G_{ij} on T and P

Whereas at our knowledge only three cases have been reported on the behaviour with temperature of G_{ij} quantities ($\partial G_{22}/\partial T$ of very dilute solutes is found positive, Matteoli and Lepori, 1984; Huot et al, 1988), the pressure effect on G_{ij} integrals, G'_{ij}, has been reported for a number of different mixtures (Lepori and Matteoli, 1988). In general, it is found that for aqueous systems, to large positive values of G_{ij} correspond large negative values of G'_{ij} and viceversa, whereas in TCM solutions G'_{ij} values are commonly very small in absolute value (see Figure 6). This means that in aqueous solutions

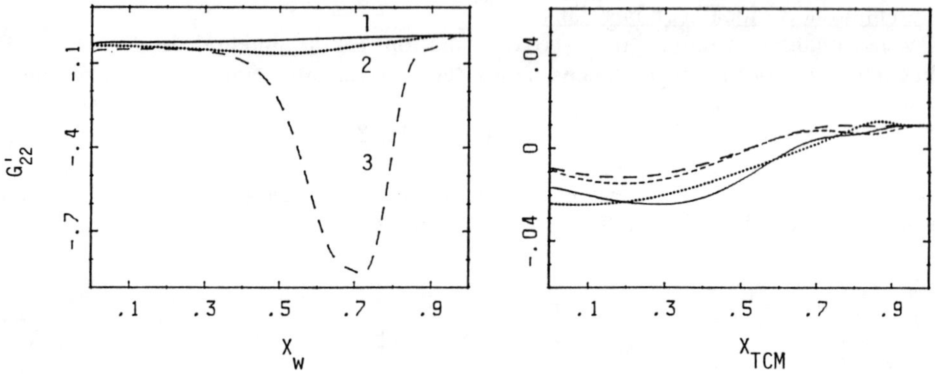

Figure 6. Pressure coefficients of G_{22}, G'_{22}, (mm^3 $mole^{-1}$ hPa^{-1}) vs. molar fraction of water, x_W, or of TCM, x_{TCM} (index 1). Other details in Figure 3.

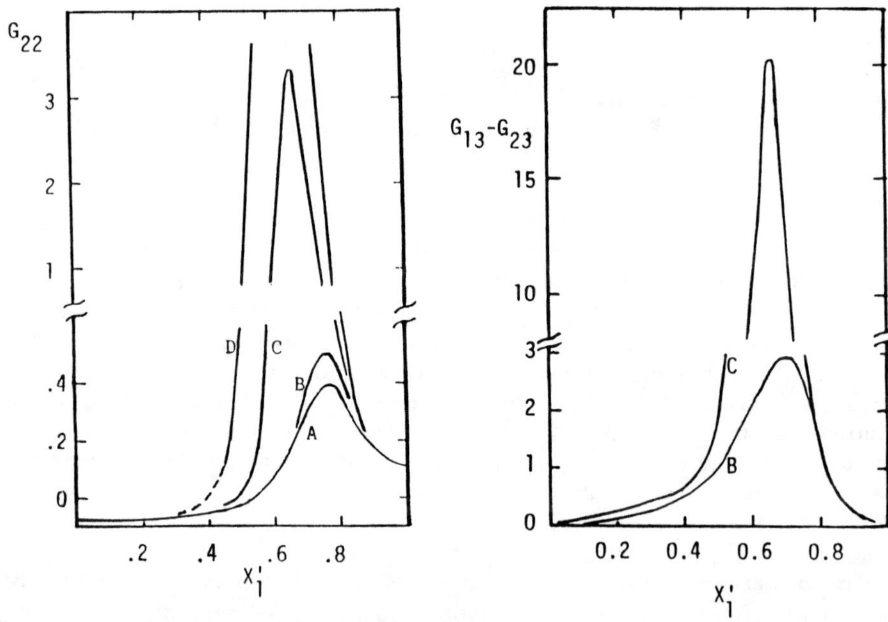

Figure 7. Plots of G_{22} and of the difference G_{31}-G_{32} for the ternary system water(1) + 1-propanol(2) + urea(3), at different values of x_3. $x'_1=x_1/(x_1+x_2)$. (A), x_3=0; (B), x_3=0.01; (C), x_3=0.05; (D), x_3=0.09. T=298.15 K.

an increase of pressure brings about a decreased strength of associative interactions among molecules of the same species, so that we can estimate that the extrema shown by all G_{ij} parameters are strongly reduced at pressures as high as 100 MPa; in the TCM-containing systems the behaviour observed is practically insensitive to pressure.

G_{ij} results for the ternary system water+propanol+urea

By using activity and density data determined in our laboratory for the ternary system water(1)+propanol(2)+urea(3), through Equations A1-A10 we have calculated all G_{ij} functions for this system (Matteoli and Lepori, 1989). By looking at Figure 7, where some of these results are reported, it can be seen that addition of urea brings about a strengthening of the alcohol-alcohol affinity, which is very high in correspondence to the maxima. This is due probably, as the large values of the difference G_{31}-G_{32} suggest, to the much larger affinity of urea for water than for propanol: the addition of urea to water+propanol mixtures promotes subtraction of water molecules from the solvation sphere of the hydrophobic moiety of propanol, favouring interactions through direct contact among them.

APPENDIX

Calculation of G_{ij} quantities for a ternary system

As said above, for a ternary system it is more convenient to use Equations (2)-(4) directly to calculate G_{ij} values. The following equations, for which only four derivatives of the chemical potentials are needed, can be used to calculate diagonal elements μ_{ii} from known expressions of $\ln\gamma_1=f(x)$ and $\ln\gamma_2=g(x)$:

$$\mu_{11} = (1-x_1)\left(\frac{\partial\mu_1}{\partial x_1}\right)_{x_2} - x_2\left(\frac{\partial\mu_1}{\partial x_2}\right)_{x_1} \quad (A1)$$

$$\mu_{22} = (1-x_2)\left(\frac{\partial\mu_2}{\partial x_2}\right)_{x_1} - x_1\left(\frac{\partial\mu_2}{\partial x_1}\right)_{x_2} \quad (A2)$$

$$\mu_{33} = x_1\left(\frac{\partial\mu_3}{\partial x_1}\right)_{x_2} + x_2\left(\frac{\partial\mu_3}{\partial x_2}\right)_{x_1} =$$

$$= \frac{1}{x_3}\left[x_1^2\left(\frac{\partial\mu_1}{\partial x_1}\right)_{x_2} + x_1x_2\left(\left(\frac{\partial\mu_1}{\partial x_2}\right)_{x_1} + \left(\frac{\partial\mu_2}{\partial x_1}\right)_{x_2}\right) + x_2^2\left(\frac{\partial\mu_2}{\partial x_2}\right)_{x_1}\right] \quad (A3)$$

$$\left(\frac{\partial\mu_1}{\partial x_1}\right)_{x_2} = RT\left[\frac{1}{x_1}\left(\frac{\partial\ln\gamma_1}{\partial x_1}\right)_{x_2}\right] \quad (A4)$$

$$\left(\frac{\partial\mu_1}{\partial x_2}\right)_{x_1} = RT\left(\frac{\partial \ln\gamma_1}{\partial x_2}\right)_{x_1} \tag{A5}$$

$$\left(\frac{\partial\mu_2}{\partial x_2}\right)_{x_1} = RT\left[\frac{1}{x_2} + \left(\frac{\partial \ln\gamma_2}{\partial x_2}\right)_{x_2}\right] \tag{A6}$$

$$\left(\frac{\partial\mu_2}{\partial x_1}\right)_{x_2} = RT\left(\frac{\partial \ln\gamma_2}{\partial x_1}\right)_{x_2} \tag{A7}$$

Off-diagonal elements can be calculated by the following relationships, obtained using the Gibbs-Duhem equation:

$$\mu_{12} = \frac{c_3^2\mu_{33} - c_1^2\mu_{11} - c_2^2\mu_{22}}{2c_1c_2} \tag{A8}$$

$$\mu_{13} = \frac{c_2^2\mu_{22} - c_1^2\mu_{11} - c_3^2\mu_{33}}{2c_1c_3} \tag{A9}$$

$$\mu_{23} = \frac{c_1^2\mu_{11} - c_2^2\mu_{22} - c_3^2\mu_{33}}{2c_2c_3} \tag{A10}$$

The partial molal volumes can be obtained by proper differentiation of the function relating the excess molar volume of the mixture V^E to the composition:

$$V_i = V_i^0 + V^E + (1 - x_i)\left(\frac{\partial V_E}{\partial x_i}\right)_{x_j} - x_j\left(\frac{\partial V^E}{\partial x_j}\right)_{x_i} \tag{A11}$$

where i=1,2,3 and j=2,3,1, respectively.

REFERENCES

Ben-Naim, A. 1977. Inversion of the Kirkwood-Buff theory of solutions: Application to water-ethanol system. J. Chem. Phys 67: 4884-4890.

Berti, P., L. Lepori and E. Matteoli. 1989. Excess volumes of cyclo-hexane + a linear ether at 298.15K. Fluid Phase Equil. 44: 285-294.

Hout, J-Y., E. Battistel, R. Lumry, G. Villeneuve, J-F. Lavallee, A. Anusiem and C. Jolicoeur. 1988. A comprehensive thermodynamic investigation of water-ethylene glycol mixtures at 5, 25, and 45°C. J. Solution Chem. 17: 601-636.

Kirkwood, J.G. and F. P. Buff. 1951. The statistical mechanical theory of solutions:I. J. Chem. Phys.19: 774-777.

Lepori, L. and E Matteoli. 1986. Excess volumes of (tetrachloromethane + an alcohol or + a cyclic ether) at 298.15K. J. Chem. Thermodyn. 18: 13-19.
Lepori, L., E. Matteoli and B. Marongiu. 1988. Isothermal liquid-vapour equilibria of mixtures containing organic compounds. 3. Excess Gibbs free energies of cyclohexane + a linear ether. Fluid Phase Equil. 42: 229-240.
Lepori, L. and E. Matteoli. 1988. Solute-solute-solvent interactions studied through the Kirkwood-Buff theory. 3. Effect of pressure on G_{ij} for some aqueous and non-aqueous mixtures. J. Phys Chem 92: 6997-7001.
Matteoli, E. and L. Lepori. 1982. Solute-solute interactions. Pairwise virial coefficients of organic compounds in water and in non-aqueous solvents. J. Phys.Chem 86: 2994.
Matteoli, E. and L. Lepori. 1984. Solute-solute interactions in water. II. An analysis through the Kirkwood-Buff integrals for 14 organic solutes. J. Chem. Phys. 80: 2856-2863.
Matteoli, E and L. Lepori. 1986. Excess Gibbs free energies of (tetrachloromethane + an alcohol or a cyclic ether) at 298.15K. J. Chem. Thermodyn. 18: 1065-1076.
Matteoli, E. and L. Lepori. 1989. Interactions in mixtures studied through the Kirkwood-Buff theory: the ternary system water+1-propanol+urea at 298.15K. Proc. Intern. Confer. on Solution Chemistry. p. 62. Jerusalem, Israel.
Matteoli, E. and L. Lepori. 1989a. Unpublished results.
Zaitsev, A. L., V.E. Petrenko and Yu. M. Kessler. 1989. Solution structure and Kirkwood-Buff theory: informativity and sensitivity to specific interactions. J. Solution Chem. 18: 115-130.

Computer Simulations of a 2-dimensional Aqueous Solution-like System with Large Concentration Fluctuations

Yosuke Kataoka
Department of Chemistry
Kyoto University, Kyoto 606, Japan

ABSTRACT
A Monte Carlo and molecular dynamics study of a 2-dimensional aqueous solution-like system is carried out by making use of the two-dimensional model of water. The static properties are studied over a wide region of temperature, density and concentration. The Kirkwood-Buff parameters are obtained by the integration of the radial distribution functions. The concentration fluctuation is calculated as a function of temperature and concentration. The mean square concentration fluctuation has a maximum as a function of temperature. The mutual diffusion coefficient, the kinetic diffusion coefficient, and the velocity correlation between different molecules are obtained. The reduced kinetic diffusion coefficient has a maximum at an intermediate temperature under a constant pressure. This feature is consistent with the temperature dependence of the mean square concentration fluctuation. The thermodynamic and dynamical properties are, at least qualitatively, in agreement with alcohol-water mixtures.

INTRODUCTION
Some binary mixtures of alcohol and water have the lower critical solution temperature (LCST) (Franks and Reid, 1973), while others show a tendency to have LCST (Rowlinson and Swinton, 1982). The inhomogeneity and the time dependent structure in the microscopic scale are the interesting problems in such solutions (de Visser et al., 1977; Roux et al., 1978; Iwasaki and Fujiyama, 1977,1979; Donkersloot, 1979; Roux et al., 1980; Ito et al., 1981ab; Ito et al., 1983) where the concentration fluctuates to a considerable extent.

Because the molecular interaction in such systems is very complicate, we are going to use computer simulations in order to get microscopic information on solutions in detail. The Monte Carlo (MC) method (e.g. Binder, 1987) is used to get static properties for simplicity, although such properties are also accessible by molecular dynamics (MD) method (e.g. Ciccotti and Hoover, 1986). After the confirmation of an interesting static properties of the model system by MC, dynamical properties will be studied by MD on the basis of the same model.

Usually it costs too much to perform computer simulations on realistic models of aqueous solutions, even if a super computer is available. For this reason we are going to adapt a very simple model to alcohol-water mixture with large concentration fluctuations. This is our first step in a systematic study of binary mixtures of alcohol and water by means of computer simulations on the basis of continuous potentials.

As the solvent, we assume a 2-dimensional (2D) water model, with which the well-known anomalous behavior of liquid water is reproduced at least in a semiquantitative way (Okazaki et al., 1981; Kataoka, 1984a; Kataoka, 1986a). The 2D solute molecule we will study is also a disk-like one which moves around in a plane and at the same time can rotate uniaxially (Kataoka, 1984b). The pair potential between water and solute molecule is anisotropic as shown in the next section.

The merits of low dimension are as follows. (1) We can save computing time and memory size. This point is important because we must perform many simulations of solution at various concentrations, temperatures and densities. (2) The linear dimension of the unit cell in the simulation is larger than that of a 3-dimension ($576=24^2\approx8.3^3$) provided that the total number of molecules is fixed. Thus, the effect of the periodic boundary condition on the local concentration fluctuations would be smaller in 2D case than that in 3D calculation with the same number of molecules.

The static properties of this model solution were studied by Monte Carlo (MC) simulations (Kataoka, 1984b). Dynamical properties were also examined by means of molecular dynamics calculations on the basis of the same model (Kataoka, 1986b). The static and dynamic properties of the model solution will be compared with those of real alcohol-water mixtures qualitatively.

MODEL

A model for binary mixture of alcohol and water is designed on the basis of the following properties. (i) The solute molecule has a stable 'clathrate hydrate type structure' at low temperatures and at optimum concentration due to the anisotropic interaction between solute and water molecules. (ii) The solute molecule has a liquid state of about the same range of temperature and density as 2D water. (iii) The size of the solute molecule is not so different from that of 2D water for the sake of simplicity.

The pair potential U_{AB} between molecules A and B are constructed after 2D water model (Suffix A and B will be used to express water or solute, generally):

$$U_{WW}=U_{LJ}^{WW}(R_{WW})+U_{ANISO}^{WW}(R_{WW},\varphi_{W1},\varphi_{W2}), \tag{2.1}$$

$$U_{WS}=U_{LJ}^{WS}(R_{WS})+U_{ANISO}^{WS}(R_{WS},\varphi_{W},\varphi_{S}), \tag{2.2}$$

$$U_{SS}=U_{LJ}^{SS}(R_{SS}). \tag{2.3}$$

Here, U_{WW} is the water-water interaction energy and is given elsewhere (Okazaki et al., 1981). And R_{AB} denotes the distance between two molecules A and B and φ_A and φ_B the angle of the respective molecules as shown in Figure 1. The first terms in Eqs. (2.1)- (2.3) $U_{LJ}^{AB}(R_{AB})$ are the Lennard-Jones potential functions of the intermolecular distance R_{AB}.

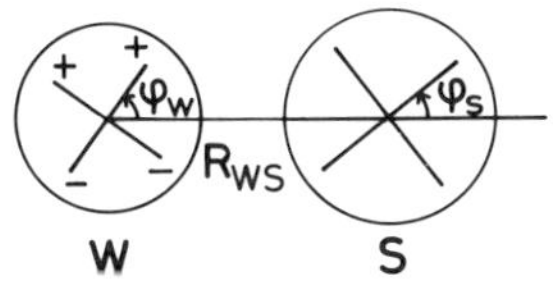

Figure 1. Two-dimensional model of water(W) and solute(S) molecules (Kataoka, 1984b).

The parameters in pair potentials are decided from the properties (ii) and (iii) described at the beginning of this section. The Lorentz-Berthelot rule is referred for the Lennard Jones parameters in U_{LJ}^{WS}. The essential part of the present model is the anisotropic interaction energy U_{ANISO}^{WS} between water and solute, and this is given as follows:

$$U_{ANISO}^{WS}(R_{WS},\varphi_W,\varphi_S)=\varepsilon_{ANISO}^{WS}S(R_{WS};R_1,R_2,R_3)(1-\cos 4\varphi_W)(1-\cos 4\varphi_S)/4. \qquad (2.4)$$

The switching function $S(R;R_1,R_2,R_3)$ is defined in Eq. (2.2) of Okazaki et al., (1981).

In order to realize the 'clathrate hydrate type structure' in 2-dimension, we are forced to use the low density ice type structure in 2D water (i.e. a solute molecule is located interstitially in the cavity of low density type structure of water molecules), although the clathrate hydrate structure is different from the low density ice in the real system. Twenty or more water molecules are located as the first neighbor of the solute in the clathrate hydrate. This seems to be the reason why there are many anomalies at low concentrations ($\fallingdotseq$0.05) in the binary mixtures of alcohol and water. The number of the first neighbor water in the 'clathrate hydrate type structure' is only 4 in our model. As the centered square lattice may be possible in this case (see Figure 5 of Okazaki et al., (1981)), the characteristic concentration may be about 0.5, where anomalies may be seen in our solution. Because of technical reasons the high concentration of 0.5 is preferable for simulations.

Further comments are added to the solute-solute potential. As the present model is spherical, it is too far from the real case. A different potential might affect the value of the Kirkwood-Buff parameters and also dynamic properties. We assume the present simple model as we believe the WW and WS interactions are more important than SS one. Some more realistic model is preferable in future.

As in the 2D water, we shall use the following reduced units. Temperature T is measured in units of (ε_0/k_B) (k_B is the Boltzmann constant), the area per molecule A in units R_0^2, and the pressure P in units of ε_0/R_0^2. Here ε_0(=24.68 kJ/mol) is the strength of water-water interaction at its absolute minimum, and $R_0(=2.76\times10^{-10}m)$ is the length of hydrogen-bond in U_{WW}. The molecular distance is measured in units of R_0.

METHOD OF MONTE CARLO CALCULATION

The number of molecules N_{tot} is 576 in a square unit cell. Periodic boundary conditions are adopted. The initial configurations at temperature T=0.169, 0.202, 0.236 and 0.270ε_0/k_B and at area per molecule A=0.957R_0^2 are made from the pure 2D water sample by the redefinition of the species as solute at random to obtain the aimed concentrations. We use the mole fraction of solute x_S to express the concentration. Then the MC simulations are carried out according to the routine procedure.

The radial distribution functions $g_{AB}(R)$ are obtained by the above MC simulations. The Kirkwood-Buff parameters G_{AB} are calculated as follows (see Figure 2);

$$G_{AB}=\frac{1}{b-a}\int_a^b G_{AB}(R_M)\,dR_M, \tag{3.1}$$

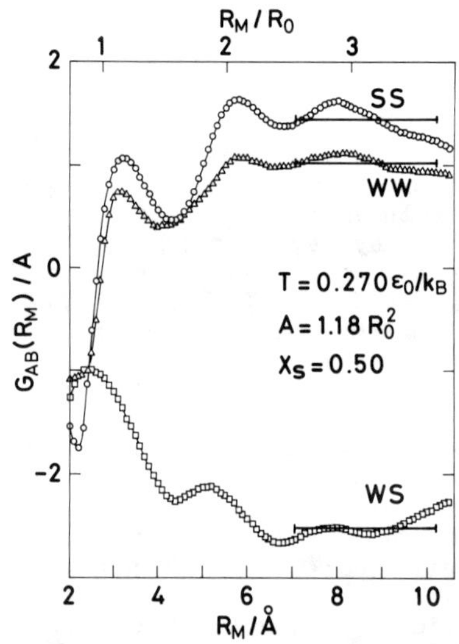

Figure 2. Integral $G_{AB}(R_M)$ vs R_M plots (Kataoka, 1984b)

$$G_{AB}(R_M)\equiv 2\pi\int_0^{R_M}(g_{AB}(R)-1)\,R\,dR, \tag{3.2}$$

$$a=7.1\times10^{-10}m,\quad b=10.2\times10^{-10}m. \tag{3.3}$$

Generally speaking, a and b should satisfy the following inequalities;

$$a\gg A^{1/2}\approx 2.76\times10^{-10}m, \tag{3.4}$$

$$b\ll (N_{tot}A)^{1/2}\approx 66\times10^{-10}m. \tag{3.5}$$

Our choice of a and b is arbitrary and the derived thermodynamic quantities will be checked in the next section.

As Ben-Naim does (Ben-Naim, 1974), it will be convenient to write

$$\eta=\rho_A+\rho_B+\rho_A\rho_B(G_{AA}+G_{BB}-2G_{AB}), \tag{3.6}$$

$$\zeta=1+\rho_A G_{AA}+\rho_B G_{BB}+\rho_A\rho_B(G_{AA}G_{BB}-G_{AB}^2). \tag{3.7}$$

Then, isothermal compressibility κ_T and the mean square concentration fluctuation $N\langle(\Delta x)^2\rangle$ are expressed in terms of G_{AB},

$$\kappa_T=\frac{\zeta}{k_BT\eta}, \tag{3.8}$$

$$N\langle(\Delta x)^2\rangle=x_S x_W A\eta. \tag{3.9}$$

where N is the average number of molecules in the region where the fluctuations are considered and A is the area per molecule.

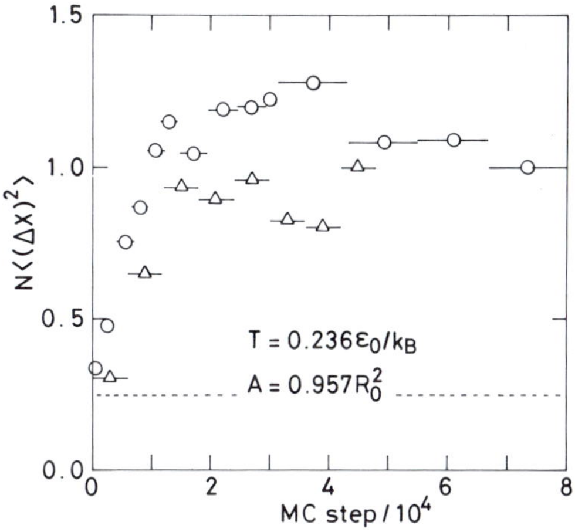

Figure 3. Convergence characteristics of the mean square concentration fluctuation $N\langle(\Delta x)^2\rangle$. One MC step means 576 successive moves. The two runs are shown. The bar means the range where the average is obtained (Kataoka, 1984b).

CONVERGENCE AND CHECK

The convergence characteristic of mean square concentration fluctuation is shown in Figure 3. We call 576 successive moves (or tries) one MC step. About 12000 MC steps seem sufficient for the equilibration in our study. For this reason, the first 12000 MC steps are thrown away and the next 30000 MC steps are usually used for averaging. The average of the difference of the position vectors at the initial MC step and at the final MC step is about $5\times10^{-10}m$. Thus, the mixing in the solution due to the MC moves seems sufficient in the region with the radius 7–$10\times10^{-10}m$.

In Figure 4, the isothermal compressibility κ_T derived from G_{AB} (open triangle) is compared with that calculated from the pressure at different densities (solid triangle). The agreements are beyond our expectation. It has been shown that this quantity has the value of about R_0^2/ε_0 in 2D water from the equation of state (Okazaki et al., 1981).

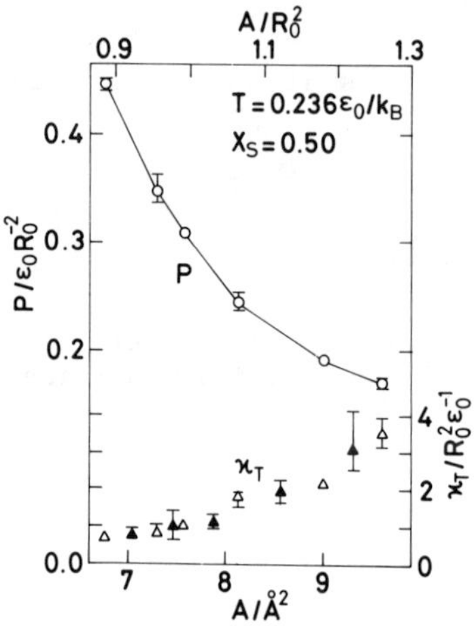

Figure 4. Pressure P vs area per molecule A (upper part) and the isothermal compressibility κ_T vs A plot (lower part). The open triangle means k_T calculated by Eq. (3.8). The solid triangle means k_T obtained from P–A plot (Kataoka, 1984b).

CONCENTRATION FLUCTUATIONS

In this section, it will be shown that the mean square concentration fluctuation of the model solution has a reasonable temperature and concentration dependence.

Figure 5 illustrates the concentration dependence of the mean square concentration fluctuation $N\langle(\Delta x)^2\rangle$, where the area per molecule A is constant at each temperature. (Hereafter in this section, the area A is constant when the concentration x_S is a variable, unless otherwise mentioned.) The concentration fluctuation has the maximum value around x_S=0.50, which is about 1.0 and larger than that of the ideal mixing case (dotted line).

Figure 6 depicts two aspects of the temperature dependence of the mean square concentration fluctuation at x_S=0.50. One of them is the constant area plot (A=0.957R_0^2). The other is the constant pressure plot, P=(0.1±0.1)ε_0/R_0^2. The temperature dependence of $N\langle(\Delta x)^2\rangle$ in Figure 6 is very interesting. In the region T<0.236ε_0/k_B, the mean square concentration fluctuation is an increasing function of temperature. This dependence is well comparable with that in the solution below LCST. Such examples are seen in Iwasaki and Fujiyama, (1979) and Ito et al, (1983). On the other hand, the concentration fluctuation decreases as temperature increases in the region T>0.236ε_0/k_B. This behavior is just the same as that of solution above UCST. In Figure 7 the Kirkwood-Buff parameters G_{AB} are plotted against x_S at T=0.236ε_0/k_B. Note that G_{WW} is maximum around x_S=0.75 and that G_{SS} is maximum around x_S=0.25.

Thus, in this solution similar species have the tendency of getting together. Overall behavior of G_{AB} is well comparable with the experimental results (Ben-Naim, 1974; Ben-Naim, 1977; Kato, 1982).

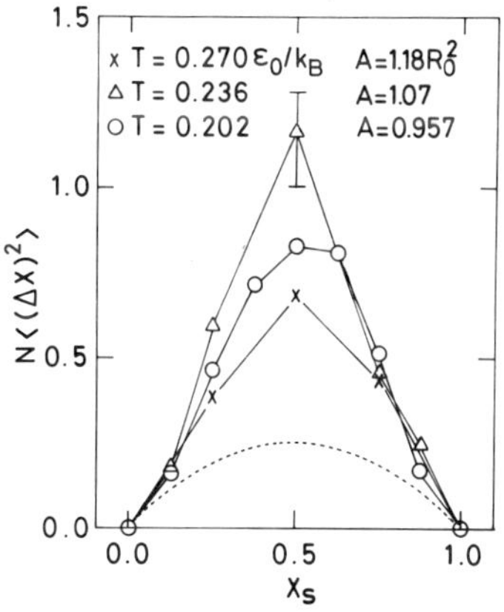

Figure 5. Mean square concentration fluctuation vs concentration x_S plot. The dotted curve means the case of ideal mixing solution (Kataoka, 1984b).

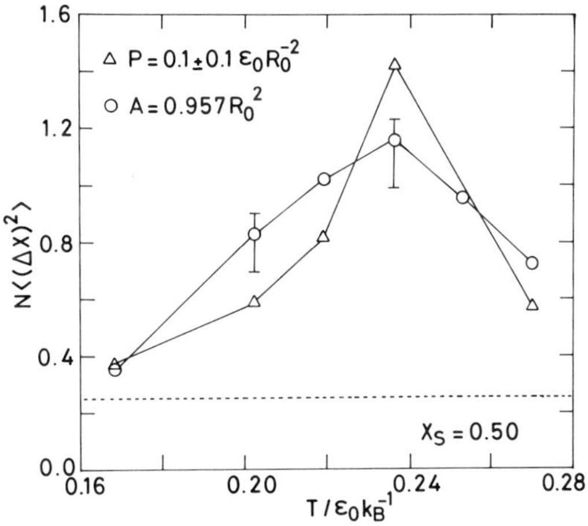

Figure 6. Mean square concentration fluctuation is plotted against temperature T. The dotted line means the case of ideal mixing solution (Kataoka, 1984b).

Chemical potential was obtained by Widom's method and the tendency for phase separation was discussed elsewhere (Kataoka, 1984b).

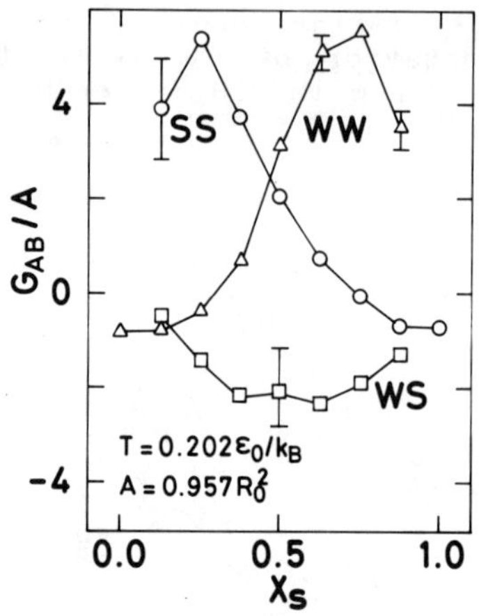

Figure 7. Kirkwood-Buff parameters G_{AB} vs x_S plots.

MUTUAL DIFFUSION COEFFICIENT

The mutual diffusion coefficient is interesting in solutions with local concentration fluctuations. The mutual diffusion coefficient D can be obtained from the mean square displacement of the center of mass of all water molecules $\langle R_{WCM}(t)^2 \rangle$ as follows (Jolly and Bearman, 1980):

$$D=\frac{Q}{4N_{tot}c_W c_S}\left\{\left(\frac{M_W}{M_S}\right) c_W+c_S\right\}^2\lim_{t\to\infty}\frac{\partial}{\partial t}\langle R_{WCM}(t)^2\rangle, \tag{6.1}$$

where c_A is the concentration (number density) of the species A, the velocity of the center of the mass of the system is fixed as 0, N_{tot} is the total number of molecules in the system, and M_A is the mass of the species A. The quantity Q is often called the thermodynamic factor and is defined as:

$$Q=\frac{x_W x_S}{N\langle(\Delta x)^2\rangle}. \tag{6.2}$$

The constant-temperature and constant-pressure MD(T-P-MD) method (Andersen, 1980; Nosé, 1984ab) was used, because this method is more convenient for obtaining the temperature dependence of the system under constant pressure. It was proved numerically that the mutual diffusion coefficient obtained by T-P-MD is not very different from that calculated by micro canonical MD (Kataoka, 1986b).

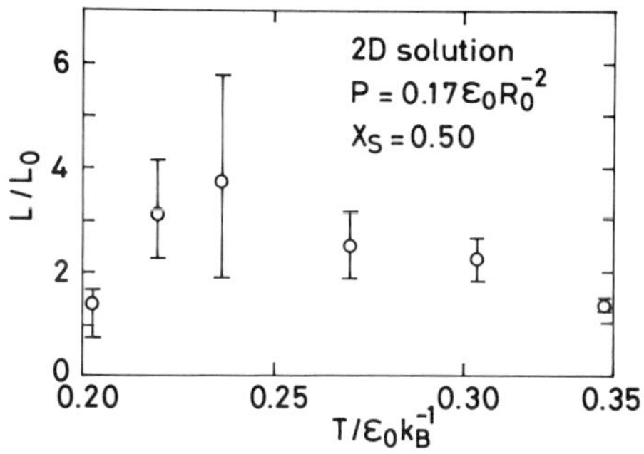

Figure 8. Plot of the reduced kinetic diffusion coefficient L/L_0 vs temperature T, at $P=0.17\varepsilon_0/R_0^2$, $x_S=0.5$ (Kataoka, 1986b).

As the mutual diffusion coefficient D has the thermodynamic factor Q, the kinetic diffusion coefficient L is more directly related to velocity correlation functions between different molecules (McCall and Douglass, 1967; Kato, 1985):

$$D=QL, \tag{6.3}$$

with

$$L=L_0+x_Wx_S(D_{WW}+D_{SS}-2D_{WS}), \tag{6.4}$$

$$L_0=x_SD_W+x_WD_S, \tag{6.5}$$

$$D_{AB}=\frac{N}{2}\int_0^\infty \langle v_{Ai}(t)\cdot v_{Bj}(0)\rangle \, dt, \qquad (i\neq j) \tag{6.6}$$

where v_{Ai} is the velocity of an i-th molecule of the species A. D_A is the self-diffusion coefficient of the species A. By the use of Eq. (6.4), the trend of the cooperative translation in the binary mixture can be seen from the comparison of L with L_0, because the second term in Eq. (6.4) comes from the velocity correlations between different molecules. At the same time, we would like to discuss the temperature dependence of L by means of the reduced quantity L/L_0, since the first term L_0 in Eq. (6.4) also depends on the temperature. The temperature dependence of L/L_0 is shown in Figure 8. The reduced kinetic diffusion coefficient L/L_0 has a maximum value around $T=0.24\varepsilon_0/k_B$. The mean square concentration fluctuation also has a maximum value around the same temperature (Figure 6). The second point to be mentioned about Figure 8 is that the value of L/L_0 is larger than 1.

The experimental values in a binary solution of t-butyl alcohol and water are as follows: L/L_0 = about 2 at 24°C and about 5 at 63°C (Ito et al.,1981b). The order of magnitude of L/L_0 shown in Figure 8 is comparable to the experimental results (Ito et al., 1981b). The temperature dependence below about $0.24\varepsilon_0/k_B$ in Figure 8 is qualitatively in agreement with such results (Ito et al., 1981b).

As the value of L/L_0 is larger than 1, the general tendency for the same species to move in the same direction is seen in this solution. In order to see this more directly, the correlation functions of velocities between the different molecules are shown in Figure 9. The correlation functions of velocities $f_{AB}(t)$ are defined as follows:

$$f_{AB}(t)=\frac{\langle v_{Ai}(t)\cdot v_{Bj}(0)\rangle}{\langle v_{Ai}(0)\cdot v_{Bj}(0)\rangle}, \qquad (i\neq j) \tag{6.7}$$

The normalization constants at temperature $T=0.236\varepsilon_0/k_B$ are given below:

$$\langle v_{Wi}(0)\cdot v_{Wj}(0)\rangle=-0.7\times10^6 m^2 s^{-2}, \tag{6.8}$$

$$\langle v_{Wi}(0)\cdot v_{Sj}(0)\rangle=-0.8\times10^6 m^2 s^{-2}, \tag{6.9}$$

$$\langle v_{Si}(0)\cdot v_{Sj}(0)\rangle=-0.7\times10^6 m^2 s^{-2}, \tag{6.10}$$

From Eqs. (6.7)- (6.10) and Figure 9 it may be seen that D_{WW} and D_{SS} are positive, while D_{WS} is negative. The tendency mentioned above for the same species generally to move in the same direction may be seen again in Figure 9.

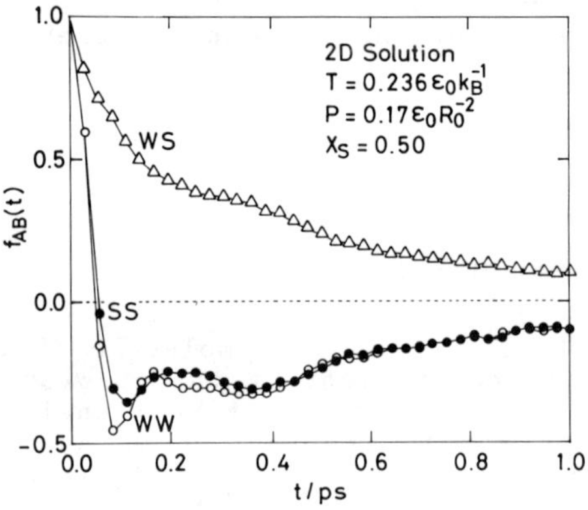

Figure 9. The correlation function of velocity vs time t. $T=0.236\varepsilon_0/k_B$, $P=0.17\varepsilon_0/R_0^2$, $x_S=0.5$ (Kataoka, 1986b).

The lifetime of the local structure is estimated from the cross-correlation function of displacement, which is computed from the time-averaged configurations, where the average period is several times longer than the velocity-autocorrelation time (Kataoka, 1986b). The correlation length of the local concentration fluctuations is obtained and compared with the snapshots of the configurations at several temperatures (Kataoka, 1986b). The dynamical properties are qualitatively in agreement with those of an alcohol-water mixture.

ACKNOWLEDGMENTS

This work is supported in part by a Grant-in-Aid for Scientific Research (Nos. 01540398 and 63430023) from the Ministry of Education, Science and Culture. This is also supported in part by the Japan Society for the Promotion of Science and Japan/U.S. research cooperation. The author thanks the Computer Center of the Institute for Molecular Science for the use of HITAC M-680H and

S-820 /80 computers. The computation is also done by FACOM VP-400E and M-780 computers at Kyoto University Data Processing Center.

REFERENCES

Andersen,H.C. 1980. Molecular dynamics simulations at constant pressure and/or temperature. J. Chem. Phys. 72: 2384-2393.

Ben-Naim,A. 1974. Water and Aqueous Solutions, New York: Plenum.

Ben-Naim.A. 1977. Inversion of the Kirkwood-Buff theory of solutions: Application to the water-ethanol system. J. Chem. Phys. 67: 4884-4890.

Binder, K. 1987. Applications of the Monte Carlo method in statistical physics. 2nd ed., Berlin: Springer.

Ciccotti, G. and W.G. Hoover. 1986. Molecular-dynamics simulations of statistical-mechanical system. Amsterdam: North-Holland.

Donkersloot.M.C.A. 1979. The structure of binary liquids. The Kirkwood-Buff theory of liquids mixtures, illustrated on the basis of the systems water/methanol, water/ethanol, and cyclohaxane/2,3-dimethylbutane, as a link between thermodynamic data and x-ray and neutron scattering results. J. Solution Chem. 8: 293-307.

de Visser.C., G.Perron, and J.E.Desnoyers. 1977. The heat capacities, volumes, and expansibilities of tert-butyl alcohol-water mixtures from 6 to 65 ℃. Can. J. Chem. 55: 856-862.

Franks,F., and D.S.Reid. 1973. Thermodynamic properties. In Water, A Comprehensive Treatise, Vol. 2, ed. F.Franks, 323-380. New York: Plenum.

Ito,N., K.Saito, T.Kato, and T.Fujiyama. 1981. Observation of mutual diffusion coefficients and cooperative motions in binary solutions of t-butyl alcohol-water and 2-butoxyethanol-water. Bull. Chem. Soc. Jpn. 54: 991-997.

Ito,N., T.Kato, and T.Fujiyama. 1981. Determination of local structure and moving unit formed in binary solution of t-butyl alcohol and water. Bull. Chem. Soc. Jpn. 54: 2573-2578.

Ito,N., T.Fujiyama, and Y. Udagawa. 1983. A study of local structure formation in binary solutions of 2-butoxyethanol and water by Rayleigh scattering and Raman spectra. Bull. Chem. Soc. Jpn. 56: 379-385.

Iwasaki.K., and T.Fujiyama. 1977. Light-scattering study of clathrate hydrate formation in binary mixtures of tert-butyl alcohol and water. J. Phys. Chem. 81, 1908-1912.

Iwasaki.K., and T.Fujiyama. 1979. Light-scattering study of clathrate hydrate formation in binary mixtures of tert-butyl alcohol and water. 2. Temperature effect. J. Phys. Chem. 83. 463-468.

Jolly,D.L., and R.J.Bearman. 1980. Molecular dynamics simulation of the mutual and self diffusion coefficients in Lennard-Jones liquid mixtures. Mol. Phys. 41: 137-147.

Kataoka,Y. 1984a. Studies of liquid water by computer simulations. III. Dynamical properties of a 2-D model. Bull. Chem. Soc. Jpn. 57: 1522-1527.

Kataoka,Y. 1984b. A Monte Carlo study of concentration fluctuations in a 2D aqueous solution. J. Chem. Phys. 80: 4470-4477.

Kataoka,Y. 1986a. Studies of liquid water by computer simulations. IV. Transport properties of a 2-D model. Bull. Chem. Soc. Jpn. 59: 1425-1431.

Kataoka.Y. 1986b. A molecular dynamical study of the mutual diffusion coefficient and cooperative motion in a 2-dimensional aqueous solution. Bull. Chem. Soc. Jpn. 59: 3341-3346.

Kato, A., T. Fujiyama, and H. Nomura, 1982. Estimation of parameters, G_{11}, G_{22}, and G_{12} in the Kirkwood-Buff solution theory on the basis of the concentration fluctuation data obtained from Rayleigh scattering. Bull. Chem. Soc. Jpn. 55: 3368-3372.
Kato, T. 1985. Determination of velocity correlation coefficients in aqueous solutions of 2-butoxyethanol by quasi-elastic light scattering, pulsed-gradient FT-NMR, and Rayleigh-Brillouin scattering. J. Phys. Chem. 89: 5750-5755.
McCall, D.W. and D.C. Douglass. 1967. Diffusion in binary solutions. J. Phys. Chem. 71: 987-997.
Nosé, S. 1984a. A molecular dynamics method for simulations in the canonical ensemble. Mol. Phys. 52: 255-268.
Nosé, S. 1984b. A unified formulation of the constant temperature molecular dynamics method. J. Chem. Phys. 81: 511-519.
Okazaki, K., S. Nosé, Y. Kataoka, and T. Yamamoto. 1981. Study of liquid water by computer simulations. I. Static properties of a 2D model. J. Chem. Phys. 75: 5864-5874.
Roux, G., G. Perron, and J.E. Desnoyers. 1978. Model systems for hyrophobic interactions: Volumes and heat capacities of n-alkoxyethanols in water. J. Solution Chem. 7: 639-654.
Roux. G., D. Roberts, G. Perron, and J.E. Desnoyers. 1980. Microheterogeneity in aqueous-organic solutions: heat capacities, volumes and expansibilities of some alcohols, aminoalcohol and tertiary amines in water. J. Solution Chem. 9: 629-647.
Rowlinson, J.S., and F.L. Swinton. 1982. Liquids and Liquid Mixtures, 3rd ed. London, Butterworths.

Theoretical Calculations of Concentration Fluctuation in Liquid Alkali Alloys

Kiyoshi Arakawa
Research Inst. of Applied Electricity
Hokkaido Univ., Sapporo, Japan

ABSTRACT

We proposed a statistical thermodynamic theory of binary liquid mixtures, where a liquid mixture was assumed to be a mixture of hard spheres immersed in a uniform background potential resulting from the attractive forces (Tokiwano et al. 1977). The formula for the hard-sphere term is given by the analytical solution of the generalized Percus-Yevick equation for a hard-sphere mixture which was given first by Lebowitz. The total free energy is expressed as a sum of terms for the hard sphere fluid mixture and a term from the uniform background potential. According to the theoretical model, the concentration fluctuation $S_{cc}(0)$ for binary liquid alkali alloys which is the long wavelength limit of the concentration-concentration structure factors has been calculated. It has turned out that the theory explains experimental data from various sources successfully.

INTRODUCTION

Liquid alkali alloys are very interesting objects from the viewpoint of the theory of liquids, because the alloys are safely regarded as mixtures composed of rigid spherical particles differing in sizes, and recent remarkable progress in experimental techniques has made it possible to obtain concentration fluctuation data of the liquid alloys form various sources.

We proposed a statistical theory of liquid mixtures based on the integral equation approach formerly (Tokiwano et al. 1977), and we here apply the theory to explain the concentration fluctuation behaviors of liquid alkali alloys.

THEORY OF LIQUID MIXTURES

Liquid Model

For the formulation of the theory, the liquid mixture which is composed of various species is regarded as a mixture of hard spheres immersed in a uniform background potential resulting from overall attractive interactions within the liquid. The model has been applied to pure liquids (Longuet-Higgins et al. 1964; Arakawa et al. 1975) and solutions (Tokiwano et al. 1977; Snider et al. 1967; Arakawa et al. 1977), being combined with the theoretical achievements according to the integral equation approach (Lebowitz et al. 1964, 1965). In the model, hard-sphere particles are moving within the uniform background potential, and the formula for the moving particle term is given by the exact analytical solution of the generalized Percus-Yevick equation for a mixture composed of hard spheres of various sizes, which was given first by Lebowitz (Lebowitz et al. 1964, 1965).

Theoretical Framework

For a r-component mixture, a uniform background potential is assumed to be proportional to the n-th power of its volume V as as follows,

$$\langle U_N \rangle_{\text{av.}} = -\frac{N^2}{V^n}\left(\sum_{i=1}^{r}\sum_{j=1}^{r} x_i x_j a_{ij}\right), \tag{2.1}$$

where $x_i = N_i / N$ is the mole fraction of the i-th species (N_i : the number of atoms for the i-th species and $N = \sum_{i=1}^{r} N_i$: the total number of atoms within the system), a_{ii} the energy constant for the pure component and a_{ij} ($i \neq j$) the one for the interaction between the i-th and j-th species.

The Helmholtz free energy F for the system at volume V and temperature T is given as a sum of the Lebowitz formula for hard spheres (Lebowitz et al. 1964, 1965) and the uniform background potential term, and, further, transformed to the following form by the authors (Tokiwano et al. 1977; Arakawa et al. 1975).

$$F = F_{\text{id.mix}} - NkT\ln(f_m) - \frac{N^2}{V^n}\left(\sum_i \sum_j x_i x_j a_{ij}\right). \tag{2.2}$$

In the Eq. (2.2), the first term $F_{\text{id.mix}}$ is the free energy of the ideal gas mixture, the second one $-NkT\ln(f_m)$ is the contribution from the repulsive interactions between hard spheres in thermal motions and the third one is that from the background potential. f_m, which is defined as the equivalent free volume fraction by the authors, is given in the form (Tokiwano et al. 1977; Arakawa et al. 1975).

$$f_m=(1-\xi)\exp\left[-\frac{9V}{\pi N}\frac{2(1-\xi)XY+X^3}{(1-\xi)^2}\right], \tag{2.3}$$

and ξ is the packing fraction for the mixture which is given by

$$\xi=\sum_{i=1}^{r}\frac{\pi N_i}{6V}R_i^3,\quad X=\sum_{i=1}^{r}\frac{\pi N_i}{6V}R_i^2 \text{ and } Y=\sum_{i=1}^{r}\frac{\pi N_i}{6V}R_i, \tag{2.4}$$

(R_i : the diameter of hard spheres for the i-th species).

Differentiating Eq.(2.2) with respect to V at a given T, the equation of state is given as

$$pV=NkT\chi_m(\xi)-n\frac{N^2}{V^n}(\sum_i\sum_j x_ix_ja_{ij}). \tag{2.5}$$

In Eq.(2.5) χ_m is the compressibility factor given by Lebowitz,

$$\chi_m(\xi)=\frac{1}{(1-\xi)^3}\left\{(1-\xi)^2+\frac{18V}{\pi N}[(1-\xi)XY+X^3]\right\}. \tag{2.6}$$

and the second term comes from the background potential.

Eqs.(2.1)~(2.6) constitute the basic framework of the theory which we have applied to the elucidation of thermodynamic behaviors of liquid alkali alloys (Ohtomo et al.1982; Tanaka et al. 1983).

The expression for the concentration fluctuation $S_{cc}(0)$ for binary mixtures, which is the long wavelength limit of the concentration -concentration structure factors, (Bhatia et al. 1970) is given as

$$S_{cc}(0)=N_AkT\left(\frac{\partial^2G}{\partial c^2}\right)^{-1}_{T,p,N}, \tag{2.7}$$

where G(= F+ pV) is the Gibbs free energy.

PROCEDURES OF CALCULATION

Background Potential for Liquid Metal and Alloy

To perform calculation of thermodynamic variables using Eqs.(2.1)~(2.7) given in the preceding section, we must determine the magnitude of n in Eq.(2.1). The value of the parameter n was 1 in the original van der Waals theory, which was proved exactly by Kac et al.(Kac et al. 1963) as the limiting form for fluids with long range interactions. In the liquid metals and alloys the situation differs, of course.

In the present series of studies, n=1/3 for liquid metals and alloys has been assigned in consideration of the fact that the classical Coulomb and the exchange energy form the main part of cohesive interaction potential to be proportional to the −(1/3)-power of volume (Ohtomo et al. 1982).

Determination of Energy Parameters a_{ii} and a_{ij}

To perform the calculation, the parameters a_{ii} and a_{ij} ($i \neq j$) in Eq. (2.1) must be given. We have determined the parameter a_{ii} for a pure liquid metal from the observed heat of vaporization $\Delta H_{vap,i}$ and the molar volume V_i at a given temperature using the following relation,

$$a_{ii}=\frac{V_i^{1/3}}{N_A^2}(\Delta H_{vap,i}-N_A kT). \qquad (3.1)$$

The values of a_{ii} at various temperatures are tabulated in Table 1 for liquid Na, K, Rb and Cs.

Table 1. Values of a_{ii}.

	m.p. (K)	T (K)	$\Delta H_{vap,i}$ [a] (kJ/mol)	V_i (cm^3/mol) [b,c,d,e]	a_{ii} (kJ·cm/mol$^{4/3}$)
Na	371.0	373	105.0	24.8	297
		384	104.6	24.9	297
		400	104.6	25.0	297
		500	103.8	25.5	293
K	336.7	373	86.6	47.5	302
		384	86.2	47.8	302
		400	86.2	48.0	301
		500	84.9	49.0	296
Rb	312.0	313	81.6	57.4	305
		373	79.5	58.5	297
		384	79.1	58.7	295
		400	78.7	59.0	293
		500	77.8	61.7	290
Cs	301.7	303	77.8	72.5	315
		373	75.7	74.1	305
		384	75.3	74.3	303
		400	74.9	74.7	301
		500	73.6	77.1	296

[a] Hultgren et al. 1963 [b] Webber and Stephens 1963

[c] Breitling and Eyring 1972
[d] Huijben et al. 1975
[e] Huijben et al. 1976

Though a very small descending trend of a_{ii} is seen with increasing temperature in Table 1 (Tanaka et al. 1983), the values of a_{ii} can be regarded as practically constant in this range of temperature. And further, the values in temperature range from 313 to 400 K are found to be 301± 6 kJ·cm/mol$^{4/3}$ for all alkali metals. Thus, throughout the following calculations, we can safely assign the value 301 kJ·cm/mol$^{4/3}$ to all the a_{ii} for Na, K, Rb and Cs. This situation clearly supports the reasonableness of the assignment of the magnitude 1/3 to value of n in the uniform background potential.

Concerning the values of a_{ij} in binary liquid alkali alloys, the energy interaction parameter a_{12} is safely assumed to be between a_{11} and a_{22} ,as in the usual solution theories (Tokiwano et al. 1977), for the cases of the volume ratio $V_2/V_1 \leqq 2$ (Na– K, K– Rb, K– Cs and Rb– Cs systems), and then, we take the values of a_{12} to be 301 kJ·cm/mol$^{4/3}$ ($a_{11} = a_{12} = a_{22}$) in those cases.

For the alloys with larger volume ratio $V_2/V_1 > 2$, the cases of Na – Rb and Na– Cs systems, the value of a_{12} is expected to be larger, as suggested by the behaviors of concentration fluctuations $S_{cc}(0)$ which show fairly noticeable peaks, as will be seen in a later section. Then, for Na– Rb and Na– Cs systems, the magnitude of a_{12} is taken to be 310 and 322 kJ·cm/mol$^{4/3}$ respectively.

Packing Fraction and Effective Hard Sphere Diameters for Liquid Metals

Packing fraction ξ is the most important parameter in the equilibrium theory of liquids. In order to determine the magnitude of ξ for pure liquids, a fitting procedure of the calculated structure factors with the observed diffraction data has been made so far, using the Percus-Yevick equations generally(Ohtomo et al. 1982; Takenaka et al. 1983).

We applied the procedure to determine the values of ξ for liquid alkali metals (Na, K, Rb and Cs), here. With respect to the calculation of the Percus-Yevick structure factors we used the formula given by Enderby and North (Enderby et al. 1968) as well as that by Ashcroft and Langreth(Ashcroft et al. 1967) on the basis of the Lebowitz solution(Lebowitz et al. 1964, 1965). The values of ξ_i at 384 K, thus determined, are given in Table 2 (Tanaka et al. 1983) together with the values of T/T_m (T_m : melting point); the magnitudes of the effective hard sphere diameter R_i calculated from ξ through Eq. (2.4) are also shown in the table.

Table 2. Packing Fraction ξ_i and Effective Hard Sphere Diameter R_i for Liquid Alkali Metals at 384 K.

	T/T_m	ξ_i	R_i (nm)
Na	1.035	0.455	0.330
K	1.140	0.434	0.404
Rb	1.231	0.415	0.426
Cs	1.273	0.402	0.456

The magnitudes of effective hard sphere diameters R_i thus determined are compared with those, which were determined by various experimental and computer simulation techniques, in Table 3 for liquid Na and K (Ohtomo et al. 1982).

Table 3. Effective Hard-sphere Diameters R_i Determined from X-ray(X) and Neutron(N) Diffraction Studies and from Monte Carlo (MC) and Molecular Dynamics(MD) Calculations, and Those Obtained by The Authors.

	R_i (nm)	T(K)	Method	Ref.		R_i (nm)	T(K)	Method	Ref.
Na	0.32	387	X	i	K	0.40	353, 387	X	i
	0.325	373	N	ii		0.40	373	N	ii
	0.320	373, 423	X	iii		0.392	373, 423	X	iii
	0.33		MD	iv		0.41	338	MD	iv
	0.326	373	MC	v		0.403	408	MC	vi
						0.415	360	MD	vii
this work									
	0.330	384				0.404	384		

i) Orton et al. 1960

ii) Gingrich and Heaton 1961

iii) Huijben and Lugt 1979

iv) Kim and Letcher 1971
v) Murphy and Klein 1973
vi) Murphy 1977
vii) Sjogren 1979

Our values obtained by solving the equation of state are seen to be compatible with those by other means(Ohtomo et al. 1982) .

The Percus-Yevick structure factors for the values of ξ_i in Table 2 are shown in Figure 1(Tanaka et al. 1983) in comparison with X-ray diffraction data

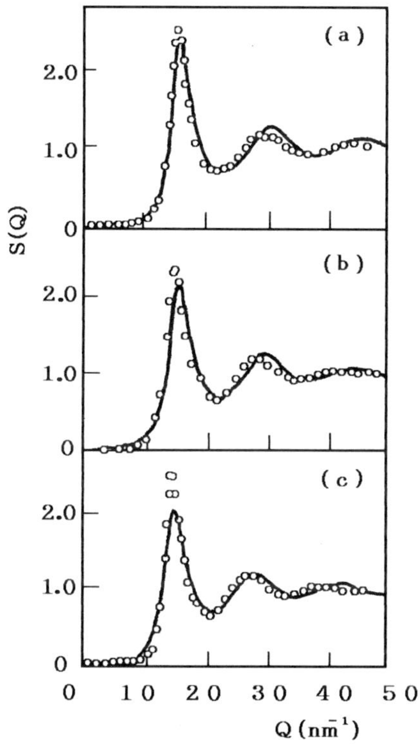

Figure 1. (a), (b) and (c) structure factors S(Q) for liquid K, Rb and Cs, respectively. Solid curves correspond to the present results at T=384 K. Open circles are experimental points: (a)T=408 K, Greenfield et al. 1971, (b) T=363 K, and (c) T=373 k, Zei 1981.

EXCESS THERMODYNAMIC FUNCTIONS AND STRUCTURE FACTORS FOR LIQUID ALKALI ALLOYS

From Eqs. (2.1)~(2.6), excess thermodynamic functions are expressed as,

$$V^E = V - \sum_i x_i V_i, \tag{4.1}$$

$$H^E = -\left(\frac{N_A^2}{V^{1/3}} \sum_i \sum_j x_i x_j a_{ij} - \sum_i x_i \frac{N_A^2}{V_i^{1/3}} a_{ii}\right) \tag{4.2}$$

$$S^E = N_A k \sum_i x_i \ln\left(\frac{f_m V}{f_i V_i}\right), \tag{4.3}$$

where f_i is the equivalent free volume fraction for pure liquid.

Calculated values of ξ, V^E, H^E and S^E are given in Table 4 (Tanaka et al. 1983) in comparison with the observed data. In performing calculations, the volume V of the mixture was obtained by solving the state Eqs. (2.5) and (2.6) at p=.10133 MPa for the equimolar mixture (x=0.5).

Table 4. Excess Thermodynamic Functions at x=0.5.

1	2	V_2/V_1	ξ	V^E (%)		H^E (J/mol)		S^E (J/K·mol)	
				Present	Obsd.	Present	Obsd.	Present	Obsd.
Na	K	1.93	0.448	−1.75	−1.4	1607	728	0.21	−0.13
Na	Rb	2.36	0.446	−4.5		992	1218	−0.21	
Na	Cs	2.99	0.445	−6.7	−5.5	615	912	−0.21	−0.75
K	Rb	1.23	0.425	−0.4	−0.2	67	126	−0.04	
K	Cs	1.55	0.420	−2.0	−1.0	310	109	−0.38	
Rb	Cs	1.27	0.415	−1.3		−105	−134	−0.63	

We calculated the Percus-Yevick structure factors for liquid alkali alloys using the effective hard sphere diameter given in Table 2. The S(Q) curve for Na−Cs alloys at x_{Na} =0.752 is compared with neutron and X-ray data (Huijben et al. 1979) in Figure 2 (Tanaka et al. 1983). The agreement of the calculated curve with neutron diffraction data is quite satisfactory in the range $Q \geqq 1$. The deviation in the region $Q < 1$ is ascribed to the feature of Percus-Yevick structure factor. Slight deviation is observed in the case of X-ray data. The agreement is good on the whole.

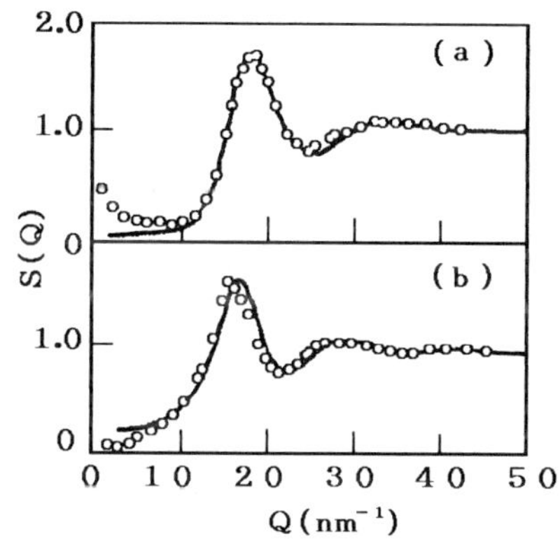

Figure 2. (a) and (b) structure factors S(Q) for liquid Na– Cs alloys at x_{Na}= 0.752. Solid curves correspond to the present results at T= 384 K. Open circles are experimental points at T= 373 K given by Huijben et al. 1979 : (a) neutron and (b) X-ray diffraction experiments.

CONCENTRATION FLUCTUATION IN LIQUID ALKALI ALLOYS

We have performed calculation of various thermodynamic quantities and structure factors for liquid alkali metals and alloys on the basis of the liquid model(Tokiwano et al. 1977; Arakawa et al. 1975 ;Arakawa et al. 1977) expressed as Eqs.(2.1)～(2.6) so far(Ohtomo et al. 1982; Tanaka et al. 1983). The results concerning excess thermodynamic functions and structure factors given in the preceding section clearly show the usefulness of the theoretical model to explain thermodynamic behavior of liquid alkali alloys.

We performed further calculations of concentration fluctuation structure factors $S_{cc}(0)$ given as Eq.(2.7) for binary alkali alloys (Ohtomo et al. 1982; Tanaka et al. 1983) along the line of the preceding section.

For liquid alkali alloys, various experimental data have been obtained near 384 K just above the melting point of liquid sodium (Huijben et al. 1975; Huijben et al.1976; Huijben et al. 1979; Kim et al. 1971), and then the present calculations are required to be carried out at this temperature.The melting point of alkali metal goes down from Na to Cs as shown in Table 2. Calculation of $S_{cc}(0)$ was performed for the systems given in Table 4.

Concentration Fluctuation $S_{cc}(0)$ for Liquid Na–K Alloys

We have calculated the concentration fluctuations $S_{cc}(0)$ for liquid Na–K alloys after some lengthy algebraic rearrangement to obtain the second derivative of G with respect to x by using Eqs. (2.1)~(4.1). The $S_{cc}(0)$ vs. x curve(Ohtomo et al. 1982) is given in Figure 3 including the curve for the ideal solution, and compared with the experimental data(Hultgren et al. 1973; Cafasso et al. 1967; Albas et al. 1980). The ideal solution values of $S_{cc}(0)$ are smaller. In consideration of the spread of the scattered data, our curve is at least qualitatively in agreement with the experimental data. The important feature of our results is the presentation of the asymmetry of the $S_{cc}(0)$ vs. x curve being skewed to the Na side, which is supposed to reflect the size effect due to the volume ratio ($V_2/V_1 = 1.93$) as shown in Tables 2 and 4. The situation will become more marked in the cases $V_2/V_1 > 2$, that is, Na–Rb and Na–Cs systems.

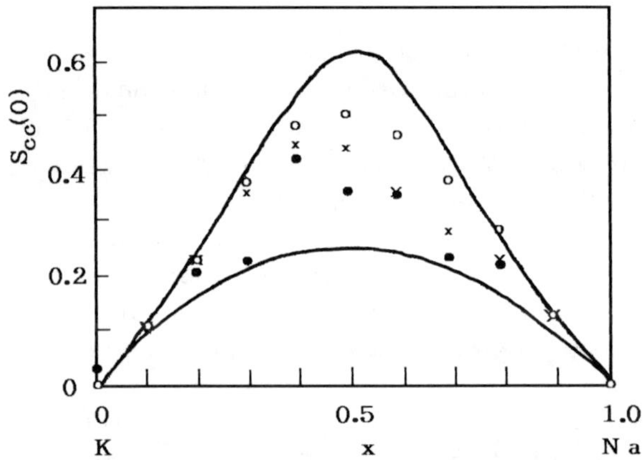

Figure 3. Concentration fluctuations $S_{cc}(0)$ for liquid Na–K alloys. Full circles are experimental points obtained by small angle X-ray diffraction measurement at T=373 K (Albas et al. 1980), open circles experimental points by the measured activity (Cafasso et al. 1967) and crosses experimental points by the measured activity (Hultgren et al. 1973). Solid curve corresponds to the present result and thin solid curve to ideal solution.

Concentration Fluctuation $S_{cc}(0)$ for Liquid K–Rb, K–Cs and Rb–Cs Alloys

For the liquid alloys, K– Rb, K– Cs and Rb– Cs systems, we performed calculations of $S_{cc}(0)$ quite similarly to the case of the Na– K alloys(Tanaka et al. 1983).

In all the cases stated above $a_{11} = a_{22} = a_{12} = 301$ kJ·cm/mol$^{4/3}$, and $V_2/V_1 < 2$. The calculated curves are given in Figure 4, and as seen for the K– Cs system, the agreement with experimental values is good, though experimental data is scarce. For the present three systems (K– Rb, K– Cs and Rb– Cs), $V_2/V_1 \leqq 1.5$ and the skewed feature is absent in the $S_{cc}(0)$ vs. x curves as seen for the Na– K system.

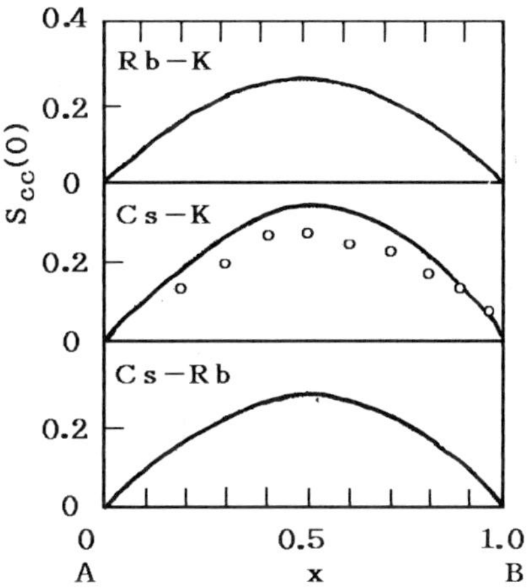

Figure 4. Concentration fluctuations $S_{cc}(0)$ for the liquid alkali alloys (K– Rb, K– Cs and Rb– Cs). Solid curves correspond to the present results at T=384 K. Open circles are experimental points at T=373 K given by Albas et al. 1981.

Concentration Fluctuation $S_{cc}(0)$ for Liquid Na– Rb and Na– Cs Alloys

For the two systems, Na– Rb and Na– Cs alloys, $V_2/V_1 > 2$ (Na– Rb: $V_2/V_1 = 2.36$ and Na– Cs: $V_2/V_1 = 2.99$) as shown in Table 4. In these two cases, $a_{11} = a_{22} \neq a_{12}$, and to the value of a_{12} we assigned 310 kJ·cm/mol$^{4/3}$ for the Na– Rb system and 322 kJ·cm/mol$^{4/3}$ for the Na– Cs system.

In Figures 5 and 6 the calculated values of $S_{cc}(0)$ are given(Tanaka et al.1983).

With respect to liquid Na– Rb alloys, the assignment of a_{12} = 310 kJ·cm/mol$^{4/3}$ has been made to obtain the best fit of calculated excess enthalpy with experimental data(Tanaka et al. 1983). In the case of the Na– Cs system the value of 322 kJ·cm/mol$^{4/3}$ has been assigned and a good fit of calculated thermodynamic functions with observed data on the whole is seen in Table 4(Tanaka et al. 1983).

In these two cases, the presence of a sharp peak at a concentration is seen as a common feature. The sharp peak of $S_{cc}(0)$ appears at the composition near $x = V_2/(V_1 + V_2)$. There has been a good deal of experimental $S_{cc}(0)$ data for the Na– Cs system from various sources, and on the contrary, no reliable experimental data for the Na– Rb system.

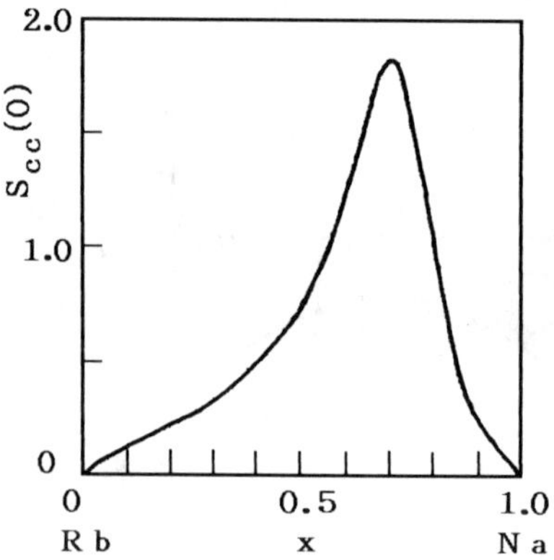

Figure 5. Calculated concentration fluctuations $S_{cc}(0)$ of the liquid Na– Rb alloys at T= 384 K.

With respect to the Na– Cs system, several authors have reported noticeable features in ultrasonic(Kim et al. 1971), chemical potential(Ichikawa et al. 1974), and X-ray and neutron diffraction data(Huijben et al. 1979). Kim and Letcher(Kim et al. 1971) pointed out from their ultrasonic data that a very pronounced absorption peak occurs near the composition at which an inflection point is present in the phase diagram(Ott et al. 1971) and stated that the peak indicates the existence of Na_3Cs as an associated species in liquid phase. However, a fairly large increase in the ultrasonic absorption is usually observed in solutions in the vicinity of its critical composition, and is supported to be parallel with the increase in concentration fluctuations(Takenaka et al. 1974; Oda et al. 1975). Through electromotive force measurements, Ichikawa et al. (Ichikawa et al. 1974) reported the existence of a pronounced peak of $S_{cc}(0)$ in the vicinity of $x_{Na} = 0.8$ at 383 K. They interpreted the marked feature as an incipient phase separation(Ichikawa et al. 1974). Thereafter, Huijben et al. also confirmed the presence of the marked feature of $S_{cc}(0)$ by neutron and X-ray diffraction experiments(Huijben et al. 1979).

Thus, the noticeable behaviors of liquid Na− Cs alloys have offered attractions to physico-chemical reseachers. Bhatia and March(Bhatia et al. 1975) attempted to interpret this behavior of $S_{cc}(0)$ in the Na− Cs system by the application of the Flory-Huggins theory for polymer solutions, and thereafter, Viser et al. (Viser et al. 1980) reported a modification of the treatment along the line.

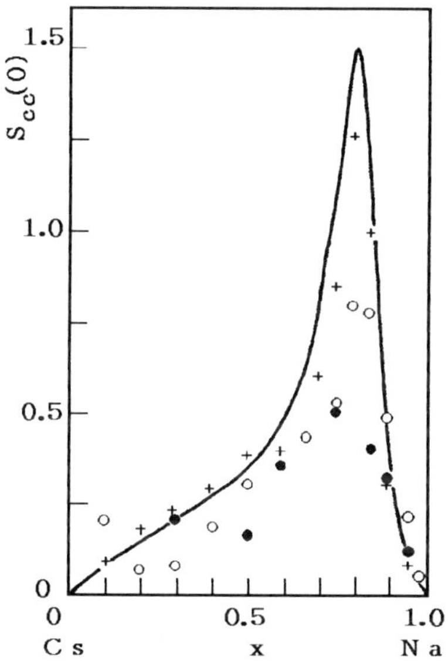

Figure 6. Concentration fluctuations $S_{cc}(0)$ for the liquid Na− Cs alloys as a function of the Na concentration. Solid curve corresponds to the present result at T=384 K. Crosses are experimental points at T=383 K given by Ichikawa et al. (Ichikawa et al. 1974) Open and full circles are experimental points by X-ray and neutron diffraction experiments at T=373 K given by Huijben et al. (Huijben et al. 1979), respectively.

The application of the Flory-Huggins theory to liquid metals is considered to be quite unreasonable from the viewpoint of its essential nature as a lattice theory, and the apparent success of its application is only fortuitous and due to arbitrary use of adjustable parameters.

As seen in Figure 6, the coincidence between the calculated curve with experimental data from various sources is very good. The marked peak observed in the $S_{cc}(0)$ vs. x curve is considered to be caused by the large difference in sizes of constituent ions, because the deviation of the energy parameter a_{12} from $a_{11} = a_{22}$ is very small ($a_{12} - a_{11} < 0.08a_{11}$). This conclusion is reasonable of course for the systems in Figures 3 and 4, where $a_{11} = a_{22} = a_{12}$, as well as for the Na-Rb system (Figure 5 : $a_{12} - a_{11} < 0.03a_{11}$). Thus, the former suggestion of the presence of an associated complex like Na_3Cs(Kim et al. 1971) should be rejected. Within the scope of the present liquid model (Eqs. (2.1)~(2.6)), the striking features in the Na-Cs system, such as negatively large V^E (Table 4) and fairly rapid increase of $S_{cc}(0)$ at about $x_{Na} \sim 0.8$, are reasonably explained.

And further, it should be noted that at the peak concentration, which appears in the vicinity of $x_1 \sim V_2/(V_1 + V_2)$, the total volume ratio between the constituents becomes nearly 1:1 . This situation is seen for all other liquid alkali alloys (Figures 3~5).

CONCLUDING REMARKS

In the present study, based on the liquid model Eqs. (2.1)~(2.6) with $n = 1/3$, we have succeeded in explaining the behaviors of concentration fluctuations $S_{cc}(0)$ (Figures. 3~6) as well as those of excess thermodynamic quantities (V^E, H^E and S^E : Table 4) including structure factors (Figure 2) for liquid alkali alloys. The striking feature of concentration fluctuation $S_{cc}(0)$, especially observed in the Na-Cs system, has been reasonably explained within the scope of the liquid model.

It has turned out that the behavior of the concentration fluctuation $S_{cc}(0)$ is primarily governed by the difference in sizes in constituent ions in excess of the ideal mixing contribution.

The success of the present approach confirms the adequacy of the assignment $n = 1/3$ in the background potential for liquid metals as well as the usefulness of the present liquid model.

REFERENCES

Albas, B. P. and W. van der Lugt. 1980. Small-angle X-ray scattering from Na-K alloys. J. Phys. **F10** :531-543

Albas, B. P., W. van der Lugt, O. Mensies and C. van Dijk. 1981. Structure of liquid potasium cesium alloys. Physica, **106B**:22-32

Arakawa, K. and K. Kojima. 1975. Investigation of the concept of free volume. Bull. Chem. Soc. Jpn. **48** :26-28

Arakawa, K., K. Tokiwano and K. Kojima. 1977. Statistical thermodynamic theory of liquid water. Bull. Chem. Soc. Jpn. **5** :65-75

Ashcroft, N. W. and D. C. Langreth. 1967. Structure of binary liquid mixtures. Phys. Rev. **156** :685-691

Bhatia, A. B. and D. E. Thornton. 1970. Structural aspects of the electrical resistivity of binary alloys. Phys. Rev. **B2** :3004-3012

Bhatia, A. B. and N. H. March. 1975. Size effects in liquid alloys. J. Phys. **F5** :1100-1106

Breitling, S. M. and H. Eyring. 1972. Liquid Metals, Chap. 5, ed. S. Z. Beer, 213 , New York: Marcel Dekker.

Cafasso, F. C., V. M. Khanna and H. M. Feder. 1967. Thermodynamic properties and ordering in liquid Na-K alloys. Adv. Phys. **16** :535-543

Enderby, J. E. and D. M. North. 1968. Percus-Yevick structure factors for liquid alloys. Phys. Chem. Liquids **1** :1-11

Gingrich, N. S. and L. Heaton. 1961. Struture of alkali metals in the liquid state. J. Chem. Phys. 34:873-878

Greenfield, A. J., J. Wellendorf and N. Wiser. 1971. X-ray determination of the static structure factor of liquid Na and K. Phys. Rev. **A4** :1607-1616

Hafner, J. 1977. Structure and thermodynamics of liquid metals and alloys. Phys. Rev. **A16** :351-364

Huijben, M. J., H. Klaucke, J. Hennehof and W. van der Lugt. 1975. Density of liquid sodium-cesium alloys. Scri. Metall. **9** : 653-656

Huijben, M. J., J. Ph. van Hasselt, K. van der Weg and W. van der Lugt. 1976. Density of liquid Na-K and K-Rb alloys. Scri. Metall. **10** :571-574

Huijben, M. J., W. van der Lugt, W. A. M. Reimert, J. Th. M. de Hosson and C. van Dijk. 1979. Structure of liquid Na-Cs alloys. Physica, **97B** :338-364

Huijben, M. J., W. van der Lugt. 1979. Acta Cryst. A35: 431.

Hultgren, R., P. D. Desai, D. T. Hawkins, M. Gleiser and K. K. Kelley. 1973. Selected Values of the Thermodynamic Properties of Binary Alloys Am. Soc. for Metals, New York: John Wiley.

Hultgren, R., R. L. Orr, P. D. Anderson and K. K. Kelley. 1963. Selected Values of the Thermodynamic Properties of Metals and Alloys New York: John Wiley

Ichikawa, K., S. M. Granstaff, Jr., and J. C. Thompson. 1974. Thermodynamics of Na-Cs alloys. J. Chem. Phys. **61** :4059-4062

Kac, G. E., G. E. Uhlenbeck and P. C. Hemmer. 1963. Van der Waals theory. J. Math. Phys. **4** :216-228

Kim, M. J. and S. V. Letcher. 1971. Ultrasonic velocity in K-Rb and Na-Cs.

J. Chem. Phys. **55** :1164-1170

Lebowitz, J. L. 1964. Percus-Yevick equation for mixture of hard spheres. Phys. Rev. **A133** :895-899

Lebowitz, J. L. and J. S. Rowlinson. 1964. Thermodynamic properties of mixtures. J. Chem. Phys., **41** :133-138

Lebowitz, J. L., E. Helfand and E. Praestgaard. 1965. Scaled particle theory of fluid mixtures. J. Chem. Phys. **43** :774-779

Longuet-Higgins, H. C. and B. Widom. 1964. A rigid model for the melting Argon. Mol. Phys. **8** :549-556

Murphy, R. D. and M. L. Klein. 1973. Radial distribution function of liquid sodium. Phys. Rev. A8:2640-2644

Murphy, R. D. 1977. Radial distribution function of liquid potassium. Phys. Rev. A15:1188-1190

Oda, K., R. Hayakawa and Y. Wada. 1975. Hypersonic relaxation in t-butanol-water mixtures. Jpn. J. Appl. Phys., **14** :1113-1125

Ohtomo, N. and K. Arakawa. 1982. Statistical thermodynamic theory of liquid Na-K alloys. J. Phys. Soc. Jpn. **51** :1282-1289

Orton, B. R., B. A. Shaw and G. I. Williams. 1960. Structure of liquid Na-K alloys. Acta Metal. 8:177-186

Ott, J. B., J. R. Goates and D. E. Oyler. 1971. Sodium- cesium alloy system. Trans. Faraday Soc, **67** :31-34

Sjogren, L. 1979. Short-wavelength density fluctuation in liquid metals. J. Phys. C12:425-430

Snider, N. S. and T. M. Herrington. 1967. Hard-sphere model of binary liquid mixtures. J. Chem. Phys. **47** :2248-2255

Takenaka, N. and K. Arakawa. 1974. Ultrasonic study on aqueous systems of ethers. Bull. Chem. Soc. Jpn. **47** :566-570

Tanaka, Y., N. Ohtomo and K. Arakawa. 1983. Statistical thermodynamics of liquid alkali alloys. J. Phys. Soc. Jpn. **52** :2093-2101

Tokiwano, K and K. Arakawa. 1977. Statistical thermodynamics of liquid mixtures. Bull. Chem. Soc. Jpn. **50** : 2217-2222

Viser, E. G., W. van der Lugt and J. Th. M de Hosson. 1980. Thermodynamic calculations for liquid alloys. J. Phys. **F10** :1681-1692

Webber, G. M. B. and R. W. B. Stephens. 1963. Physical Acoustics, Vol. 4, Part B, ed W. P. Mason, 53. (Academic Press, New York: Academic Press.

Yokokawa, T. and O. J. Kleppa. 1964. Heat of mixing in binary-liquid-alkali-metal mixtures. J. Chem. Phys. **40** :46-54

Zei, M. S. 1981. X-ray analysis of the structure of liquid Rb and Cs. Phys. Rev., **B24** :4509-4514

Generalized Transport Theory and Its Application to Binary Mixtures

Enrique Peacock-López
Department of Chemistry
Williams College, Williamstown, MA

Luis Woodhouse
Department of Chemistry
University of California, Davis, CA

ABSTRACT
Using the statistical theory of nonequilibrium thermodynamics, we analyze the mass and heat transport in a multicomponent system arbitrarily away from equilibrium. First, we obtain the generalized Onsager's coefficients which satisfy reciprocal relations. Second, we consider an isothermal nonelectrolyte binary mixture, and for the simplest elementary process, we obtain an explicit expression for the mutual diffusion coefficient. Finally, we consider the benzene-cyclohexane and carbon tetrachloride-cyclohexane mixtures and compare our theoretical values with the experimental coefficients and the Bearman-Kirkwood theory. We find that our theory is in good agreement with the Bearman-Kirkwood theory and both give a good qualitative and quantitative description of the experimental behavior.

1.- INTRODUCTION

Differences in concentrations and temperature between regions in space are reduced spontaneously by mass and heat flows. These phenomena are fundamental examples in the discussion of irreversible processes and the second law of thermodynamics. A general theory of irreversibility and the second law for transport processes has been developed for systems at or near equilibrium. As an example of the widely accepted theory of processes at or near equilibrium, we will refer the reader to the seminal work by de Groot and Mazur (1962). On the other hand, transport processes, like mass diffusion, are likely to be driven away and far from equilibrium. The description of systems far from equilibrium lacks a general accepted theory. Instead several approaches have been developed to deal with systems at steady states far from equilibrium. One of these approaches is the fluctuation-dissipation (F-D) theory also known as the mechanistic statistical theory of nonequilibrium thermodynamics (Keizer 1987). This theory generalizes the near-equilibrium linear theory of Onsager and Machlup (1953) and Fox and Uhlenbeck (1970a) which provides a thermodynamic description of the extensive variable fluctuations.

According to Onsager's theory (1931a and 1931b) of irreversible processes near-equilibrium, the fluxes are linearly related to the mass and temperature gradients. In this case, the coupling coefficients, which appear in the linear equations, are related through phenomenological reciprocal relations. One completes the description of the process with the Onsager and Machlup's fluctuation theory (Onsager 1953 and Machlup 1953) of linear irreversible thermodynamics. Here, one assumes that the fluctuations around equilibrium behave in accordance to a set of stochastic equations. These stochastic equations are obtained by adding a Gaussian random term to each of the linear relations between fluxes and gradients.

Originally Onsager and Machlup's theory was used to describe the near-equilibrium transport of mass, heat and charge. Later, Fox and Uhlenbeck extended and applied the theory to other linear phenomena near equilibrium, such as the linearized hydrodynamical equations (Fox 1970a), and the Boltzman equation (Fox 1970b). Only recently has a generalization of Onsager's theory to systems far from equilibrium been outlined by Keizer (1975a, 1976a, 1976b, and 1987). The generalization, known as the mechanistic statistical theory of nonequilibrium thermodynamics or the fluctuation-dissipation (F-D) theory, determines completely the microscopic equations for the extensive variables and the decay of fluctuations in terms of elementary molecular mechanisms. This theory, which reduces to the previous near-equilibrium formalisms, has been applied to a variety of phenomena such as: the effect of diffusion on chemical reaction in solution (Keizer 1975b, 1983a and 1983b), the coupling of density and energy fluctuations by isothermal chemical reactions (Keizer 1984 and 1987), the effect of diffusion in one dimensional reactive systems (Peacock-López 1988), receptor-mediate endocytosis (Keizer 1985 and Peacock-López 1986), non-local fluctuating hydrodynamics (Medina-Noyola 1981, Keizer 1982 and Peacock-López 1985), and isothermal diffusion in multicomponent fluids (Woodhouse 1984).

In this paper, the fluctuation-dissipation theory is used to analyze a multicomponent system arbitrarily away from equilibrium, where mass and heat transport are the only dissipative processes. In Section 2, we calculate the generalized coupling coefficients, which satisfy generalized reciprocal relations. Mass and energy fluctuations around stationary states are discussed in Section 3. In Section 4, a nonelectrolyte binary solution under isothermal conditions is considered, and the mutual diffusion coefficient dependence on activities is calculated. In Section 5, simple elementary processes which include the solvent are considered for the explicit calculation of the mutual diffusion coefficient. The results are compared with the theoretical Bearman-Kirkwood (1958) equations and experimental values of the mixtures benzene-cyclohexane (Kamal 1966) and carbon tetrachloride-cyclohexane (Horrocks 1961). A summary and final comments are included in Section 6.

2.- GENERALIZED DIFFUSION

In this section we analyze the mass and the energy transport of an n-component system in the context of the fluctuation-dissipation theory (F-D) (Keizer 1975a, 1976a, and 1976b). According to this theory, one first divides the system into small cells, where N_{ip} denotes the amount of the ith extensive variable in the pth cell. For simplicity, we consider an isotropic system and analyze the elementary processes related to the pth cell along the x-axis. We want to emphasize that not all of the simplifications are essential. Their purpose is to facilitate the calculations and emphasize the physical ideas.

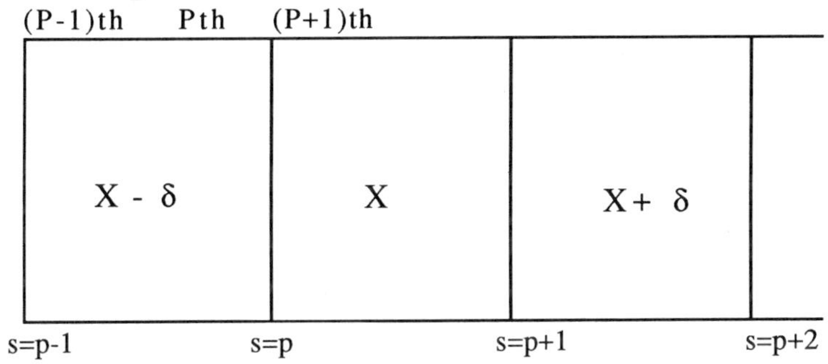

Figure 1. The pth cell centered at x. d is the size of the box, and s denotes the boundary between cells.

Figure 1 depicts the pth cell where δ is the length of a side of the box, s denotes the boundary between cells and x is the position of the pth cell. The discrete elementary processes are represented by k_τ and s. Here k_τ denotes a position independent elementary process, τ stands for either mass (M) or thermal (T) transport, and s represents the boundary at which transport is taking place. These elementary processes are associated with the molecular amounts $n_i^{+r}(k_\tau)$ and $n_i^{+l}(k_\tau)$ that participate in the process (Keizer 1975a)

$$\sum_{i=1} n_i^{+l}(k_\tau) + \sum_{i=1} n_i^{+r}(k_\tau) \rightleftarrows \sum_{i=1} n_i^{-l}(k_\tau) + \sum_{i=1} n_i^{-r}(k_\tau) \tag{2.1}$$

The superscript +(-) stands for the forward (backward) direction and l(r) stands for the left (right) side of a boundary. Since no reactions or sources have been considered, we have a conservation relation

$$n_i^{+l}(k_\tau) + n_i^{+r}(k_\tau) = n_i^{-l}(k_\tau) + n_i^{-r}(k_\tau) \tag{2.2}$$

On the other hand, the density $n_{ip} = N_{ip}/\delta^3$ is related to molecular changes $n_{ip}^{\pm}(k_\tau,s)$. In the one-dimensional case, the $n_{ip}^{\pm}(k_\tau,s)$ are given by

$$n_{ip}^{+}(k_\tau,s) = n_i^{+r}(k_\tau)\,\delta_{s\,p} + n_i^{+l}(k_\tau)\,\delta_{s\,p+1}\,, \tag{2.3}$$

where δ_{sp} is a Kroneker delta. Following the F-D theory, (Keizer 1987) one constructs the change of n_{ip} due to the (k_τ, s) elementary process. This net change is given by (Keizer 1975a)

$$w_{ip}(k_\tau,s) = n_{ip}^{-r}(k_\tau,s) - n_{ip}^{+r}(k_\tau,s)\,. \tag{2.4}$$

If we define $w_i(k_\tau) = n_i^{-r}(k_\tau) - n_i^{+r}(k_\tau)$, Eq. (2.4) reduces to

$$w_{ip}(k_\tau,s) = w_i(k_\tau)\left[\delta_{s\,p} - \delta_{s\,p+1}\right] \tag{2.5}$$

The next step is the construction of the forward and backward rates for each elementary process. According to the F-D theory (Keizer 1975a and 1976a), the rates have the following canonical form (for details see Appendix A)

$$V_{k_\tau s}^{\pm} = \Omega_{k_\tau}^{\pm}\exp\left[-\sum_{j\,m} F_j(m)\, n_{jm}^{\pm}(k_\tau,s)/k_B\right], \tag{2.6}$$

where the $\Omega^{\pm}_{k_\tau}$'s are intrinsic rates associated with each transport process, and k_B is the Boltzman constant. The $F_j(m)$'s are the local conjugate intensive variable to the $n_{j\,m}$'s, namely

$$F_j(m) = \left(\frac{\partial S^{le}(m)}{\partial n_{j\,m}}\right), \tag{2.7}$$

with $S^{le}(m)$ the local entropy in the mth cell. The first postulate in the F-D theory states (Keizer 1975a) that the rate of change of the extensive variable, $n_{j\,m}\delta^3$, due to dissipative processes, is given by

$$\left[\frac{d}{dt}\left(n_{i\,p}\delta^3\right)\right]_{diss} = \sum_{k_\tau s} w_{i\,p}(k_\tau s)\left[V^{+}_{k_\tau s} - V^{-}_{k_\tau s}\right] \tag{2.8}$$

Using equations (2.5) and (2.6) in (2.8), one gets

$$\left[\frac{d}{dt}\left(n_{i\,p}\delta^3\right)\right]_{diss} = \sum_{k_\tau} w_i(k_\tau)\Omega_{k_\tau}\cdot$$

$$\cdot\left[R^{+}_{k_\tau}(p)L^{+}_{k_\tau}(p-1)+R^{-}_{k_\tau}(p+1)L^{-}_{k_\tau}(p)-R^{-}_{k_\tau}(p)L^{-}_{k_\tau}(p-1)-R^{+}_{k_\tau}(p+1)L^{+}_{k_\tau}(p)\right] \tag{2.9}$$

where we have assumed microscopic reversibility, i.e. $\Omega^{+}_{k_\tau} = \Omega^{-}_{k_\tau}$,

and defined the following functions

$$R^{\pm}_{k_\tau}(p) = \exp\left[-\sum_j F_j(p)\, n^{\pm r}_j(k_\tau)/\, k_B\right] \tag{2.10a}$$

$$L^{\pm}_{k_\tau}(p) = \exp\left[-\sum_j F_j(p)\, n^{\pm l}_j(k_\tau)/\, k_B\right]. \tag{2.10b}$$

Recalling that the pth cell has been centered around x and expanding Eq. (2.9) to leading order in $\delta^2 \Omega_{k_\tau}$, one gets

$$\left[\frac{d}{dt}\left(n_{ip}\delta^3\right)\right]_{diss} = - \sum_{k_\tau j} w_i(k_\tau)w_j(k_\tau)\Omega_{k_\tau}\delta^2 \cdot$$

$$\left\{ \frac{\partial}{\partial x}\left[\overset{+}{R}_{k_\tau}(x)\, \overset{+}{L}_{k_\tau}(x) \frac{1}{k_B}\frac{\partial F_i}{\partial x}\right]\right\} + O\left(\delta^3\, \Omega_{k_\tau}\right), \qquad (2.11)$$

where we have used the equality $\overset{+}{R}_{k_\tau}(x)\, \overset{+}{L}_{k_\tau}(x) = \overset{-}{R}_{k_\tau}(x)\, \overset{-}{L}_{k_\tau}(x)$. In order to

obtain the macroscopic equations, we divide Eq. (2.11) by δ^3 and take the limit of small δ. In the continuum we have

$$\lim_{\delta \to 0} \frac{d}{dt} n_{ip} = \frac{\partial}{\partial t} \bar{n}_i(x),$$

where $\bar{n}_i(x)$ is the macroscopic number density of the ith species, and $\lim_{\delta \to 0}$

$\Omega_{k_\tau}/\delta = \bar{\Omega}_{k_\tau}$ Finally, we obtain

$$\left[\frac{\partial}{\partial t}\, \bar{n}_i(x)\right]_{diss} = \frac{\partial}{\partial x} J_i(x)\,, \qquad (2.12)$$

where we have identified the J_is as

$$J_i(x) = -\frac{1}{k_B}\sum_{k_\tau j} w_i(k_\tau)w_j(k_\tau)\bar{\Omega}_{k_\tau}\exp\left[- \sum_{j'} F_{j'}(x)\, [n_{j'}^{+r}(k_\tau)+n_{j'}^{+l}(k_\tau)]/\, k_B\right]$$

$$\times \; \frac{\partial F_j(x)}{\partial x}\,. \qquad (2.13)$$

From Eq. (2.13) the generalized Onsager's coefficients are defined as

$$L_{ij}=\sum_{k_\tau j} w_i(k_\tau)w_j(k_\tau)\bar{\Omega}_{k_\tau}\exp\left[-\sum_{j'} F_{j'}(x)[n_{j'}^{+r}(k_\tau)+n_{j'}^{+l}(k_\tau)]/\,k_B\right]. \quad (2.14)$$

For an isotropic system, the extension to three dimensions is straightforward. Therefore, one gets

$$\left[\frac{\partial}{\partial t}\bar{n}_i(\mathbf{x})\right]_{diss} = -\nabla\cdot\sum_j L_{ij}\nabla\left(\frac{F_j}{k_B}\right) \quad (2.15)$$

or

$$\left[\frac{\partial}{\partial t}\bar{n}_i(\mathbf{x})\right]_{diss} = -\nabla\cdot\sum_j D_{ij}\nabla\bar{n}_j \quad , \quad (2.16)$$

where we have defined the diffusion coefficients as

$$D_{ij} = \sum_s L_{is}\,\frac{\partial F_s/k_B}{\partial \bar{n}_j} \quad . \quad (2.17)$$

Notice that one can replace $\bar{\mathbf{n}}$ by $\bar{\mathbf{n}}^{eq}$ and recover Onsager's theory near equilibrium, or one can use $\bar{\mathbf{n}} = \bar{\mathbf{n}}^{ss}$, where the superscript refers to an asymptotically stable nonequilibrium stationary state. In this case, the extension of Onsager's theory to nonequilibrium steady states is obtained.

For mass and energy transport in an n-component system, we label the species i = 1,2....s, and, for energy, i = e. Using equations (2.14) and (2.15), we get the following expressions

$$\left[\frac{\partial}{\partial t}\bar{n}_i(\mathbf{x})\right]_{diss} = -\nabla\cdot\left\{\sum_{j=1}^{s}\sum_{k_\tau} w_i(k_\tau)w_j(k_\tau)\bar{\Omega}_{k_\tau}P_{k_\tau}\nabla(-\beta\mu_j)\right\}$$

$$-\nabla\cdot\left\{\sum_{k_\tau} w_i(k_\tau)w_e(k_\tau)\bar{\Omega}_{k_\tau}P_{k_\tau}\nabla\beta\right\} \quad (2.18)$$

$$\left[\frac{d}{dt}\,\bar{n}_e(\mathbf{x})\right]_{diss} = -\,\nabla\cdot\left\{\sum_{j=1}^{s}\sum_{k_\tau} w_e(k_\tau)w_j(k_\tau)\bar{\Omega}_{k_\tau}P_{k_\tau}\,\nabla(-\beta\mu_j)\right\}$$

$$-\,\nabla\cdot\left\{\sum_{k_\tau} w_e(k_\tau)w_e(k_\tau)\bar{\Omega}_{k_\tau}P_{k_\tau}\,\nabla\beta\right\}, \tag{2.19}$$

where μ_i is the chemical potential for the ith species, T is the absolute temperature, $\beta = 1/k_BT$, and we have defined P_{k_τ} as

$$P_{k_\tau} = \exp\left[-\sum_{j'} F_{j'}(\mathbf{x})\,[n_{j'}^{+r}(k_\tau)+n_{j'}^{+l}(k_\tau)]/\,k_B\right]. \tag{2.20}$$

The sums, over k_τ in Eq. (2.18) and (2.19), split in two. The first sum is over the elementary process due to mass gradients (k_M), and the second sum is over k_T due to temperatures gradients. For example, one can make the following identification:

$$n_i^{+l}(k_T) = m_i^l(k_T) \tag{2.21a}$$

$$n_i^{-l}(k_T) = \bar{m}_i^l(k_T) \tag{2.21b}$$

$$n_e^{+l}(k_T) = e^l(k_T) \tag{2.21c}$$

$$n_e^{-l}(k_T) = \bar{e}^l(k_T) \tag{2.21d}$$

with similar expressions for the right-hand-side of the boundary surface. In this case, the elementary processes k_T are represented as

$$\sum_{i=1}^{s}\left[m_i^r(k_T) + m_i^l(k_T)\right] + e^r(k_T) + e^l(k_T) \quad \rightleftarrows$$

$$\sum_{i=1}^{s}\left[\bar{m}_i^{\,r}(k_T)+\bar{m}_i^{\,l}(k_T)\right]+\bar{e}^r(k_T)+\bar{e}^l(k_T)\ . \tag{2.22}$$

Each m_i represents the number of ith particles and e the amount of energy involved in the elementary process k_T. In a similar way, in the presence of a mass gradient, we redefine the n_i 's as

$$n_i^{+1}(k_M) = m_i^l(k_M) \tag{2.23a}$$

$$n_i^{-1}(k_M) = \bar{m}_i^l(k_M) \tag{2.23b}$$

$$n_e^{+1}(k_M) = e^l(k_M) \tag{2.23c}$$

$$n_e^{-1}(k_M) = \bar{e}^l(k_M) \tag{2.23d}$$

with similar expressions for the right-hand-side of the boundary. Using equations (2.21) and (2.23) in (2.20), we define the following functions

$$P_{k_M}(\mathbf{n}) = \Omega_{k_M} \exp\left[\sum_{j=1}^{s}[m_j^r(k_M)+m_j^l(k_M)]\ \beta\mu_j\right]\exp\left[-[e^r(k_M)+e^l(k_M)]\ \beta\right] \tag{2.24}$$

$$P_{k_T}(\mathbf{n}) = \Omega_{k_T} \exp\left[\sum_{j=1}^{s}[m_j^r(k_T)+m_j^l(k_T)]\ \beta\mu_j\right]\exp\left[[e^r(k_T)+e^l(k_T)]\ \beta\right], \tag{2.25}$$

where $\mathbf{n} = (n_1,\ldots,n_i,\ldots,e)$. Finally, we rewrite equations (2.18) and (2.19) in the following form

$$\left[\frac{\partial}{\partial t}\,\bar{n}_i(\mathbf{x})\right]_{diss} = -\,\nabla\cdot\left\{\sum_{j=1}^{s} L_{ij}\nabla(-\beta\mu_j) + L_{ie}\nabla\beta\right\} \tag{2.26}$$

$$\left[\frac{\partial}{\partial t}\,\bar{e}(\mathbf{x})\right]_{diss} = -\,\nabla\cdot\left\{\sum_{j=1}^{s} L_{ej}\nabla(-\beta\mu_j) + L_{ee}\nabla\beta\right\}\,. \tag{2.27}$$

In the presence of mass and temperature gradients, the generalized Onsager coefficients are given by

$$L_{ij} = \sum_{k_M} w_i(k_M)w_j(k_M)\,P_{k_M} + \sum_{k_T} w_i(k_T)w_j(k_T)\,P_{k_T} \tag{2.28a}$$

$$L_{ie} = L_{ei} = \sum_{k_M} w_i(k_M)w_e(k_M)\,P_{k_M} + \sum_{k_T} w_i(k_T)w_e(k_T)\,P_{k_T} \tag{2.28b}$$

$$L_{ee} = \sum_{k_M} w_e(k_M)w_e(k_M)\,P_{k_M} + \sum_{k_T} w_e(k_T)w_e(k_T)\,P_{k_T}\,. \tag{2.28c}$$

It is easily verified that the coefficients satisfy the reciprocal relations, i.e. $L_{ij} = L_{ji}$. Eqs. (2.26) - (2.28) also contain the Soret effect (mass flow due to a temperature gradient) and the Dufour effect (a heat flow due to a mass gradient) as result of the terms $w_i(k_T)$ and $w_i(k_M)$ in L_{ie}. Eqs (2.26)-(2.28) involve nonlinear terms and, for isotropic systems, extend the linear theory of heat and mass flow in to the nonlinear domain. In the case of anisotropic systems, one obtains equivalent equations.

In this section we have cast the macroscopic mass and energy transport equations of an n-component system within the fluctuation-dissipation theory. As prescribed for a continuum system, this is done by a discretization of the system. This procedure allows us to apply the F-D theory, which yields a generalization of Onsager's theory of irreversible processes to far from equilibrium steady states.

3.- MASS AND ENERGY FLUCTUATIONS

The study of mass and energy is completed by a consistent fluctuation theory. The characterization of fluctuations in the extensive variables around their conditional averages is contained in the second postulate of the fluctuation-dissipation theory (Keizer 1975a, 1976a, and 1976b). This postulate states that the fluctuations, $\delta\mathbf{n} = \bar{\mathbf{n}}(t) - \mathbf{n}(t)$, satisfy a Langevin-type equation obtained by the linearization of the macroscopic transport equations, e.g.,

$$\frac{\partial\,\delta\mathbf{n}(t)}{\partial\,t} = \mathbf{H}\left(\bar{\mathbf{n}}(t)\right)\delta\mathbf{n}(t) + \tilde{\mathbf{f}} \tag{3.1}$$

where $\mathbf{H}\left(\bar{\mathbf{n}}(t)\right)$, the relaxation matrix, is given by

$$H_{ij}\left(\bar{\mathbf{n}}(t)\right) = \left[\frac{\partial}{\partial\,n_j}\left(R_i(\mathbf{n}) - S_i(\mathbf{n})\right)\right]_{\mathbf{n}=\bar{\mathbf{n}}} . \tag{3.2}$$

$R_i(\mathbf{n})$ represents the dissipative contribution to the macroscopic equation i.e., Eq. (2.16), and $S_i(\mathbf{n})$ are streaming terms i.e., conservative contributions. The term $\tilde{\mathbf{f}}$ is a purely random Gaussian process with vanishing average. Its covariance $\langle\tilde{\mathbf{f}}(t)\,\tilde{\mathbf{f}}^T(t')\rangle$ is given by (Keizer 1975a, 1976a, and 1976b)

$$\langle\tilde{\mathbf{f}}(t)\,\tilde{\mathbf{f}}^T(t')\rangle = \gamma\left(\bar{\mathbf{n}}(t)\right)\delta(t-t') , \tag{3.3}$$

where $\delta(t-t')$ is the Dirac delta function. The strength of the fluctuations, γ, is given by the third postulate of the F-D theory (Keizer 1975a, 1976a, and 1976b), namely

$$\gamma_{ij}\left(\bar{\mathbf{n}}(t)\right) = \sum_{k_\tau} w_i\left(k_\tau\right) w_j\left(k_\tau\right)\left[V_{k_\tau}^{+} + V_{k_\tau}^{-}\right], \tag{3.4}$$

where the conditional covariance matrix $\sigma_{ij}(\bar{\mathbf{n}}(t)) = \langle\delta n_i \delta n_j\rangle$ satisfies a generalized fluctuation-dissipation theorem. Using the relaxation matrix, the noise strength and the conditional covariance matrix, the generalized fluctuation-dissipation theorem is given by

$$\frac{d\sigma(\bar{\mathbf{n}}(t))}{dt} = \mathbf{H}(\bar{\mathbf{n}})\sigma(\bar{\mathbf{n}}) + \sigma(\bar{\mathbf{n}})\mathbf{H}^T(\bar{\mathbf{n}}) + \gamma(\bar{\mathbf{n}}) , \tag{3.5}$$

here $\mathbf{H}^T$ denotes the transpose of **H.** For asymptotically stable stationary states, one has to solve the macroscopic equation to find $\bar{\mathbf{n}}^{ss}$. In this case, Eq. (3.5) reduces to

$$\mathbf{H}(\bar{\mathbf{n}}^{ss})\sigma(\bar{\mathbf{n}}^{ss}) + \sigma(\bar{\mathbf{n}}^{ss})\mathbf{H}^T(\bar{\mathbf{n}}^{ss}) = -\gamma(\bar{\mathbf{n}}^{ss}) , \tag{3.6}$$

where

$$\sigma(\bar{\mathbf{n}}^{ss}) = \langle \delta\mathbf{n}\, \delta\mathbf{n}^T \rangle^{ss}, \tag{3.7}$$

and $\langle ... \rangle^{ss}$ is an average over the steady state ensemble.

To obtain the γ_{ij} in the continuum limit, first we consider its action as integral operator, and second, we discretize the integral, namely

$$\int \gamma_{ij}(x,x',t)\, G(x')\, dx' \longrightarrow \sum_{p'} \gamma_{ijpp'} G_{p'} \delta^3 \quad , \tag{3.8}$$

where $G_p = G(x_p)$. Combining the results of previous section with Eq. (3.4) , one finds that

$$\gamma_{ijpp'} = \frac{1}{\delta^3} \sum_{k_\tau s} w_{ip}\left(k_\tau s\right) w_{js}\left(k_\tau s\right) \left[V^+_{k_\tau s} + V^-_{k_\tau s} \right] . \tag{3.9}$$

Using Eq (2.5) and expanding (3.9) to leading order in δ, one gets

$$\int \gamma_{ij}(x,x',t)\, G(x')\, dx' \longrightarrow (-1) \sum_{k_\tau} w_i\left(k_\tau\right) w_j\left(k_\tau\right) \left(\Omega_{k_\tau}/\delta\right)$$

$$\mathrm{x}\left\{ 2\left(R^+_{k_\tau} L^+_{k_\tau} \right) \frac{\partial^2 G}{\partial x^2} + 2\frac{\partial}{\partial x}\left(R^+_{k_\tau} L^+_{k_\tau} \right) \frac{\partial G}{\partial x} \right\} + O\left(\Omega_{k_\tau}\right) . \tag{3.10}$$

In order to find the final expressions in the continuum limit, we take the limit δ —> 0 in Eq. (3.10) and recall Eq. (2.14). In this case we obtain an expression for γ_{ij}, as an integral operator in the continuum i.e.,

$$\int \gamma_{ij}(x,x',t)\, G(x')\, dx' \;=\; -2\left[\frac{\partial}{\partial x}\, L_{ij}\, \frac{\partial G}{\partial x}\right] . \tag{3.11}$$

The extension to a three dimensional isotropic system is given by

$$\gamma_{ij}(\mathbf{r},\mathbf{r}',t) \;=\; -2\left[\nabla_{\mathbf{r}}\cdot(L_{ij}\nabla_{\mathbf{r}})\right]\delta(\mathbf{r}-\mathbf{r}') , \tag{3.12}$$

where $\delta(\mathbf{r}-\mathbf{r}')$ is the 3-dimensional Dirac delta function.

In this section we have considered the behavior of the fluctuations of the extensive variables and their connection with the generalized Onsager coefficients. Eq. (3.12) is the main result in this section and constitute, with Eqs. (2.16) and (2.14), the basis for the generalized diffusion theory. The application of the theory to nonelectrolyte binary mixtures is discussed in the following section.

4.- MUTUAL DIFFUSION IN ISOTHERMAL BINARY MIXTURES

In this section, we will consider an isothermal nonelectrolyte binary mixture, where matter diffuses independently of energy transport,i.e., $w_i(k_T) = w_e(k_M) = 0$. For this system, we will calculate the mutual diffusion coefficient in the volume-fixed frame.

In practice, we have to select the elementary processes from a physical description of the molecular events involved in the macroscopic process under study. In the present case, we picture solvated solute particles diffusing in the solvent. Based on this solvation effect, which operates when either of the species is in low concentration, we extend the picture and consider also groups of solvated solute molecules diffusing. For illustration, we define the following elementary processes which are responsible for the diffusion of the solute (molecule 1) and the solvent (molecule 2). For the first set of elementary process, we consider that molecule 1 is solvated by β_2 molecules 2. These processes can be described by:

$$m_1^{1}(\kappa_{1n}) = n \qquad \bar{m}_1^{\,1}(\kappa_{1n}) = 0 \tag{4.1a}$$

$$m_2^{1}(\kappa_{1n}) = n\beta_2 \qquad \bar{m}_2^{\,1}(\kappa_{1n}) = n(\beta_2 + \nu_2) \tag{4.1b}$$

$$m_1^{r}(\kappa_{1n}) = 0 \qquad \bar{m}_1^{\,r}(\kappa_{1n}) = n \tag{4.1c}$$

$$m^r_2(\kappa_{1n}) = n(\beta_2 + \nu_2) \quad \bar{m}^r_2(\kappa_{1n}) = n\beta_2 \quad . \tag{4.1d}$$

In this case, n=1 represents one solvated molecule and n>1 represents groups of molecules diffusing. A second set of elementary processes is described by:

$$m^1_1(\kappa_{2n}) = n\beta_1 \qquad \bar{m}^1_1(\kappa_{2n}) = n(\beta_1 + \nu_1) \tag{4.1e}$$

$$m^1_2(\kappa_{2n}) = n \qquad \bar{m}^1_2(\kappa_{2n}) = 0 \tag{4.1f}$$

$$m^r_1(\kappa_{2n}) = n(\beta_1 + \nu_1) \qquad \bar{m}^r_1(\kappa_{2n}) = n\beta_1 \tag{4.1g}$$

$$m^r_2(\kappa_{2n}) = 0 \qquad \bar{m}^r_2(\kappa_{2n}) = n \tag{4.1h}$$

The quantities ν_1 and ν_2 are required by conservation of momentum and their values are determined below (c.f. Eq. (4.9))

Our choice of elementary processes for the mutual diffusion is dictated by the need to represent the molecular interactions that exist in real systems, and, also, to keep the equations as simple as possible. In the first case, we consider n molecules of kind 1, i.e. $m^1_1(\kappa_{1n}) = n$, surrounded by $n\beta_2$ molecules of kind 2, i.e. $m^1_2(\kappa_{1n}) = n\beta_2$, diffusing from the left side of the pth surface (see Fig. 1) to the right side. To complete the description of the elementary process, one has to include the diffusion of $n(\beta_2+\nu_2)$ molecules 2 ($m^r_1(\kappa_{1n}) = 0$ and $m^r_2(\kappa_{1n}) = n(\beta_2+\nu_2)$) from the right side of the pth surface to the left side. This process will dominate when molecule 1 is the solute and molecule 2 is the solvent. For the second set of elementary processes, we have exchanged the roles of solute and solvent. In this case, we have n molecules of kind 2 , solvated by $n\beta_1$ molecules of kind 1, moving from left to right, and $n(\beta_1+\nu_1)$ molecules of kind 1 diffusing in the opposite direction.

Certainly, many other elementary processes, which include size shape and aggregation, can be selected. But, for the purpose of illustration, we have chosen only the simplest elementary processes. In this case, the net changes for these two elementary processes are found from Eq. (2.4) to be

$$w_1(\kappa_{1n}) = n \qquad w_2(\kappa_{1n}) = -n\nu_2 \tag{4.2a}$$

$$w_1(\kappa_{2n}) = -n\nu_1 \qquad w_2(\kappa_{2n}) = n \quad , \tag{4.2b}$$

where n varies from one to infinity. Therefore in our description of diffusion, we are including an infinite number of similar elementary processes.

Next, we use Eqs. (2.7) and (2.13) to construct the mass fluxes. In general, we have

$$\mathbf{J}_i = \sum_{\kappa_{1n} j} w_j(\kappa_{1n}) w_i(\kappa_{1n}) V^+_{\kappa_{1n}} \nabla(\beta\mu_j) + \sum_{\kappa_{2\nu} j} w_j(\kappa_{2n}) w_i(\kappa_{2n}) V^+_{\kappa_{2n}} \nabla(\beta\mu_j) , \tag{4.3}$$

where the $V^+_{\kappa_{in}}$'s are given by

$$V^+_{\kappa_{1n}} = \bar{\Omega}_{\kappa_{1n}} \exp\left[\sum_{j'} \beta\mu_{j'} \left(n^{+r}_{j'}(\kappa_{1n}) + n^{+1}_{j'}(\kappa_{1n})\right)\right] \tag{4.4a}$$

$$V^+_{\kappa_{2n}} = \bar{\Omega}_{\kappa_{2n}} \exp\left[\sum_{j'} \beta\mu_{j'} \left(n^{+r}_{j'}(\kappa_{2n}) + n^{+1}_{j'}(\kappa_{2n})\right)\right] \tag{4.4b}$$

For example, if i=1, we get

$$\begin{aligned} \mathbf{J}_1 = & \sum_{\kappa_{1n}} w_1(\kappa_{1n}) w_1(\kappa_{1n}) \; \bar{\Omega}_{\kappa_{1n}} P_{\kappa_{1n}} \nabla(\beta\mu_1) + \\ & \sum_{\kappa_{1n}} w_1(\kappa_{1n}) w_2(\kappa_{1n}) \; \bar{\Omega}_{\kappa_{1n}} P_{\kappa_{1n}} \nabla(\beta\mu_2) + \\ & \sum_{\kappa_{2n}} w_1(\kappa_{2n}) w_1(\kappa_{2n}) \; \bar{\Omega}_{\kappa_{2n}} P_{\kappa_{2n}} \nabla(\beta\mu_1) \\ & + \sum_{\kappa_{2n}} w_1(\kappa_{2n}) w_2(\kappa_{2n}) \; \bar{\Omega}_{\kappa_{2n}} P_{\kappa_{2n}} \nabla(\beta\mu_2) , \end{aligned} \tag{4.5}$$

where the $P_{\kappa_{in}}$'s have been defined as:

$$P_{\kappa_{1n}} = \exp\left[n\beta\left(\mu_1 + (\nu_2 + 2\beta_2)\mu_2\right)\right] \tag{4.6a}$$

$$P_{\kappa_{2n}} = \exp\left[n\beta\left((\nu_1+2\beta_1)\mu_1+\mu_2\right)\right] . \qquad (4.6b)$$

In the center of the mass frame (C.M.) the fluxes satisfy the following condition

$$m_1 \mathbf{J}_1 + m_2 \mathbf{J}_2 = 0 , \qquad (4.7)$$

where the m_i's are the molecular masses. Eq. (4.7) leads to

$$\mathbf{J}_2 = - \nu_2 \mathbf{J}_1 , \qquad (4.8)$$

with

$$\nu_2 = \frac{m_1}{m_2} = \frac{1}{\nu_1} . \qquad (4.9)$$

For isobaric and isothermal conditions, we use the Gibbs-Duhem equation (Moore 1972)

$$n_1 \nabla(\beta\mu_1) + n_2 \nabla(\beta\mu_2) = 0 , \qquad 4.10)$$

where n_i's are the number densities of the ith species, to rewrite Eq. (4.5). Namely, we have

$$\mathbf{J}_1^{CM} = \sum_{\kappa_{1n}} w_1(\kappa_{1n})\, \bar{\Omega}_{\kappa_{1n}} \left[w_1(\kappa_{1n}) - \frac{n_1}{n_2} w_2(\kappa_{1n})\right] P_{\kappa_{1n}} \nabla(\beta\mu_1)$$

$$+ \sum_{\kappa_{2n}} w_1(\kappa_{2n})\, \bar{\Omega}_{\kappa_{2n}} \left[w_1(\kappa_{2n}) - \frac{n_1}{n_2} w_2(\kappa_{2n})\right] P_{\kappa_{2n}} \nabla(\beta\mu_1) \qquad (4.11)$$

Let us choose the solvent frame where $(\mathbf{J}_2)_{sol} = 0$. In this frame one has (Kirkwood 1960)

$$\mathbf{J}_i^{SOL} = \mathbf{J}_i^{CM} + n_i \mathbf{v}_{SOL}^{CM} , \qquad (4.12)$$

where $\mathbf{v}^{CM}_{SOL}$ is the relative velocity between the CM frame and the solvent frame. Using the condition $\mathbf{J}_2^{SOL} = 0$, one gets an expression for the relative velocity which relates $\mathbf{J}_1^{SOL}$ with $\mathbf{J}_1^{CM}$, i.e.,

$$\mathbf{J}_1^{SOL} = \left[1 + \frac{n_1 m_1}{n_2 m_2} \right] \mathbf{J}_1^{CM} \quad . \tag{4.13}$$

Combining Eqs. (4.11) and (4.13), the flux in the solvent-fixed frame can be written as

$$\begin{aligned} \mathbf{J}_1^{SOL} = & \left(1 + v_2 \frac{n_1}{n_2} \right) X \\ & X \{ \sum_{\kappa_{1n}} w_1(\kappa_{1n}) \left[w_1(\kappa_{1n}) - \frac{n_1}{n_2} w_2(\kappa_{1n}) \right] \bar{\Omega}_{\kappa_{1n}} P_{\kappa_{1n}} + \\ & \sum_{\kappa_{2n}} w_1(\kappa_{2n}) \left[w_1(\kappa_{2n}) - \frac{n_1}{n_2} w_2(\kappa_{2n}) \right] \bar{\Omega}_{\kappa_{2n}} P_{\kappa_{2n}} \} \frac{\partial (\beta\mu_1)}{\partial n_1} \nabla n_1 \end{aligned} \tag{4.14}$$

From Eq. (4.14), the diffusion coefficient D^{SOL} is given by:

$$\begin{aligned} D^{SOL} = & \left(1 + v_2 \frac{n_1}{n_2} \right) X \{ \\ & \sum_{\kappa_{1n}} w_1(\kappa_{1n}) \left[w_1(\kappa_{1n}) - \frac{n_1}{n_2} w_2(\kappa_{1n}) \right] \bar{\Omega}_{\kappa_{1n}} P_{\kappa_{1n}} + \\ & \sum_{\kappa_{2n}} w_1(\kappa_{2n}) \left[w_1(\kappa_{2n}) - \frac{n_1}{n_2} w_2(\kappa_{2n}) \right] \bar{\Omega}_{\kappa_{2n}} P_{\kappa_{2n}} \} \frac{\partial (\beta\mu_1)}{\partial n_1} . \end{aligned} \tag{4.15}$$

In this frame, the diffusion coefficient is related to the volume-fixed frame, D^V , by the relation (Kirkwood 1960)

$$D^V = \frac{\rho_M x_2 \bar{V}_2}{M_1 x_1 + M_2 x_2} D^{SOL} \quad , \tag{4.16}$$

where $\bar{V}_2$ is the partial molar volume of species 2, ρ_M is the overall density of the solution, M_i and x_i are the molar mass and the mole fraction of the ith specie. Moreover, if the partial specific volumes are independent or almost independent of concentration, the mutual diffusion coefficient is given by D^V (Kirkwood 1960). Therefore for the case discussed in this section, the mutual diffusion coefficient for a binary mixture with zero mixing volume is

$$D^V = \frac{\rho_M x_2 \bar{V}_2\left[1 + \nu_2\left(n_1/n_2\right)\right]}{M_1(x_1 + \nu_1 x_2)} \left\{ \sum_{n=1}^{\infty} n^2 \left[1 + \nu_2 \frac{n_1}{n_2}\right] \bar{\Omega}_{\kappa_{1n}} P_{\kappa_{1n}} + \sum_{n=1}^{\infty} n^2 \left[\nu_1 + \frac{n_1}{n_2}\right] \bar{\Omega}_{\kappa_{2n}} P_{\kappa_{2n}} \right\} \frac{\partial \ln x_1}{\partial n_1} \frac{\partial \ln a_1}{\partial \ln x_1} \quad , \tag{4.17}$$

where we have used the relation between activities, a_i and chemical potential i.e. $\mu_i = \mu_i^o + \beta^{-1} \ln a_i$. Using this relation and Eqs. (4.6a-4.6b), one can further reduce Eq. (4.17), i.e.,

$$D^V = \frac{\rho_M \bar{V}_2}{M_1 \nu_1^2} \frac{\left(x_1 + \nu_1 x_2\right)}{x_2} \left\{ \sum_{n=1}^{\infty} n^2 \bar{\Omega}_{\kappa_{1n}} \left(e^{f_1^o} a_1 \, a_2^{\left(\nu_2 + 2\beta_2\right)}\right)^n + \right.$$

$$\left. + \sum_{n=1}^{\infty} n^2 \nu_1^2 \bar{\Omega}_{\kappa_{2n}} \left(e^{f_2^o} a_1^{\left(\nu_1 + 2\beta_1\right)} a_2\right)^n \right\} \frac{\partial \ln x_1}{\partial n_1} \frac{\partial \ln a_1}{\partial \ln x_1} \quad , \tag{4.18}$$

where we have defined

$$f_1^0 = \beta\left[\mu_1^o + \left(\nu_2 + 2\beta_2\right)\mu_2^o\right] \tag{4.19a}$$

$$f_2^0 = \beta\left[\left(\nu_1+2\beta_1\right)\mu_1^o + \mu_2^o\right] \quad . \tag{4.19b}$$

By expressing the number density, n_1, as a function of the mole fraction, x_1, i.e.,

$$n_1 = \frac{N_1}{V_M} = \frac{\rho_M x_1}{\left(m_1-m_2\right)x_1 + m_2} \quad , \tag{4.20}$$

where V_M is the total volume of the mixture, one is able to rewrite Eq. (4.18) as

$$D^V = \frac{m_1\bar{V}_2}{M_1\nu_1^2}\frac{\left(x_1 + \nu_1 x_2\right)^3}{\nu_1 + x_1\left(x_1 + \nu_1 x_2\right)\frac{d\ln\rho_M}{d\,x_1}}$$

$$x\left\{\sum_{n=1}^{\infty} n^2\,\bar{\Omega}_{\kappa_{1n}}\, e^{f_1^o} a_2^{\left(\nu_2+2\beta_2-1\right)}\left(e^{f_1^o} a_1\, a_2^{\left(\nu_2+2\beta_2\right)}\right)^{n-1} + \right.$$

$$\left.\sum_{n=1}^{\infty} n^2\nu_1^2\bar{\Omega}_{\kappa_{2n}}\, e^{f_2^o} a_1^{\left(\nu_1+2\beta_1-1\right)}\left(e^{f_2^o} a_1^{\left(\nu_1+2\beta_1\right)}\, a_2\right)^{n-1}\right\}\frac{a_1a_2}{x_1x_2}\frac{\partial\ln a_1}{\partial\ln x_1} \tag{4.21}$$

where we have used Eq. (B.4).

At this point, certain assumptions have to be taken. For example, it is reasonable to assume that the larger the number of solvated molecules the harder it is to diffuse. Explicitly, we assume that the Ω's are inversely proportional to the number molecules diffusing, i.e.

$$\bar{\Omega}_{\kappa_{in}} = \frac{\Omega_{\kappa_i}}{n} \tag{4.22}$$

In this case, the sums in Eq.(55) can be carried out, and we get

$$D^V = \frac{m_1\bar{V}_2}{M_1\nu_1^{\,2}} \; \frac{\left(x_1 + \nu_1 x_2\right)^3}{\nu_1 + x_1\left(x_1 + \nu_1 x_2\right)\dfrac{d\ \ln\rho_M}{d\ x_1}}$$

$$\times\left[\frac{\Omega_{\kappa_1}\ e^{f_1^o}\, a_2^{\left(\nu_2+2\beta_2-1\right)}}{\left(1 - e^{f_1^o}\, a_1\, a_2^{\left(\nu_2+2\beta_2\right)}\right)^2} + \frac{\nu_1^2\ \Omega_{\kappa_2}\ e^{f_2^o}\, a_1^{\left(\nu_1+2\beta_1-1\right)}}{\left(1 - e^{f_2^o}\, a_1^{\left(\nu_1+2\beta_1\right)}\, a_2\right)^2}\right]$$

$$\left(\frac{a_1 a_2}{x_1 x_2}\right)\frac{\partial \ln\ a_1}{\partial \ln\ x_1}. \tag{4.23}$$

Eq. (4.23) is the main result obtained in this section. The molar volume, density, and activities have to be obtained from vapor-liquid equilibrium experiments. The ν_is are related to the conservation of momentum in the center of mass frame. Explicitly, these two parameters are given by Eq. (4.9) as molecular mass ratios. Thus, we are left with an equation with four parameters to be determined i.e. Ω_{κ_i} and β_i.

As it has been observed elsewhere, (Medina-Noyola 1981,Keizer 1987), one has to find relations between the microscopic theoretical parameters, i.e., Ω_{κ_i} and β_i, and the macroscopic experimental transport coefficients. In the case of Eq. (4.23), we consider the infinite dilution diffusion coefficients, and we find the corresponding limits of Eq. (4.23). Namely,

$$\lim_{x_1 \to 0} D^V = D_1^\infty = \Omega_{\kappa_1}\, e^{f_1^o}\, \frac{m_1\bar{V}_2^0}{M_1}\, A_0 \tag{4.24}$$

$$\lim_{x_1 \to 1} D^V = D_2^\infty = \Omega_{\kappa_2}\, e^{f_1^o}\, \frac{m_2\bar{V}_2^0}{M_1}\, \frac{A_1}{B_1}, \tag{4.25}$$

where we have defined the following constants:

$$A_o = \lim_{x_1 \to 0}\left(\frac{a_1 a_2}{x_1 x_2}\right) \tag{4.26}$$

$$A_1 = \lim_{x_1 \to 1} \left(\frac{a_1 a_2}{x_1 x_2} \right) \tag{4.27}$$

$$B_1 = \lim_{x_1 \to 1} \left(\frac{1}{v_1 + x_1 \left(x_1 + v_1 x_2 \right) \frac{d \ln \rho_M}{d x_1}} \right). \tag{4.28}$$

Using Eqs. (4.24)-(4.25) the final expression for the mutual diffusion coefficient is given by

$$D^V = \frac{1}{v_1^2} \frac{\left(x_1 + v_1 x_2 \right)^3 \left(1 + f(x_1) \right)}{v_1 + x_1 \left(x_1 + v_1 x_2 \right) \frac{d \ln \rho_M}{d x_1}}$$

$$x \left[\frac{\left(D_1^\infty / A_o \right) a_2^{\left(v_2 + 2\beta_2 - 1 \right)}}{\left(1 - e^{f_1^o} a_1 \, a_2^{\left(v_2 + 2\beta_2 \right)} \right)^2} + \frac{\left(v_1^2 B_1 D_2^\infty / A_1 \right) a_1^{\left(v_1 + 2\beta_1 - 1 \right)}}{\left(1 - e^{f_2^o} a_1^{\left(v_1 + 2\beta_1 \right)} a_2 \right)^2} \right]$$

$$\left(\frac{a_1 a_2}{x_1 x_2} \right) \frac{\partial \ln a_1}{\partial \ln x_1}, \tag{4.29}$$

where we have used the following relation

$$\bar{V}_2 = \bar{V}_2^o \left(1 + f(x_1) \right), \tag{4.30}$$

where $\bar{V}_2^o$ is the partial molar volume of the pure component 2.

In this section we applied our generalized transport theory to binary nonelectrolyte mixtures, and in particular, we calculate the mutual diffusion coefficient. Our final result, Eq. (4.29), expresses the mutual diffusion coefficient of a nonelectrolyte binary solution without mixing under isobaric and isothermal conditions as a function of the mole fraction. The v_is are given by Eq. (4.9) and depend on the molecular mass. The infinite dilution diffusion coefficients, density and the activities are obtained from vapor-liquid equilibrium experiments. If the standard free energies of the pure substance, μ_i^o, are known, Eq. (4.29) is a two independent parameter equation. These parameters, β_1 and β_2, are related to a particular set of elementary processes.

5.- COMPARISON BETWEEN EXPERIMENTS AND THEORY.

In this section, we compare our results, Eq. (4.29), with the experimental measurements (Horrocks 1962 and Kamal 1966). For this purpose, the parameters β_1 and β_2, which can be related to solvation, have to be determined. This can be done in two ways: a) One can select two experimental values of the mutual diffusion coefficient at two different mole fractions. Namely, we calculate the parameters by matching the theoretical initial (x_1 = .1) and final (x_1 = .9) values of D^V with the experimental diffusion coefficients. In this case, Eq. (4.29) yields a set of simultaneous equations, which have to be solved. b) One can try different values of β_1 and β_2 and select those that best fit the experimental values of D_v from x_1 = 0 to x_1 =1.

Since we need vapor-liquid equilibrium information, the binary mixtures of the carbon tetrachloride-cyclohexane (Horrocks 1966) and benzene-cyclohexane (Kamal 1966) are selected to test our predictions based on the elementary processes that are considered in the previous section. All the quantities appearing in Eq. (4.29) can be obtained from experimental measurements of vapor-liquid equilibrium and solution densities of Scatchard et al (1938 and 1939). Even though these experiments are performed at temperatures greater than 303K, we can extrapolate the phenomenological expressions to 298K as it has been done before by others (Horrocks 1966 and Kamal 1966).

As it has been shown in the previous section, the intrinsic rates, Ω_{κ_i}, are related to the infinity dilution diffusion coefficients, D_i^∞, and the β_is. These relations are given by Eqs. (4.24)-(4.25) and (4.19a)-(4.19b). The ν_is are given by Eq. (4.9). Since the standard molar free energies are not available for the mixtures consider in the present discussion, we define the following parameters:

$$\lambda_1 = e^{f_1^o} \qquad (5.1)$$

$$\lambda_2 = e^{f_2^o}\ , \qquad (5.2)$$

where the f_i^os are defined by Eqs. (4.19a)-(4.19b). These new parameters change Eq. (4.29) to a four parameter equation. For convenience in notation, we define another functions

$$E(x_1) = \frac{y(x_1)}{A_o} \frac{a_2^{(\nu_2+2\beta_2-1)}}{\left(1 - \lambda_1 a_1 a_2^{(\nu_2+2\beta_2)}\right)^2} \left(\frac{a_1 a_2}{x_1 x_2}\right), \tag{5.3}$$

$$P(x_1) = \frac{y(x_1)\nu_1^2 B_1}{A_1} \frac{a_1^{(\nu_1+2\beta_1-1)}}{\left(1 - \lambda_2 a_1^{(\nu_1+2\beta_1)} a_2\right)^2} \left(\frac{a_1 a_2}{x_1 x_2}\right), \tag{5.4}$$

with $y(x_1)$ defined by

$$y(x_1) = \frac{1}{\nu_1^2} \frac{\left(x_1 + \nu_1 x_2\right)^3 \left(1 + f(x_1)\right)}{\nu_1 + x_1\left(x_1 + \nu_1 x_2\right)\frac{d\ \ln\rho_M}{d\ x_1}} . \tag{5.5}$$

Hence, D^V can be written as

$$D^V(x_1) = \left(\frac{\partial \ln a_1}{\partial \ln x_1}\right)\left[E(x_1)D_1^\infty + P(x_1)D_2^\infty\right] \tag{5.6}$$

Finally, we notice that Eq. (5.6) formally resembles the expression obtained by Bearman and Kirkwood for the mutual diffusion coefficient of a binary isothermal system of molecules (Bearman 1958). Namely, if one assumes spherically symmetric force fields, a composition independent radial distribution function, zero volume change of mixing, and a ratio of thermal expansion to isothermal compressibility independent of composition, the Bearman-Kirkwood (B-K) equation reduces to (Horrocks 1962 and Kamal 1966)

$$D_{BK}^V(x_1) = \left(\frac{\partial \ln a_1}{\partial \ln x_1}\right)\left[\left(\eta_1^o \frac{1 - x_1}{\eta(x_1)}\right)D_1^\infty + \left(\eta_1^o \frac{V_1}{V_2} \frac{x_1}{\eta(x_1)}\right)D_2^\infty\right], \tag{5.7}$$

where η_1^o is the viscosity of species 1 in the pure state, $\eta(x_1)$ is the viscosity of the mixture at composition x_1, and v_1 and v_2 are the molecular volumes. Notice that the general result, Eq.(5.6), which was obtained using the F-D theory, reduces to an equation similar to the B-K equation if we consider a particular set of coefficients. Namely, if $\nu_i + 2\beta_i = 2$, one recovers Eq. (5.7) with

$$\frac{\eta_1^o}{\eta(x_1)} = \frac{y(x_1)}{A_o} \frac{\gamma_1 \gamma_2^{\,2}}{\left(1 - e^{f_1^o} a_1\, a_2^2\right)^2} \tag{5.8}$$

$$\frac{\eta_1^o}{\eta(x_1)} \frac{v_1}{v_2} = \frac{y(x_1) v_1^2 B_1}{A_1} \frac{\gamma_1^{\,2} \gamma_2}{\left(1 - e^{f_2^o} a_1^2\, a_2\right)^2}, \tag{5.9}$$

where the γs are the activity coefficients associated with the activities, which are function of the mole fraction.

For the benzene-cyclohexane mixture one finds that m_1=78.12, m_2=84.16 and $v_1=1.08=1/v_2$. The activities and density have been reported by Scatchard (1938). In Table 1, we calculate the mutual diffusion coefficient for the benzene-cyclohexane mixture at 298K. The values for D^V are obtained using Eq. (5.6). Since the phenomenological expressions of the activities and densities are extrapolated to 298K, we calculated the βs and the λs by the second method, which is described at the beginning of the this section. In the present work, we first consider n=1 and estimate the βs by solving the set of of transcendental equations, as described at the beginning of the section. Next, using Eq. (5.6), which includes an infinite number of elementary processes, i.e., $n \in [1,\infty)$, we adjust the values of the βs and λs such that the values of the mutual diffusion coefficient are within 2% of the experimental values. Using this approach, we find $\beta_1=\beta_2=0.5$, $\lambda_1 = 0.321$, and $\lambda_2 = 0.454$. The experimental points and the theoretical values predicted by Eq. (5.7) are taken from Kamal (1966). From the values obtained using the F-D theory, we observe a decrease in the mutual diffusion coefficient for the mole fraction near zero, with a minimum near $x_1 = 0.5$.

Table 1 Mutual Diffusion Coefficient of the Benzene-Cyclohexane Mixture			
Benzene Mole Fraction	Experimental (Kamal 1962) D_{exp} $x10^9(m^2/sec)$	F-D Theory D_{PLW} $x10^9(m^2/sec)$	B-K Theory (Kamal 1962) D_{BK} $x10^9(m^2/sec)$
0.0	1.880	1.88	1.88
0.1	1.859	1.87	1.86
0.2	1.833	1.84	1.86
0.3	1.809	1.80	1.84
0.4	1.796	1.77	1.85
0.5	1.797	1.76	1.86
0.6	1.815	1.77	1.90
0.7	1.847	1.82	1.96
0.8	1.903	1.90	2.03
0.9	1.991	2.00	2.15
1.0	2.101	2.10	2.29

This behavior agrees qualitatively with the experimental results. Quantitatively, our theoretical values are within 2% of the experimental coefficients. This is a good result if one considers that we have used the simplest elementary processes and no experimental viscosity is needed. In contrast, the Bearman-Kirkwood (B-K) theory, which requires the experimental viscosity of the mixture, yields diffusion coefficients close to the experimental values, but not better than the D-F theory results, only for low mole fractions and cannot reproduce the value of the diffusion coefficient at $x_1 = 1$, i.e., pure benzene. Figure 2 depicts the differences between the theoretical calculations and the experimental values of the mutual diffusion coefficient for the mixture benzene-cyclohexane. For this mixture, notice that the F-D theory results are better than the B-K theory.

As it has been observed in other cases (Peacock-López 1986 and Keizer 1987), results obtained with the fluctuation-dissipation theory are sensitive to the experimental parameters needed to describe the system under consideration. Better numerical results can be obtained by including other elementary processes, experimental measurements in the range $x_1 < .1$ and $x_1 > .9$ and experimental measurements of the vapor-liquid equilibrium at 298K. In particular, measurements of the mutual diffusion coefficient in the mole fraction range $x_1 < .1$ and $x_1 > .9$ are required for a better matching of the slopes. In table 2, we calculate the mutual diffusion coefficient for the carbon

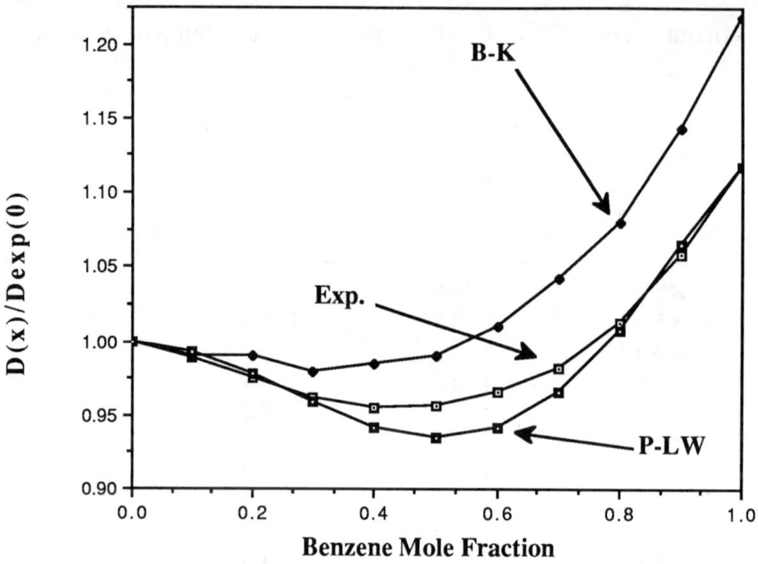

Figure 2. Mutual diffusion divided by the diffusion coefficient of pure cyclohexane. The system is a mixture of benzene-cyclohexane at 298K. Experimental values taken from Kamal (1966); B-K values obtained using Eq. (5.7)(Kamal 1966); P-LW values obtained from Eq. (5.6) with $\beta_1 = \beta_2 = 0.5$, $\lambda_1 = 0.321$ and $\lambda_2 = 0.454$.

Table 2 Mutual Diffusion Coefficient of the Carbon Tetrachloride-Cyclohexane Mixture

CCl_4 Mole Fraction	Experimental (Horrocks 1962) D_{exp} $x10^9(m^2/sec)$	F-D Theory D_{PLW} $x10^9(m^2/sec)$	B-K Theory (Horrocks 1962) D_{BK} $x10^9(m^2/sec)$
0.0	1.486	1.486	1.486
0.1	1.474	1.474	1.473
0.2	1.460	1.458	1.456
0.3	1.444	1.443	1.433
0.4	1.426	1.432	1.413
0.5	1.405	1.421	1.395
0.6	1.382	1.405	1.373
0.7	1.357	1.375	1.360
0.8	1.329	1.329	1.348
0.9	1.298	1.276	1.332
1.0	2.265	1.265	1.321

tetrachloride-cyclohexane mixture. For this mixture one finds that m_1=153.82, m_2=84.16 and v_1=0.55=1/v_2. Using these values and the method described earlier, we find that β_1 = 0.75, β_2 = 0.40, λ_1 = 0.05, and λ_2 = 0.75 yield values of the mutual diffusion coefficient within 2% of the experimental values. The experimental points and the theoretical values predicted by Eq. (5.6) are taken from Horrocks (1962).

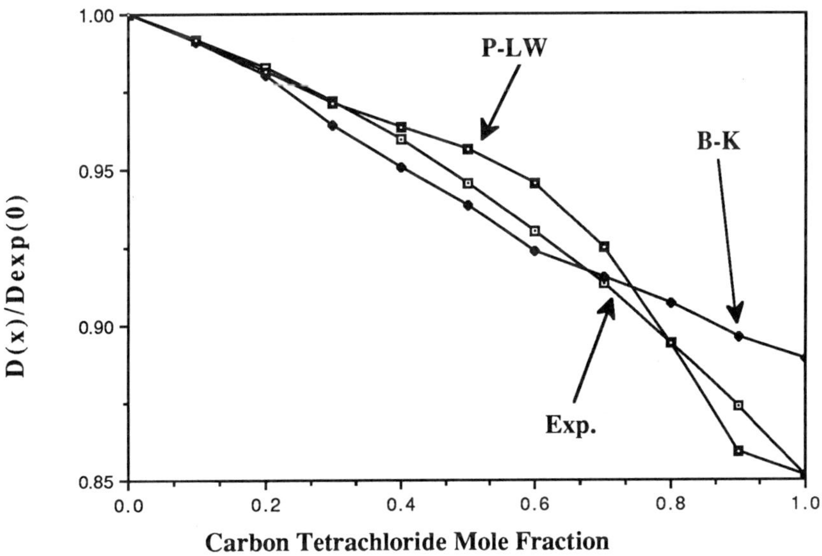

Figure 3. Mutual diffusion divided by the diffusion coefficient of pure cyclohexane. The system is a mixture of carbon tetrachloride-cyclohexane at 298K. Experimental values taken from Horrocks (1962); B-K values obtained using Eq. (5.7)(Horrocks 1962); P-LW values obtained from Eq. (5.6) with β_1 = β_2 = 0.4, λ_1 = 0.05 and λ_2 = 0.75.

For this mixture, we are not able to simultaneously match the values of D^V_{exp} and the shape of the experimental curve. This problem can be overcome by including other elementary processes. But, we do not consider other processes for three reasons. First, the lack of experimental measurements for mole fractions x_1 < 0.05 and x_1 > 0.95, second, the lack of experimental measurements of vapor-liquid equilibrium and solution densities below 303K, and third, we are considering binary mixtures as an example to assess the applicability of the generalized theory of diffusion developed in previous sections using the Fluctuation-Dissipation theory. Nonetheless, we are able to match the experimental values of the diffusion coefficient within a 2%. Figure 3 depicts the differences between the theoretical calculations and the experimental values of the mutual diffusion coefficient for the mixture carbon tetrachloride-cyclohexane. For this mixture, notice that the F-D theory results are comparable with the B-K theory.

Based on our description, the fact that in one case we get a minimum in the diffusion coefficient for the benzene-cyclohexane, but not for the other mixture can be explained. In the case of the benzene-cyclohexane mixture, the decrease in the diffusion constant is related to an increase in the size of the diffusing particles. As we increase the solute concentration, we change the intermolecular forces. The effect of this second factor, which is related to the strength of the interaction between the molecules and its dependence on the mole fraction, is reflected by the viscosity of the mixture. In the case of the benzene-cyclohexane mixture, the value of the viscosity of cyclohexane is greater that the value for benzene. But, the diffusion coefficient is inversely proportional to the radius of the diffusing particle and the viscosity. Therefore, if an increase in particle size upsets the increase in viscosity, the diffusion coefficient decreases in value. According to our model, the size of the diffusing particles has a maximum value around x=0.5. As we reverse the roles of the solvent and the solute, the size of the diffusing particle goes through a maximum. And as the viscosity decreases the diffusion coefficient increases. For the the carbon tetrachloride-cyclohexane mixture, the value of viscosity of cyclohexane is less than value for carbon tetrachloride. In this case, even though the size goes through a maximum value as we reverse the roles of solvent and solute, the value of the viscosity increases. The combined changes yield a decreasing diffusion coefficient as one changes x_1 from zero to one.

6. SUMMARY

Using the F-D theory, we have analyzed mass and heat transport in multicomponent systems. And we have obtained the generalized Onsager coefficients. This coefficients depend in a nonlinear way on the composition of the system under observation. As an application of the theory, we have selected an isothermal nonelectrolyte binary mixture. For this system, we have calculated the diffusion coefficient in the center of mass frame and the solvent frame. The expression obtained for these coefficients depend in a nonlinear way on the composition and are valid for systems with nonzero mixing volume. In principle, if one uses Kirkwood's relation (1960) between the volume-fixed diffusion coefficient and the coefficient in the solvent frame, one is able to obtain the experimental mutual diffusion coefficients. Difficulties are encountered when analytical expressions of the diffusion coefficient in the volume-fixed frame are desired for a mixture with nonzero mixing volume. As we mention before, numerically this does not represent a problem. But if we want an analytical expression, we have to solve a complicated integral. For mixtures with zero mixing volume, one finds a simple relation between the diffusion coefficients in different frames. For this case we have obtained expression for the mutual diffusion coefficient that depends on the infinite dilute diffusion coefficients, molecular masses, density, activities, standard free energies and two independent parameters. In the case of the benzene-cyclohexane and carbon tetrachloride mixtures, the standard free energies are not available. This difficulty has been overcome by the definition of two more parameters. By adjusting these four parameters, we obtain values of the mutual diffusion coefficient within 2% of the experimental values.

Finally, we have compared our theory with the Bearman-Kirkwood theory that has been developed from first principles. Also, the B-K theory has been applied to the same mixtures under similar approximations. The B-K equation, which requires of the experimental mixture's viscosity, gives good numerical results for $x_1 < 0.7$, but it cannot reproduce the value of the mutual diffusion coefficient for pure benzene or pure CCl_4. In particular, the B-K equation cannot reproduce the overall behavior of these binary mixtures because the relation

$$\eta_2^o (v_1/v_2) D_1^\infty = \eta_1^o D_2^\infty , \tag{6.1}$$

predicted by the Bearman-Kirkwood theory, is not satisfied by the benzene-cyclohexane and carbon tetrachloride-cyclohexane mixtures. This restriction that is not imposed on our results allows us to get slightly better results. Namely, if one considers the simplest elementary processes, our model of binary mixtures reproduces qualitatively the experimental behavior of the mutual diffusion coefficient as a function of the mole of fraction, and quantitatively, our results are within 2% from the experimental values without the knowledge of the experimental viscosity as is required by the B-K theory.

ACKNOWLEDGMENTS

The authors would like to thank Joel Keizer for uncountable discussions on the theory of mixtures and the Fluctuation-Dissipation theory. One of us (EP-L) would like to thank Drs. Murphy and Allan Mallinger for their hospitality during his summer visits to the University of California at San Diego, where part of the work on mixtures was developed. Finally, support to one of us (EP-L) from the National Science Foundation (CHE 83-11583 and CHE 86-18647), the Universitywide Energy Research Group of the University of California and the Bronfman Science Center is gratefully acknowledged.

APPENDIX A

For extensive variables, n_i, the Fluctuation-Dissipation theory first postulates that the dissipative part of the macroscopic equation for the average values of the extensive variables, $\bar{n}_i$, can be written in the following form:

$$\left(\frac{d\,\bar{n}_i}{dt}\right)_{diss} = \sum_{\kappa} w_i(\kappa)\left[V_\kappa^+ - V_\kappa^-\right] = R_i \;, \qquad (A.1)$$

where

$$w_i(\kappa) = n_i^+(\kappa) - \bar{n}_i(\kappa) \;, \qquad (A.2)$$

and we have defined

$$V_\kappa^\pm = \Omega_\kappa^\pm \exp\left[- \sum_{j=1}^{s} F_j \, n_j^\pm(\kappa)/k_B \right] . \qquad (A.3)$$

The second postulate states that the fluctuation around an asymptotic stable steady state, $\delta n_i = n_i(t) - \bar{n}_i$, satisfy a stochastic differential equation. Namely,

$$\frac{d\,\delta n_i}{d\,t} = \sum_{j=1}^{s} \left(\frac{\partial\left(R_i + S_i\right)}{\partial\, n_j}\right)_{n=\bar{n}} \delta n_j + f_i \;, \qquad (A.4)$$

where S_j represents the conservative terms of the macroscopic equations, f_i is a zero centered Gaussian white noise. In other words, the noise have the following statistical properties:

$$< f_i(\mathbf{r},t) > = 0 \qquad (A.5)$$

$$< f_i(\mathbf{r},t)\; f_j(\mathbf{r}',t') > = \gamma_{ij}(\mathbf{r})\, \delta(\mathbf{r}-\mathbf{r}')\, \delta(t-t') \;. \qquad (A.6)$$

Finally, the γ_{ij} is given by the third postulate. In this case we have

$$\gamma_{ij}(\mathbf{r}) = \sum_{\kappa} w_i(\kappa)\, w_j(\kappa) \left[V_\kappa^+ + V_\kappa^- \right] . \qquad (A.7)$$

APPENDIX B

For the most general case, one notices that the flux $\mathbf{J}_1$ in the center of mass frame is

$$J_1^{CM} = \left(L_{11} - L_{12} \frac{x_1}{x_2} \right) \nabla (\beta \mu_1)$$

(B.1)

in the isobaric and isothermal case. From B-1 it follows that J_1^{SOL} is given by

$$J_1^{SOL} = \left(1 + \frac{1}{v_1} \frac{x_1}{x_2} \right) \left(L_{11} - L_{12} \frac{x_1}{x_2} \right) \frac{\partial (\beta \mu_1)}{\partial n_1} \nabla n_1 \qquad \text{(B.2)}$$

which implies that D^v is

$$D^v = \frac{\rho_M \bar{V}_2}{M_1 v_1} \left(L_{11} - L_{12} \frac{x_1}{x_2} \right) \frac{1}{x_1} \frac{\partial x_1}{\partial n_1} \frac{\partial \ln a_1}{\partial \ln x_1} \ . \qquad \text{(B.3)}$$

Using the relation

$$\frac{n_1}{x_1} \frac{\partial x_1}{\partial n_1} = \frac{x_1 + v_1 x_2}{v_1 + x_1 \left(x_1 + v_1 x_2 \right) \dfrac{d \ln \rho_M}{d x_1}} \qquad \text{(B.4)}$$

and the expression for the L_{ij}, we get Eq. (4.23).

REFERENCES

Bearman, R. J., and J. G. Kirkwood, 1958. Statitistical mechanics of transport processes. XI. Equation of transport in multicomponent systems J. Chem. Phys. 28: 136-145.

de Groot, S. R., and P. Mazur, 1984. Non-equilibrium Thermodynamics. North Holland, Amsterdam; reprinted by Dover, 1984.

Fox, R.F.,and G.E. Uhlenbeck, 1970a. Contributions to non-equilibrium thermodynamics. I. Phys. Fluids 13: 1893-1902.

Fox, R.F., and G.E. Uhlenbeck, 1970b. Contributions to non-equilibrium thermodynamics. II. Phys. Fluids 13: 2881-2890.

Horrocks, J.K., and E. McLaughlin, 1962. Transport properties of binary regular solutions. Trans. Farad. Soc. 58: 1357-1362.
Kamal, I., and E. McLaughlin, 1966. Self-diffusion in binary non-electrolyte mixtures. Trans. Farad. Soc. 62: 1762-1768.
Keizer, J. 1975a. A theory of spontaneous fluctuations in macroscopic systems. J. Chem. Phys. 63: 398-403.
Keizer, J. 1975b. Nonequilibrium statistical thermodynamics and the effect of diffusion on chemical reaction rates. J. Phys. Chem. 86: 5052-5067.
Keizer, J. 1976a. Dissipation and fluctuations in nonequilibrium thermodynamics. J. Chem. Phys. 64: 1679-1687.
Keizer, J. 1976b. On the kinetic meaning of the second law of thermodynamics. J. Chem. Phys. 64: 4466-4474.
Keizer, J. and M. Medina-Noyola, 1982. Spatially nonlocal fluctuation theories: hydrodynamic fluctuations for simple fluids. Physica 115A: 301-338.
Keizer, J. 1983a. Trapping by static sinks: An exact result based on fluctuation nonequilibrium thermodynamics. J. Chem. Phys. 79: 4877-4881.
Keizer, J. 1983b. Nonlinear fluorescence quenching and the origin of positive curvature in Stern-Volmer plots. J. Am. Chem. Soc. 105: 1494-1498.
Keizer, J. 1984. The coupling of density and internal energy fluctuations by isothermal chemical reactions. J. Chem. Phys. 80: 4185-4192.
Keizer, J. Ramirez, and E. Peacock-López, 1985. The effect of diffusion on the binding of membrane-bound receptors to coated pits. Biophys. J. 47: 79-87.
Keizer, J. and E. Peacock-López, 1987. Energy transport effects on rapid bimolecular chemical reactions. Physica 147A: 61-76.
Keizer, J. 1987 Statistical Thermodynamics of Nonequilibrium Processes Springer-Verlag: New York.
Kirkwood, J.G., R.L. Baldwin, P.J. Dunlop, L.J. Gosting, and G. Kegeles, 1960. Flow equations and frames of references for isothermal diffusion in liquids. J. Chem. Phys. 33: 1505-1513.
Machlup, S., and L. Onsager, 1953. Fluctuations and irreversible processes. II. Phys. Rev. 91: 1512-1515.
Medina-Noyola, M. and J. Keizer, 1981. Spatial correlation in nonequilibrium systems: the effect of diffusion. Physica 107A: 437-463.
Moore, W.J. 1972. Physical Chemistry 4th ed. Prentice-Hall, Englewoods.
Onsager,L. 1931a. Reciprocal relations in irreversible processes. I. Phys. Rev. 37: 405-426.
Onsager,L. 1931b. Reciprocal relations in irreversible processes. II. Phys. Rev. 38: 2265-2279.
Onsager,L., S. Machlup, 1953. Fluctuations and irreversible processes. I. Phys.Rev. 91:1505-1512.
Peacock-López, E., and J. Keizer, 1985. Hydrodynamic calculation of static correlation functions for homogeneous shear. Phys. Lett. 108A: 85-90.
Peacock-López, E. and J. Ramirez, 1986. The effect of diffusion on the trapping of membrane-bound receptors by localized coated pits. Biophys. Chem. 25: 117-125
Peacock-López, E., and J. Keizer, 1988. One-dimensional reactive systems: the effect of diffusion on rapid bimolecular processes. J. Chem. Phys. 88: 1997-2003.

Scatchard, G., S.E. Wood, and J.M. Mochel, 1938. Vapor-liquid equilibrium III. J. Phys. Chem. 43: 119-130.

Scatchard, G., S.E. Wood, and J.M. Mochel, 1939. Vapor-liquid equilibrium. IV. Carbon tetrachloride-cyclohexane mixtures. J. Am. Chem. Soc. 61: 3206-3210.

Woodhouse, L. 1984. Nonequilibrium Studies on Diffusion, doctoral dissertation, University of California, Davis.

On the Infinite-Dilution Solute Chemical Potential: Theory and Applications

Daniel Jonah
Univ. Research and Development Services Bureau
Freetown, Sierra Leone, W. Africa

ABSTRACT

Both a reciprocal temperature- and density-expansion are presented for Henry's Constant. These have been derived by perturbing an infinitely dilute binary solution about the pure solvent.

The second order perturbation in the reciprocal temperature -expansion is a fluctuation term, which unlike the other leading terms, is not thermodynamically accessible. Two thermodynamic approximations with adjustable parameters are suggested for this term. Tests against Molecular Dynamics data for binary dilute mixtures of Lennard-Jones (6-12) molecules, with equal well depths and varying size ratios, show quite good agreement; but these same parameters reproduce simulation data for molecules of different well depths rather poorly.

The reciprocal temperature series is next compared with the closed expression for the Henry's Constant obtained from the one-fluid conformal solution theory. Though our expression does'nt invoke mixing rules in its derivation, it shows that there is nevertheless a subtle and very interesting relationship between this expression and that from one-fluid conformal solution theory through the concept of mixing rules: briefly, the effect of the "correct" mixing rules (which are presently not known) is to telescope the remainder of the reciprocal temperature series into its leading term- that is, the first order perturbation term .

Finally applications to low-pressure vapor-liquid equilibria and solution thermodynamics are considered.

1. INTRODUCTION

Recently, there has been a lot of interest shown in the residual solute chemical potential at infinite dilution [(Shing and Gubbins; Shing,Gubbins and Lucas 1983,1988); (Hamad and Mansoori 1987);(Meyer 1988);(Lofti and Fischer 1989)]. Efforts have been directed along two main fronts: simulation and analytical development in the form of perturbation expansion. Knowledge about this thermodynamic function is basic to the understanding of solubility thermodynamics, which finds several applications in processes of industrial interest, as well as being of intrinsic interest in itself as a testing ground for combining and mixing rules.It is therefore not surprising that it has constituted an area of so much interest and activity.

Traditionally, perturbation treatment of dilute solutions usually starts with the mixture at some finite composition, perturbing its Helmholtz free energy about an equivalent hard sphere system, followed by the limiting operation to infinite dilution. Such a procedure inevitably brings in, in addition to the solute-solvent and solvent-solvent interactions, also the solute-solute interactions. Intuitively it seems strange to have a formulation of dilute solution theory involving solute-solute interactions of which thermodynamic properties ought to be independent. As a matter of fact, such a consideration has formed the basis of one of the consistency tests in the recent simulation (Lofti and Fischer 1989) of the residual solute chemical potential at infinite dilution. It seems therefore reasonable to have alternative formulations from which the solute-solute interactions are absent. A perturbation treatment using the pure solvent as the reference substance, and the solute-solvent interactions as the perturbation, provides this kind of formulation; here,of course,the contribution from the solute-solute interactions to the total potential energy is assumed negligible and ignored from the start.

An attempt at such a formulation was made by (Jonah,Shing and Gubbins 1981) to include only the first order perturbation, and by (Jonah 1986) up to the second order perturbation. An interesting feature about this expansion is the fact that the first order perturbation can be expressed in terms of pure solvent thermodynamic properties for the case of Lennard-Jones pair potentials. For molecules of the same size, but different interaction energies, comparison with simulation data indicated reasonable convergence up to the first order perturbation; however,the convergence was poor for molecules differing significantly in size.

The second order perturbation term is a fluctuation term in the difference between "private" solute-solvent and solvent-solvent interactions. Though this term is not thermodynamically accessible, it's form is suggestive of thermodynamic approximations in terms of pure solvent properties, using fluctuation theory within the canonical ensemble theory.

Mention must be made of the expression obtained through 1-fluid conformal solution theory. Here we have a closed expression for the solute residual chemical potential at infinite dilution. It has two satisfying features: (1) solute-solute interactions are absent for standard mixing rules like the 1-fluid Van der Waals (VDW1) mixing rules;(2) the zeroth order term is the same as that for a perturbation about the pure solvent component.

In this paper we try to do the following: we present two expansions for the residual solute chemical potential at infinite dilution— an inverse temperature expansion, and a density-expansion; the latter expansion is shown to yield the correct low density limit up to the third virial coefficient,as well as the inverse temperature expansion. Next,we present two thermodynamic approximations for the second order perturbation term in the inverse temperature-expansion, both of which have been motivated by fluctuation theory,and contain two adjustable parameters.In the next section ,the 1-fluid comformal solution theory is introduced, and a subtle relationship between this and the inverse temperature expansion in the context of mixing rules is elucidated.

Finally we consider some applications of the preceding theory to both simulation and experimental data. The two thermodynamic approximations are tested against simulation data for Henry's

Constant for a Lennard-Jones(6-12) fluid, as presented by Shing et al.,(1988),and Lofti and Fischer,(1989).Methods for correlating low-pressure vapor-liquid equilibria data, and solid solubilities in supercritical fluids according to 1-fluid conformal theory, are outlined.

2. EXPRESSIONS FOR INFINITE DILUTION CHEMICAL POTENTIAL

In this section, we obtain various expressions for the infinite dilution solute chemical potential.

We consider for simplicity only binary mixtures; the extension to multi-component mixtures will be quite straight-forward.

Our system of interest is an infinitely dilute binary mixture with N_b solvent molecules belonging to the component *b*, and one solute molecule of the second component *a*, in the thermodynamic state characterized by temperature T, and density ρ_b of the pure solvent.

Let us define the "private" energies of a solute and solvent molecule, Ψ and Ψ_b as follows:

$$\Psi = \sum_{j=1}^{N} u_{ab}(\mathbf{r}_o,\mathbf{r}_j) \equiv \sum_{j=1}^{N} u_{ab}(o,j) \quad , \tag{2.1a}$$

$$\Psi_b = \sum_{j=1}^{N} u_{bb}(\mathbf{r}_o,\mathbf{r}_j) \equiv \sum_{j=1}^{N} u_{bb}(o,j) \quad . \tag{2.1b}$$

In (2.1a),the position vector $\mathbf{r}_o$, is associated with the solute molecule, whilst in (2.1b), it is associated with a solvent *b*-molecule, which has replaced the solute molecule.

The basic expression[1] from which the others are derived is the following: (see appendix A)

$$\left.\begin{aligned} &\exp\{-\beta(\mu^{\infty}_{ar}-\mu^{\bullet}_{br})\}=\langle\exp(-\beta\Delta\Psi)\rangle_{ps}, \\ &\Delta\Psi = \Psi - \Psi_{b} \ . \end{aligned}\right\} \tag{2.2}$$

μ^{∞}_{ar} denotes the residual infinite dilution chemical potential of the solute component *a*, whilst $\mu^{\bullet}_{br}$ is the corresponding thermodynamic function for the pure solvent component *b*, both at temperature T, and density ρ ($=\rho_b$); $\beta \equiv 1/(k_B T)$, where k_B denotes the Boltzmann constant. $\langle\ \rangle_{ps}$ indicates a canonical average over the *pure* solvent component.

Explicitly,

$$\langle\exp(-\beta\ \Delta\Psi)\rangle_{ps} \equiv \frac{\int\exp(-\beta\ \Delta\Psi\ )\exp(-\beta\ U^{solv}_{N+1})d\{N+1\}}{\int\exp(-\beta\ U^{solv}_{N+1})\ d\{N+1\}} \tag{2.2a}$$

where in standard notation,

$$d\{N+1\} \equiv \prod_{j=1}^{N+1} dr_j$$

whilst U^{solv}_{N+1} denotes the total potential energy of the pure solvent (N+1)-molecule system.

We may observe at this stage, that according to (2.2a), only solvent-solvent pair correlation function $g_{bb}(r)$ appears in the relation (2.2). This situation is not inconsistent with the Kirkwood's expression for μ^{∞}_{ar} , namely:

---

1) It was brought to the author's attention (prior to the publication of their paper), that Stecki and Samborski (1989) have made equation (2.2) the basis of their computer simulation.

$$\frac{\mu_{ar}^{\infty}}{RT} = \beta\rho \int_0^1 d\xi \int dr_{12} u_{ab}(1,2)\; g_{ab}^{\infty}(1,2;\xi) \tag{2.3}$$

where ξ is the Kirkwood's coupling parameter,— the solute *a-molecule* interacts with the solvent *b*-molecules only to the extent ξ, $(0 \le \xi \le 1)$ — whilst $g_{ab}^{\infty}(r;\xi)$ denotes the solute-solvent pair correlation function in the infinitely dilute system; this correlation function can be expressed in terms of the corresponding solvent-solvent pair correlation function by perturbing about the pure solvent component.

The next observation about the relation (2.2) is that it satisfies both of the Lofti-Fischer consistency tests (Lofti, Fischer 1989); the satisfaction of the second is of course automatic, by virtue of the fact that the mixture is considered at infinite dilution.

From equation (2.2), it is possible to derive two expansions — a reciprocal temperature- and a density-expansion— for μ_{ar}^{∞} with coefficients which are molecular integrals over the solvent-solvent correlation funtions.But before we do this, let us re-write (2.2) in terms of the Henry's Constant H_{ab}, and the pure solvent fugacity $f_b^{\bullet}$; that is ,

$$\ln (H_{ab}/f_b^{\bullet}) = -\ln\langle\exp(-\beta\, \Delta\Psi)\rangle_{ps} . \tag{2.4}$$

We note the striking similarity between the above relation, and that resulting from perturbing the Helmholtz free energy A about an arbitrary reference A_{ref}:

$$\left.\begin{aligned} &\beta(A-A_{ref}) = -\ln\langle\exp(-\beta\, \Delta W_N)\rangle_{ref} , \\ &\Delta W_N = \Sigma_{i,j}[U(i,j) - U_{ref}(i,j)]. \end{aligned}\right\} \tag{2.4a}$$

<u>Reciprocal Temperature Expansion</u>

Expansion of the exponential term on the right hand side of (2.2) in powers of β leads at once to:

$$-\beta(\mu^{\infty}_{ar} - \mu^{\bullet}_{br}) = \beta <\Delta\Psi>_{ps} - \frac{\beta^2}{2}<(\Delta\Psi - <\Delta\Psi>)>^2_{ps} + O(\beta^3) \quad . \tag{2.5}$$

Alternatively the left hand side of the above relation may be replaced by that of (2.4). In this form, up to the first term (first order perturbation), (2.5) has been previously derived (Jonah, Shing and Gubbins, 1981); (2.5) is also the limiting form ($x_a \to 0$) of an equation [numbered (5)] in a more recent publication (Jonah 1986).

For pair potentials of the Lennard-Jones form— that is, for u(r) of the form,

$$u(r) = A_1\phi_1(r) + A_2\phi_2(r),$$

and such that,

$$r\frac{du}{dr} = B_1\phi_1(r) + B_2\phi_2(r),$$

(where A_1, B_1 etc. are coefficients independent of r), $<\Delta\Psi>_{ps}$ can be expressed in terms of pure solvent thermodynamic properties. Thus for the L-J (6-12) potential, we have:

$$<\Delta\psi>_{ps} = \frac{\varepsilon_{ab}}{\varepsilon}\left(\frac{\sigma_{ab}}{\sigma}\right)^3\left\{\left(\frac{\sigma_{ab}}{\sigma}\right)^6 - 1\right\}RT\ z^{\bullet}_{br}$$

$$-2\left\{1 + \frac{\varepsilon_{ab}}{\varepsilon}\left(\frac{\sigma_{ab}}{\sigma}\right)^3\left[\left(\frac{\sigma_{ab}}{\sigma}\right)^6 - 2\right\}\ U^{\bullet}_{br}, \tag{2.6}$$

where $U^{\bullet}_{br}$ and $z^{\bullet}_{br}$ are respectively, the residual internal energy and compressibility factor of the pure solvent; ε_{ab}, σ_{ab}, $\varepsilon(\equiv\varepsilon_{bb})$, $\sigma(\equiv\sigma_{bb})$ denote the usual solute-solvent and solvent-solvent molecular parameters. Similar relations to (2.6) for other potentials are not obvious to the author.

The resulting expression from (2.5) for the Henry's Constant, when only the first order perturbation is considered, bears a striking resemblance to that obtained through 1-Fluid conformal solution theory; this similarity will be taken up later in the course of this paper.

As has been demonstrated previously (Jonah,et.al, 1981), comparison with simulation data for H_{ab} shows quite good agreement for molecules of the same size, in the range $0.5< \varepsilon_{ab}/\varepsilon <2$, when calculated from equation (2.5); the comparison for variation with size ratios σ_{ab}/σ ($\varepsilon_{ab}/\varepsilon =1$) however, showed poor agreement.Thus, the convergence quality of (2.5) for mixtures of similar-sized molecules would seem to be quite good, using only the first order perturbation; to describe molecules differing significantly in size, we may need to go beyond the first order perturbation.

The second order perturbation is represented by the canonical average of fluctuations in the difference between solute and solvent "private" energies in the infinitely dilute mixture. Though this term cannot, like the first order perturbation, be expressed exactly in terms of pure solvent thermodynamic properties, we may hope to be able to use fluctuation theory in finding such approximate expressions. In the next section, such considerations will be taken up. The possibility of approximating the Henry's Constant in terms of pure solvent thermodynamic properties, is quite attractive, and deserving of attention.

Density Expansion

The reciprocal temperature expansion as given in (2.5) is clearly unsuitable for hard-core potentials. For such potentials , the exponential on the r.h.s. of (2.2) needs to be expanded in terms of the Mayer's f-function; that is, we write:

$$\exp(-\beta\ \Delta\Psi) = \prod_{J=1}^{N}[1 + \Delta f(o,j)] \quad , \tag{2.7}$$

$$\Delta f(r) = \exp\{-\beta[u_{ab}(r) - u_{bb}(r)]\} - 1 \quad . \tag{2.7a}$$

A more convenient representation for $\Delta f(r)$ is the following:

$$\left.\begin{aligned} \Delta f(r) &= \omega(r) \exp[\beta u_{bb}(r)], \\ \omega(r) &= f_{ab}(r) - f_{bb}(r) \quad , \\ f_{ij}(r) &= \exp[\beta u_{ij}(r)] - 1 . \end{aligned}\right\} \tag{2.7b}$$

Substitution of (2.7) into (2.2) now yields the desired density expansion:

$$\exp\{-\beta(\mu^{\infty}_{ar} - \mu^{\bullet}_{br})\} = 1 + \langle\Sigma' \Delta f(o,j)\rangle_{ps} + \langle {}_{j,k}\Sigma' \Delta f(o,j)\Delta f(o,k)\rangle_{ps} + .. \quad (2.8)$$

where the prime on ${}_{j,k}\Sigma'$ indicates that $j \neq k$.

The averages in the equation (2.8) may be expressed in terms of the background correlation functions, $y^{(2)}(1,2)$, $y^{(3)}(1,2,3)$ etc. defined by (see for example, Hill 1956),

$$y^{(2)}(1,2) = g^{(2)}(1,2) \exp[\beta\, u(1,2)], \quad (2.9a)$$

$$y^{(3)}(1,2,3) = g^{(3)}(1,2,3) \exp[\beta\{ u(1,2) + u(2,3) + u(1,3)\}]. \quad (2.9b)$$

Thus for the terms indicated in (2.8), we have:

$$\langle \sum_j \Delta f(o,j)\rangle_{ps} = \frac{\rho}{V}\int d1\, d2\, \omega(1,2)y^{(2)}(1,2) , \quad (2.10a)$$

$$\langle {}_{j,k}\Sigma' \Delta f(o,j)\Delta f(o,k)\rangle_{ps}$$

$$= \frac{\rho^2}{2V} \int d1d2d3\, \omega(1,2)\omega(1,3)e^{-\beta u(2,3)}y^{(3)}(1,2,3) , \quad (2.10b)$$

where it is to be understood that the correlation functions $y^{(2)}$, $y^{(3)}$ refer to the *pure* solvent component, and $\omega(r)$ represents the difference between the Mayer f-function for solute-solvent, and solvent-solvent interactions.

The general term in (2.8), can be readily written down, and expressed in terms of higher order correlation functions $y^{(n)}$, ($n > 3$), but there is no need for this.

By substituting (2.10a) and (2.10b) in (2.8), we may write down the density expansion in the form:

$$-\beta(\mu^{\infty}_{ar} - \mu^{\bullet}_{br}) = \rho \frac{S_2}{V} + \frac{1}{2} \frac{\rho^2}{2V}(S_3 - S_2^2) + O(\rho^3), \quad (2.10c)$$

$$S_2 = \int d1\ d2\ \omega(1,2) y^{(2)}(1,2) \ , \tag{2.10d}$$

$$S_3 = \int d1\ d2\ d3\ \omega(1,2)\ \omega(1,3) e^{-\beta u(2,3)} y^{(3)}(1,2,3) \ . \tag{2.10e}$$

Two ways in which we can check the correctness of (2.10c) are as follows:

(1) It should be possible to recover the reciprocal temperature-expansion (2.5) from (2.10c); this can be shown to be the case;

(2) the correct low-density limit should also be recoverable; this can also be readily verified up to the third virial coefficient [see appendix B].

3.THERMODYNAMIC APPROXIMATIONS FOR SECOND ORDER PERTURBATION TERM

In this section, we shall consider two approximations for the second order perturbation term, both of which have been motivated by fluctuation theory. The expressions will be in terms of pure solvent thermodynamic properties, with coefficients which are functions of the molecular parameters.

The First Approximation

We consider a Lennard-Jones pair potential u(r) of the form:

$$u(r) = ar^{-n} - br^{-m}.$$

Denoting the solvent total potential by U, we have:

$$U - \langle U \rangle = a \sum_{ij} (r_{ij}^{-m} - \langle r_{ij}^{-m} \rangle) - b \sum_{ij} (r_{ij}^{-n} - \langle r_{ij}^{-n} \rangle) \tag{3.1}$$

Or in more convenient notation:

$$\left.\begin{aligned} U - \langle U \rangle &= a \sum_i \chi_i^{(m)} - b \sum_i \chi_i^{(n)} , \\ \chi_i^{(l)} &= \sum_{j>i} (r_{ij}^{-l} - \langle r_{ij}^{-l} \rangle), \qquad (l = m,n). \end{aligned}\right\} \tag{3.1a}$$

In the same notation, we have:

$$\Delta\Psi - <\Delta\Psi> = \Delta a\ \chi_1^{(m)} - \Delta b\ \chi_1^{(n)} \tag{3.2}$$

$$\left.\begin{aligned} \Delta a &= \varepsilon_{ab}\sigma_{ab}^{\ m} - \varepsilon\ \sigma^{m} \\ \Delta b &= \varepsilon_{ab}\sigma_{ab}^{\ n} - \varepsilon\ \sigma^{n}. \end{aligned}\right\} \tag{3.2a}$$

We shall assume that $\Sigma\ \chi_i^{(l)}$ $(l = m,n)$ is simply a multiple of $\chi_1^{(l)}$, and also that the contribution from one of the two terms on the r.h.s. of (3.2) dominates that of the other; it then seems plausible to write either,

$$\Delta\Psi - <\Delta\Psi> \propto \frac{\Delta a}{a}\ (\ a \sum_i \chi_i^{(m)} - b \sum_i \chi_i^{(n)}) \quad , \tag{3.2b}$$

or

$$\Delta\Psi - <\Delta\Psi> \propto \frac{\Delta b}{b}\ (\ a \sum_i \chi_i^{(m)} - b \sum_i \chi_i^{(n)}) \quad , \tag{3.2c}$$

depending on which term is the dominant term in (3.2a).

Probably better still, we may take a weighted average of (3.2b) and (3.2c), so that we have (using equation (3.1a)):

$$\Delta\Psi - <\Delta\Psi> \propto (c_1\frac{\Delta a}{a} + c_2\frac{\Delta b}{b})\,(U - < U>) \tag{3.3}$$

c_1 and c_2 being adjustable parameters.

Now making use of the standard fluctuation relation,

$$<(U - <U>)^2> = k_B T^2 C_V \ , \tag{3.4}$$

where C_V is the constant volume heat capacity, we have as our first approximation,

$$\beta^2 <(\Delta\Psi - <\Delta\Psi>)^2> = \frac{C_V}{R}(k_1\frac{\Delta a}{a} + k_2\frac{\Delta b}{b})^2, \tag{3.5}$$

where k_1 and k_2 again, are adjustable parameters.

Specializing to the L-J (6-12) potential, we have:

$$\left.\begin{aligned} \frac{\Delta a}{a} &= \frac{\varepsilon_{ab}}{\varepsilon}\left(\frac{\sigma_{ab}}{\sigma}\right)^{12} - 1 , \\ \frac{\Delta b}{b} &= \frac{\varepsilon_{ab}}{\varepsilon}\left(\frac{\sigma_{ab}}{\sigma}\right)^{6} - 1 . \end{aligned}\right\} \tag{3.6}$$

<u>A Second Approximation</u>

We may express the fluctuation in the differences in "private" energies for the L-J (n,m) pair potential, as follows:

$$<(\Delta\Psi - <\Delta\Psi>)^2> = (\Delta a)^2 X' + 2\,\Delta a\,\Delta b\, Y' + (\Delta b)^2 Z' \tag{3.7}$$

where X'etc. are given by:

$$\left.\begin{aligned} X' &\equiv <(\chi_1^{(m)})^2>, \\ Y' &\equiv <(\chi_1^{(m)})(\chi_1^{(n)})> \\ Z' &\equiv <(\chi_1^{(n)})^2> \quad . \end{aligned}\right\} \tag{3.7a}$$

Similarly, we may write the fluctuation in the total potential energy:

$$< (U - <U>)^2> = a^2X + 2abY + b^2Z \tag{3.8}$$

where,

$$\left.\begin{aligned} X &\equiv <(\sum_i \chi_i^{(m)})^2>, \\ Y &\equiv <(\sum_i \chi_i^{(m)})(\sum_i \chi_i^{(n)})>, \\ Z &\equiv <(\sum_i \chi_i^{(n)})^2>. \end{aligned}\right\} \tag{3.8a}$$

To arrive at the second approximation, we need to find two other fluctuation relations involving X, Y and Z; now that we have three linear equations in these variables, these can be solved for the three unknowns in terms of thermodynamic properties. The next step then consists in finding approximate relations between X and X′, Y and Y′, Z and Z′.

We identify the three fluctuation relations for the L-J (n-m) potential as the following (Rowlinson 1959):

$$<(U - <U>)(\mathcal{V} - <\mathcal{V}>)> = k_B T\{ TV(\partial P/\partial T)_V - Nk_B T \} \equiv h_1,$$

$$<(U - <U>)^2> = k_B T^2 C_V \equiv h_2 ,$$

$$<(\mathcal{V} - <\mathcal{V}>)^2> = T\{- \frac{nm}{9} U_r + \frac{(n + m + 3)}{3} PV - \frac{(n + m)}{3} NkT - V/\kappa\} \equiv h_3; \quad (3.9)$$

$\mathcal{V}$ is the virial function defined by:

$$\mathcal{V} = -(1/3) \sum_{i,j} r_{ij} du/dr_{ij} ,$$

whilst κ denotes the isothermal compressibility coefficient. Expressed in terms of X, Y and Z, the set of equations (3.9) assumes the form:

$$\left.\begin{aligned} ma^2X - ab(m + n)Y + nb^2Z &= h_1 , \\ a^2X - 2abY + b^2Z &= h_2 , \\ m^2a^2X - 2mnabY + n^2b^2Z &= h_3. \end{aligned}\right\} \quad (3.9a)$$

The solution to the above set is readily verified to be given by:

$$X = \frac{-1}{a^2(n-m)^3}\{2n(m-n)h_1 + (m+n)(h_2n^2 - h_3) + 2n(h_3 - mnh_2)\},$$

$$Y = \frac{1}{b(n-m)^3}\{m(n^2h_2 - h_3) + n(h_3 - m^2h_2) + h_1(m^2 - n^2)\}, \quad (3.10)$$

$$Z = \frac{-1}{b^2(n-m)^3}\{2m(n-m)h_1 + (m+n)(h_2m^2 - h_3) + 2m(h_3 - mnh_2)\} .$$

What we have done up to now is exact; our approximation is going to be introduced through the relationship we shall suggest between X and X$'$ etc.

For this, let us assume a simple proportionality relationship between each pair; we shall use the same proportionality constant in each case, except for the relationship between Z and Z′; we have used a different proportionality constant for this pair in an attempt to give unequal weightings to the effects of the attractive and repulsive parts of the potential.

Based on these assumptions, our second approximation to the second order perturbation term takes the form:

$$\beta^2\langle(\Delta\Psi - \langle\Delta\Psi\rangle)^2\rangle = K_1\{-(st^4-1)^2(h_1/3 - h_2 - h_3/36)$$

$$+ 2(st^4-1)(st^2-1)(h_1/2 - 2h_2 - h_3/36)\}$$

$$+ K_2(st^2-1)^2(-2h_1/3 + 4h_2 + h_3/36)\ , \qquad (3.11)$$

where,

$$\left.\begin{aligned} h_1 &= \alpha V/(\kappa R) - 1, \\ h_2 &= C_V/R\ , \\ h_3 &= -8U_r/(RT) + 6(z-1) + V(P-\kappa^{-1})/(RT). \end{aligned}\right\} \qquad (3.11a)$$

We must emphasize the empirical nature of the above approximations, which have been partly motivated by some results in fluctuation theory, and partly by mathematical convenience.

4. RELATION TO 1-FLUID CONFORMAL SOLUTION THEORY

There is a very interesting, though subtle, connection between expansions obtained in the preceding sections, and corresponding expressions based on 1-fluid conformal solution theory. We shall now address this feature.

Through 1-fluid conformal solution theory, we have for the residual solute chemical potential at infinite dilution the following expression — see for example Hamad and Mansoori (1987):

$$\frac{\mu^\infty_{ar}}{RT} = \frac{\mu^\bullet_{br}}{RT} + \left(\frac{N}{\varepsilon_m}\partial\varepsilon_m/\partial N_a\right)^\infty \frac{U^\bullet_{br}}{RT} + \left(\frac{N}{\sigma^3_m}\partial\sigma^3_m/\partial N_a\right)^\infty z^\bullet_{br}\ ; \qquad (4.1)$$

ε_m, σ_m are the molecular parameters for the pseudo-single component fluid; all thermodynamic properties on the r.h.s. of the above equation pertain to the pure solvent component, which in this case has been taken as the reference fluid.The derivation of this equation may be based, either on the residual Gibbs free energy, or the Helmholtz free energy .

Equation (4.1) provides us with an expression for the residual solute chemical potential at infinite dilution in closed form, that is exact for conformal mixtures; the main limitation however to its usefulness, is the lack of knowledge of the *correct* mixing rules. Comparison of this equation with either equation (2.5) or or (2.10c) provides us in principle with a closed expression for the remainders in the density- and reciprocal temperature-expansion of μ^{∞}_{ar}, beyond the zeroth order term . Thus knowledge of the correct mixing rules, is tantamount to knowledge about the sum of the remainders in these two expansions.

Let us consider more closely the effect of mixing rules. We shall re-write (4.1) in a slightly different, but more convenient form:

$$\left.\begin{aligned} \frac{\mu^{\infty}_{ar}}{RT} &= \frac{\mu^{\bullet}_{br}}{RT} + F_u(s,t;T,\rho)\frac{U^{\bullet}_{br}}{RT} + F_z(s,t;T,\rho)\, z^{\bullet}_{br} \\ F_u &\equiv \left(\frac{N}{\varepsilon_m}\,\partial\varepsilon_m/\partial N_a\right)^{\infty}, \\ F_z &\equiv \left(\frac{N}{\sigma^3_m}\,\partial\sigma^3_m/\partial N_a\right)^{\infty}; \end{aligned}\right\} \qquad (4.1a)$$

and s and t are used to denote the molecular parameter ratios according to:

$$s \equiv \varepsilon_{ab}/\varepsilon, \qquad t \equiv (\sigma_{ab}/\sigma)^3.$$

We have purposely suggested in the above notation that F_u and F_z are state dependent, in recognition of the fact that any realistic mixing rules would have to be state-dependent; as a matter of fact subsequent considerations in this section will confirm this view.

Now, for the VdW1 mixing rules F_u and F_z are functions of the molecular parameter ratios alone (see equation (5.4a)); and then equation (4.1a) assumes a form that is remarkably similar to that of equation (2.5) truncated up to the first order perturbation term, and making use of the L-J (6-12) - expression in (2.6) for this term.

But as we have already observed, (4.1) may be regarded as the sum of the infinite series in (2.5), assuming the convergence of the latter, and the appropriate forms of F_u and F_z through the use of the correct mixing rules. It appears then, that if the coefficients of $U^{\bullet}_{br}/(RT)$ and $z^{\bullet}_{br}$ in (2.6) are suitably modified - in particular made to be state-dependent - then contributions from the remainder of the infinite series in (2.5) beyond the first order term can be accounted for. These remarks must of course be understood within the context of conformality.

It should be possible to use this kind of considerations for a systematic improvement of the VDW1 mixing rules which are known to be correct at low densities for hard spheres (Lebowitz 1964), and also for soft spheres, provided the temperature is sufficiently high. Such investigations have been pursued elsewhere (Jonah and Deiters).

5. APPLICATIONS

<u>Henry's Constant From Equation (2.5)</u>

Recently, Shing, Gubbins and Lucas (1988), and Lofti and Fischer (1989), generated simulation data for the Henry's Constant based on the L-J (6-12) pair potential at reduced temperature $T^*(\equiv k_B T/\varepsilon) = 1.2$, and reduced density $\rho^*(\equiv \rho\sigma^3) = 0.7$. We shall test equation (2.5) with the two thermodynamic approximations (3.5) and (3.11) for the second order perturbation term. We define additional reduced variables as follows:

$$P^* \equiv P\sigma^3/\varepsilon, \qquad U^*_r \equiv U^{\bullet}_r / N\varepsilon, \qquad \mu^*_r \equiv \mu^{\bullet}_{br} / N\varepsilon .$$

Expressed in terms of the above variables, we have for these approximations:

First Approximation

$$\ln\left(\frac{H_{ab}}{\rho k_B T}\right) = \left[\ln\left(\frac{H_{ab}}{\rho k_B T}\right)\right]_1 - \frac{1}{2}\frac{\partial U^*}{\partial T^*}\{ k_1(st^4-1)+k_2(st^2-1)\}^2 \qquad (5.1)$$

$$\left[\ln\left(\frac{H_{ab}}{\rho k_B T}\right)\right]_1 = \frac{\mu^*_{br}}{T^*} - \frac{2U^*_{br}}{T^*}\{1 + st^2(t^2-2)\} + st^2(t^2-1)z^*_{br} \qquad (5.1a)$$

$$z^*_{br} = \frac{P^*}{\rho^* T^*} - 1. \qquad (5.1b)$$

Second Approximation.

$$\ln\left(\frac{H_{ab}}{\rho k_B T}\right) = \left[\ln\left(\frac{H_{ab}}{\rho k_B T}\right)\right]_1 - \frac{\beta^2}{2}\langle(\Delta\Psi - \langle\Delta\Psi\rangle)^2\rangle \qquad (5.2)$$

with the second term given by (3.11), but with h'_1 (replacing h_1) etc.now given by,

$$\left.\begin{aligned} h'_1 &= \frac{1}{\rho^*}\frac{\partial P^*}{\partial T^*} - 1 , \\ h'_2 &= \frac{\partial U^*}{\partial T^*} , \\ h'_3 &= -\frac{8U^*}{T^*} + 6z + \frac{1}{T^*}\left\{\frac{P^*}{\rho^*} - \left(\frac{\partial \rho^*}{\partial P^*}\right)^{-1}\right\} . \end{aligned}\right\} \qquad (5.2a)$$

All thermodynamic properties of the pure solvent have been calculated from the equation of state for L-J (6-12) fluids according to Nicolas et. al.(1979); these are listed in Table 1.

Table 1. Thermodynamic Properties for L-J (6-12) Fluids at Reduced Temperature T^*=1.2, and Reduced Density ρ^*=0.7 (Nicolas et al. 1979).

U_r^*	z_r^*	$\partial U_r^*/\partial T^*$	$\partial P^*/\partial T^*$	$(\partial \rho^*/\partial T^*)^{-1}$
-4.768	-0.243	0.712	3.099	8.550

Figure 1 compares the simulated values (Lofti and Fischer 1989) with those calculated from equation (5.1), for the cases s=1, 1.5 and 0.5; values calculated from (5.2) differ very little from those of (5.1). The adjustable parameters in either of these two equations were evaluated by constraining the curve to pass through the points (2, -2.14) and (1.5, -1.98) (s=1); in this way we obtain k_1= .01093, k_2= 4.5748, K_1= .0133, K_2= -6.087. The poor agreement for the cases s=1.5 and 0.5 clearly indicates that the parameters are not independent of well-depth ratios.

It is interesting to observe at this point, that the use of the 1-fluid conformal expression for the Henry's Constant based on the VDW1 mixing rules, - equations (4.1) and (5.4a) - provides values which are in very good agreement with the simulation data, except for the small region t < 0.5, s = 1 .

Low-Pressure Vapor Liquid Equilibria (VLE) and Solid Solubilities in Supercritical Fluids

In the preceding sections,we have presented several expressions for the residual solute chemical potential at infinite dilution, and have elucidated the role of mixing rules in relating the closed 1-fluid form to the series expansion. In this section, we shall consider the application of this important thermodynamic function to the study of low-pressure VLE on the one hand, and solid solubilities in supercritical fluids for sparingly soluble substances, on the other.Instead of using the truncated series, in which the second order perturbation term has been approximated - this approach has been previously demonstrated (Jonah 1986; 1989)-

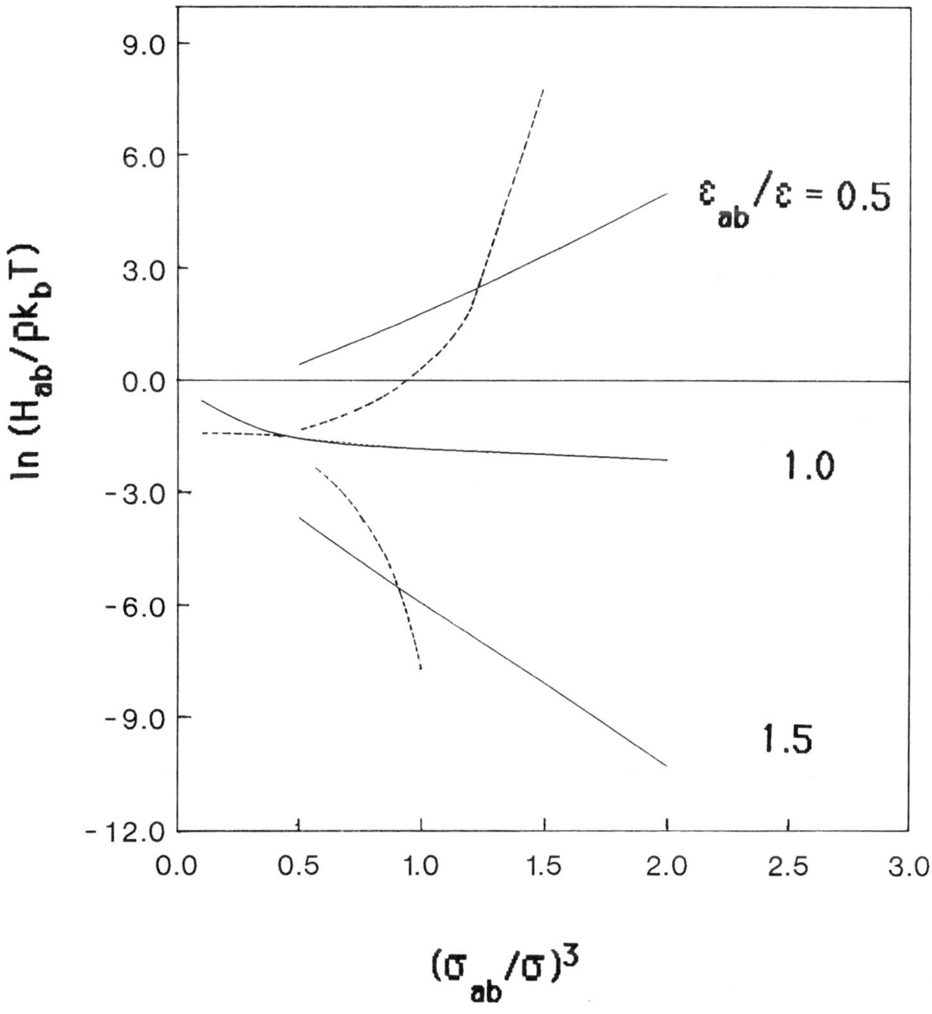

Figure 1. Variation of Henry's Constant with size ratio (σ_{ab}/σ) for various values of $(\varepsilon_{ab}/\varepsilon)$, for L-J(6-12) mixtures at reduced temperature $T^*=1.2$ and reduced density $\rho^*=0.7$. —— expt. data [Lofti and Fisher 1989]; - - - - calculated from equations (5.1) and (5.2) .

we shall outline the procedure for the 1-fluid closed expression. The application to low-pressure VLE will be based on VDW1 mixing rules; the application to solubilities of solids in supercritical fluids, on the other hand ,will invoke no particular mixing rules, but will assume a certain density-dependence for the coefficients F_u and F_z.

Low-pressure vapor-liquid equilibria. Activity coefficients at infinite dilution play an important role in the prediction of VLE at low pressures, when such studies are based on model equations whose adjustable parameters are expressible in terms of appropriate infinitely dilute solution properties.In principle,all the two-parameter models for activity coefficients as functions of composition do have this property; the difference between various equations of this kind arises from the fact that for some of them, the relationship between the parameters and the infinite dilution properties is simpler than for others.

Some three- and four- parameter models are expressible not only in terms of limiting activity coefficients,but also in terms of these and the limiting first composition derivatives. So, if we can predict these limiting quantities from molecular considerations (or otherwise), we have available in principle, expressions for the excess Gibbs free energy isotherms as functions of composition.

For the limiting activity coefficients γ_a^∞ in the symmetric standard state convention, and the limiting first composition derivative, $(\partial \ln \gamma_a / \partial x_a)^\infty$, the 1-fluid conformal solution theory (according to VDW1 mixing rules) gives:

$$\ln\left(\frac{\gamma_a^{\infty} f_a^{\bullet}}{f_b^{\bullet}}\right) = F_u \frac{U_{br}^{\bullet}}{RT} + F_z z_{br}^{\bullet}, \tag{5.3}$$

$$\left(\frac{\partial \ln\gamma_a}{\partial x_a}\right)^{\infty} = -\frac{F_u}{R}\left(\frac{\partial \mu_{br}^{\bullet}}{\partial T} + F_u \frac{\partial U_{br}^{\bullet}}{\partial T}\right) + \frac{\rho F_z}{RT}\left(\frac{\partial \mu_{br}^{\bullet}}{\partial \rho} + F_u \frac{\partial U_{br}^{\bullet}}{\partial \rho}\right) + \left(\frac{\partial F_u}{\partial x_a}\right)^{\infty} \frac{U_{br}^{\bullet}}{RT} - \left(\frac{\partial F_z}{\partial x_a}\right)^{\infty} \tag{5.4}$$

where,

$$\mathbb{F}_u \equiv \frac{N}{\varepsilon_m}\frac{\partial \varepsilon_m}{\partial N_a}, \qquad \mathbb{F}_z \equiv \frac{N}{\sigma_m^3}\frac{\partial \sigma_m^3}{\partial N_a};$$

$$F_u = 2\left(\frac{\sigma_{ab}}{\sigma}\right)^3\left(\frac{\varepsilon_{ab}}{\varepsilon} - 1\right), \qquad F_z = 2\left[\left(\frac{\sigma_{ab}}{\sigma}\right)^3 - 1\right]; \tag{5.4a}$$

$$\left(\frac{\partial \mathbb{F}_u}{\partial x_a}\right)^{\infty} = F_u\left\{1 - \left(\frac{\sigma_{ab}}{\sigma}\right)^3 - 2\frac{\varepsilon_{ab}}{\varepsilon}\left(\frac{\sigma_{ab}}{\sigma}\right)^3\right\} + 2\left(\frac{\sigma_{aa}}{\sigma}\right)^3\left(\frac{\varepsilon_{aa}}{\varepsilon} - 1\right), \tag{5.4b}$$

$$\left(\frac{\partial \mathbb{F}_z}{\partial x_a}\right)^{\infty} = -\frac{1}{2}F_z^2 - F_z\left(\frac{\sigma_{ab}}{\sigma}\right)^3 + 2\left\{\left(\frac{\sigma_{aa}}{\sigma}\right)^3 + 1 - 2\left(\frac{\sigma_{ab}}{\sigma}\right)^3\right\} .$$

In computing the limiting composition derivative of the activity coefficients, we have neglected the derivative $\partial z^{\bullet}_{br}/\partial x_a$ which must be small at liquid densities. The temperature and density derivatives of the residual pure solvent chemical potential and internal energy are readily obtained by standard thermodynamic manipulations.

At liquid densities, $z^{\bullet}_{br} \simeq -1$, and (5.3) may conveniently be written as:

$$\ln\left(\frac{\gamma_a^{\infty} f_a^{\bullet}}{f_b^{\bullet}}\right) = 2\left(\frac{\sigma_{ab}}{\sigma}\right)^3\left(\frac{\varepsilon_{ab}}{\varepsilon} - 1\right)\frac{U^{\bullet}_{br}}{RT} - 2\left[\left(\frac{\sigma_{ab}}{\sigma}\right)^3 - 1\right] \tag{5.5}$$

Given γ_a^{∞} at various temperatures, it is now a simple matter to deduce effective values for the molecular parameter ratios σ_{ab}/σ and $\varepsilon_{ab}/\varepsilon$, by plotting,

$$\ln\left(\frac{\gamma_a^{\infty} f_a^{\bullet}}{f_b^{\bullet}}\right) \quad \text{versus} \quad \frac{U^{\bullet}_{br}}{RT}, \tag{5.5a}$$

to obtain a straight line, whose gradient and intercept will yield two linear equations, from which these parameters may be determined; for examples of such plots see Jonah et al. (1981). With these values of the parameters, $(\partial \ln \gamma_a/\partial x_a)$ at infinite dilution may also be computed as a function of temperature.

Because of the linear plot resulting from (5.5a), it means that in practice we can use experimental values of γ_a^∞ at only two temperatures to estimate the effective molecular parameter ratios, and hence $(\partial \ln\gamma_a/\partial x_a)^\infty$ at various temperatures. Once these infinite dilution properties are available as functions of temperature, appropriate empirical models for activity coefficients as functions of composition may be used to compute excess Gibbs free energy- and heats of mixing-isotherms as functions of composition. This kind of exercise had been carried out before based on a semi-empirical thermodynamic approximation for the second order perturbation term in (2.5); we do not repeat the procedure here, but would refer the interested reader to this publication (Jonah 1988).

Solid solubilities in supercritical fluids. Assuming negligible solubility of the gas in the solid phase, we have the equilibrium condition:

$$f_a^{vap} = f_a^{\bullet solid}, \tag{5.6}$$

where the component *a* is the solid, and the component *b*, the supercritical gas ; $f_a^\bullet$ represents the fugacity of the pure component *a*.

For a sparingly soluble solid, we may invoke a generalized Henry's law, and write,

$$f_a^{vap} = K_{ab} \times \text{(solute mole fraction)}, \tag{5.6a}$$

K_{ab} denoting the generalized Henry's Constant. K_{ab} obviously consists of contributions from the solvent-solvent, and solute-solvent interactions; in recognition of this fact we shall write,

$$K_{ab} = f_b^{\bullet vap} \exp\{-\beta(\mu_{ar}^\infty - \mu_{br}^\bullet)\} \tag{5.6b}$$

$f_b^{\bullet vap}$ denoting the fugacity of the pure solvent gas.

In writing (5.6b), we have simply used the thermodynamic identities:

$$\left.\begin{aligned} \ln\left(\frac{K_{ab}}{\rho k_B T}\right) &= \beta\ \mu^{\infty}_{ar}, \\ \ln\left(\frac{f^{\bullet}_{b}}{\rho k_B T}\right) &= \beta\ \mu^{\bullet}_{br}. \end{aligned}\right\} \tag{5.6c}$$

We now have for our equilibrium condition, on using (5.6a) and (5.6b) in (5.6):

$$f^{\bullet solid}_{a} = y_a f^{\bullet vap}_{b} \ \exp\{\beta(\mu^{\infty}_{ar} - \mu^{\bullet}_{br})\}, \tag{5.7}$$

where y_a denotes the mole fraction of the dissolved solid.

Since all the quantities (other than y_a) involved in (5.7) are mole fraction-independent,we may readily solve this equation for y_a to obtain:

$$y_a = (f^{\bullet solid}_{a} / f^{\bullet vap}_{b})\exp\{-\beta(\mu^{\infty}_{ar} - \mu^{\bullet}_{br})\}. \tag{5.8}$$

We may now use one of the several expressions presented for μ^{∞}_{ar} in the preceding sections, to compute the solubility y_a.

We had in a previous publication considered the semi-empirical correlation of y_a with certain pure solvent thermodynamic properties (Jonah 1989); such a treatment was based on equation (2.5) with a thermodynamic approximation for the second order perturbation term, different from either of those presented in this paper. Here we shall use the more exact form as given by 1-fluid conformal solution theory in equation (4.1a).

Using the Prausnitz expression (1986) for $f^{\bullet solid}_{a}$ namely,

$$f^{\bullet solid}_{a} = f^{\bullet sat}_{a} \exp\{v^{\bullet solid}_{a}(P - P^{sat}_{a})/(RT)\} \tag{5.9}$$

(where $v^{\bullet solid}_{a}$ is the molar volume of the pure solid at saturation) we may re-write the latter equation in the more convenient form:

$$\ln\left(\frac{P_a^{sat}}{y_a f_b^{\bullet vap}}\right) + \frac{Pv_a^{\bullet solid}}{RT} = F_u \frac{U_{br}^{\bullet}}{RT} + F_z z_{br}^{\bullet}. \tag{5.10}$$

In writing (5.10), we have neglected P_a^{sat} in comparison with the system pressure P in (5.9). F_u and F_z are functions of the molecular parameter ratios s and t; for realistic mixing rules we expect F_u and F_z to be also functions of temperature and density.

In the present context, it is not difficult to verify a very weak temperature-dependence for F_u and F_z. To do this, we first define the functions,

$$\left.\begin{aligned} \mathbf{Y} &\equiv \left\{\ln\left(\frac{P_a^{sat}}{y_a f_b^{\bullet vap}}\right) + \frac{Pv_a^{\bullet solid}}{RT}\right\} / z_{br}^{\bullet}, \\ \mathbf{U} &\equiv \left(\frac{U_{br}^{\bullet}}{RT}\right) / z_{br}^{\bullet}. \end{aligned}\right\} \tag{5.11}$$

Now according to (5.10), if **Y** is plotted against **U** at constant density, we would expect to have linear graphs with gradient F_u, and intercept F_z when the temperature dependence is very weak. Experimental data for several solids dissolved in supercritical ethylene and carbon dioxide have been examined in this way, and found to yield linear isochores; the systems examined were 2,3- and 2,6-Dimethyl Naphthalene in CO_2 and C_2H_4; benzoic acid, phenanthrene and adamantane in CO_2. Figure 2 illustrates such linear isochores for adamantane/CO_2. Because of this very weak temperature-dependence, it becomes possible to extrapolate and interpolate sparse experimental data along constant density - paths. For example, if we have available solubilities at two temperatures T_1 and T_2, and the same density ρ, these data will suffice for extrapolating to, or interpolating at, some other temperature T and the same density ρ. Table 2 illustrates the effectiveness of such a procedure for adamantane solubilities in supercritical carbon dioxide; for comparison, the result of direct linear extrapolation and interpolation are given in the last column.

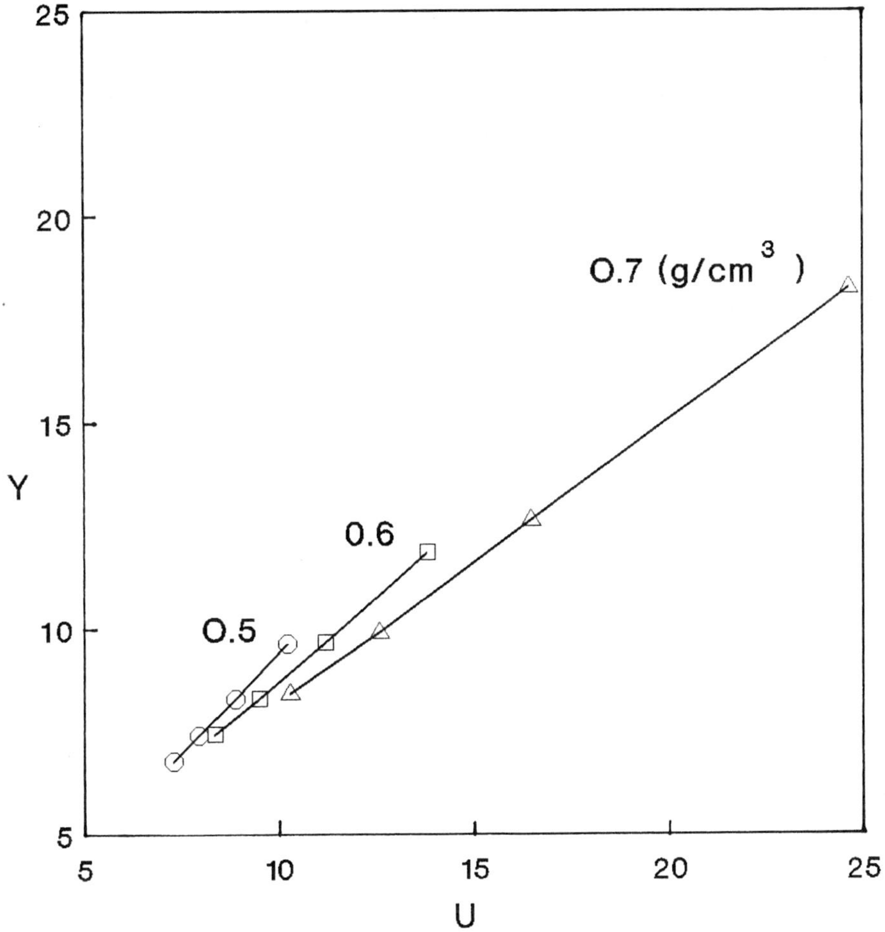

Figure 2. Linear isochores for the solubilities of adamantane in carbon dioxide [Swaid et al.1985]; points are experimental values. Y and U are defined in equation (5.11).

The density-dependence of the coefficients F_u and F_z, on the other hand, is quite marked; for the systems examined, this dependence appears to be at least quadratic, in contrast to the Meyer's form (Meyer 1988) which suggests a linear dependence. However, our findings confirm the Meyer's prediction of temperature-independent F_u and F_z .Without committing ourselves to any particular mixing rules, it is possible to make use of equation (5.10) in an empirical way to construct a solubility-temperature-density (y-T-ρ) surface. To accomplish this, we fit F_u and F_z from various isochores to the functions:

$$\left.\begin{aligned} F_u &= F_u^{(0)} + F_u^{(1)}\rho + F_u^{(2)}\rho^2, \\ F_z &= F_z^{(0)} + F_z^{(1)}\rho + F_z^{(2)}\rho^2. \end{aligned}\right\} \qquad (5.12)$$

So we now have for the solubility y_a:

$$y_a = \frac{P_a^{sat}}{f_b^{\bullet vap}} \exp\left\{\frac{Pv_a^{\bullet solid}}{RT} - F_u(\rho)\frac{U_{br}^{\bullet}}{RT} - F_z(\rho)z_{br}^{\bullet}\right\} \qquad (5.12a)$$

with $F_u(\rho)$ and $F_z(\rho)$ given by (5.12); ρ is the density of the pure supercritical solvent gas.

By making use of isochores at $\rho(g/cm^3)$ = 0.45, 0.5, 0.6, 0.7, 0.85, 0.9, for the system adamantane/CO_2 in the temperature range 343K ≤ T ≤ 402K, we have been able to construct such a surface with parameter values:

$$F_u^{(0)} = 2.417533 \qquad F_z^{(0)} = -10.9927$$

$$F_u^{(1)} = -3.893446\ cm^3g^{-1} \qquad F_z^{(1)} = 31.20106\ cm^3 g^{-1}$$

$$F_u^{(2)} = 2.004615\ cm^6g^{-2} \qquad F_z^{(2)} = -19.55385\ cm^6g^{-2}$$

Table 2. Extrapolation and Interpolation of Solid Solubilities in Supercritical CO_2 along Isochoric Paths for the system Adamantane/CO_2 (Swaid et al. 1985). Data Given at (T_1,ρ) and (T_2,ρ); Extrapolation/Interpolation Made at T(K).

ρ(g/cc)	T_1(K)	T_2(K)	T(K)	Expt. ($10^3 c_a$)	Extrap. $10^3 c_a$ Theory	Extrap. $10^3 c_a$ Linear
0.5	343	362.5	382	19.7	17.6	14.9
			402	33.3	28.5	19.4
0.7	343	362.5	382	48.3	47.2	40.4
			402	76.34	76.2	52.8
					Interp. $10^3 c_a$ Theory	**Interp. $10^3 c_a$ Linear**
0.5	343	402	362.5	10.5	11.0	15.1
			382.0	19.7	19.4	24.1
0.7	343	402	362.5	28.3	28.2	36.1
			382.0	48.3	47.1	56.0

Figure 3 shows the comparison between experimental and calculated solubilities for four isotherms in the density range 0.45 $g/cm^3 \le \rho \le 0.9 g/cm^3$ for which the fit was made. The properties of CO_2 were obtained from Angus et al. (1976); in Table 3 we have listed these properties at certain selected states, as well as the relevant properties of the pure solid. The experimental solubiities for adamantane in CO_2 are reported in terms of concentration c, instead of mole fraction; we use the following equation to convert from y_a to c:

$$c = y_a \rho M_a / [M_b(1-y_a)],$$

M_a, M_b denoting respectively, the molecular weights of the solute and the solvent gas. The solid solubilities as calculated from

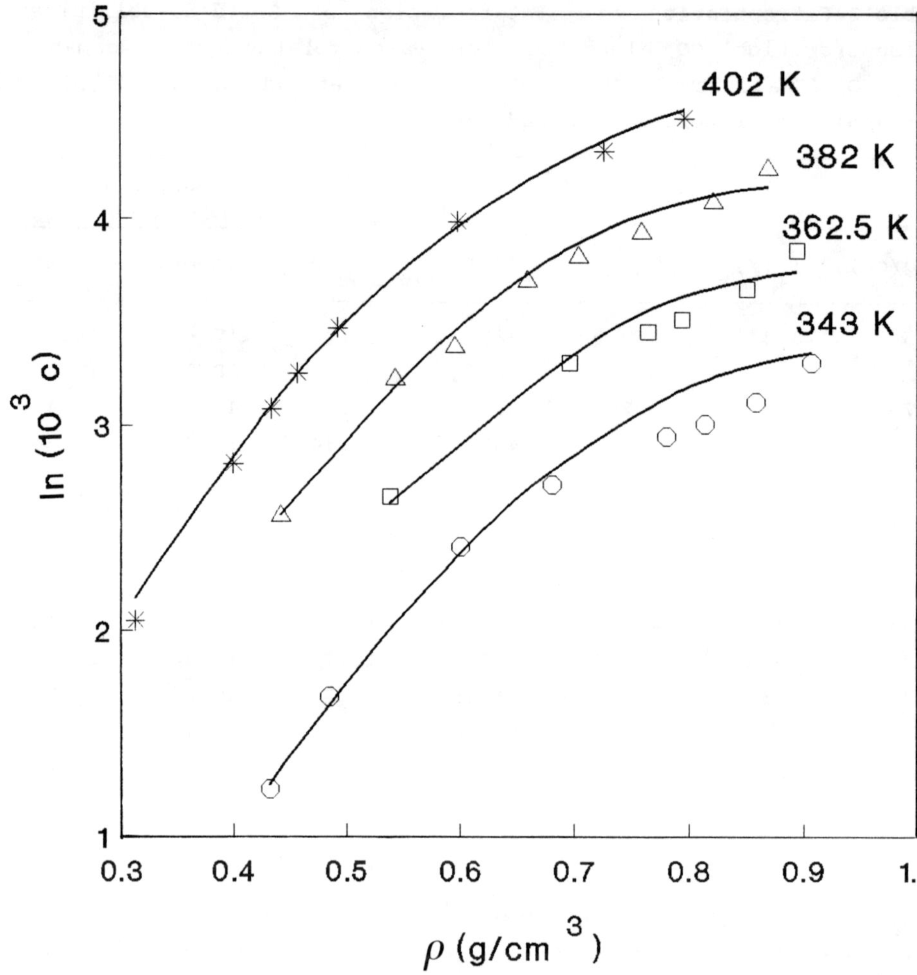

Figure 3. Comparison between experimental and calculated solubities in carbon dioxide according to equations (5.12) and (5.12a).Curves indicate experimental values [Swaid et al 1985]; the points denote calculated values.

the $(y_a\text{-}T\text{-}\rho)$ surface are seen in good agreement with the experimental values.

It is interesting to consider the possibility of extracting the second and third interaction virial coefficients, B_{ab} and C_{abb} respectively, from solid solubilities in supercritical gases using the above procedure for data reduction. This we can do by considering the low-density limit of equation (4.1a), which takes the following form:

$$2(B_{ab}-B)\rho + \frac{3}{2}(C_{abb}-C)\rho^2 + O(\rho^3)$$
$$= (F_u^{(0)}+\rho F_u^{(1)}+\rho^2 F_u^{(2)}+..)\{-T(\rho\frac{dB}{dT}+\frac{\rho^2}{2}\frac{dC}{dT}+.....)\}$$
$$(F_z^{(0)}+\rho F_z^{(1)}+\rho^2 F_z^{(2)}+..)(B\rho + C\rho^2+......), \qquad (5.13)$$

B and C denoting respectively, the second and third virial coefficients of the pure solvent gas.

Equating coefficients of like powers of ρ on the l.h.s.and r.h.s. of (5.13), we deduce at once expressions for the interaction virial coefficients B_{ab} and C_{abb}, together with their temperature derivatives. Thus, we have,

$$B_{ab} = \frac{1}{2}F_u^{(0)}T^{-1}\frac{dB}{dT^{-1}} + \frac{B}{2}(F_z^{(0)} + 2), \qquad (5.14a)$$

$$C_{abb} = C(1+\frac{2}{3}F_z^{(0)}) + \frac{1}{3}F_u^{(0)}T^{-1}\frac{dC}{dT^{-1}} + \frac{2}{3}(F_u^{(1)}T^{-1}\frac{dB}{dT^{-1}}+BF_z^{(1)}) \qquad (5.14b)$$

The main problems however, in the application of (5.14a) and (5.14b) consist in estimating reliably,

(a) the coefficients $F_u^{(0)}$, $F_z^{(0)}$ etc.,

(b) the temperature derivatives of B and C.

(b) will perhaps be the more difficult for dC/dT^{-1}, due to the usual large scatter in experimental data for C. However we can obtain quite close fits for B expressed as a cubic polynomial in reciprocal temperature, from which we may obtain reasonable values of its temperature derivatives.

It is interesting to observe in connection with equation (5.14a), that if $F_u^{(0)}$, and $F_z^{(0)}$ are identified with the values of F_u, and F_z associated with the VDW1 mixing rules, then we recover correctly the leading terms in an expansion for B_{ab} about $\varepsilon_{ab}/\varepsilon = 1$, valid for any two-parameter conformal pair potential. These considerations are developed further elsewhere (Jonah and Deiters).

Table 3. Solvent (CO_2) and Solute (Adamantane) Properties at Selected States.

		CARBON DIOXIDE				ADAMANTANE	
T /K	ρ /g cm^{-3}	P /bar	f /bar	$\frac{-U_r}{RT}$	$-z_r$	v /liter	$10^3 P^{sat}$ /bar
343	.680	210	102.5	5.108	.5233	0.127	4.33
343	.906	500	177.5	6.648	.1485		
362.5	.695	290	143.5	4.819	.3907	0.127	12.09
362.5	.8926	600	251.2	6.095	.0184		
382	.442	200	129.9	3.000	.3730	0.127	30.35
382	.703	370	194.2	4.534	.2708		
382	.821	550	268.9	5.244	.0714		
402	.457	240	160.0	2.867	.3086	0.127	71.06
402	.725	475	268.0	4.353	.1369		

It may be observed at this point, that the procedure utilized above for solid solubilities, is also applicable to liquid solubilities in supercritical fluid solvents, provided the liquid is sparingly soluble; the important modification to the theory however,consists in the proper modelling of the liquid phase to account for the non-negligible solubility of the gas in the liquid. Once this has been achieved, the left hand side of equation (5.7) is now simply replaced by the fugacity of the solute component *a* in the liquid phase, which is no longer assumed to be pure.

SUMMARY

We have presented an alternative to the traditional formulation for the residual solute chemical potential at infinite dilution, which de-emphasizes the role of solute-solute interactions, as is appropriate for this level of dilution. Our procedure leads to two kinds of perturbation expansions, both using the pure solvent as reference: an inverse temperature-expansion, and a density-expansion. Internal consistency between the two forms is established through the fact that the inverse temperature-expansion can be recovered from the density-expansion, which in turn leads to the correct low-density limit.

For pair potentials of the Lennard- Jones form, the first and second order perturbation terms are shown to be expressible in terms of the pure solvent thermodynamic properties - the first order term exactly, whilst the second order term may be so expressed approximately by making use of fluctuation theory within the canonical ensemble theory.Two such approximations have been suggested, and shown to reproduce closely simulation data for the Henry,s Constant for L-J (6-12) molecules of equal well depths, in the range $0.3 < (\sigma_{ab}/\sigma) < 2$ when two adjustable parameters are suitably chosen ; but these parameters are unfortunately dependent on the well depths ratio. It is true that these approximations are semi-empirical in nature, but we believe they do help us in identifying useful pure solvent properties in terms of which the second order perturbation term may be expressed.

The relationship between the above two expansions and the corresponding expression from 1-fluid conformal solution theory is elucidated within the context of mixing rules.

Finally, we have outlined how 1-fluid conformal solution theory may be used for estimating the adjustable parameters in certain model equations for the activity coefficients as functions of composition; we have also illustrated the use of this theory in calculating solid solubilities in supercritical fluids.In the first application, specific mixing rules (VDW1) are invoked;

whilst in the second application, we have a demonstration of how 1-fluid theory may be used semi-empirically without committing ourselves to any particular mixing rules.

The results of applying 1-fluid conformal theory to solid solubilities in supercritical fluids have been particularly encouraging, especially in the production of linear isochores which may be used for the extrapolation and interpolation of sparse data, as well as the construction of solubility-temperature-density surfaces.The possibility of extracting values for the second and third interaction virial coefficients is also attractive.

ACKNOWLEDGEMENT

The author acknowledges with deep gratitude many helpful discussions with Dr.J.M.H.Levelt-Sengers and Prof.W.R.Smith.The kindness of Drs.A.Lofti and J.Fischer in making available a preprint of their paper is also gratefully acknowledged.

The author thanks very much the Department of Mathematics and Statistics, University of Guelph for their kind hospitality, and the Stiftung Volkswagenwerk for partial support of this work.

The assistance of Dr.Edmundo de Azevedo and Mr. S.Blonski in the word-processing of the manuscript is deeply appreciated.

APPENDICES

A. Derivation of Basic Relation

We consider an infinitely dilute binary mixture consisting of one solute molecule (*a*-species) amongst N solvent molecules (species *b*). We have then the standard relation:

$$\mu^{\infty}_{ar} = -\beta^{-1}(\ln Q^{mix}_{N+1} / Q^{solv}_{N}), \qquad (A1)$$

$$Q_N = \int_V \exp(-\beta\, U_N)\; d\{N\}; \qquad (A2)$$

the superscripts "mix" and "solv" denote respectively, mixture pure solvent quantities.

Let us consider $Q^{mix}_{N+1}/ Q^{solv}_{N}$:

$$\frac{Q^{mix}_{N+1}}{Q^{solv}_{N}} = \frac{\int \exp(-\beta\psi_a)\exp(-\beta\ U^{solv}_{N})\ d\{N+1\}}{\int \exp(-\beta U^{solv}_{N})\ d\{N\}}$$

$$= \frac{\int \exp(-\beta\psi_b)\exp(-\beta\Delta\Psi)\exp(-\beta U^{solv}_{N})\ d\{N+1\}}{\int \exp(-\beta\Psi_b)\exp(-\beta U^{solv}_{N})d\{N+1\}}$$

$$\times \frac{\int \exp(-\beta\psi_b)\exp(-\beta\ U^{solv}_{N})\ d\{N+1\}}{\int \exp(-\beta U^{solv}_{N})\ d\{N\}} \qquad \text{(A3)}$$

The second factor on the r.h.s. of (A3) is simply,

$$(Q_{N+1}/Q_N)^{(solv)} ,$$

which is $\exp(-\beta\mu^{\bullet}_{br})$; the first factor is $\langle\exp(-\beta\Delta\Psi)\rangle_{ps}$.So (A1) - (A3) leads at once to the basic relation (2.2).

B. <u>Low-density Limit of Equation (2.10c)</u>

To recover the iow-density limit, we use the approximation:

$$\left.\begin{aligned} y^{(2)}(1,2) &\cong 1+ \rho\int f(1,3)f(2,3)\ d(3), \\ y^{(3)}(1,2,3) &\cong 1. \end{aligned}\right\} \qquad \text{(B1)}$$

Using (B1) in the equation,

$$-\beta(\mu^{\infty}_{ar}- \mu^{\bullet}_{br}) = \rho\ \frac{S_2}{V} +\frac{1}{2}\ \frac{\rho^2}{2V}\ (S_3-S_2^2)+ O(\rho^3)$$

we have after some cancellations,

$$-\beta\mu^{\infty}_{ar} = -\beta\mu^{\bullet}_{br} - [\frac{\rho}{V}\int f(1,2)\ d(1)d(2)$$

$$+ \frac{\rho^2}{2V}\int f(1,3)f(2,3)f(1,2)\ d\{3\}]$$

$$+\frac{\rho}{V}\int f_{ab}(1,2)d(1)d(2) + \frac{\rho^2}{2V}\int f(2,3)f_{ab}(1,2)f_{ab}(1,3)d\{3\}$$

$$+ O(\rho^3) \quad ; \qquad \text{(B3)}$$

the Mayer f-functions without any suffixes are understood to pertain to the pure solvent.

The term in the square brackets [ ] is simply $-\beta\mu^{\bullet}_{br}$, which cancels the first term; B(3) thus reduces to the virial expansion for $-\beta\mu^{\infty}_{br}$ up to terms of $O(\rho^2)$.

REFERENCES

Angus,S.,B.Armstrong,K.M.Reuck,W.Featherstone,and M.R.Gibson.1976. International Tables of the Fluid State. (IUPAC) Carbon Dioxide.

Hamad,E.Z. and G.A.Mansoori.1987. Dense fluid theory of mixtures. J.Chem.Phys.87: 6046-6052.

Hill,T.L.1956. Statistical Mechanics. McGraw-Hill Book Company Inc.,New York.

Jonah,D.A., K.S.Shing, and K.E.Gubbins.1981.Theory of dilute solutions. In Proc. Eighth Symp. on Thermophysical Properties, Vol. 1, ed .Jan V. Sengers, 335-342. Amer. Soc. Mech. Eng., Washington,D.C.

Jonah,D.A.1986.Consideration of some dilute phenomena based on an expression for the Gibbs free energy. Internat.J.Thermophys. 7:935- 948.

Jonah,D.A.1988. Interpolation between two excess Gibbs free energy -composition isotherms. Chem.Eng.Comm.65: 95-108; Heats of mixing-composition isotherms deduced from only two (g^E-x) isotherms. Fluid Phase Equilibria 39:267-291.

Jonah,D.A.1989. A linear correlation for solid solubilities in supercritical gases. Chem.Eng.Comm.79:1-13.

Jonah,D.A.,and U.K.Deiters.To be published.

Kirkwood, J.G. 1935. Statistical Mechanics of liquid mixtures . J.Chem.Phys.3:300-313.

Lebowitz,J.L.1984 Exact solution of the generalized Percus-Yevick equation for a mixture of hard spheres.Phys.Rev.A,133:895-899.

Lofti,A.,and J.Fischer.1989. Chemical potential of model and real dense fluid mixtures from perturbation and simulation. Molec. Phys.66: 199-219.

Meyer,E.C.1988. A one-fluid mixing rule for hard spheres. Fluid Phase Equilibria 41:19-29.

Nicolas,J.J.,K.E.Gubbins,W.B.Streett,and D.J.Tildesley.1979.Equation of state for Lennard-Jones fluids. Molec.Phys.37: 1429 -1454.

Prausnitz,J.M., R.N.Lichtenthaler,and E.G.de Azevedo.1986. Molecular Thermodynamics of Fluid Phase Equilibria, Second Edition. 34 - 35. Prentice-Hall, Englewood Cliff.

Rowlinson,J.S.1959.Liquids and Liquid Mixtures,First Edition.267 -268.Butterworths Pub.Ltd.London.

Shing,K.S.,and K.E.Gubbins.1983. The chemical potential in non-ideal liquid mixtures: computer simulation and theory.Molecular Phys.49: 1121- 1138.

Shing,K.S., K.E.Gubbins,and K.Lucas.1988. Henry's Constant in non-ideal fluid mixtures: computer simulation and theory. Molec.Phys.65: 1235- 1252.

Stecki,J.,and A.Samborski.1989. Solute molecules at infinite dilution and the Henry's Constant.Fluid Phase Equilibria 48: 141-160.

Swaid,I.,D.Nickel,and G.M.Schneider.1985. NIR-Spectroscopic investigation on phase behavior of low-volatile organic substances in supercritical carbon dioxide. Fluid Phase Equilibria 21: 95-112.

NOMENCLATURE

A	Helmholtz free energy.
B	Second pure solvent virial coefficient.
C	Third pure solvent virial coefficient.
c	Concentration of solute in supercritical solvent.
C_v	Constant volume heat capacity.
f	Fugacity.
f(r)	Mayer f-function.
H	Henry's Constant
K	Generalized Henry's Constant.
M	Molecular weight.
N	Avogadro's number.
P	System pressure.
Q	Configurational partition function.
R	Universal gas constant.
s,t	Molecular parameter ratios.
T	System temperature.
U	Total potential energy.
u	Pair potential energy.
U	Dimensionless quantity in linear isochoric plots.
V	System volume.
V	Virial function.
y	Mole fraction in gas phase.
z	Compressibility factor.

<u>Greek</u> <u>Letters</u>

α	Isobaric coefficient of volume expansion.
β	$\equiv 1/(k_B T)$
γ	Activity coefficient.
Δ	Difference between functions pertaining to solute-solvent and solvent-solvent interactions.
ξ	Kirkwood's coupling parameter.
ε	Potential well depth.
Ψ	Private potential energy.
κ	Isothermal coefficient of compressibility.
μ	Chemical potential.
ρ	Number density.
σ	Collision diameter.

Superscripts

$\cdot$	Pure component.
*	Reduced quantity.
sat	Saturation value.
solid	Solid phase.
vap	Vapor phase.
mix	Mixture.
∞	Infinite dilution quantity.

Subscripts

a,b	Solute and solvent component respectively.
N	Pertaining to a system containing N molecules.
r	Residual thermodynamic function.
v	Constant volume.

Application of Kirkwood-Buff Theory to Electrolyte Solutions

Kenneth E. Newman
Dept. of Chemistry, The King's College
Edmonton, Alberta T5H 2M1, Canada

ABSTRACT

The application of Kirkwood-Buff theory to electrolytic solutions is complicated by the fact that the various ion - ion and ion - solvent correlations are not independent being connected by electroneutrality conditions. In addition, the long-range inter-ion electrostatic potential (which manifests itself in the Debye-Hückel limiting law) has major implications for the form of the Kirkwood-Buff integrals. Using the Kirkwood-Buff equations for *non-electrolytic* solutions and treating the solute-solute interactions as a suitable linear combination of cation-cation, cation-anion and anion-anion interactions, equations can be developed for a variety of constant temperature thermodynamic properties which have the correct thermodynamic form, i.e. are ion additive or give the correct limiting form at infinite dilution. The Debye-Hückel limiting slopes arise not from the first term in the exponential expansion of the Poisson-Boltzmann distribution (as in conventional theory) but from the second term. The theory has been applied to the interpretation of salt partial molar volumes in both single and mixed solvents, to free energies of transfer from one solvent to a binary mixture and to salt activity coefficients in single solvents.

1. INTRODUCTION

In their original work, Kirkwood and Buff (1951) mentioned in passing that their theory was an alternative approach towards obtaining electrolyte activity coefficients from the Poisson-Boltzmann equation rather than the commonly used charging cycles. Their suggestion was never subsequently explored in depth and since that time most of the effort directed at understanding electrolyte solutions has been concerned with developing cluster expansion approaches (Rasaiah and Friedman, 1968, 1969; Waisman and Lebowitz, 1972 a,b). The reasons for this are clear. McMillan-Mayer theory (1945) shows that, if we consider only solutions in osmotic equilibrium with pure solvent [the so-called McMillan-Mayer (MM) standard state] then the thermodynamic properties depend only on the solute-solute intermolecular potentials and not on the solute-solvent terms. The solute-solvent interaction appears only in the way it modifies the solute-solute term. Within the MM standard state, most thermodynamic properties involve the product of the intermolecular potential (or some derivative thereof) and the radial distribution function (rdf) integrated over all space. The intermolecular potential and the the radial distribution function are clearly causally linked, a strongly attractive potential implies a strongly positive contribution rdf. However, the connection is not obvious and most theories have been concerned with forging such a link generally by using cluster expansion-type approaches. KB theory as well as being entirely rigourous, is one of the very few theories available which does not require knowledge of the intermolecular potential and it can be formulated in either the MM standard state or the more conventional "Lewis-Randall" (LR) standard state. It is also exact. The major disadvantage of the theory, obviously inherent in any fluctuation theory, is that the radial distribution function *per se* is not available, only is its integral G_{ij} defined as:

$$G_{ij} = \int 4\pi r^2 [g_{ij}(r) - 1] dr \tag{1.1}$$

In view of the importance of these terms we shall refer to them as "Kirkwood-Buff (or KB) integrals". It is the purpose of this paper to review recent work on the application of Kirkwood-Buff theory to understand the thermodynamic properties of electrolyte solutions in both pure and mixed solvents (Cheng and Newman, 1990a,b; Covington and Newman, 1988; Mindermann and Newman, 1990; Newman, 1988a,b; 1989 ; 1990).

2. ELECTRONEUTRALITY CONSIDERATIONS

Electrolytic solutions differ from non-electrolytic solutions in that the ion-ion interactions are extremely long-range and from a classical thermodynamic viewpoint, only thermodynamic parameters of neutral combinations of ions are available (Guggenheim, 1929). In a later section, we will discuss this point in more detail and following Hall (1972) we will show that, under certain circumstances, a KB definition of a single ion thermodynamic parameter is possible. Fluctuation theory in general, and, more specifically, KB theory, is developed via the Grand Canonical Ensemble. When such an ensemble is applied to electrolytic solutions, it is important to note that *each member* of the ensemble must be neutral. This implies that cation-cation, cation-anion and anion-anion correlations cannot be independent (Beeby, 1973). In addition, it turns out that neither can cation-solvent and anion-solvent interactions. For an electrolyte which consists of $\nu+$ cations M of charge z+ and $\nu-$ anions X of charge z-, i.e. has the formula, $M_{\nu+}X_{\nu-}$, we have the condition:

$$(\nu+)(z+) + (\nu-)(z-) = 0 \tag{2.1}$$

The electroneutrality conditions (Beeby, 1973) require that for a salt number density n_s:

$$(z+) + (z+)(\nu+)n_s G_{MM} + (z-)(\nu-)n_s G_{MX} = 0 \tag{2.2}$$

This equation simply reads that the charge on the cation plus the charge on the ion atmosphere due to both cations and anions must be zero. In combination with equation (2.1), (2.2) becomes:

$$1/(\nu+) + n_s(G_{MM} - G_{MX}) = 0 \tag{2.3}$$

There is a similar equation which relates anion-anion and cation-anion interactions and thus we readily see that:

$$1/(\nu+) + n_s G_{MM} = 1/(\nu-) + n_s G_{XX} \qquad (2.4)$$

With regard to ion-solvent interactions, the total correlation equals the immediate correlation of solvent around the ion plus the solvent correlated to the ions (both positive and negative) within the ion atmosphere. In the spirit of the Ornstein-Zernike equation(see e.g. Watts and McGee, 1976), we may write (Cheng and Newman, 1990b):

$$h_{MW}(r) = h^*_{MW}(r) + \int n_s(\nu+) h^*_{MM}(s) h_{MW}(r) ds + \int n_s(\nu-) h_{MX}(s) h^*_{XW}(r) ds \qquad (2.5)$$

where:

$$h_{ij}(r) = g_{ij}(r) - 1 \qquad (2.5b)$$

The starred term (*) implies the immediate correlation between i and j. A similar equation applies to the anion-solvent interactions. If we now take the Fourier transform of equation (2.5), making use of the fact that the r.h.s. of the equation contains a convolution integral (Watts and McGee, 1976), we obtain:

$$\tilde{h}_{MW}(q) = \tilde{h}^*_{MW}(q)[1 + n_s(\nu+)\tilde{h}_{MM}(q)] + \tilde{h}^*_{XW}(q) n_s(\nu-)\tilde{h}_{MX}(q) \qquad (2.6)$$

From the basic properties of Fourier transforms it is clear that:

$$G_{MW} = G^*_{MW}[1 + n_s(\nu+)G_{MM}] + n_s(\nu-)G_{MX}G^*_{XW} \qquad (2.7)$$

where again the starred term implies the corresponding integral of the immediate correlation function. A similar equation applies to anion-solvent interactions. If equation (2.7) and its anion analogue are combined with equation (2.3) then we readily obtain:

$$G_{MW} = n_s G_{MX}[(\nu-)G^*_{MW} + (\nu-)G^*_{XW}] = G_{XW} \qquad (2.8)$$

With regard to an electrolyte in a binary solvent mixture $\alpha+\beta$, similar arguments apply and equations analogous to (2.8) are readily obtained with $W = \alpha$ and β.

3. POISSON-BOLTZMANN DISTRIBUTION

The Poisson-Boltzmann distribution for the ion j around ion i is given by (Pitzer, 1977; Newman, 1989):

$$g_{ij}(r) = \exp[-q_{ij}(r)] \qquad (3.1)$$

where:

$$q_{ij}(r) = \frac{z_i z_j e^2 \exp(\kappa a_{ij}) \exp(-\kappa r)}{4\pi\varepsilon_o \varepsilon_r k_B T(1 + \kappa a_{ij}) r} \qquad (3.2)$$

with a_{ij} the distance of closest approach between i and j, and:

$$\kappa^2 = (e^2/\varepsilon_0\varepsilon_r k_B T)\sum_i z_i n_i^2 \tag{3.3}$$

where all the terms have their normal significance. The exponential of equation (3.1) can be expanded out to give:

$$g_{ij}(r) = 1 - q_{ij}(r) + q_{ij}^2/2! - q_{ij}^3/3! + \ldots.. \tag{3.4}$$

We may readily obtain the KB integral from equation (3.4) after noting that in the range 0 to a_{ij}, $g_{ij}(r)$ equals -1:

$$G_{ij} = z_i z_j / \sum_i z_i{}^2 n_i - 4\pi a^2/3 + G'_{ij} = G^{\circ}_{ij} - 4\pi a^3/3 + G'_{ij} \tag{3.5}$$

where the last term represents the as yet unconsidered higher terms in the exponential expansion. Table 3.1 gives the values of the first term on the right hand side of equation (3.5) for various charge types. Turning our attention to the higher order term as obtained in the integrating of equation (3.4), we obtain from the square term the expression (Newman, 1989):

$$G'_{ij} = z_i^2 z_j^2 \left(\frac{e^6}{256\pi^2 \varepsilon_0^3 \varepsilon_r^3 k_B^3 T^3 \sum_i z_i^2 n_i} \right)^{0.5} \frac{1}{(1 + \kappa a_{ij})^2} \tag{3.6}$$

or in terms of the salt number density n_s:

$$G'_{ij} = \left(\frac{e^6}{256\pi^2 \varepsilon_0^3 \varepsilon_r^3 k_B^3 T^3 n_s} \right)^{0.5} \left(\frac{z_i^4 z_j^4}{(\nu+)(z+)^2 + (\nu-)(z-)^2} \right)^{0.5} \frac{1}{(1 + \kappa a_{ij})^2} \tag{3.7}$$

Table 3.1 also summarizes values for the second term on the right hand side of equation (3.7) in square brackets. It is clear that since the first term on the r.h.s. of (3.5) obeys the electroneutrality conditions (2.3) and (2.4), then the cation-cation, cation-anion and anion-terms given by equation (3.7) must be equal and this is only true for symmetrical electrolytes. We thus conclude that the inclusion of the second term in the exponential expansion of the Poisson-Boltzmann distribution is not valid for asymmetric salts. This is disturbing since the Debye-Hückel slope comes from this term. This point has been made before with regard to conventional Debye-Hückel treatments. However, in practice this limitation is not too serious and a treatment of asymmetric as well as symmetric electrolytes is still possible. With regard to higher order terms in the expansion, it is clear that the odd terms must always be

Table 3.1. Values of $n_s G^\circ_{ij}$ and $z_i^4 z_j^4/[(\nu+)(z+)^2+(\nu+)(z+)^2] = f_{ij}$ for various electrolyte charge types

Salt $z_+:z_-$	$n_s G^\circ_{MM}$	$n_s G^\circ_{XX}$	$n_s G^\circ_{MX}$	f_{MM}	f_{XX}	f_{MX}
1:1	-1/2	-1/2	1/2	1/2	1/2	1/2
1:2	-1/6	-4/6	2/6	1/6	$2^8/6$	$2^4/6$
1:3	-1/12	-9/12	3/12	1/12	$3^8/12$	$3^4/12$
1:4	-1/20	-16/20	4/20	1/20	$4^8/20$	$4^4/20$
2:1	-4/6	-1/6	2/6	$2^8/6$	1/6	$2^8/6$
2:2	-1/2	-1/2	1/2	$2^8/8$	$2^8/8$	$2^8/8$
2:3	-4/30	-9/30	6/30	$2^8/30$	$3^8/30$	$2^4 3^4/30$
2:4	-1/6	-4/6	2/6	$2^8/24$	$4^8/24$	$2^4 4^4/24$
3:1	-9/12	-1/12	3/12	$3^8/12$	1/12	$3^4/12$
3:2	-9/30	-4/30	6/30	$3^8/30$	$2^8/30$	$3^4 2^4/30$
3:3	-1/2	-1/2	1/2	$3^8/18$	$3^8/18$	$3^8/30$
4:1	-16/20	-1/20	4/20	$4^8/20$	1/20	$4^4/20$
4:2	-4/6	-1/6	2/6	$4^8/24$	$2^8/24$	$4^4 2^4/24$
4:3	-16/84	-9/84	12/84	$4^8/84$	$3^8/84$	$4^4 3^4/84$
4:4	-1/2	-1/2	1/2	$4^8/32$	$4^8/32$	$4^8/32$

discounted whereas the even terms only obey electroneutrality for the symmetric salts.

Returning again to equation (3.5), we also see that the electroneutrality conditions are only ever valid under the Poisson-Boltzmann distribution providing the distance of closest approach term is the same for cation-cation, cation-anion and anion-anion distances. We shall thus drop the subscripts from the term a. It is not at this stage clear what implications this limitation has on the treatment of salts with vastely different ion sizes. We should also add that, at the present time, the interpretation of the experimentally determined distance of closest approach is by no means clear since they show no obvious relationship with published ionic radii data; the derived distances are however well defined by the data and the treatment.

4. FREE ENERGIES OF TRANSFER

The chemical potential of a solute (at constant temperature and pressure) varies due to changes in its concentration and/or solvent composition; non-ideality contributions to the former effect are generally discussed in terms of activity coefficients (see section 6 below).The latter effect is conveniently discussed in terms of the free energy of transfer of the solute from one solvent (α) to another (β), $\Delta G^\circ(\beta \leftarrow \alpha)$. Thus:

$$\Delta G^{\circ}(\beta \leftarrow \alpha) = \mu_s^{\circ}(\beta) - \mu_s^{\circ}(\alpha) \tag{4.1}$$

or in differential form:

$$\Delta G^{\circ}(\beta \leftarrow \alpha) = \int_{\alpha}^{\beta} d\mu_s^{\circ} \tag{4.2}$$

Within the Hall (1971) formulation of Kirkwood-Buff theory, the chemical potentials in a three component solution of *non-electrolyic* solute S in a binary solvent mixture $\alpha + \beta$, are related as follows:

$$k_B T \, d \ln n_s - (1 + n_s G_{ss}) d\mu_s + \sum_{\gamma=\alpha,\beta} G_{s\gamma} d\mu_\gamma \tag{4.3}$$

For a neutral solute, as $n_s \rightarrow 0$, then so does $n_s G_{ss}$ and we obtain:

$$d\mu_s^{\circ} = \lim_{n_s \rightarrow 0} (d\mu_s - k_B T \ln n_s) = - \sum_{\gamma=\alpha,\beta} n_{s\gamma} G_{s\gamma} d\mu_\gamma \tag{4.4}$$

It is not immediately obvious how to apply such equations to electrolyte solutions since, a) it is not clear whether we can apply an equation for a ternary mixture to a mixture of cations, anions plus two solvent components; b) the solute-solute KB integrals do not go to zero as the concentration tends to zero; c) in the electrolyte analogue of equation (4.4), the logarithmic term should be multiplied by $[(\nu+)+(\nu-)]$. We should perhaps note at this stage that, in very dilute solution, only the first term on the r.h.s. of equation (3.5) will be important and thus for "infinite dilution" thermodynamic properties, these are the only ion-ion terms we need to include. Our initial work on 1:1 electrolytes showed that it was simply necessary to substitute in the non-electrolyte equation for the solute-solute and solute-solvent terms, the cation-cation and cation-solvent terms respectively and the resulting equation had the correct form automatically (Newman, 1989). In anticipation of discussion of activity coefficients below, this procedure also gave the correct DH limiting law. For the unsymmetrical electrolytes, it would appear that some linear combination of cation-cation, anion-anion and cation-anion terms is the correct procedure. We seek a combination such that $1 + n_s G_{ss} = 1/[(\nu+) + (\nu-)]$; the required choice is:

$$G_{ss} = (\nu+) G_{MM} + (\nu-) G_{XX} + G_{MX} \tag{4.5}$$

Final justification of this choice will be the fact that it gives equations which show the required ion-additivity relationships (where necessary), give the required behaviour at infinite dilution and yields the Debye-Hückel limiting law. It is worth noting that in the limit of infinite dilution (see equation 3.5):

$$1 + n_s G_{ss} = n_s G_{MX} = 1/[(\nu+) + (\nu-)] \tag{4.6}$$

and in practice it is easier to simply substitute according to equation (4.6). Substituting into equation (4.2) yields:

$$\begin{aligned} d\mu_s^\circ &= \lim_{n_s \to 0} \{d\mu_s - [(\nu+) + (\nu-)]k_B T \, d \ln n_s\} \\ &= - \sum_{\gamma=\alpha,\beta} [(\nu+) + (\nu-)] n_\gamma G_{s\gamma} d\mu_\gamma \end{aligned} \tag{4.7}$$

The necessary substitution for $G_{s\gamma}(\gamma=\alpha,\beta)$ is simply:

$$G_{s\gamma} = G_{M\gamma} = G_{X\gamma} \tag{4.8}$$

After substituting equations (2.7), (4.6) and (4.8) into (4.7), we obtain:

$$d\mu_s^\circ = - \sum (\nu+) n_\gamma G^*_{M\gamma} d\mu_\gamma - \sum (\nu-) n_\gamma G^*_{M\gamma} d\mu_\gamma \tag{4.9}$$

This equation originally developed by Hall (1972) very satisfyingly has the required ion-additivity relationship and is directly analogous to the corresponding equation for the neutral solute (Hall, 1971):

$$d\mu_s^\circ = - \sum_{\gamma=\beta,\alpha} n_\gamma G_{s\gamma} d\mu_\gamma \tag{4.10}$$

As Hall (1972) has pointed out, equation (4.9) allows definition of a single ion (infinite dilution) chemical potential:

$$d\mu_M^\circ = - \sum_{\gamma=\alpha,\beta} n_\gamma G^*_{M\gamma} d\mu_\gamma \tag{4.11}$$

Attempts to estimate the immediate ion-solvent KB integrals for single ions and hence single ion free energies of transfer will be discussed below. The two solvent potentials are not independent being related (at constant temperature and pressure) by the Gibbs-Duhem expression:

$$n_\alpha d\mu_\alpha + n_\beta d\mu_\beta = 0 \tag{4.12}$$

Combination of equations (4.11) and (4.12) readily yields:

$$d\mu_M^\circ = n_\alpha (G^*_{M\beta} - G^*_{M\alpha}) d\mu_\alpha \tag{4.13}$$

The variation of solvent chemical potentials with solvent composition is readily available from vapour pressure measurements and thus from free energy of transfer measurements the variation of the bracketed term in equation (4.13) with composition is obtainable. We must note that a direct analysis of thermodynamic data yields only the sum of ion-solvent terms for neutral combinations of ions.

5. ELECTROLYTE PARTIAL MOLAR VOLUMES

According to Kirkwood-Buff theory (1951), the partial molar volume of a solute S in a single solvent component W is given by:

$$\overline{v}_S = \frac{1 + n_W(G_{WW} - G_{SW})}{n_S + n_W + n_S n_W(G_{SS} + G_{WW} - 2G_{SW})} \tag{5.1}$$

For a non-electrolyte solute, the limiting value of the partial molar volume as n_S approaches zero is:

$$\lim_{n_S \to 0} (\overline{v}_S) = 1/n_W + (G_{WW} - G_{SW}) \tag{5.2}$$

The water-water KB integral for pure water readily relates to the isothermal compressibility of the pure solvent:

$$\kappa_T k_B T = 1/n_W + G_{WW} \tag{5.3}$$

and we readily obtain for the solute partial molar volume:

$$\lim_{n_S \to 0} (\overline{v}_S) = \kappa_T k_B T - G_{SW} \tag{5.4}$$

Adopting a similar procedure to that adopted for free energies of transfer, we substitute equations (4.6), (4.8) and (2.7) into (5.1) and obtain in the limit that n_S goes to zero:

$$\lim_{n_S \to 0} (\overline{v}_S) = [(\nu+) + (\nu-)]\kappa_T k_B T - (\nu+)G^*_{MW} - (\nu-)G^*_{XW} \tag{5.5}$$

which has the required ion-additivity relationships. If we return to the Hall derivation (1971) of KB theory [equation (4.3)], we note that for a neutral solute S in a single solvent W:

$$k_B T \, d \ln n_S = (1 + n_S G_{SS}) d\mu_S + n_W G_{SW} d\mu_W \tag{5.6}$$

The pressure derivative of this equation gives (in the limit that the salt concentration tends to zero) equation (5.4) as we expect. However, we may note that the (molar standard state) infinite dilution *standard* partial molar volume is given by:

$$\overline{v}^\circ_S = - G_{SW} \tag{5.7}$$

since the solvent compressibility contribution has been cancelled by the pressure derivative of the concentration in the equation analogous to (4.4). In general, infinite dilution apparent molar volumes are given in the literature and the solvent compressibility term must be included.

6. ELECTROLYTE ACTIVITY COEFFICIENTS

The concentration dependence of the non-ideality of an electrolyte solution at constant temperature and pressure is generally given in terms of an activity coefficient; throughout this work we have been dealing with the number density standard state and thus we use the activity coefficient $y_{\pm}$ defined as:

$$\mu_S = \mu_S^{\circ} + k_B T \ln n_M^{\nu+} n_X^{\nu-} + k_B T \ln y_{\pm}^{\nu} \qquad (6.1)$$

where

$$\nu = (\nu+) + (\nu-) \qquad (6.2)$$

For an electrolyte solution, the Debye-Hückel limiting law gives us information about the limiting behaviour of $\ln y_{\pm}$ as the concentration tends to zero. Thus:

$$\ln y_{\pm} = (1/2^{0.5})(z+)(z-)A\left[\sum_i z_i n_i^2\right]^{0.5} \qquad (6.3)$$

with the summation i over all ionic species and where:

$$A = (e^6/32\pi^2\varepsilon_0^3\varepsilon_r^3k_B^3T^3)^{0.5} \qquad (6.4)$$

In combination with equation (2.1), (6.3) readily arranges to give:

$$\frac{d \ln y_{\pm}}{d n_S} = \frac{A}{(8n_s)^{0.5}} \{-(z+)^3(z-)^3[(\nu+) + (\nu-)]\}^{0.5} \qquad (6.5)$$

The definition for A differs slightly from that given normally (Pitzer, 1977) since it is defined for a number density scale and thus involves the Boltzmann constant rather than the gas constant and Avagadro's number and it does not involve the solvent density. We should note immediately that A is very closely related to equation (3.7) obtained above. It is in fact useful to rewrite this equation in terms of the DH slope (after making use of the salt neutrality condition (2.1):

$$G'_{MX} = \frac{A}{(8n_S)^{0.5}}\left(\frac{z^3 z^3}{\nu_+ + \nu_-}\right)^{0.5} \frac{1}{(1 + \kappa a)^2} \qquad (6.6)$$

The variation for the chemical potential of a neutral solute S with number density of S in a solvent W is given by:

$$\frac{1}{k_B T}\frac{d\mu_S}{dn_S} = \frac{1}{n_S} + \frac{G_{SW} - G_{SS}}{1 + n_S(G_{SS} - G_{SW})} \qquad (6.7)$$

In section 4 on free energies of transfer of electrolytic solutes from one solvent to another, we concluded that, in order to generate an equation suitable for an electrolytic solution, the correct substitution for G_{SS} and G_{SW} were:

$$1 + n_S G_{SS} = n_S G_{MX} = 1/[(\nu+)+(\nu-)] + n_S G'_{MX} - n_S V(ex) \tag{6.8}$$

and

$$G_{SW} = G_{MW} = G_{XW} \tag{6.9}$$

where we have substituted the expression $4\pi a^2/3$ in equation (3.5) with V(ex), the excluded volume. Substitution of equations (6.8) and (6.9) into (6.7) yields:

$$\frac{1}{k_B T}\frac{d\mu_S}{dn_S} = \frac{(\nu+)+(\nu-)}{n_S} + \frac{[(\nu+)+(\nu-)]^2[G_{MW}-G'_{MX}+V(ex)]}{1+n_S[(\nu+)+(\nu-)][G'_{MX}-G_{MW}-V(ex)]} \tag{6.10}$$

Thus it is clear that:

$$\frac{d \ln y_{\pm}}{dn_S} = \frac{[((\nu+)+(\nu-)][G_{MW}-G'_{MX}+V(ex)]}{1+[(\nu+)+(\nu-)]n_S[G'_{MX}-G_{MW}-V(ex)]} \tag{6.11}$$

Substitution of equation (6.6) into (6.11) readily yields the Debye-Hückel limiting law [equation (6.5)]. We have thus generalized our previous treatment (Newman, 1989) to any electrolyte charge type. Our justification for the substitutions (6.8) and (6.9) is that they give the correct form of equation for the free energies of transfer (see section 4), partial molar volumes (section 5), the correct limiting behaviour for the chemical potential (see section 4 and this section above) and they produce the Debye-Hückel limiting law. For asymmetric salts, the correct DH slope is also produced even though the electroneutrality conditions have been breached. Perhaps it is the cation-cation and anion-anion terms terms that are incorrectly obtained from the Poisson-Boltzmann distribution since the cation-anion term yields the correct limiting behaviour.

It is perhaps also useful to consider the form of the cation-solvent KB integral. We do not currently have any experimental information on the form of this term beyond the fact that at infinite dilution it relates closely to the salt partial molar volume. However, from our consideration of electroneutrality conditions, we also know that it depends on ion-ion terms.Thus we have:

$$G_{MW} = n_S G_{MX}[(\nu+)G^*_{MW} + (\nu-)G^*_{XW}] \tag{6.12}$$

and we have already had an expression for the cation-anion term [equation (6.8)]. The second term on the r.h.s. of equation (6.12) would likely vary with concentration but presumably only slowly. The infinite dilution value of this term is known:

$$(\nu+)G^{*}_{MW} + (\nu-)G^{*}_{XW} = [(\nu+)+(\nu-)]\kappa_T k_B T - \lim_{n_s \to 0} (\bar{v}_s) \qquad (6.13)$$

Throughout the above development, we have used a number density concentration scale and standard state and the chemical potentials imply free energy per molecule. Conversion to a molar concentration scale simply requires change of Boltzmann's constant to the gas constant; the chemical potentials are then in units of free energy per mole; this implies that the Kirkwood-Buff integrals now have the units of volume per mole (i.e. have been multiplied by Avagadro's number).

Much of the experimental work on electrolytic solution activity coefficients has been obtained through isopiestic measurements of osmotic coefficients (Robinson and Stokes, 1959) and it is perhaps useful to underline that the differential nature of KB theory allows one to switch from solute to solvent chemical potentials with the greatest of ease through use of the constant temperature and pressure Gibbs-Duhem expression:

$$n_s d\mu_s + n_W d\mu_W = 0 \qquad (6.14)$$

Thus we obtain from equation (6.7):

$$\frac{1}{k_B T}\frac{d\mu_W}{dn_s} = -\frac{1}{n_W(1 + n_s(G_{SS} - G_{SW})} \qquad (6.15)$$

The osmotic coefficient is defined as (Robinson and Stokes, 1959):

$$\phi = -\frac{\ln a_W}{M_W[(\nu+)+(\nu-)]m_s} \qquad (6.16)$$

with M_W the molar mass of W and m_s molality. Noting that:

$$\frac{1}{k_B T}\frac{d\mu_W}{dn_s} = \frac{d \ln a_W}{dn_s} \qquad (6.17)$$

we obtain:

$$M_W[(\nu+)+(\nu-)]\left(m_s\frac{d\phi}{dn_s} + \phi\frac{dm_s}{dn_s}\right) = \frac{1}{n_W[1 + n_s(G_{SS} - G_{SW})]} \qquad (6.18)$$

We can substitute equations (6.8) and (6.9) into (6.18) and obtain an exactly analogous equation for osmotic coefficients to that for activity coefficients (6.11):

$$\frac{d(\phi m_S)}{dn_S} = \frac{1}{M_W n_W[(\nu+)+(\nu-)]\{1+[(\nu+)+(\nu-)]n_S[G'_{MX}-G_{MW}-V(ex)]\}} \qquad (6.19)$$

7. SEPARATION OF KIRKWOOD-BUFF PARAMETERS

One of the earlier uses of KB theory was to take three of the constant temperature thermodynamic data sets (e.g. vapour pressures, densities and compressibilities) for a binary mixture $\alpha + \beta$ and by suitable combination to obtain the three KB integrals for the α-α, α-β and β-β interactions (Ben-Naim, 1977). It was disappointing that the three parameters showed complex behaviour with composition and thus did not shed as much light on solution structure as had been hoped. With regard to solutions of solute S plus solvent W, the situation is interesting. There are four constant temperature thermodynamic properties, namely activity coefficient, osmotic pressure, partial molar volume and compressibility although only three of the four can be mutually independent. In terms of KB theory this simply translates into there being three KB parameters which determine the thermodynamics. The four equations are:

$$\frac{1}{k_B T}\frac{\partial \pi}{\partial n_S} = \frac{1}{1 + n_S G_{SS}} \qquad (7.1)$$

$$\bar{v}_S = \frac{1 + n_W(G_{WW} - G_{SW})}{n_S + n_W + n_S n_W(G_{SS} + G_{WW} - 2G_{SW})} \qquad (7.2)$$

$$\kappa_T k_B T = \frac{1 + n_W G_{WW} + n_S G_{SS} + n_S n_W(G_{SS}G_{WW} - G^2_{SW})}{n_S + n_W + n_S n_W(G_{SS} + G_{WW} - 2G_{SW})} \qquad (7.3)$$

and equation (6.7). It is a shame that osmotic pressures are not available for most electrolyte solutions since this would greatly enhance our ability to separate the three terms. They are available for a great number of large biomolecules and polymers and also a variety of sugars. We will explore the application to aqueous solutions of sugars in a later section. It is obvious from equation (6.7) that the variation of solute chemical potential with solute number density allows evaluation of the difference between the Kirkwood-Buff solute-solute and solvent-solvent integrals; through the symmetry of KB theory, it is clear that there exists an equation similar to equation (6.7) but with S replaced by W and *vice versa*. It is however easier to obtain information about the solvent-solvent integrals through the KB equation for partial molar volumes (5.1) to obtain:

$$\frac{\bar{v}_S}{\bar{v}_W} = \frac{1 + n_W(G_{WW} - G_{SW})}{1 + n_S(G_{SS} - G_{SW})} \qquad (7.4)$$

Partial molar volumes are related by the expression;

$$n_W\bar{v}_W + n_S\bar{v}_S = 1 \tag{7.5}$$

and using the Gibbs-Duhem expression for partial molar volumes, it is easy to show that:

$$\frac{dn_S}{dn_W} = -\frac{\bar{v}_W}{\bar{v}_S} \tag{7.6}$$

From density data, we can obtain the l.h.s. of equation (7.6) which equals the r.h.s. of equation (7.4). Thus a combination of solute activity coefficients and densities allows both $G_{SS}-G_{SW}$ and $G_{WW}-G_{SW}$ to be obtained. Finally, defining:

$$D_{ij} = G_{ii} - G_{ij} \qquad (\textit{note } D_{ij} \neq D_{ji}) \tag{7.7}$$

we obtain from the equation for compressibility (7.3) after a certain amount of tedious rearrangement:

$$G_{SW}=\frac{\kappa_T k_B T[n_S+n_W+n_S n_W(D_{SW}-D_{WS})]-1-n_s D_{SW}-n_W D_{WS}+n_S n_W D_{SW} D_{WS}}{n_S + n_W + n_S n_W(D_{SW} + D_{WS})} \tag{7.8}$$

We can thus obtain all three KB integrals from the combination of activity coefficients, density and compressibility data.

8. COMPARISON WITH EXPERIMENTAL RESULTS

Within the field of thermodynamics of electrolytic solutions, two major goals are apparent. The first is the parametrization of experimental data in a precise and yet concise algebraic form so that the information may be used in other disciplines in the areas of both science and engineering. The other goal is the understanding of the various intermolecular forces, how they effect the structure of the solution and how these two effects impinge on the thermodynamic properties. Ideally, we would like to be able to work in any direction so that by starting with any one of the three (intermolecular forces, solution structure or thermodynamic parameters), we could obtain quantitative information about the other two. In practice, this second goal is extremely difficult to achieve due to the complexity of the solutions themselves, the complexity and shortcomings of the statistical mechanical approaches it is necessary to use, and difficulties in obtaining and interpreting structural data on solutions. As mentioned in the introduction, Kirkwood-Buff theory is formulated without reference to the various intermolecular forces and thus allows a rigourous connection to be drawn between thermodynamics and parameters relating directly to solution structure. A disadvantage of the theory, as mentioned in the introduction, is that only the integrals of the radial distribution functions are obtainable. Since radial distribution functions are relatively long-range and oscillatory in nature, some concern has been expressed (Friedman, 1988) that interpretation of the KB integrals in a simplistic way may not be easy. This problem has been partially addressed in an empirical way (Newman, 1988c) and will be addressed further below. It is not the goal of this chapter to discuss in great detail the analysis of the vaste array of thermodynamic data available in the literature; this task is still proceeding. Our aim will be simply to highlight certain areas in order to show the power of the approach.

a. Free Energies of Transfer

In section 4 above, the effect of change of solute chemical potential on solvent composition was formulated in terms of free energies of transfer. Other equivalent formulations include the primary medium effect and salting-in/salting-out coefficients and conversion between them is straightforward. One of the earliest realizations that an ion in a binary solvent mixture will sort solvent molecules such that molecules with the better solvating ability will be found preferentially close to the ion, was made by Debye in 1927. He developed a dielectric continuum model whereby the sorting is the balance between two factors, one being the decrease in free energy of charging the ion due to finding the solvent component with the higher relative permittivity (dielectric constant) closer to the ion , the other being the increase in free energy due to demixing the solvent. An alternative way of looking at this problem is that in a mixed solvent, the Born ion charging process should be performed at constant solvent component chemical potential and not at constant solvent mixture composition as is normally done (Koenig, 1937; Frank, 1955). It is relatively straightforward to calculate the solvent sorting caused by a charged sphere immersed in a dielectric continuum binary mixture and thence to use Kirkwood-Buff theory to obtain the free energy of transfer (Covington, Hall and Newman, 1990). In more recent years spectroscopic measurements have been made of the composition of solvent close to an ion in a binary solvent mixture. The reason for the great interest in this area arises from the fact that spectroscopic properties are obtainable for individual ions uncomplicated by effects due to counter ions whereas thermodynamic measurements are only available for neutral combinations of ions. Initial attempts at interpreting the spectroscopic data were based on solvation shell chemical equilibrium models (Covington, Lilley, Newman and Porthouse, 1973; Covington, Newman and Lilley, 1973; Covington and Newman, 1979) but more recently a Kirkwood-Buff approach has been developed (Covington and Newman, 1988). This latter approach arose in response to the gradual realization that (with the exception of transition metal ions and some other 3+ ions), it is unreasonable to treat the solvation shell around the ion as a distinct chemical entity since the lifetimes of the solvent molecules within the shell are so short, the solvation shell structure is so diffuse and is frequently very poorly defined. A treatment based on the ion-solvent radial distribution functions completely circumvents this problem. As was shown in Section 4 above, Kirkwood-Buff theory allows rigourous definition of single ion free energies of transfer whose "measurement" has been an important goal for many years. The interested reader is referred to the original work for more details (Covington and Newman, 1988).

The alternative procedure is to start with the experimental thermodynamic parameters and manipulate them mathematically so as to obtain the two solute-solvent Kirkwood-Buff integrals directly. Such integrals are simply the sum of cation and anion terms. For a solute S in binary solvent mixture $\alpha+\beta$, evaluation of both solute-α and solute-β integrals requires two sets of thermodynamic data most generally free energies of transfer and solute partial molar volumes over the entire solvent composition range; these latter are surprisingly sparse in the literature. It is not proposed to discuss in detail the results of this type of analysis; the interested reader is referred to the original work (Newman, 1988a). It has often been argued that higher temperature derivatives of thermodynamic properties are more sensitive to "structural changes" occurring and we have also recently reported a KB study of enthalpies of transfer and solute expansibilites which allow estimation of the temperature dependences of the two KB integrals (Newman, 1988b). Experimental solute expansibilites in mixed solvents are also very sparse in the literature.

b. Electrolyte Partial Molar Volumes

Through the years most theoretical ideas about partial molar volumes have touched upon three main ideas. The first is that salt partial molar volumes at infinite dilution are ion additive and much effort has related to trying to obtain single ion values. It is clear from inspection of equation (5.6) that a single ion infinite dilution partial molar volume of ion M in solvent W may be written as:

$$\bar{V}^{\circ}_{M} = \kappa_{T} k_{B} T - G^{*}_{MW} \quad (8.1)$$

The other two ideas relate to using either continuum or discrete models for the ion-solvent interaction. In the continuum model, solvent close to the ion suffers electrostriction (reduction in volume due to the high electric fields); hence the highly negative volumes for multicharged ions. The discrete model considers explicitly the volume occupied by water molecules within the solvation shells around the ion. This latter approach is clearly related in some sense to KB theory. The simplest KB model that we can envisage treats the solvation as a certain number of water molecules contained within the first solvation shell. Under this model we can write (Newman, 1989):

$$G^{*}_{MW} = N/n_{W} - (4\pi/3)r^{3} + \text{long-range contribution} \quad (8.2)$$

where N is the number of water molecules within the solvation shell, r is the radius of this shell and the "long-range contribution" arises from wiggles in the radial distribution function beyond the first maximum. The value of N is obtainable from the area under the first maximum of the radial distribution function as derived from neutron or X-ray diffraction data and since the extent of the first maximum is somewhat arbitrary then so is the value of N; similar arguments clearly apply to the value of r as well. In order to obtain some insight into this model we calculate the value of r on the assumption that the long-range contribution is negligible, from the values of partial molar volumes (assuming that a split between cations and anions is available) and the value of N derived from neutron/X-ray diffraction data. The results for a whole variety of ions are summarized in Table 8.1 which also includes distances between the ion and the oxygen of the nearest neighbour water molecules r(MO) also derived from from neutron/X-ray measurements, and the differences $\Delta r[=r-r(MO)]$. The values of N used were in general obtained from X-ray or neutron diffraction data (or for the first row transition metal ions the well established values of six were used). For some ions where there is uncertainty about the values of N, a range of possible values was chosen. The calculations are very insensitive to the exact values of $\overline{v}^{\circ}$ but do show strong sensitivity to N. For the singly and doubly charged cations, the difference Δr between the two distances is remarkably constant at about 1Å although there is some evidence that the values are slightly less for the triply charged cations. This constancy at the sort of value expected for this naïve model is strong evidence that large long-range contributions to the ion-water KB integrals are are unimportant at least for inorganic ions and the long-range oscillations in the radial distribution functions must largely cancel out. For anions, uncertainties in the values of N preclude definitive conclusions although the value of Δr appears constant at a value somewhat less than that of the cations.

c. Electrolyte Activity Coefficients

In Section 6, an expression for the molar standard state activity coefficient of an electrolyte was obtained from the Poisson-Boltzmann distribution and electroneutrality considerations. If it is assumed that immediate ion-solvent correlation terms do not vary with concentration, then they can be obtained from the salt infinite dilution partial molar volume [equation (6.13)]. The resulting expression for the activity coefficient would then have only one parameter to be determined, the distance of closest approach, a. Our initial work (Newman, 1989) did not consider the effect of G_{MX} on G_{MW} (equation 6.12) and simply treated

$$G_{MW} + V(ex) = C \tag{8.3}$$

with C a fitting parameter (see equation 6.11). The method was able to fit the activity coeffients of 23 1:1 electrolytes in water to reasonable precision with two parameters, a and C, and to high precision with three parameters, a, C_0 and C_1 where:

$$C = C_0 + C_1 c_s^{0.5} \tag{8.4}$$

The reader is referred to the original work for further details (Newman, 1989); Cheng and Newman, 1990a,b). In more recent work (Newman, 1990), we have used equation (6.12) to represent the ion-solvent KB integral with the term in parentheses held constant at the value obtained from the infinite dilution partial molar volume. The term V(ex) is then formally replaced by a fitting parameter E in equation (6.11):

$$V = E = E_0 + E_1 c^{0.5} + E_2 c + \ldots\ldots \tag{8.5}$$

Table 8.1. Partial Molar Volumes, Solvation numbers and derived Properties for Various Ions

Ion	Volume[a] cm^3/mol	N	r Å	r(MO) Å	Δr Å
Li^+	-6	4.0[b]	2.95	1.90[b]	1.05
Li^+	-6	5.5[c]	3.32	1.95[c]	1.37
Na^+	-6	8.0[d]	3.79	2.50[d]	1.29
K^+	+4	8.0[d]	3.88	2.70[d]	1.18
Cs^+	+16	8.0[b]	4.00	2.95[b]	1.05
NH_4^+	+13	10[e]	4.2	3.2[e]	1.0
NMe_4^+	+85	20[e]	5.6	4.6[e]	1.0
Mg^{2+}	-31	6.0[f]	3.11	2.11[g]	1.00
Ca^{2+}	-28	10.0[c]	3.91	2.46[c]	1.45
Mn^{2+}	-28	6.0[f]	3.15	2.22[g]	0.93
Fe^{2+}	-35	6.0[f]	3.05	2.12[g]	0.93
Co^{2+}	-34	6.0[f]	3.07	2.09[g]	0.98
Ni^{2+}	-34	6.0[c]	3.07	2.07[c]	1.00
Cu^{2+}	-38	6.0[c]	3.01	1.97[c]	1.04
Zn^{2+}	-32	6.0[f]	3.10	2.08[g]	1.02
Cd^{2+}	-30	6.0[f]	3.12	2.34[g]	0.78
Al^{3+}	-57	6.0[f]	2.70	1.88[g]	0.82
Cr^{3+}	-55	6.0[f]	2.74	1.99[g]	0.75
Fe^{3+}	-59	6.0[f]	2.67	2.04[g]	0.63
La^{3+}	-54	8.0[f]	3.28	2.51[g]	0.77
La^{3+}	-54	9.0[f]	3.49	2.51[g]	0.98
Nd^{3+}	-58	8.5[c]	3.34	2.48[c]	0.86
F^-	+4	6.0[f]	3.53	2.74[g]	0.79
Cl^-	+23	6.0[f]	3.72	3.12[g]	0.60
Cl^-	+23	6.0[c]	3.72	3.25[c]	0.47
Cl^-	+23	6.0[h]	3.72	3.10[h]	0.62
Br^-	+30	6.0[f]		3.33[g]	
Br^-	+30	6.0[h]	3.79	3.21[h]	0.58
I^-	+41	6.0[f]	3.89	3.61[g]	0.27
I^-	+41	8.0[f]	4.18	3.61[g]	0.57
NO_3^-	+34	2.4[c]	3.11	2.65[c]	0.46

[a]Millero (1971) with $\bar{v}^o(H^+)=-5 cm^3/mol$; [b]Ohtomo et al (1979); [c]Enderby (1983); [d]Ohtomo et al (1980); [e]Finney and Turner (1988); [f]Assumed; [g]Marcus (1983); [h]Ohtomo et al (1981).

This is equivalent to adding a virial term (or more) to the cation-anion integral. The newer two parameter fits (a and E) are considerably better than the earlier ones, indeed it appears that the new two parameter fits are of similar quality to the earlier three parameter ones. Before we discuss the fits in more detail it is useful to consider the functional form of the term which determines activity coefficients (i.e. the second term in square braces in equation 6.11), in more detail. Figure 8.1 shows the variation of this term for sodium chloride, calcium chloride and cadmium sulphate derived by direct non-linear least squares curve fitting to the experimental activity coefficient data. All three

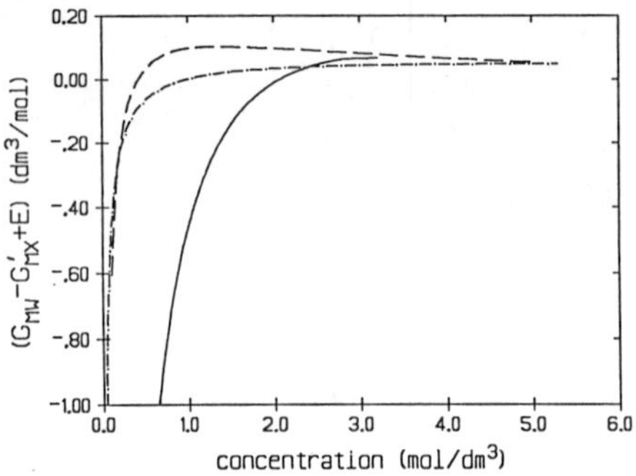

Figure 8.1. Variation of the term $[G_{MW}-G'_{MX}-V(ex)]$ with concentration for aqueous solutions of the salts NaCl —·—·— , $CaCl_2$ — — — , and $CdSO_4$ ——— at 298.15K.

salts show vastely different behaviours with respect to their activity coefficients and yet the functional forms of this term are all similar. The function is obviously a rapidly decreasing term in combination with a term which is either constant or only slowly varying with concentration; this is precisely the behaviour predicted by the treatment in section 6. The rate of initial decrease varies from salt to salt being determined by the value of the distance of closest approach. However, for cadmium sulphate and presumably also other 2:2 electrolytes, as discussed frequently in the literature (see e.g. Pitzer and Mayorga, 1974) the low concentration data appear anomalous in that the initial slope of log of the activity coefficient with square root of the concentration is larger than the DH value. Under this treatment this would imply a negative distance of closest approach; a value suitable to fit the dilute (0.1 - 0.2M) data however causes the term $(1+\kappa a)$ to pass through zero at high concentration which obviously destroys the fit. We have resolved this problem in a *de facto* manner simply by allowing the Debye-Hückel slope to vary. The quality of fits so obtained is remarkably good. The value of the plateau region is determined by the values of E and the value of the cation-solvent integral, the latter term showing some concentration dependence [see equation (6.12). Figure 8.2 shows the separation of the three contributions to the function for sodium chloride. Figure 8.3 shows the fits of activity coefficient against molar concentration for the three salts discussed earlier; the corresponding fitting parameters are given in Table 8.2 together with some indication of the quality of the fits. Work is proceeding in the analysis of other electrolytic solutions.

The majority of salt activity coefficients available in the literature are given in the molal standard state and it is in this

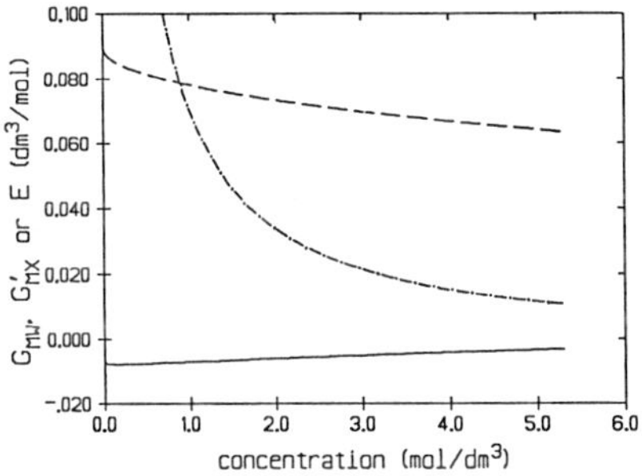

Figure 8.2. Variation of the three terms G'_{MX} —·—·— , G_{MW} ——— , and V(ex) — — — with concentration for aqueous solutions of NaCl at 298.15K.

Table 8.2. Molar Standard State Activity Coefficient Parameters for Various Salts in Water at 298.15K

Salt	m^a_{max} / mol/kg	a / Å	10^2E_0 / M^{-1}	10^2E_1 / $M^{-1.5}$	10^2E_2 / M^{-2}	e^b_{max}	m^c_e / mol/kg
$NaCl^d$	6.0	3.19±0.03	+8.94±0.2	-1.12±0.1	-	5.5	6.0
NaF^e	1.0	3.07±0.02	-1.71±0.1	-	-	0.73	0.05
$CaCl_2^f$	6.0	3.02±0.02	+36.1±0.4	-20.2±0.4	+3.20±0.1	6.8^g	3.0
$CdSO_4^h$	3.5	-	+321±3	-88.9±1.5	-	0.35	0.1

[a] Maximum concentration in fitting procedure.
[b] 10^3 times maximum difference between observed and calculated activity coefficients.
[c] Concentration at which maximum error occurs.
[d] Data from Lobo and Quaresma (1981 and 1984); logarithmic fit with equal weights.
[e] Data from Robinson and Stokes (1959); logarithmic fit with equal weights;two parameter fit.
[f] Data from Robinson and Stokes (1959); at 3.5m and above, activity coefficient given to only ±0.01; linear fit with appropriate weights; four parameter fit.
[g] Largest difference for data below 3.5m[f].
[h] Data from Robinson and Stokes (1959); logarithmic fit with equal weights; distance of closest approach set to zero; DH slope varied in fitting procedure to yield 1.395±0.004 $(mol/dm^3)^{-0.5}$.

concentration scale that they are often most useful. We have thus attempted to use the equations developed here directly with this sort of data. The parameters so obtained are not immediately relateable to KB integrals but the equations are remarkably successful at parametrizing an enormous volume of data. Table 8.3 gives parameters for a large selection of the data (mainly 1:1 electrolytes) given in in the appendix of Robinson and Stokes' standard text (1959). In many instances, the fits are within 0.001

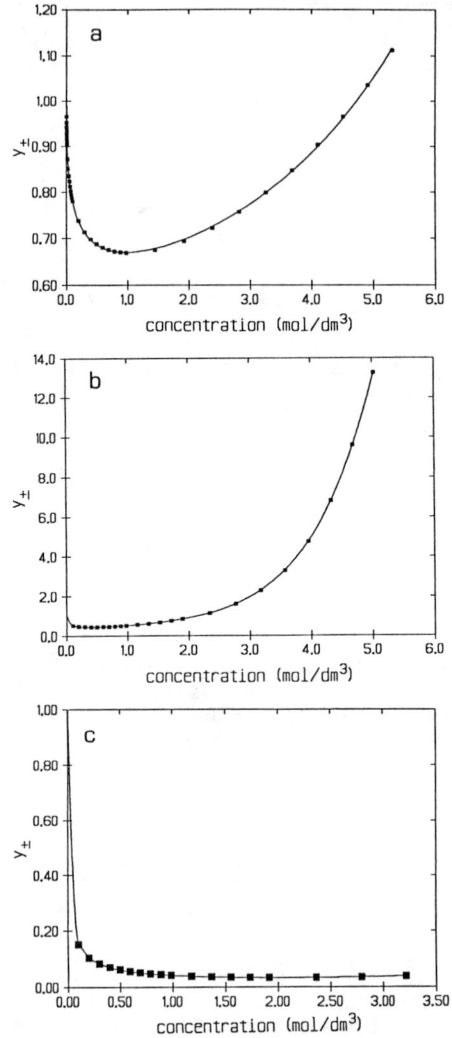

Figure 8.3. Observed and calculated molar standard state activity coefficients for aqueous solutions of: a) NaCl, b) $CaCl_2$ and c) $CdSO_4$ at 298.15K.

Table 8.3. Molal Standard State Activity Coefficient Parameters[a] derived from the Tabulations of Robinson and Stokes (1959)

Salt	m^{b}_{max}	a / Å	10^3E_0 / $(mol/kg)^{-1}$	10^3E_1 / $(mol/kg)^{-1.5}$	e^{c}_{max}	m^{d}_{e}
LiCl[e]	6.0	3.58±0.03	+168.5±1.3	-43.1±0.6	4.6	0.2
LiBr[e]	6.0	3.63±0.05	+196.6±1.8	-53.0±0.8	8.6	2.0
$LiClO_4$	4.0	4.85±0.04	+219.9±1.4	-67.3±0.8	4.3	0.1
NaF	1.0	3.09±0.02	-18.2±1.4	f	0.7	0.1
NaCl	6.0	3.21±0.03	+76.9±2.0	-10.6±0.9	4.7	6.0
NaBr	4.0	3.34±0.02	+106.6±1.3	-19.9±0.7	0.2	1.6
NaI	3.5	3.63±0.06	+144.2±1.9	-34.3±1.9	3.4	0.1
$NaClO_4$	3.5	3.45±0.02	+52.3±0.9	-4.9±0.4	1.6	0.4
$NaNO_3$	6.0	3.09±0.03	-41.2±2.0	+16.7±0.1	2.3	0.1
NaFor	3.5	3.45±0.06	+80.7±4.2	-21.2±2.5	2.3	0.2
NaAc	3.5	3.66±0.07	+172.7±3.9	-53.8±2.2	3.5	0.9
NaProp	3.0	3.76±0.03	+225.1±1.2	-79.3±1.2	1.9	1.2
KF	4.0	2.91±0.02	+74.7±1.8	-8.2±1.0	2.1	0.1
KCl	4.5	3.09±0.03	+29.4±1.1	+3.7±1.1	1.4	0.1
KBr	5.5	3.06±0.02	+53.9±1.4	-6.6±0.7	1.4	6.0
KI	4.5	3.36±0.03	+80.1±1.1	-17.0±1.1	1.3	0.2
RbCl	5.0	2.61±0.01	+38.1±0.2	f	1.9	0.1
RbBr	5.0	2.64±0.01	+34.6±0.2	f	1.4	0.1
RbI	5.0	2.56±0.01	+41.9±0.3	f	1.8	3.5
CsCl	6.0	1.74±0.06	+61.1±6.1	-7.2±2.5	6.2	0.1
CsI	3.0	2.10±0.12	+24.6±12.8	+1.1±7.3	3.7	0.1
$Mg(NO_3)_2$[g]	2.0	3.39±0.05	+325.6±7.9	-145±6	5.9	0.1
$ZnSO_4$	3.5	h	+3529±12	-908±19	6.4	0.1

[a] Logarithmic fit except where noted.
[b] Maximum concentration (mol/kg) fitted; except where noted, this is the highest concentration given.
[c] 10^3 times maximum difference between observed and calculated activity coefficent.
[d] Concentration (mol/kg) at which maximum error occurs.
[e] Linear fit weighted according to number of significant figures in data.
[f] Two parameter fits.
[g] Original data extend to 5.0m.
[h] Distance of closest approach set to zero; DH slope varied in fitting procedure to yield 1.384±0.004 $(mol/kg)^{-.05}$.

Table 8.4. Observed[a] and Calculated Molal standard State Activity Coefficients of Aqueous $ZnSO_4$ at 298.15K

m / mol/kg	$\gamma_\pm$(obs)	$\gamma_\pm$(cal)
0.1	0.150	0.156
0.2	0.104	0.106
0.3	0.0835	0.0840
0.4	0.0714	0.0711
0.5	0.0630	0.0625
0.6	0.0569	0.0564
0.7	0.0523	0.0518
0.8	0.0487	0.0487
0.9	0.0458	0.0455
1.0	0.0435	0.0432
1.2	0.0401	0.0399
1.4	0.0378	0.0377
1.6	0.0363	0.0363
1.8	0.0356	0.0356
2.0	0.0357	0.0354
2.5	0.0367	0.0370
3.0	0.0408	0.0415
3.5	0.0480	0.0475

[a]Taken from Robinson and Stokes (1959)

(the precision given in the tables) over the entire concentration range with three parameters (Mindermann and Newman, 1990). This work is still far from completion but the utility of the approach is already clear. The fitting is in general most successful for activity coefficients that decrease monotonically with concentration. For salts whose values increase rapidly at very high concentrations, four parameters are sometimes needed out to the very highest concentration. For 2:2 electrolytes, similar comments apply to those given above about the apparent value of the Debye-Hückel slope. Table 8.4 gives the values of observed and calculated molal activity coefficients for zinc sulphate obtained by treating the DH slope as variable with the value of a constrained to zero and with E given by an equation analogous to (8.5) but in terms of molality. The corresponding parameters are given in Table 8.3.

d. Separation of Kirkwood-Buff Integrals

As mentioned in section 7, the analysis of three constant temperature thermodynamic properties for a binary mixture S+W as a function of concentration allows separate determination of the three S-S, S-W and W-W Kirkwood-Buff parameters (Ben-Naim, 1977).It is clear that such a separation for electrolytic solutions would shed much insight into the detailed interpretation of the thermodynamic properties most particularly activity coefficients. Such a separation has not yet been completed but is under way; it is indeed a pity that osmotic pressures are not available in general for salt solutions since they would clearly have simplified this task. We have completed a separation for aqueous solutions of sugars where reliable (although very old) osmotic pressure measurements are available. These were analysed along with activity coefficients and density data as outlined in section 7 using non-linear least squares methods on the differential equations. The values of the S-S integral derived in the temperature range 273.15 to 353.15K showed but very slight variation over the concentration range 0-1 molal and in this simplified treatment were assumed constant; the results are given in Figure 8.6. The difference between the S-S and S-W integrals at 298.15K derived from activity coefficient data shows some variation over the much wider concentration range 0-6 molal ; the results are also summarized in Figure 8.4. Finally from density data at 298.15K, the differential in equation (7.6) was obtained and in combination with equation (7.4), the W-W integral was evaluated in the concentration range 0-1 molal. The variation of the three KB parameters with concentration in the range 0-1 molal (see Fig.8.4). Ben-Naim (1977) has tentatively suggested that the

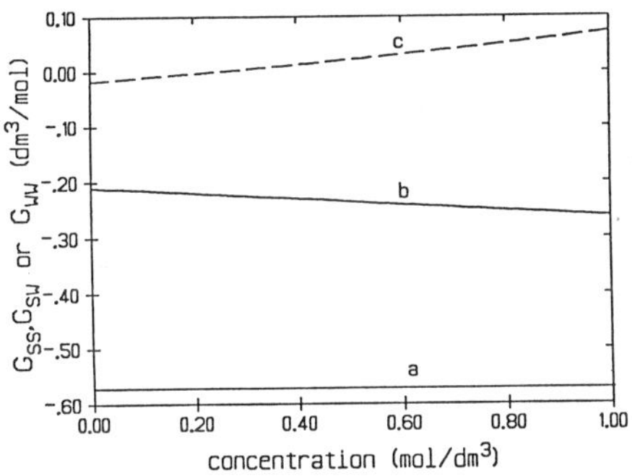

Figure 8.4. Calculated values of G_{SS} (a), G_{SW} (b) and G_{WW} (c) for aqueous solutions of sucrose at 298.15K as a function of concentration.

rapid increase in the value of G_{ww} for ethanol+water mixtures is a manifestation of the well-known water structure enhancement of alcohols in dilute solutions. In light of a similar finding for the very different non-hydrophobic solute sucrose, this conclusion may be subject to some revision

9. CONCLUDING REMARKS

In this paper we have given an outline of the development of the application of Kirkwood-Buff theory to electrolytic solutions. We have placed strong emphasis on the electroneutrality conditions and on the Poisson-Boltzmann distribution. A simple salt solution cannot be treated as a ternary mixture of two ions plus solvent due to the completely correlated nature of the various ion-ion and ion-solvent KB integrals. The development of the corresponding KB equations for such a solution has been obtained in an entirely consistent but non-rigourous way whose justification arises from the fact that the procedure adopted produces equations that give correct ion-additivity relationships, the correct form for the ideal behaviour of salts and reproduce the Debye-Hückel limiting law. It is clear that a more rigourous development of the theory as applied to salt solutions is however desirable. The work of interpreting the vaste array of thermodynamic data has only just begun. Further work is clearly necessary in order to resolve the question of the extent of long-range contributions to the KB integrals. In addition, systematic separation of the three integrals, for a whole variety of salts is needed. A major area within the field of electrolytic solutions not discussed in the paper is that of mixed salt systems. Work is proceeding in this direction and will be reported at some future date. Finally, there is a need to form connections between KB theory and some of the other approaches frequently used; examples would be the chemical equilibria approaches to free energies of transfer (Covington, Newman and Lilley 1973) and the integral equation appraoches widely used in the interpretation of activity coefficients (Rasaiah and Friedman, 1968 and 1969; Waisman and Lebowitz, 1972a,b).

ACKNOWLEDGEMENTS

We wish to thank Ms. Paulina Mindermann for help in the analysis of experimental data. We gratefully acknowledge partial support for this work by the Natural Sciences and Engineering Research Council (NSERC) of Canada.

REFERENCES

Beeby, J.L. 1973. Neutron scattering from aqueous solutions: the long wavalength limit. J. Phys. C: Solid State Phys. 6: 2262-2268

Ben-Naim, A. 1977. Inversion of the Kirkwood-Buff theory of solutions: Application to the water-ethanol system. J. Chem. Phys. 67: 4884-4890.

Cheng, Y. and K. E. Newman. 1990a. Application of Kirkwood-Buff theory to electrolyte solutions. III. Activity coefficients of aqueous NaCl between 273.15 and 573.15K. J. Solution Chem. 19: (in press)

Cheng, Y. and K. E. Newman. 1990b. Application of Kirkwood-Buff theory to electrolyte solutions. III. Activity coefficients of 1:1 electrolytes at 298.15 Can. J. Chem. (Submitted).

Covington, A.K., D.G. Hall and K.E. Newman. 1988. Application of Kirkwood-Buff theory to free energies of transfer of electrolytes from one solvent to another. J. Chem. Soc., Faraday Trans. 1. 84: 1393-1404.

Covington, A.K., D.G. Hall, and K.E. Newman. 1990. Work in progress.

Covington, A.K., T.H. Lilley, K.E. Newman and G.A. Porthouse. 1973. Nuclear Magnetic Resonance studies of preferential solvation. 1. Hydrogen-peroxide + water. J. Chem. Soc., Faraday Trans. 1. 69: 963-972.

Covington, A.K., K.E. Newman and T.H. Lilley. 1973. Nuclear Magnetic Resonance studies of preferential solvation. 2.Thermodynamic treatment and application to methanol + water solvents.J. Chem. Soc., Faraday Trans. 1. 69: 973-983.

Covington, A.K. and K.E. Newman. 1979. Approaches to the problems of solvation in pure solvents and preferential solvation in mixed solvents. Pure Appl. Chem. 51: 2041-2057.

Debye, P. 1927. Das elektrische Ionenfeld und das Aussalzen. Z. Physik. Chem. 130: 56-64.

Enderby, J.E. 1983. Neutron scattering from ionic solutions. Ann. Rev. Phys. Chem. 34: 155-185.

Finney J.L. and J. Turner. 1988. Direct measurement by neutron diffraction of the solvation of polar and apolar molecules in water. The hydration of the tetramethylammonium ion. Faraday Discuss. Chem. Soc. 85: 125-135.

Frank, H.S. 1955. Thermodynamics of a fluid substance in the electrostatic field. J. Chem. Phys. 23: 2023-2032.

Friedman, H.L. 1989. General Discussion. Faraday Discuss. Chem. Soc. 85: 284.

Guggenheim, E.A. 1929. The conception of electrical potential difference between two phases and the individual activities of ions. J. Phys. Chem. 33: 842-849.

Hall, D.G. 1971. Kirkwood-Buff theory of solutions. An alternative derivation of part of it and some applications. Trans. Faraday Soc. 67: 2516-2524.

Hall, D.G. 1972. A definition of the primary medium effect for individual ionic species in solution. J. Chem. Soc., Faraday Trans. 2. 68: 25-32.

Kirkwood, J.G. and F.P. Buff. 1951.The statistical mechanical theory of solutions. I. J. Chem. Phys. 19: 774-777.

Koenig, F.O. 1937. The thermodynamics of the electric field with special reference to chemical equilibrium. J. Chem. Phys. 41: 597-620.

Lobo, V.M.M. and J.L. Quaresma. 1981 and 1984. Electrolyte solutions: literature data on thermodynamic and transport properties, Vol. 1 and 2. Portugal: University of Coimbra.

McMillan, W.G. and J.G. Mayer. 1945. The statistical mechanics of multicomponent system. J. Chem. Phys. 13: 276-305.

Marcus, Y. 1983. The cristal radii of ions. J. Solution Chem. 12: 271-279.

Millero, F.J. 1971. The molal volumes of electrolytes. Chem. Revs. 71: 147-176.

Minderman, P. and K.E. Newman. 1990. To be published.

Newman, K.E. 1988a. Interpretation of free energies of transfer and solute partial molar volumes in mixed solvents using Kirkwood-Buff theory. J. Chem. Soc., Faraday Trans. 1 84: 1387-1391.

Newman, K.E. 1988b. Application of Kirkwood-Buff theory to enthalpies of transfer and expansibilities of solutes in binary solvent mixtures. J. Chem. Soc., Faraday Trans. 1 84: 3885-3890.

Newman, K.E. 1988c. General discussion . Faraday Discuss. Chem. Soc. 85: 283 and 284.

Newman, K.E. 1989. A Kirkwood-Buff theoretical approach to Debye-Hückel theory. J. Chem. Soc., Faraday Trans. 1. 85: 485-492

Newman, K.E. 1990. Work in progress.

Ohtomo, N. and K. Arakawa. 1979. Neutron diffraction study of aqueous ionic solutions. I. Aqueous solutions of lithium chloride and caesium chloride. Bull. Chem. Soc. Japan. 52: 2755-2759.

Ohtomo, N. and K. Arakawa. 1980. Neutron diffraction study of aqueous ionic solutions. II. Aqueous solutions of sodium chloride and potassium chloride. Bull. Chem. Soc. Japan. 53: 1789-1794.

Ohtomo, N., K. Arakawa, M. Takeuchi, T. Yamaguchi, and H. Ohtaki. 1981. Neutron diffraction study of aqueous hydrochloric and hydrobromic acid solutions. Bull. Chem. Soc. Japan. 54: 1314-1319.

Pitzer, K.S. 1977. Electrolyte theory - Improvements since Debye and Hückel. Accounts Chem. Research. 10: 371-384.

Pitzer, K.S. and G. Mayorga. 1974. Thermodynamics of electrolytes. III. Activity and osmotic coefficients for 2-2 electrolytes. J. Solution Chem. 3: 539-546.

Rasaiah, J.C. and H.L. Friedman. 1968. Integral equation methods in the computation of equilibrium properties of ionic solutions. J. Chem. Phys. 48: 2742-2752.

Rasaiah, J.C. and H.L. Friedman. 1969. Integral equation computations for aqueous 1-1 electrlytes. Accuracy of the method. J. Chem. Phys. 50: 3965-3976.

Robinson, R.A. and R.H. Stokes. 1959. Electrolyte solutions. London: Butterworths.

Waisman, E. and J.L. Lebowitz. 1972a. Mean spherical model integral equation for charged hard spheres. I. Method of solution. J. Chem. Phys. 56: 3086-3093.

Waisman, E. and J.L. Lebowitz. 1972b. Mean spherical model integral equation for charged hard spheres. II. Results. J. Chem. Phys. 56: 3093-3099.

Watts, R.O. and I.J. McGee. 1976. Liquid state chemical physics, 67-68. New York: Wiley.

Index